17584

FISH PHYSIOLOGY

Volume IV

*The Nervous System,
Circulation, and Respiration*

CONTRIBUTORS

C. ALBERS
JERALD J. BERNSTEIN
GRAEME CAMPBELL
JOHN E. CUSHING
JAMES CLARKE FENWICK

MALCOLM S. GORDON
KJELL JOHANSEN
D. J. RANDALL
AUSTEN RIGGS
G. SHELTON

JOHAN B. STEEN

FISH PHYSIOLOGY

Edited by
W. S. HOAR
DEPARTMENT OF ZOOLOGY
UNIVERSITY OF BRITISH COLUMBIA
VANCOUVER, CANADA

and
D. J. RANDALL
DEPARTMENT OF ZOOLOGY
UNIVERSITY OF BRITISH COLUMBIA
VANCOUVER, CANADA

Volume IV

The Nervous System,
Circulation, and Respiration

Academic Press   New York and London   1970

COPYRIGHT © 1970, BY ACADEMIC PRESS, INC.
ALL RIGHTS RESERVED
NO PART OF THIS BOOK MAY BE REPRODUCED IN ANY FORM,
BY PHOTOSTAT, MICROFILM, RETRIEVAL SYSTEM, OR ANY
OTHER MEANS, WITHOUT WRITTEN PERMISSION FROM
THE PUBLISHERS.

ACADEMIC PRESS, INC.
111 Fifth Avenue, New York, New York 10003

United Kingdom Edition published by
ACADEMIC PRESS, INC. (LONDON) LTD.
Berkeley Square House, London W1X 6BA

LIBRARY OF CONGRESS CATALOG CARD NUMBER: 76-84233

PRINTED IN THE UNITED STATES OF AMERICA

CONTENTS

LIST OF CONTRIBUTORS	ix
PREFACE	xi
CONTENTS OF OTHER VOLUMES	xiii

1. Anatomy and Physiology of the Central Nervous System
Jerald J. Bernstein

I.	Introduction	2
II.	Ultrastructure	4
III.	Telencephalon	10
IV.	Diencephalon	26
V.	Mesencephalon	32
VI.	Cerebellum	50
VII.	Medulla Oblongata	54
VIII.	Spinal Cord	68
IX.	Myotypic Respecification of Regenerated Nerves	76
	References	78

2. The Pineal Organ
James Clarke Fenwick

I.	Introduction	91
II.	The Pineal Dilemma	91
III.	Ontogeny and Structure of the Pineal Organ	92
IV.	Physiology of the Pineal Body	96
V.	Synopsis	103
	References	104

3. Autonomic Nervous Systems
Graeme Campbell

I.	Introduction	109
II.	Cyclostomes	110

III.	Gnathostomatous Fish	114
	References	128

4. The Circulatory System
D. J. Randall

I.	Introduction	133
II.	Blood Pressures in Arteries and Veins	135
III.	The Heart	138
IV.	General Properties of the Cardiovascular System	158
V.	The Effects of Some Substances on the Circulatory System	165
	References	168

5. Acid–Base Balance
C. Albers

I.	Introduction	173
II.	Basic Concepts of Physical Chemistry	174
III.	The Transport of CO_2 in the Blood	188
IV.	The Intracellular pH	203
V.	Controlling Mechanisms of the Acid–Base Balance	204
	References	205

6. Properties of Fish Hemoglobins
Austen Riggs

I.	Introduction	209
II.	Structural Properties	210
III.	Oxygen Transport	222
	References	246

7. Gas Exchange in Fish
D. J. Randall

I.	Introduction	253
II.	Gas Exchange between Blood and Water	254
III.	Gas Exchange between Blood and Tissues	278
IV.	Methods of Analysis of the Transfer of Gases	279
V.	The Effects of Various Parameters on Gas Exchange	283
	References	286

8. The Regulation of Breathing
G. Shelton

I.	Introduction	293
II.	The Respiratory Pump	294
III.	Central Factors in the Regulation of Breathing Patterns	316
IV.	The Role of Mechanoreceptors in Respiratory Regulation	326
V.	The Chemical Regulation of Respiration	328
VI.	The Relationship between Ventilation and Perfusion	347
	References	352

9. Air Breathing in Fishes
Kjell Johansen

I.	Occurrence and Bionomics of Air-Breathing Fishes	361
II.	Nature of the Structural Adaptations for Air Breathing	363
III.	Physiological Adaptations in Air-Breathing Fishes	375
	References	408

10. The Swim Bladder as a Hydrostatic Organ
Johan B. Steen

I.	Introduction	414
II.	The Biological Significance of the Swim Bladder	414
III.	The Architecture of the Swim Bladder	415
IV.	The Performance of the Swim Bladder	423
V.	The Mechanisms of Gas Transport	428
VI.	The Realization of Countercurrent Multiplication in the Swim Bladder	432
VII.	Nervous Control of the Hydrostatic Function of the Swim Bladder	438
	References	440

11. Hydrostatic Pressure
Malcolm S. Gordon

I.	Introduction	445
II.	Fishes Living under High Pressures	450
III.	Experimental Studies	454
	References	460

12. Immunology of Fish
John E. Cushing

I.	Introduction	465
II.	Antibodies	468
III.	The Cellular Basis of the Immunological Response	479

IV.	Complement	483
V.	Blood Groups	486
VI.	Cyclostomes	490
VII.	Final Considerations	494
	References	494

AUTHOR INDEX 501

SYSTEMATIC INDEX 517

SUBJECT INDEX 523

LIST OF CONTRIBUTORS

Numbers in parentheses indicate the pages on which the authors' contributions begin.

C. ALBERS (173), *Kerckhoff Institut der Max-Planck-Gesellschaft, Bad Nauheim, Germany*

JERALD J. BERNSTEIN (1), *Department of Anatomical Sciences, and Center for Neurobiological Sciences, University of Florida, College of Medicine, Gainesville, Florida*

GRAEME CAMPBELL (109), *Department of Zoology, University of Melbourne, Parkville, Australia*

JOHN E. CUSHING (465), *Department of Biological Sciences, University of California, Santa Barbara, California*

JAMES CLARKE FENWICK (91), *Department of Biology, University of Ottawa, Ottawa, Canada*

MALCOLM S. GORDON (445), *Department of Zoology, University of California, Los Angeles, California*

KJELL JOHANSEN (361), *Department of Zoology, University of Washington, Seattle, Washington*

D. J. RANDALL (133, 253), *Department of Zoology, University of British Columbia, Vancouver, Canada*

AUSTEN RIGGS (209), *Department of Zoology, University of Texas, Austin, Texas*

G. SHELTON (293), *School of Biological Sciences, University of East Anglia, Norwich, England*

JOHAN B. STEEN (413), *Institute of Physiology, University of Oslo, Oslo, Norway*

PREFACE

In the first three chapters of Volume IV the anatomy and physiology of the nervous system are discussed and provide a background for many of the subsequent chapters in this volume and in Volume V. The major portion of this fourth volume is devoted to a review of aspects of respiration, circulation, and properties of the blood in fishes. There is a chapter on the role of the swim bladder as a hydrostatic organ and another on the effects of hydrostatic pressure on animal function. The swim bladder presents a fascinating example of the special organization of the circulatory system to meet a particular functional demand, in this case the secretion of gases against large concentration gradients. Some recent developments in the understanding of how gases are secreted into the swim bladder are reported.

The subject matter discussed has been attracting the interest of a large number of people from many disciplines, as has comparative physiology in general. Engineers, mathematicians, physicists, biochemists, zoologists, and physiologists are combining abilities to produce new approaches and fresh insight into old problems in fish physiology. Engineers are providing some of the theoretical background required for the quantitative analysis of the movement of fish through water and the movement of water over the gills. Biochemists in conjunction with physiologists are relating changes observed in the behavior of molecules to changes in the functional organization of the animal as a whole to meet a new set of internal and external environmental conditions. Mammalian physiologists are discovering a wealth of interesting adaptive problems in fish physiology. We hope that this volume will serve as a useful source of information for these people.

CONTENTS OF OTHER VOLUMES

Volume I

The Body Compartments and the Distribution of Electrolytes
    W. N. Holmes and Edward M. Donaldson

The Kidney
    Cleveland P. Hickman, Jr., and Benjamin F. Trump

Salt Secretion
    Frank P. Conte

The Effects of Salinity on the Eggs and Larvae of Teleosts
    F. G. T. Holliday

Formation of Excretory Products
    Roy P. Forster and Leon Goldstein

Intermediary Metabolism in Fishes
    P. W. Hochachka

Nutrition, Digestion, and Energy Utilization
    Arthur M. Phillips, Jr.

AUTHOR INDEX—SYSTEMATIC INDEX—SUBJECT INDEX

Volume II

The Pituitary Gland: Anatomy and Histophysiology
    J. N. Ball and Bridget I. Baker

The Neurohypophysis
    A. M. Perks

Prolactin (Fish Prolactin or Paralactin) and Growth Hormone
    J. N. Ball

Thyroid Function and Its Control in Fishes
    Aubrey Gorbman

The Endocrine Pancreas
August Epple

The Adrenocortical Steroids, Adrenocorticotropin and the Corpuscles of Stannius
I. Chester Jones, D. K. O. Chan, I. W. Henderson, and J. N. Ball

The Ultimobranchial Glands and Calcium Regulation
D. Harold Copp

Urophysis and Caudal Neurosecretory System
Howard A. Bern

AUTHOR INDEX—SYSTEMATIC INDEX—SUBJECT INDEX

Volume III

Reproduction
William S. Hoar

Hormones and Reproductive Behavior in Fishes
N. R. Liley

Sex Differentiation
Toki-o Yamamoto

Development: Eggs and Larvae
J. H. S. Blaxter

Fish Cell and Tissue Culture
Ken Wolf and M. C. Quimby

Chromatophores and Pigments
Ryozo Fujii

Bioluminescence
J. A. C. Nicol

Poisons and Venoms
Findlay E. Russell

AUTHOR INDEX—SYSTEMATIC INDEX—SUBJECT INDEX

Volume V (Tentative)

Vision: Visual Pigments
F. W. Munz

Vision: Electrophysiology of the Retina
   Tsuneo Tomita

Vision: The Experimental Analysis of Visual Behavior
   D. Ingle

Chemoreception
   Toshiaki J. Hara

Temperature Receptors
   R. W. Murray

Sound Production and Detection
   W. N. Tavolga

The Labyrinth
   O. Lowenstein

Mechanoreceptors: The Lateral Line Organ Receptors
   Ake Flock

Electroreception
   M. V. L. Bennett

Electric Organs
   M. V. L. Bennett

Volume VI (Tentative)

The Effect of Environmental Factors on the Physiology of Fish: An Examination of the Different Categories of Physiological Adaptation
   F. E. J. Fry

Action of the Environment on Biochemical Systems
   P. W. Hochachka and G. N. Somero

Freezing Resistance in Fishes
   Arthur L. DeVries

Learning and Memory
   Paul Rozin and Henry Gleitman

The Ethological Analysis of Fish Behavior
   G. P. Baerends

Biological Rhythms
   H. O. Schwassmann

Orientation and Fish Migration
   A. D. Hasler

Special Techniques
   D. J. Randall and W. S. Hoar

FISH PHYSIOLOGY

Volume IV

*The Nervous System,
Circulation, and Respiration*

1

ANATOMY AND PHYSIOLOGY OF THE CENTRAL NERVOUS SYSTEM

JERALD J. BERNSTEIN

I. Introduction	2
II. Ultrastructure	4
III. Telencephalon	10
A. Olfactory Bulb Anatomy	10
B. Telencephalic Anatomy	11
C. Neurophysiological Studies of the Olfactory System	14
D. Electroencephalography of the Telencephalon	17
E. Telencephalon and Reproductive Behavior	18
F. Specific Sensory Deficits following Telencephalic Ablation	19
G. Telencephalon and Learning	20
H. Stimulation of the Telencephalon	23
I. Regenerative Capacity of the Telencephalon	24
IV. Diencephalon	26
A. Anatomy	26
B. Photosensitivity of the Pineal Organ	28
C. Electrical Activity of Neurosecretory Cells	28
D. Regeneration of the Hypophysis (Pituitary)	31
V. Mesencephalon	32
A. Anatomy	32
B. Spontaneous Activity of the Optic Tectum	34
C. Electrophysiological Characteristics of the Visual Input	34
D. Spectral Sensitivity of Optic Tectal Units	40
E. Interhemispheric Transfer of Visual Information	41
F. The Optic Tectum and Learning	43
G. Retinotopic Respecification of the Regenerated Optic Nerve	44
H. Regeneration of the Optic Tectum	48
I. Axoplasmic Flow	49
VI. Cerebellum	50
A. Anatomy	50
B. Deficit Function	51
C. Stimulation of the Cerebellum	52
VII. Medulla Oblongata	54
A. Anatomy	54

    B. Spontaneous Activity of the Medulla Oblongata . . . 55
    C. Reticulomotor System 56
    D. Taste 58
    E. Audition 62
VIII. Spinal Cord 68
    A. Anatomy 68
    B. Caudal Neurosecretory System and Urophysis 70
    C. Electrophysiological Properties of the Neurosecretory Cells
       and Intramedullary Neurons 72
    D. Regeneration of the Spinal Cord 74
IX. Myotypic Respecification of Regenerated Nerves 76
References 78

I. INTRODUCTION

The primary function of the central nervous system is integration. Integration results in adaptive behavior which is initiated by the perception of peripheral events with all their various attributes, qualities, and amplitudes. These peripheral events are encoded; information is transmitted to different areas of the central nervous system; messages converge on and divert to the appropriate central nervous system centers; this information is transmitted to appropriate efferent centers with the resultant appropriate messages going to the efferent organs of the body, resulting in the final behavior of the animal. Anatomically, information is carried over the fiber tracts of the central nervous system to the various neural centers; physiologically, this results in the dendritic summation of excitatory and/or inhibitory effects on the neural components of the nervous system center.

Nervous system function can be viewed as a computer with digital and analog components. The information impinging on the central nervous system via physical energy from the environment is extremely reliable. This information is encoded by receptors and transmitted via axons as a series of all-or-none impulses. The digital axonal information has no known qualitative differences available except for factors such as the temporal rates of firing from the periphery (digital encoding). The digital information of the axon is centrally converted to an analog input to the neurons of termination by the propagation of graded potentials in the postsynaptic elements. The graded potential results in a more-or-less type of analog information being propagated in the dendrites and cell body of the neuron. This physiological event is recorded as the postsynaptic potential of neurons. The electrophysiological recording will indicate if the graded threshold was of sufficient amplitude to discharge the neuron resulting in an action potential or spike (excitation), or the postsynaptic membrane may become hyperpolarized owing to axonal discharge (inhibition) so that subsequent information will not generate an action potential. The interaction or the balance of inhibitory and excitatory information converging on a neuron makes up the sum total of the postsynaptic neuronal membrane characteristics. This synaptic function with all its parts may well be the anatomical area that transforms a digital input into an analog system. The information, then, is passed from the periphery into the central nervous system and to different integrative areas within the central nervous system itself. The sum total of the behavior of the animal then can be summarized by the integrative area which is established by the anatomical connections of the nervous system and the response of the neurons to the encoded messages.

There have been many experimental methods and designs developed to ascertain the integrative function of various portions of the nervous system of fish. Deficit function work has been carried out by removing parts of the nervous system in order to locate portions of the brain which are responsible for special integrative action. For the most part behavioral end points have been used in this type of experimentation. That is, following lesion, animals have been placed in certain behavioral situations to ascertain the animal's performance before and after the loss of a particular area. Stimulation of portions of the nervous system also has been utilized to ascertain the function of specific neural centers. Electrical stimulation can elicit specific behavioral patterns. In addition, electrophysiological recording from the various neural elements involved in integration has yielded valuable information on the characteristics of processing of neuronal information. These types of studies have demonstrated interactions of the various neural centers within the nervous system proper and the interaction of the nervous system with the periphery (central control of periphery). The neuroanatomical relationships of the various centers in the central nervous system have been advanced by the use of modern techniques such as the use of specific stains for the degeneration of neural pathways and the utilization of radioactive tracers for the study of axoplasmic flow. In addition, electron microscopy has revealed several unusual features of the fish nervous system.

The study of the integrative function of the central nervous system of fish is aided by one of the most remarkable features of the system which is its ability to regenerate. Thus, one can remove an area in the fish central nervous system and have the area reconstituted. This characteristic of the central nervous system makes the fish an elegant experimental animal for the study of central nervous system function. Parts of the central nervous system can be removed and be reconstituted often with the return of normal function.

Recent findings have shown that removal of particular portions of the brain of fish results in specific losses in behavioral patterns and perception. This has been a major finding since at one time it was thought the removal of a particular portion of the brain did not result in deficits in specific sensory, motor, or behavioral patterns, such as reproductive behavior, because of the loss of particular areas of the brain.

It is the integrative function within the central nervous system to which this chapter is devoted. Unfortunately, the wealth of information which has been written on the subject far exceeds the space available.

II. ULTRASTRUCTURE

The central nervous system of vertebrates contains neurons and non-neuronal elements consisting of neuroglia, endothelia, ependyma, and pericytes of blood vessel walls. In general the ultrastructure of neurons of the fish central nervous system is similar to that of neurons found in all vertebrates (Palay *et al.,* 1962). The nerve cells have a rather large nucleus with a distinct nucleolus. The nucleus is surrounded by a porous double membrane. The inner membrane is the denser of the two, and the outer membrane is continuous with the membranes of the extensive vacuolar system within the cytoplasm of the cell (endoplasmic reticulum).

Cytoplasm of the typical neuron perikaryon contains laminae of ergastoplasm (rough endoplasmic reticulum, i.e., endoplasmic reticulum lined by ribosomes) with free ribosomes interposed between the laminae. In light microscopy this organelle stains with basic dyes and is the characteristic Nissl body. Nissl bodies are not found as frequently in fish neurons as in mammalian neurons. Therefore, as can be expected, the bulk of the ribosomes are free in the cytoplasm. The Golgi apparatus is a complex specialization of the endoplasmic reticulum and is comprised of stacked cisternae and vesicles of smooth endoplasmic reticulum. Organelles such as mitochondria are more or less evenly distributed in the cytoplasm. Lysosomes may be present and are bound by a single membrane. Mitochondria in fish are double membrane bound and often contain tubular cristae. In addition, the cytoplasm of the perikaryon of the neuron contains neurofilaments and often neurotubules.

Primary dendrites are similar in ultrastructure to the perikaryon with rough endoplasmic reticulum that may extend into the basal regions of the dendrites and the occasional lamination with free ribosomes into Nissl bodies. Many dendrites contain regularly, longitudinally oriented neurotubules or canaliculi of indefinite length and about 200 Å in diameter as well as neurofilaments. In the goldfish, dendritic neurotubules can be observed in regular laminated patterns (Fig. 1). Dendrites also possess many spinous projections (dendritic spines) which are sites of increased neuronal area for synaptic contact (axodendritic synapses).

Neurofilaments are a conspicuous element of axoplasm. Smooth endoplasmic reticulum is common and rough endoplasmic reticulum is absent.

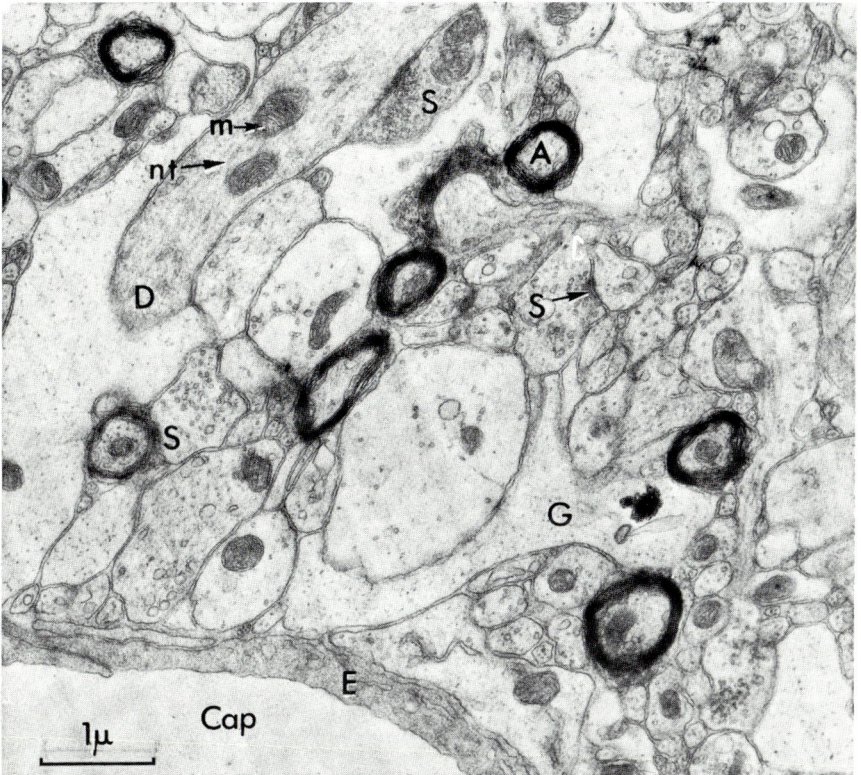

Fig. 1. Electron micrograph of the optic tectum of the goldfish. Note the typical 150-Å extracellular space and the processes of glial cells (G) between the endothelial cells (E) of the capillary (Cap) and the neural elements. Bouton terminaux are present and are the presynaptic (S) elements of the synapse. Note the synaptic bar (arrow), vesicles, and mitochondria in the synaptic profiles. Dendrites (D) containing neurotubules (nt) and mitochondria (m) are often present. In addition, myelinated (A) and unmyelinated axons can be observed. Osmium fixation, lead citrate and uranyl acetate stain. ×15,500.

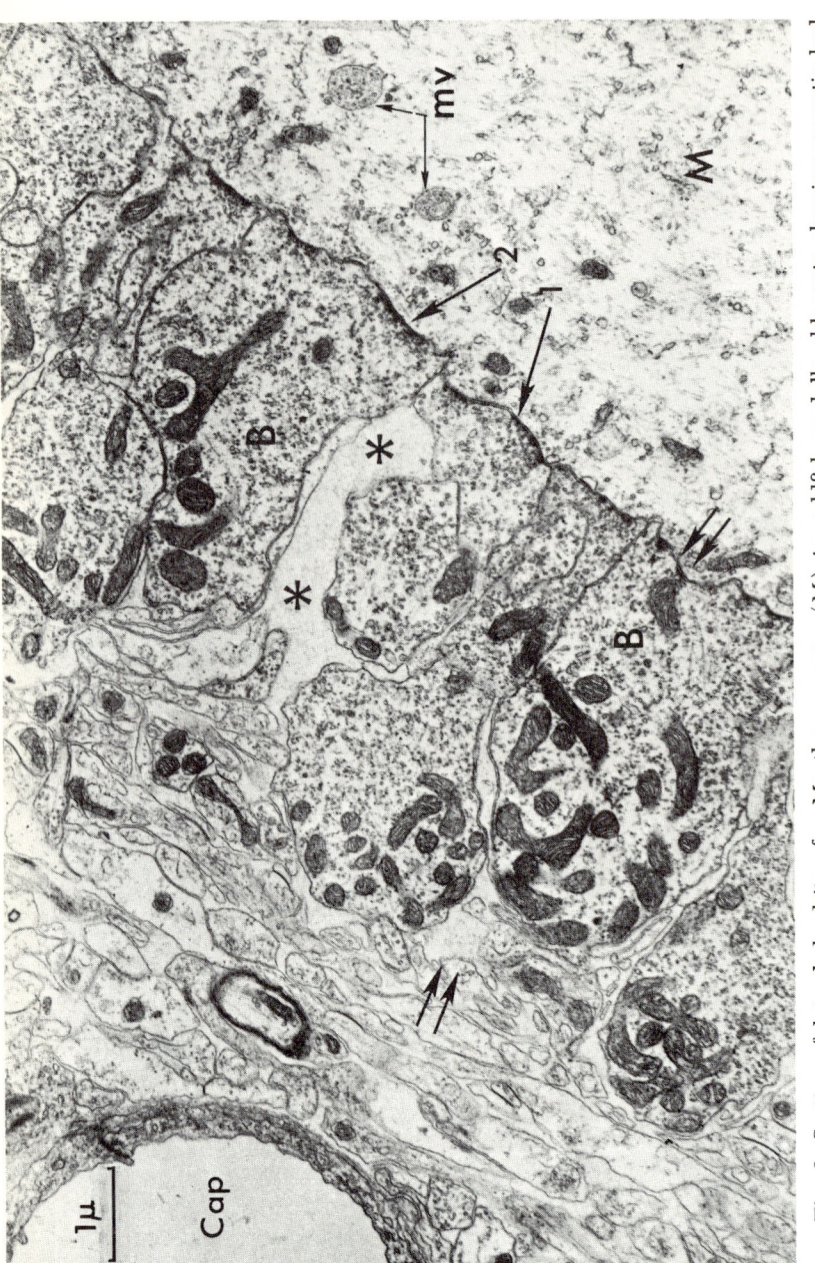

Fig. 2. Section of lateral dendrite of a Mauthner neuron (M) in goldfish medulla oblongata showing synaptic bed (between double arrows). The bouton terminaux (B) are filled with synaptic vesicles and mitochondria. Extracellular matrix material (*) is present and is a special feature of the synaptic bed of Mauthner's neurons. Accumulations of synaptic vesicles are present next to presynaptic membranes (arrows 1 and 2). Multivesicular bodies are shown in the lateral dendrite (mv). A capillary lies in the upper left (Cap). Osmium fixation, potassium nitrate and lead stained. ×10,000. After Robertson et al. (1963).

Neurotubules (200 Å in diameter) may also be present but are more frequently observed in unmyelinated axons. The axoplasm contains oval or elongated mitochondria with double membranes and often tubular patterns of cristae and occasional membrane bound vacuoles. Neurofilaments are frequent, especially in myelinated axons. Myelin sheaths of central nervous system axons have laminations showing a periodicity of major dense lines lying 120–140 Å apart, separated by intraperiod lines. Schmidt-Lanterman clefts are occasionally observed in the myelin sheath of central nervous system axons. Unlike mammalian central nervous system, nodes of Ranvier are frequent along axons of the fish central nervous system. Internodal segments are often of great length in the central nervous system of fish and the area of the node devoid of the myelin sheath may be considerable.

Bouton terminaux are characterized by numerous synaptic vesicles, mitochondria, neurofilaments, and occasional vesicles (Long et al., 1968; Pappas, 1966; Robertson, 1963; Robertson et al., 1963). The synaptic specialization has pre- and postsynaptic membrane thickenings (synaptic bars) and a 250-Å synaptic cleft (Fig. 2). Electrotonic synaptic junctions have been described in the medulla oblongata and spinal cord of puffers and electric fish (Bennett et al., 1967a–d; Nakajima et al., 1965; Pappas, 1966) and has been postulated for the club endings on Mauthner neurons (Robertson, 1963; Robertson et al., 1963).

In the primitive myxinids, *Myxine glutinosa,* nerve cells have a clear, rounded, relatively large nucleus occupying a central position in the cell body (Mugnaini, 1967). The cytoplasm contains ergastoplasm, free ribosomes, Nissl bodies, and large, heterogeneous dense bodies. Dendrites and axons contain neurotubules which pass from the perikaryon into the cell processes. Axosomatic synapses are usually absent, but axodendritic synapses occur frequently (Mugnaini, 1967).

The neurons of the tiger, *Galeocerdo cuvieri,* hammerhead, *Sphyrna zygaena,* and nurse shark, *Ginglymostoma ciriratum,* frequently have round nuclei containing a prominent nucleolus and abundant cytoplasm containing neuronal organelles typical of vertebrates (Bakay and Lee, 1966; Long et al., 1968). Special neurosecretory neurons are located in the preoptic nucleus and have been described in the goldfish (Palay, 1957, 1960) and *Zoarces viviparus* (Oztan, 1967), in the adenohypophysis of *Zoarces viviparus* (Oztan, 1966), and in the caudal neurosecretory system of teleosts (Fridberg and Bern, 1968). The neurohypophyseal neurosecretory neurons in the preoptic nucleus contain marginated Nissl substance, an eccentric folded nucleus, a central zone of Golgi apparatus, mitochondria, multivesicular bodies, and two size classes of secretory droplets, 0.1 and 1.0 μ and larger (Palay, 1957, 1960).

In teleosts bluegill, *Lepomis macrochirus*, and sand bass, *Paralabrax nebulifer*, axosomatic synapses have been observed in the optic tectum on some undifferentiated cells of the subependymal "matrix" layer, suggesting their identification as neurons (Kruger and Maxwell, 1967). These observations are in agreement with the work of Kirsche (1967) at the light microscopic level who has described continuous generation of neurons by matrix layers of the crucian carp optic tectum. Glycogen particles appear to be more widespread in lower vertebrates where they are usually present in endothelial and ependymal cells, and to a variable extent in neurons (Kruger and Maxwell, 1967). Because of fragility during fixation, neurons appear to lose glycogen granules. Since glycogen is found in some neurons these cells could be the ultimate consumers of this source of glucose (Kruger and Maxwell, 1967).

Neuroglial cells can be classified as astrocytes and oligodendrocytes. These two neuroglial cell types have been observed in the author's laboratory in the goldfish spinal cord and have been observed in the goldfish medulla oblongata (Robertson *et al.*, 1963), shark brain (Long *et al.*, 1968), and bluegill and bass brain (Kruger and Maxwell, 1967). However, two neuroglial types have not been observed in the telencephalon of the myxinid *Myxine glutinosa* (Mugnaini and Walberg, 1965), the lamprey spinal cord, *Petromyzon marinus* (Schultz *et al.*, 1956), and *Lampetra planeri, L. zanandreai, and L. fluviatilis* (Bertolini, 1964, 1966). The presence of one neuroglial cell type perhaps could have been predicted since the oligodendrocyte is the glial element that produces the myelin sheath of central nervous system axons and the myxinids and lampreys possess only an unmyelinated nervous system.

In the myxinid telencephalon, *Myxine glutinosa* L. (Mugnaini and Walberg, 1965), and lamprey spinal cord, *Petromyzon marinus* (Schultz *et al.*, 1956), glial cells are classified as one class. The processes of the neuroglial cells intrude between the nervous elements of the neuropil or encircle neuronal perikarya. The nuclei of neuroglia are often smaller, more irregular in outline, and contain denser chromatin than the nuclei of nerve cells. The nucleus and cytoplasm of myxinid neuroglial cells often contain rodlets (Mugnaini, 1967). Neuroglial cytoplasm also contains numerous mitochondria, limited ergastoplasm, fine neuroglial filaments, and heterogeneous dense bodies. Desmosomes are often observed between contiguous cells. Neuroglial cells with their cell bodies and processes form an extensive layer covering the subpial or ependymal surface (Mugnaini, 1964) and blood vessels (Mugnaini and Walberg,

1965). The pericapillary space (100–200 Å) between the endothelial cells of the blood vessel and neuroglial cell contains basement membrane. Neurons are not contiguous to endothelial cells of the capillary. Neuroglial processes lie between blood vessels and neurons (Mugnaini and Walberg, 1965). In the shark the two neuroglial types are observed (Long et al., 1968). Astrocytes are smaller and contain fewer organelles than neurons. Glial fibrils are often present in the cytoplasm. Oligodendrocytes are smaller than astrocytes, have denser cytoplasm, and possess intensely staining nuclei (Long et al., 1968). These two neuroglial cell types are also found in bluegills, sand bass, and goldfish (Kruger and Maxwell, 1967; Robertson, 1963; Robertson et al., 1963). Capillary walls in the cyclostome (Mugnaini and Walberg, 1965) appear to be nearly completely invested with astrocyte processes. In the teleost and shark this investment may be partially derived from ependymal cells (Kruger and Maxwell, 1967; Long et al., 1968).

The processes of ependymal cells of fish can extend from the ventricular to the pial surface of the brain and generally maintain an orientation perpendicular to the pial surface. The smooth endoplasmic reticulum of ependymal cells of the sand bass, *Paralabrax nebulifer,* is organized into discoid clusters of membrane at irregular intervals across the long axes of the processes (Kruger and Maxwell, 1966). The ependymal cell processes generally contain irregularly scattered small densecore vesicles resembling secretory granules. Zona occludens (tight junctions) are numerous at the subpial ependymal end-foot contacts and appear to intermix with and contact with glial processes. Ependymal and glial end-feet comprise the subpial specialization in teleosts (Kruger and Maxwell, 1967). The extensive distribution of vacuoles in the subpial processes may reflect transport in the direction of exchange of materials with the subarachnoid space (Rahmann, 1968). The extensive distribution of glycogen and the occasional contacts with capillary walls suggests that ependyma may parallel some of the functional properties of astrocytes which are the principal perivascular and glycogen-laden elements in mammalian brain (Kruger and Maxwell, 1966). Ependymal cells possess cilia and microvilli that extend into the ventricles and spinal canals (Schultz et al., 1956). The most remarkable morphological feature of teleost ependymal cells is the evidence for a specialized metabolic role for ependyma because of the large quantities of glycogen and mitochondria of unusual size. Ependymal cells of the teleost brain appear to be pluripotential, retaining the embryonic capacity to dedifferentiate and redifferentiate into spongio- (glial precursors) and neuroblasts (neuronal precursors) (Kirsche, 1967; Fridberg et al., 1966).

III. TELENCEPHALON

A. Olfactory Bulb Anatomy

In the fish, as in other animals, the sensory receptors for olfaction project to the olfactory bulb (Wilson and Westerman, 1967). The axons of the cells from the bulb are the primary efferent axons of the olfactory system and pass posteriorly into the telencephalon. In most vertebrates the olfactory bulbs are located immediately adjacent to the wall of the nasal cavities and thus the olfactory tracts are short. However, in certain teleosts the olfactory bulbs are separated from the telencephalon proper by a long olfactory tract (Fig. 3)

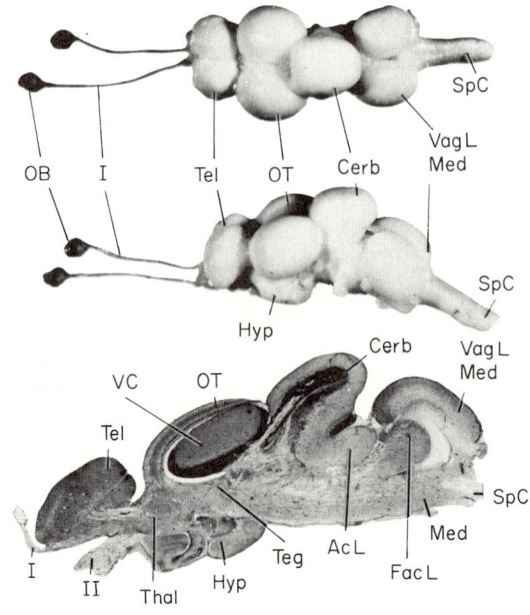

Fig. 3. Dorsal (upper), lateral (middle), and longitudinal (lower) section of a goldfish brain, *Carassius auratus*. The longitudinal section was stained with gallocyanin and periodic acid-Schiff counterstain. Abbreviations: I, olfactory nerve; II, optic nerve; Ac L, acoustic lobe of medulla oblongata; Cerb, cerebellum; Fac L, facial lobe of medulla oblongata; Hyp, hypothalamus; Med, medulla oblongata; OB, olfactory bulb; OT, optic tectum; Sp C, spinal cord; Teg, tegmentum; Tel, telencephalon; Thal, thalamus; Vag L Med, vagal lobe of medulla oblongata; and VC, valvula cerebelli.

(Schnitzlein, 1964; Nieuwenhuys, 1967b). In the gar, *Lepisosteus osseus,* the bulb can be many centimeters from the olfactory integrating centers found within the piriform area of the telencephalon. In other species such as sharks, *Scylliorhinus,* the two bulbs lie lateral to the telencephalic hemispheres.

In most fish, the mitral cells are arranged in a diffuse layer which lies immediately central to the glomerula zone of the nasal cavity. However, mitral cells can be found to occupy more peripheral positions within the bulb and often are scattered along the glomerulus in lampreys (Nieuwenhuys, 1967b). With few exceptions interglomerula association cells have been demonstrated for selachians, actinopterygians, and dipnoans. Within the olfactory bulb there are granular elements which form the periventricular bulbar layer of gray matter. This layer of gray matter is extensive, and its neurons provide distinct axons which contribute to the olfactory tract. Dendrites of the cells within the gray matter comprise a significant portion of the glomerulus. Within the deep granular layer of the glomerulus some neurons possess short axons that innervate other glomerular neurons. Thus, not all the granular cells send axons into the olfactory tract.

The olfactory bulbs are joined by an interbulbar commissure located in the brain proper. Efferent nerve fibers from the olfactory bulb have been located in the anterior commissure, habenular commissure, and other areas (Nieuwenhuys, 1967b). These anatomical observations have been corroborated by electrophysiological data which have indicated that the central interbulbar commissures are part of the neural pathway for the integration of olfactory information.

B. Telencephalic Anatomy

The anatomy of the telencephalon of fish has presented certain unique problems for neuroanatomists because structures from which homologies are drawn in higher vertebrates appear to be lacking in the fish telencephalon. For instance, there is the lack of a demonstrable lateral ventricle, the lack of a prominent ventricular sulcus, and the presence of an extensive and thin roof plate. For these reasons it is difficult to establish homology between comparable regions of the fish telencephalon and the telencephalon of other vertebrates (Droogleever Fortuyn, 1961; Nieuwenhuys, 1962a,b, 1967b; Schnitzlein, 1964). The telencephalon of bony fishes as in other vertebrates has its embryonic origin in a tube-shaped structure (Nieuwenhuys, 1959, 1960, 1962a–d, 1963, 1967b). As the telencephalon develops there appears to be only an eversion of the dorsal part of the lateral wall of the telencephalon concomitant with an expansion of the mass of the telencephalic hemispheres. This eversion is not followed, as in mammals, by evagination and involution of the hemispheres (Aronson, 1963; Nieuwenhuys, 1966, 1967b). The eversion of the telencephalon results in the following early telencephalic characteristics: The lines of attachment of the roof plate are shifted laterally and ventrally in comparison to their original dorsomedial position; the ependymal cells which only line the ventricles in mammals are found on the outer surface of the telencephalon of fish and the main blood vessels that supply the telencephalic parenchyma are found to enter only to the ventral surface (Nieuwenhuys, 1962a,b; Segaar, 1965; Segaar and Nieuwenhuys, 1963). These embryonic characteristics result in the location of the highest density of neurons at the periphery of the telencephalic wall; the nerve cells appear to be inverted in comparison to higher forms of animals. The ependymal cells possess long processes that converge toward the basal surface of the adult fish forebrain. The vascular, ependymal, and glial patterns persist into adulthood and are extremely characteristic of the eversion process (Nieuwenhuys, 1967b). However, there are many variations on the eversion theme. In the shark there are isolated islets of neurons and glia (Klatzo, 1967). In the primitive Polypteriformes, a unique actinopterygian fish, the eversion of the telencephalon has not been accompanied by the thickening of the telencephalic wall and thus the telencephalon is saclike in appearance. In contrast, in the goldfish *Carassius auratus*, the telencephalic walls are greatly thickened and in dipnoans the thickening is accompanied by slight involution of the telencephalic walls (Aronson, 1963; Nieuwenhuys, 1967b).

Therefore, it is difficult to determine if the telencephalon of the fish has evolved along separate lines or contains areas homologous to other vertebrate species. For this reason two basically different systems of nomenclature have been developed for the telencephalon (Fig. 4). There are workers who have homologized areas of the telencephalon according to certain characteristics found in other vertebrates. This represents the classic approach to comparative neuroanatomy (Ariëns Kappers *et al.*, 1960; Droogleever Fortuyn, 1961; Schnitzlein, 1964). In general three criteria have been used to establish homologies: (1) location and general relation of a zone or mass of parenchyma, (2) the histology and cytology of the zone, and (3) the efferent and afferent connections of the area. Using these criteria, it appears that primordia and thus homologs of various areas of the telencephalon (which are adequately developed in higher vertebrates) can be homologized in the fish forebrain.

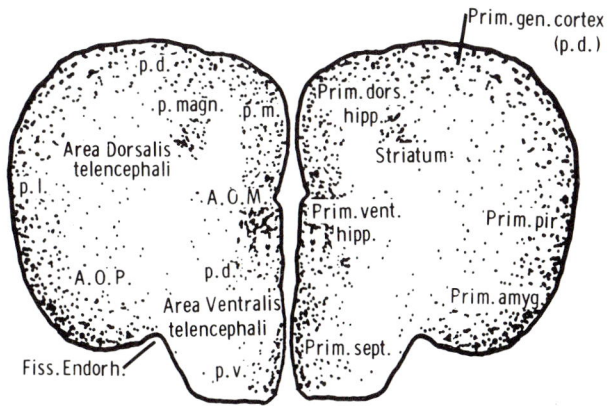

Fig. 4. Schematic of a cyprinid telencephalon. The left tectal hemisphere is labeled according to the nomenclature of Nieuwenhuys. The right tectal hemisphere is labeled according to the classic neuroanatomical nomenclature. A.O.M., area olfactorius medialis; A.O.P., area olfactorius posterior; Fiss. Endorh., endorhinal fissure; p.d., pars dorsalis; p.l., pars lateralis; p.m., pars medialis; p. magn., pars magnocellularis; Prim. amyg., primordial amygdala; Prim. dors. hipp., primordial dorsal hippocampus; Prim. gen. cortex (p.d.), primordial general cortex (pars dorsalis); Prim. pir., primordial piriformis; Prim. sept., primordial septum; Prim. vent. hipp., primordial ventral hippocampus; and p.v., pars ventralis.

Primordia of various structures of the cortex or pallium that are seen in higher animals have been defined (Droogleever Fortuyn, 1961; Schnitzlein, 1964). The olfactory bulb and striatum of the telencephalon make up the bulk of the hemisphere. By definition, cortex is the bark or peripheral structure of the striatum. Peripheral zones of the hemispheres have been the morphological criterion used for corresponding structures of the higher vertebrate cortex. Therefore, the primordial septum is on the ventral medial surface and the primordial hippocampus on the medial and dorsal medial surface of the hemisphere facing the common ventricle. This area can then be traced around to the dorsal surface of the telencephalon and is then designated as the primitive general cortex (Ariëns Kappers et al., 1960). The lateral portion of the telencephalic lobe has been designated as the primordial piriform area and the lateral ventral surface the primordial amygdala (Fig. 4).

The areas of primordial hippocampus, entorhinalis, and portions of the nucleus amygdala as well as some portions of the septal areas (Segaar, 1965) may belong to the telencephalic portion of the limbic system. The telencephalon was formerly thought to be exclusively olfactory in function; consequently, both the olfactory bulb and telencephalic limbic system were combined in the term "rhinencephalon."

Recent studies in mammals and in fish have demonstrated that areas designated as limbic portions of the telencephalon may be involved in emotion (Segaar, 1965).

Carefully following the development of the telencephalon, Nieuwenhuys (1963, 1966, 1967b) has concluded that the telencephalon of actinopterygians only everts and thickens. Therefore, he has not homologized the areas of the teleostian telencephalon to higher vertebrate nervous systems (Nieuwenhuys, 1967b), but has divided the telencephalon into two principal regions, the area ventralis and the area dorsalis telencephali (Fig. 4). The ventral area of the telencephalon receives secondary olfactory fibers and is a rather constant structure throughout the whole subclass of bony fishes. However, there appears to be a progressive differentiation of the dorsal telencephalic area. In the Chondrostei, Holostei, and Teleostei neurons are located throughout the entire mass of the telencephalic wall. These neurons comprise the nuclear masses of telencephalic gray that are identified as the telencephalic nuclei. In the Holostei the caudal area of the area dorsalis appears to receive secondary olfactory fibers. However, in the higher teleost fishes these fibers can be traced almost exclusively to the caudal ventral part of the area dorsalis. In the evolutionary sequence through Chondrostei, Holostei, and Teleostei there appears to be a decrease in the influence of the olfactory area on the area dorsalis of the telencephalon as sequentially increasing connections of the dorsal area of the pars dorsalis of the telencephalon with the ventral portion of the diencephalon are observed. In the most advanced bony fishes, the area dorsalis telencephali and the ventral thalamus possess increasing numbers of connections. In the teleost we also find an advance in differentiation by an increase in the mass of the area dorsalis telencephali, the lateral forebrain bundle, and other areas. The forebrain of the bony fishes appears to have developed into a highly differentiated complex center that processes much more than olfactory information. These neural centers have developed strong diencephalic connections (Nieuwenhuys, 1963, 1967b).

C. Neurophysiological Studies of the Olfactory System

Neuroanatomical and behavioral studies suggest that olfaction plays a major role in the behavior of fishes (Kleerekoper, 1967; Hasler, 1957, 1966; Healey, 1957). It is obvious from the anatomical data that the influence of the telencephalic and diencephalic areas subserving olfaction is species dependent (Nieuwenhuys, 1967b). The spontaneous activity of the neural elements in the olfactory bulb of fish also appears to differ from species to species. In goldfish, *Carassius auratus*, the spontaneous activity of the surface of the olfactory bulb was consistently of relatively high amplitude (Hara and Gorbman, 1967). Frequencies of 14–16 Hz with amplitudes of 70–100 μV peak to peak were obtained from the olfactory bulb. Infusion of salt solutions into the nasal cavity evoked synchronous waves of comparatively high amplitude (150–200 μV). This synchronous pattern of discharge decreased and was subsequently followed by desynchronous wave patterns with relatively higher frequencies and lower amplitudes (Hara and Gorbman, 1967). However, spontaneous activity in the olfactory bulb of homing and juvenile salmon of the species *Oncorhynchus tshawytscha* and *O. kisutch* had discharges of relatively lower frequency, 7–9 Hz with amplitudes of 35–65 μV (Hara et al., 1965). In contrast to this, young hatchery-reared salmon of the same species exhibited consistently higher frequencies of 8–10 Hz with amplitudes of 35–65 μV (Hara et al., 1965).

The evoked response of the olfactory bulb increased in magnitude (linearly) with the infusion of the nasal cavity with sodium chloride solutions of increasing concentrations (Hara and Gorbman, 1967). Transection of the ipsilateral medial bundle of the olfactory tract resulted in augmentation of the synchronized electrical activity of the olfactory bulb in response to sodium chloride infusion. If the lateral bundle of the ipsilateral olfactory tract was sectioned there was no change either in the spontaneous discharge pattern or in the evoked synchronized activities within the bulb. When the olfactory bulb was stimulated electrically, transection of one of the olfactory tracts augmented the efferent-induced activity and accelerated the excitability cycle, resulting in reduction of the threshold for electrically evoked potentials in the olfactory bulb. These data indicate that centrifugal tonic depressive influences are normally exerted on the excitability of the olfactory bulb and that this is mediated through the medial bundle of the olfactory tract (Hara and Gorbman, 1967).

Repetitive electrical stimulation to the ipsilateral olfactory bulb depressed spontaneous activity in the contralateral olfactory bulb (Hara and Gorbman, 1967). This activity was also depressed after the application of strong olfactory stimuli but was facilitated and synchronized by weaker olfactory stimuli. These phenomena indicate that there is an inhibitory action derived from the activity of the contralateral bulb via the olfactory tract. Furthermore, since no recognizable change occurred after sectioning of the contralateral olfactory tract, it is likely that the inhibitory influence in the contralateral bulb is not of a tonic nature but may be triggered only when the stimulation to the ipsilateral bulb is of relatively greater magnitude in comparison to the stimulation of the contralateral olfactory bulb. In addition, the patterns of electrical stimulation of the contralateral olfactory bulb to both the physiologically and electrically induced discharges were depressed when the anterior commissure of the telencephalon was electrically stimulated (Hara and Gorbman, 1967). This confirms electrophysiologically the earlier neuroanatomical observation that interbulbar connections carrying olfactory information are mediated through the anterior commissure (Nieuwenhuys, 1967b).

Stimulation of the posterior portion of the telencephalon produced characteristic changes in the synchronous electrical activity of the olfactory bulb (Hara and Gorbman, 1967). These changes varied with changing frequencies of stimulation. Stimulation of the anterior part of the telencephalon only slightly depressed the intrinsic electrical activity of the bulb and then only at high frequencies and intensities of stimulation. Stimulation of the preoptic area of the diencephalon facilitated responses in the ipsilateral bulb. It would appear that the neurons of the preoptic nucleus were directly affected by stimulation of the olfactory bulb. These data indicate that there are olfactory centrifugal systems that exert a tonic inhibitory influence upon the olfactory bulb of the goldfish. This influence is mediated along centrifugal nerve fiber bundles in the medial olfactory tract. The results of the telencephalic stimulation studies strongly suggest that the bulb can be influenced by activity originating in the ipsilateral hemisphere. There is central regulation of afferent transmission within the nervous system of fish (central control of periphery). This appears to be a common mode of neural integration (Hara and Gorbman, 1967).

The electrophysiological properties of the nerve fibers of the olfactory tract of the burbot, *Lota*, were studied by Döving and Gemne (1965, 1966). The rostral portion of the olfactory tract was stimulated and potentials recorded from the transected caudal stump. The action potentials of each of four classes of nerve fibers showed two initial components and a third late component. Three components were evident in the action potentials recorded from the olfactory tract nerve fibers. These components varied in conduction velocity and refractory period in the medial and lateral divisions of the olfactory tract. The units recorded from the olfactory tract had a mean absolute refractory period of 3.7 msec. The relative refractory period for the units are 16–20 msec for the first component, 28–39 msec for the second component, and about 120 msec for the third component. Neurons in the olfactory bulb were found to be spontaneously active and efferent impulses could be evoked by stimulation of the ipsi- and contralateral olfactory tracts (Döving and Gemne, 1966). The postetanic potentiation of the response together with other evidence in the burbot indicate that as in the goldfish (Hara and

Gorbman, 1967) there is a system of efferent nerve fibers in the olfactory tract derived from the telencephalon proper. The cell bodies of these efferent neurons were activated both by ipsi- and contralateral stimulation of the olfactory tract and demonstrated long periods of suppressed excitability when single pulses of electrical stimulation were applied to the opposite tract (Döving and Gemne, 1966). It is also interesting to note that the spontaneous activity of the second-order neurons in the olfactory bulb were inhibited by touch stimuli to the skin. There is then a possibility of extraolfactory influences ascending from lower brain centers that can influence olfactory bulb activity. In addition, the second-order neurons of the olfactory system could be affected by physiological stimulation of the olfactory epithelium with various solutions. After transecting the olfactory tract recordings could be taken from single units in the olfactory bulb for several hours (Döving and Gemne, 1966). Using a large variety of olfactory stimuli it was found that 30% of the stimuli produced increased activity in the second-order neurons of the olfactory bulb and approximately 20% caused inhibition of the spontaneous discharge of these units. Approximately half of the stimulating solutions failed to produce any change in the discharge rate. These data demonstrate the influence of rhinencephalic centers of the telencephalon as well as the diencephalic preoptic area upon the discharge rate of the olfactory bulb and thus central control of periphery.

The rate of olfactory bulb discharge can develop a differential discharge pattern (Hara et al., 1965; Ueda et al., 1967). Neurons in the olfactory bulb of Pacific coho and Chinook salmon, *O. kisutch* and *O. tshawytscha*, responded differentially to water from the pool in which it was known the animals would spawn and to water from different watersheds. High amplitude responses were recorded from home pond water infused into the nasal cavity but not to water from different breeding ponds. Weaker responses could be evoked in the olfactory bulb by waters traversed by the spawning salmon when swimming toward the spawning site. Water was taken from a branch of the stream that passed by the spawning site or from water above the spawning site (Ueda et al., 1967). This is remarkable since the adult pacific salmon had not been exposed to these precise olfactory stimuli for many years. Therefore, the response appears to be "imprinted" in the central nervous system of the young salmon (neuronal imprinting).

D. Electroencephalography of the Telencephalon

The EEG's of the telencephalon of several species of sharks (Hodgson et al., 1967) and bony fishes have been recorded. Recordings from surface electrodes on the telencephalon of goldfish, *Carassius auratus,* reveal a dominant electrical pattern of spontaneous low amplitude electrical potentials of 9–14 Hz in a quiet, dark room (Schadé, 1959; Schadé and Weiler, 1959). These waves tended to be accentuated by general illumination levels or moderate sounds in the room. Occasionally a telencephalic alpha rhythm was present in the animals, demonstrating what has been classified as "arousal" in higher vertebrates. Although the spontaneous electrical activity of the fish telencephalon can be influenced by visual and auditory stimuli, evoked responses cannot be elicited in the telencephalon after visual or auditory stimuli (Karamyan, 1956). Electroencephalograms recorded from the telencephalon of codfish, *Gadus callarias,* show a dominant rhythm of low amplitude 8–13 Hz rhythm (Enger, 1957). However, slower rhythms of less than 7 Hz were most pronounced in the telencephalon, but they were relatively unaffected by changes in the environment.

E. Telencephalon and Reproductive Behavior

The telencephalon plays an important but varied role in the reproductive behavior of fishes. Removal of both hemispheres of the telencephalon of *Hemichromis bimaculatus* and *Betta splendens* resulted in complete loss of all sexual behavior. However, complete removal of the telencephalon of *Tilapia macrocephala* only resulted in the loss of specific portions of the mating behavior (Aronson, 1948). There were differential losses in courtship behavior such as a decrease in nest building activity and deficits in the ability of males to fertilize the eggs. Subtotal extirpation of the telencephalon of the guppy, *Lebistes reticulatus,* inhibited but did not prevent normal mating and aggressive behavior whereas complete removal of the telencephalon in *Xiphophorus maculatus* resulted in a depression of the frequency of copulation with no indication that any sexual pattern within the courtship of the animal could be completely eliminated (Kamrin and Aronson, 1954). Since removal of the telencephalon did not eliminate the behavior it was suggested that there is a facilitative action of the telencephalon on lower centers and that the telencephalon does not play a direct role in the organization of sexual behavior pattern. Thus, Aronson (1967) characterized the telencephalon as a general sensitizor or facilitator of mechanisms present in lower portions of the brain (Kamrin and Aronson, 1954).

Removal of portions of the telencephalon of the stickleback, *Gasterosteus aculeatus,* resulted in losses of specific aspects of reproductive behavior and weakening of sequential patterns of its complex series of reproductive behavioral responses (Segaar, 1965). In a series of experiments, Segaar removed discrete portions of the stickleback telencephalon and found differential suppression of aggressive, sexual, and parental behavior (Segaar, 1956, 1961, 1965; Segaar and Nieuwenhuys, 1963). After removal of either the rostral or the lateral portion of the telencephalon aggression was greatly suppressed, sexual activity was short in duration, and parental care was rather extensive. However, after ablation of the medial caudal portion of the telencephalon these patterns appeared to be reversed, with elevated aggression, continuous sex drive, and low parental activity. Following medial lesions of the telencephalon, sticklebacks demonstrated rising aggressive behavior and depression of sex drive, but otherwise they appeared to be normal. Lesions in the extreme medial part of the caudal medial and dorsal medial areas of the telencephalon resulted in disinhibition of parental activity while aggression and sex drive were normal (Segaar, 1965).

These data show that specific telencephalic lesions result in disruption of the balance of specific behavioral responses such as aggressive behavior, sexual activity, and parental behavior in sticklebacks (Segaar, 1965). The telencephalon appears to function as an organizer and integrator of segments of innate (DNA contingent) behavior and maintains aggression, sexual activity, and parental behavior of fishes in the proper balance to insure successful reproduction of the species.

F. Specific Sensory Deficits following Telencephalic Ablation

Color vision in goldfish has been studied using cardiac deceleration, a conditioned autonomic response, as the measure of discrimination (McCleary and Bernstein, 1959). After establishing the relative brightnesses of two pairs of red and green stimulus patches as perceived by the fish, these same stimulus patches were used to determine whether a subsequent discrimination by the fish was based on hue or brightness. Stimulus generalization was used as a method of evaluating what aspect of the stimulus was used by the fish in learning a given discrimination. Goldfish were found to be capable of red–green hue discrimination under experimental conditions which ruled out the possibility that brightness cues were being employed. The goldfish was also found to be capable of stimulus generalization involving either hue or brightness characteristics of the stimuli (McCleary and Bernstein, 1959).

Forebrain extirpation resulted in a selective suppression of the color visual discrimination in goldfish. Immediately following complete bilateral telencephalic ablation, animals trained to red and green stimuli of equal brightness and tested with red and green stimuli of differing brightnesses were found to respond only to the brightness differences of the test stimuli (Bernstein, 1961b). If normal fish learned to discriminate stimuli that differed in both hue and brightness, their ability to discriminate was impaired when they were subsequently tested with stimuli having identical hues but reversed brightness. Such a reversal test did not confuse the operated fish since they persisted in reacting to the brightness differences despite the confounding of hue–brightness relationships (Bernstein, 1961b).

Forebrain-ablated goldfish trained to gray stimuli of varying brightnesses and tested on black and white stimuli were found to make brightness discriminations whereas normal fish did not make this discrimination. Furthermore, fish lacking a forebrain were able to generalize to a different brightness problem (Bernstein, 1961a).

Interocular transfer of a hue discrimination was made by normal goldfish and goldfish with the forebrain ablated contralateral to the trained eye. However, ablation of the forebrain homolateral to the trained eye or bilateral forebrain ablation and subsequent immediate training and testing (within 10 min) resulted in a loss in ability to make an interocular transfer of a hue discrimination. Training and testing preceded by a 4-hr interval after complete telencephalic ablation resulted in spontaneous recovery of color visual function and interocular transfer of an acquired hue discrimination (Bernstein, 1962, 1963). These results demonstrate that the telencephalon is not essential for color vision but does partially function in the integration of color visual information. This color visual integrating function of the telencephalon can be assumed by lower brain centers. Spontaneous recovery of nervous system function is a well-known phenomenon in vertebrates (Ruch, 1958).

G. Telencephalon and Learning

Ablation of the telencephalon of goldfish greatly impaired or prevented the acquisition of an instrumental response (one-way conditioned avoidance), abolished retention of the response in animals when the task was previously learned, and greatly reduced resistance to extinction in one-way avoidance learning (Hainsworth *et al.,* 1967). However, forebrainless goldfish eventually do learn the response. *Tilapa macrocephala* have been trained to escape in conditioned avoidance training by passing through a 4-cm hole to avoid shock [unconditioned stimulus (US)] in response to the onset of a light [conditioned stimulus (CS)]. Animals were trained to avoid the shock within 2.5 sec CS–US interval

(Kaplan and Aronson, 1967). After the animals had learned the response, the olfactory bulbs were ablated. Removal of olfactory bulbs before or after conditioning had no effect on the animal's ability to make the conditioned avoidance response (Aronson and Kaplan, 1963; Kaplan and Aronson, 1967). Following telencephalic ablation there was no obvious sensory or motor deficit. However, there was a marked decline in performance. The ablation resulted in a rapid drop in the percentage of avoidances and an increase in latency to avoidance. Forebrain-lesioned fish exhibited varying degrees of improvement with continued training; however, the preoperative level of performance was not attained after ablation. If the telencephalon was removed prior to training, the forebrain-ablated fish were unable to learn the avoidance response (Aronson and Kaplan, 1963; Aronson, 1967; Kaplan and Aronson, 1967).

Goldfish were also trained to avoid shock after the onset of the conditioned stimulus (light) by swimming through an area which was previously covered by a partition (Hainsworth et al., 1967). The animals were then returned to the starting box with a net (one-way conditioned avoidance). Normal and sham-operated animals showed no significant differences in learning. In contrast, forebrain removal resulted in marked impairment of acquisition of the avoidance response. Following acquisition of the response, forebrainless fish exhibited rapid extinction leading to the conclusion that this type of learning in forebrainless fish is an unstable association. If the animals were trained and then the telencephalon removed, the operation resulted in a complete loss of the previously learned avoidance response (Hainsworth et al., 1967). However, unlike *Tilapia* (Kaplan and Aronson, 1967) the forebrainless goldfish did eventually relearn the avoidance response to preoperative levels. The animals required as many trials to relearn the problem without the telencephalon as they did with the telencephalon. Removal of the forebrain resulted in the complete loss of the previously learned response since the animals demonstrated no savings from previous training when compared with the performance of previously naive forebrainless subjects (Hainsworth et al., 1967). Fish spontaneously recovered the ability to acquire and retain a conditioned avoidance response if sufficient time was allowed after forebrain removal before subsequent retraining. These data indicate that removal of the telencephalon results in the loss of ability to acquire instrumental conditioning, greatly impairs or prevents the acquisition of instrumental avoidance conditioning, abolishes the retention of a previously learned response, and greatly reduces the resistance to extinction (Hainsworth et al., 1967). The specific and permanent deficits found in the forebrainless goldfish in this series of experiments suggest that in fish the telencephalon subserves complex learned and emotional behavior and provides evidence for functional homologies between the mammalian limbic system and parts of the forebrain of fish. These data are in support of Segaar's hypothesis (1965) that the function of the telencephalon is partially limbic in nature in reproductive behavior in fish.

In keeping with the findings that forebrainless fish conditioned to light form unstable associations, Kholodov (1960) and Baru (1951, from Karamyan, 1957) found that conditioning to sound or magnetic fields was not affected by forebrain ablation whereas conditioned responses to light were not stable. Crucian carp, ruff, shark, and ray were trained to discriminate a light and a bell. Following removal of the telencephalon animals required as many trials to learn the response as normal fish of the same species. In lamprey, ganoids, and plagiostomes, this conditioned response was unstable since animals did not respond one day after initial acquisition of the task (Baru, 1951, from Karamyan, 1957).

Aronson and Herberman (1960) trained *Tilapia* to an operant task. Animals were rewarded with food for striking a small Plexiglas target. After subjects were trained to criterion, normal, sham-operated, and fish with both olfactory bulbs ablated continued to respond with average latency. Animals with complete forebrain ablation responded appropriately, but their average latency of response and variability in latency of response increased considerably. Some animals with complete telencephalic ablation ceased responding after the seventh and eighth trial on the first day postoperative.

The effect of telencephalic lesions on learning in paradise fish, *Macropodus opercularis*, was tested with a series of *Umweg* problems (mazes of increasing complexity) and discrimination reversal tasks (Warren, 1960, 1961). Telencephalic ablation was found to severely impair the animals' ability to perform in the maze. Normal paradise fish were compared with telencephalic ablated fish in their ability to learn a reversal task. Training was carried out in a T maze in which the targets presented both brightness and positional cues. The fish were to select a black alley on the right. After reaching criterion, the subjects were required to reverse the discrimination habit and to go to the white alley on the left. It was found that forebrain ablated fish were inferior to controls in learning a reversal task in the T maze (Warren, 1960, 1961).

In a continuous Y maze in which goldfish were forced to turn consistently in a specified direction, it was found that forebrainless fish learned the problem and its reversal better statistically than unoperated controls. However, in alternation learning (first the fish go right and then go left), forebrainless fish did not perform as well as controls. However, histological verification of the lesions was not available (Ingle, 1965a,b).

These data are in agreement with the findings with autonomic conditioning to brightness discrimination where forebrainless fish perform better than normals (Bernstein, 1961b). Therefore, certain tasks are performed better by forebrainless fish than normal animals. Although these results appear odd there is an increased body of knowledge in mammals that demonstrate limbic lesions often lead to improved performance in certain specified tasks. Obviously, one must take the normal animals' behavior as a base line for performance and not as maximum (or 100%) performance on given tasks.

These data support experimental results on the learning and retention of conditioned avoidance responses by goldfish in a two-way avoidance task designed to minimize the difficulty of task for forebrainless goldfish (Dewsbury and Bernstein, 1969). The fish were in an open shuttlebox with a minimal barrier at the center and had to cross the barrier at the onset of light (CS) to avoid shock (US). Forebrainless fish were less active than normal fish in both pretraining response rate and intertrial spontaneous cross rate. Forebrainless goldfish learned the avoidance response with a minimal barrier and a 10-sec light shock interval. However, conditioned avoidance learning by forebrainless fish was inferior to normals when the size of the opening for avoidance was reduced and the light shock interval was reduced to 4 sec. Removal of the telencephalon produced significant decreases in the performance of a previously learned avoidance task; however, there was evidence of partial retention (savings) following training, subsequent to forebrain ablation, and retesting. Since specific lesions of the telencephalon results in specific losses of reproductive behavior and losses in ability to perform specific learning tasks the normal function of the telencephalon has been proposed by Dewsbury and Bernstein (1969) to be best conceptualized as a facilitator for the processing of information. The concept of the forebrain as a nonspecific energizer should be reexamined in the face of more recent data.

H. Stimulation of the Telencephalon

Because of the lack of control for spread of current during stimulation, earlier studies on stimulation and behavior must be interpreted with caution (Healey, 1957; Segaar, 1965). In more recent studies, chronically implanted bipolar electrodes have been utilized to stimulate the telencephalon of free swimming goldfish (Grimm, 1960). Stimulation of the olfactory crura at 0.075–0.15 mA with pulses of 0.5 msec duration elicited complete feeding behavior. This included arousal and intention movements of the feeding dance and low intensity feeding commonly seen in normal goldfish. In contrast, stimulation of the lateral and medial olfactory regions of the telencephalon usually did not produce full intention feeding movements. Stimulation of lower brain centers such as the vagal or facial lobes of the medulla, valvula cerebelli, or cerebellum did not elicit feeding behavior (Grimm, 1960).

The brain of fish has been stimulated to ascertain if areas of the nervous system can be used as positive or negative reinforcement during operant conditioning. Portions of the telencephalon (Boyd and Gardner, 1962) were stimulated with chronically implanted monopolar electrodes while the animal was being trained. Tests were made in free operant behavioral situations. In the first series of experiments the animals were tested on a side preference test. Currents of between 5 and 150 mA were delivered to goldfish when on one side and not on the other side of an aquarium. Light was the cue for the association by the fish for the appropriate side of the aquarium during electrical stimulation. The animal was thus free to receive or avoid electrical stimulation by swimming toward or away from the side of the aquarium at which he was stimulated. With progressive current two telencephalic implants (which are difficult to locate anatomically) acted as negative (aversive) stimuli. One implant in the optic tectum acted as positive reinforcement at high levels of stimulation. After these tests, animals were trained to strike a target to get a train of electrical impulses to the brain. Animals with telencephalic implants did not learn the task. However, stimulation of the anterior optic tectum resulted in the fish hitting the target in order to produce self-stimulation (Boyd and Gardner, 1962).

Not only does stimulation of the telencephalon elicit feeding behavior and appear to be a positive or negative reinforcer, but also it appears to have other effects on what might be determined as feeding behavior. Clark *et al.* (1960) found that stimulation of the telencephalon produced movement of the barbels in the maxillary region of the catfish. These responses were extremely stable, and stimulation of the telencephalon would invariably elicit barbel movement. Repeated stimulation of the telencephalon elicited barbel movement in repeatable patterns as if a topography of representation for such movements occurred in the neural mechanisms of the telencephalon (Clark *et al.*, 1960).

I. Regenerative Capacity of the Telencephalon

The regenerative capacity of the telencephalon of fish is related to the age of the animals and to the species of fish involved. The olfactory tract appears to regenerate completely in the carp, *Cyprinus carpio* (Westerman, 1965; Westerman and von Baumgarten, 1964). Sectioning of the olfactory tract resulted in regeneration of the axons from the olfactory bulb. The regenerated axons appeared to selectively reinnervate lower centers in the telencephalon. After regeneration of the olfactory crura, a previously acquired conditioned response to olfactory stimuli was reestablished and operated animals responded appropriately upon subsequent testing with no retraining.

In *Lebistes reticulatus,* the guppy, a complete telencephalon was regenerated within 60–120 days following complete extirpation (Marón, 1963). There appeared to be no difference in the size or shape of the telencephalon in regenerates and normal animals. Segaar found that, excluding lesions of the rostral telencephalon which did not regenerate, selected portions of the telencephalon of the stickleback were reconstituted following ablation (Segaar, 1965; Segaar and Nieuwenhuys, 1963). Regeneration appeared to start from the ependymal layer of the telencephalon. This ependymal layer has been demonstrated to be a matrix zone for the regeneration of the optic tectum and telencephalon (Kirsche, 1965a,b, 1967). New neurons and nerve fibers were developed within the reconstituted telencephalic hemisphere (Segaar, 1965). The reconstituted neural units are arranged in the same cytoarchitecture as found in normal animals. This return of normal cytoarchitectonics was concomitant with the return of normal courtship patterns (Segaar, 1965).

In contrast, the telencephalon of goldfish does not reconstitute ablated areas although tracts do regenerate (Bernstein, 1967). In a series of operations such as either bilateral or unilateral ablation, removal of only the lateral portion of the hemisphere, or stab wounds in the hemisphere, it was found that the parenchyma of the goldfish telencephalon did not reconstitute the ablated mass. Following stab wounds, necrotic areas were replaced by neuroglial and ependymal cells. Although the mass of the telencephalon of goldfish was not reconstituted, new neurons were derived from the ependymal cells lining the common ventricle (Bernstein and Sadlack, 1969). Goldfish were injected with tritiated thymidine following unilateral telencephalic lesions. Neurons were found with reduced silver grains (by autoradiography) on the 8–16 postoperative day. However, there were few new neurons formed and the numbers of total cells (glial, ependymal, and neurons) dividing was low and divisions ceased by the sixteenth postoperative day. The number of total new cells formed were not adequate to reconstitute the ablated telencephalic parenchyma (Bernstein and Sadlack, 1969). Regeneration of the telencephalon is species and age specific in the fish. The nervous system of different species of fish appears to mature at differing rates.

Therefore, maturity of the nervous system of an animal cannot be determined by its sexual maturity (Bernstein, 1965). The disparity of results may be because age and species differences are critical in the ability of fish to reconstitute ablated telencephalic parenchyma.

IV. DIENCEPHALON

A. Anatomy

Although the diencephalon of fish appears to be extremely variable, it can be divided into three zones: the epithalamus, thalamus (subdivided into dorsal thalamus and ventral thalamus; Herrick, 1910), and the hypothalamus. In the elasmobranchs the dorsal thalamus is small and poorly developed. In the more advanced teleosts the dorsal and ventral thalamus are not readily separated because of the problem of locating various sulci that are used in order to determine the homologies of the different areas (Nieuwenhuys and Bodenheimer, 1966). In addition, the various nuclear masses are so varied within teleosts that the picture one could gain from homologies is quite confusing and has been reviewed elsewhere (Aronson, 1963; Nieuwenhuys and Bodenheimer, 1966; Schnitzlein, 1962).

The epithalamus consists of two parts: (1) the pineal complex and (2) the habenular nuclei. The pineal organ has been shown to be light sensitive (Healey, 1957). A conelike photoreceptor has been described in the parietal eye of the lamprey and many sharks and teleosts and is presumably the primary sensory receptor (Altner, 1966; Bertolini and Mangia, 1966; Breucker and Horstmann, 1965; Eakin, 1963; Rüdeberg, 1966, 1968ab). The pineal gland of elasmobranchs consists of three different cell types (Altner, 1966; Rüdeberg, 1968b). All authors have reported that there are sensory cells and supporting cells within the pineal body. The third type of cells, nerve cells, has not been observed in most studies, but are found in the shark, *Scyliorhinus*, associated with rodlike photoreceptors (Rüdeberg, 1968b).

The pineal organ of teleosts contains three different definable cell types. These are the sensory, supporting, and ganglion cells. The outer segments of the photoreceptors are conelike in teleost fish (Oksche and Kirschstein, 1967; Rüdeberg, 1966, 1968a). Ganglion cells have not been found in all species of teleosts but have been observed in *Salmo irideus* (Y. Morita, 1966) and *Oncorhynchus nerka* (Hafeez and Ford, 1967) and are present but difficult to demonstrate in *Mugil* (Rüdeberg, 1968a).

The outer segments of the sensory cells appear to have a structure similar to cones in that there is a laminated outer structure on which photosensitive pigments are presumed to be located. The outer segments vary in shape within each given receptor and also between different species of fish studied (Breucker and Horstmann, 1965; Oksche and Kirschstein, 1967; Rüdeberg, 1966, 1968a). In addition to photosensitivity, the pineal gland may serve an endocrine function in teleosts (Pflugfelder, 1953, 1954, 1964). Aldehyde fuchsin-positive granules have been found in the pineal organ of salmon (Hafeez and Ford, 1967) but these granules have not been found in all animals studied (Holmgren, 1959; Rasquin, 1958; Rüdeberg, 1968a).

The habenular, the second portion of the epithalamus, receives a massive afferent input from the pineal gland and a moderate input from the olfactory region and telencephalon. The efferent outflow of the habenular nuclei is mainly to other thalamic and lower centers.

A dorsal and a ventral thalamus, although present in fish, are difficult to define because of the problem of determining the sulcus medians which is the traditional boundary between these two divisions of the thalamus (Aronson, 1963; Nieuwenhuys and Bodenheimer, 1966). The difficulty homologizing many of the different thalamic nuclei has been intensified by the differences in interpretation of nuclear masses by various authors. For instance, the lateral geniculate nucleus which is found in the lateral portion of the dorsal thalamus may be small and undeveloped, relatively large and considerably developed, or contain a high degree of complexity so that the nucleus is laminated. The complexity of structure is species specific and thus this nucleus of such variable structure has not always been properly identified. In addition to the lateral geniculate we find a rotundus complex of nuclei, pretectal nuclei, which are neurosecretory as well as neural in function, and other thalamic nuclei. The dorsal and ventral thalamus send connections to the hypothalamus, midbrain, and other lower neural centers. There is also a plethora of nerve fibers of passage from the telencephalon through the thalamus to the hypothalamus.

The hypothalamus contains a number of well-differentiated cellular masses. In fish the hypothalamus appears to be a major center for the confluence of information coming from the telencephalon. This afferent input is mainly derived from the medial and lateral forebrain bundles which also terminate in the preoptic area of the thalamus (Nieuwenhuys, 1967b). Nerve fibers from gustatory regions ascend into the hypothalamus as well as fibers from the acoustico-lateralis system. Efferent pathways from the hypothalamus lead to the various parts of the telencephalon and motor centers of medulla, to portions of the dorsal thalamus, optic tectum, cerebellum, and neurohypophysis (Ariëns Kappers *et al.*, 1960).

B. Photosensitivity of the Pineal Organ

Dodt (1963) has studied the photosensitivity of the pineal organ of the rainbow trout, *Salmo irideus*, by recording with steel electrodes during photic stimulation. Stimulation by white light elicited an inhibition of the spontaneous discharge of the organ following a 30-msec delay between onset of stimulus and inhibition of discharge. Following offset of light pulses of 0.1 sec duration there was a resumption of the spontaneous discharge. Light pulses of 0.65 sec duration resulted in a transient inhibitory effect. Thus, illumination of the pineal gland elicited both inhibitory and excitatory changes within the gland. In addition, the pineal gland could also dark adapt and demonstrated a 5.0-log unit loss of threshold after 30 min of dark adaptation. The pineal gland also had a spectral sensitivity of the inhibitory response which was highest in the blue–green range and declined on either side of this spectrum. Sensitivity function of the absorption spectrum demonstrated a visual pigment with a maximal absorption of 505 mμ. Electrophysiological findings compared favorably with the two visual pigments found in the lateral eye of the rainbow trout with a maximum at 505 and 533 mμ (Bridges, 1956). It is interesting to contemplate that the only known visual pigment with an absorption spectrum at 507 mμ is rhodopsin (Dodt, 1963).

C. Electrical Activity of Neurosecretory Cells

There appears to be morphological similarity between the teleost and mammalian neuroendocrine system (Lederis, 1964; Palay, 1957, 1960). However, the systems are not identical in function. The posterior pituitary hormone differed somewhat in its structure in the two classes (Sawyer *et al.*, 1960) and did not influence renal tubular absorption of water in fish as it did in higher vertebrates (Pickford and Atz, 1957). In teleosts the posterior pituitary hormone facilitated sodium ion influx across the membranes of the gills of freshwater fish (Maetz, 1963; Maetz and Julien, 1961; Meier and Fleming, 1962). The cells studied for neuroendocrine function were located in the preoptic nucleus of fish. In fish, the preoptic nucleus does not differentiate into supraoptic and paraventricular nuclei as in higher forms. The neurons of this nuclear mass produce the hormones of the neural lobe of the pituitary in fish (Pickford and Atz, 1957). This nucleus and its outflow, the hypothalamic pituitary pathway, appears to be rather a universal feature of the neuroendocrine system of vertebrates and thus fish can serve as a model for neuroendocrine function.

Kandel (1964) studied the electrophysiological properties of neuroendocrine cells in the magnocellular area of the preoptic nucleus of goldfish. These cells usually range in size from 12 to 30 μ with occasional large cells up to 50 μ. Single cell recordings were made with glass fluid-filled pipettes. Antidromic stimulation of the preoptic cells was accomplished by stimulation of the pituitary gland proper. The average latency for antidromic activation was 6.0 msec. Since axon length was 2.8 mm, a conduction velocity of 0.46 meters/sec was calculated for the axons of the neuroendocrine cells. A similar conduction velocity for the axons of the preoptic neurons (0.5 meters/sec) was observed in *Lophius* (Potter and Loewenstein, 1955). Spike heights of the neuroendocrine cells ranged up to 117 mV and were distinguished by long durations of discharge up to 3.5 msec (Kandel, 1964).

The neuroendocrine cells demonstrated a slow spontaneous firing rate of 2–8 impulses/sec. Action potentials and afterpotentials similar to those occurring spontaneously were initiated by this stimulation. Afferent stimulation of preoptic neurons was produced by stimulation of the olfactory tract which produced a long latency depolarizing excitatory postsynaptic potential (EPSP). This response was graded and triggered an action potential when the critical discharge potential was attained. The EPSP was multiphasic and most probably polysynaptic. Suprathreshold stimuli produce a faster firing EPSP and a shorter latency spike.

Antidromic stimulation, from stimulation of the pituitary gland, activated the neurohypophysial tract and produced antidromic action potentials in the preoptic nucleus. This type of stimulation resulted in inhibitory postsynaptic potentials (IPSP) in 80% of the neurons in the preoptic nucleus. The IPSP was also a graded response and proved to be sensitive to polarizing current; that is, the potential change was increased by depolarizing current pulses and decreased or abolished by hyperpolarizing current pulses (Kandel, 1964).

Flushing the gills of goldfish with water containing as little as 0.1–0.35% sodium chloride (NaCl) produced effects localized in the preoptic nucleus. Higher percentages of saltwater tended to produce small positive extracellular potential changes in surrounding areas of the brain. Single infusions of saltwater produced little or no effect, but multiple infusions appeared to summate the response and produced significant slowing of spontaneous activity in preoptic neurons. However, infusion of seawater three, four, and five times with 10–60 sec/pulse interval will elicit a slowing of spontaneous discharge rate. Multiple infusions of saltwater resulted in a cessation of the spontaneous firing rate associated with a hyperpolarization and a resumption of the spontaneous firing rate with the return of membrane potential to its earlier level.

It is of particular interest that there was antidromic inhibition of the preoptic neuroendocrine cells (via hypophysial stimulation) and orthodromic excitation of preoptic neuroendocrine cells via olfactory nerve stimulation. The antidromic inhibition constituted electrophysiological evidence for inhibitory recurrent collateral axons synapsing on the preoptic neuron of origin. In addition, IPSP's could be produced with stimuli that elicited only a low threshold potential in some of the neurohypophysial tract fibers. It would appear then that the neurosecretory cells of the preoptic nucleus of the thalamus of the fish are not only neurosecretory in function but also serve as functional units in a neural center. It is interesting to note that Hara and Gorbman (1967) found that the olfactory system decussates within the preoptic area of the goldfish brain and that stimulation of the ipsilateral olfactory tract resulted in suppression of the contralateral olfactory bulb.

The relationship between the olfactory tract and the activation of spike potentials in the preoptic nucleus (Kandel, 1964) has led to interesting work on the function of the olfactory–preoptic relationship within the goldfish. Infusion of 0.1% NaCl over the nasal epithelium of the goldfish induced inhibition in some units in the preoptic nucleus (pars parvocellularis), excited some units in this area, and did not affect other units (Jasinski et al., 1967). Infusion of 0.1% NaCl solutions into the nasal cavity or emersion of the goldfish in a solution of this percentage for up to 7 days had little to no effect on the accumulation of neurosecretory granules in the hypothalamo-hypophysial tract. However, relatively short stimulation (60 sec) of the olfactory tract resulted in complete degranulation of the neurosecretory neurons of the nucleus preopticus. In general, neurons of specimens sacrificed immediately following stimulation contained more degranulated neurons than those fixed after a short delay. Olfactory stimulation for longer than 1 min did produce clear and consistent degranulation of the cells examined.

The axons of neurons examined were located in the nucleus preopticus (pars parvocellularis and pars magnocellularis) as well as juxtosomal axons in most specimens. The cytological change of the axons included the disappearance of the characteristic Herring bodies and loss of stainability. The fuchsin stainable granules were depleted in one-half of the axons examined. Regranulation was in the distal part of the neurosecretory axons in the hypothalamo-hypophysial tract after olfactory tract stimulation. This was not true in the neurohypophysis proper where no consistent change was found in the amount of neurosecretory substance after olfactory tract stimulation. These data indicated that degranulation of the hypothalamic neurosecretory systems may take place via olfactory stimulation to the animal. It is not known whether the olfactory epithelium is acting as an osmoreceptor if receptors in the epithelium are specific for chemoreception of sodium and/or chloride ions. However, weak NaCl solutions act in a manner which resembles other olfactory agents and can evoke neurosecretory activity and derivative phenomena (Jasinski et al., 1967).

Results similar to these have been found with work on the angler or goose fish, *Lophius*. In these animals the pituitary stalk may be 2.0 cm or longer (Bennett et al., 1968). Bursts of efferent impulses were evoked in hypothalamic neurons by tactile cutaneous stimulation and electrical stimulation of spinal, trigeminal, optic, and olfactory nerves. Antidromic stimulation of these units by stalk stimulation generated action potentials that were 65 mV in amplitude and 10 sec in duration when recorded intracellularly. Excitatory postsynaptic potentials were evoked in a single cell by stimulation of both olfactory and trigeminal nerves. These results were in agreement with the earlier work on the goldfish (Jasinski et al., 1967; Kandel, 1964). However, the units in *Lophius* did not appear to show recurrent inhibition. There appeared to be tonic ascending inhibitory influence of the units recorded from lower centers since section of the brain caudal to the hypothalamus results in many hours of spontaneous activity. Four days after stalk section all proximal axons that were in the hypophysial stalk were packed with granules close to the area of transection. This indicates that in *Lophius* as in the goldfish stalk axons are neurosecretory (Bennett et al., 1968).

The nucleus lateralis tuberis also has been found to contain neurosecretory neurons. This nucleus is also a neurosecretory neural center in the fish, *Clarias batrachus* (Dixit, 1967).

D. Regeneration of the Hypophysis (Pituitary)

A prolactinlike hormone was necessary for the survival of *Fundulus heteroclitus* in freshwater since animals died in freshwater following hypophysectomy (Pickford et al., 1965). In a series of experimental animals one hypophysectomized animal survived in freshwater. This observation, together with other data, indicated that this one surviving fish may have regenerated the hypophysis from a piece of parenchyma left following operation (Pickford et al., 1965). This proved to be the case since surrounding the hypophysial remnant was a cluster of adenohypophysial cells surrounding a stalk of neurohypophysial tissue which descended from the floor of the third ventricle (Ball, 1965). Almost all of the cells (87%) of this hypophysial regenerate are erythrosinophils. These cells were in the process of proliferation since active mitosis was observed in erythrosinophilic cells. In addition, other cell types were noted between the erythrosinophils and the neurohypophysial tissue. There were a few acidophylic and basophylic cells (3.6%) which are indicative of gonadotropic (GH) and somatotropic (STH) hormone secreting cells. Neurochemical assay revealed that some cells (8.5%) were adrenocorticotropic (ACTH) secreting cells. This would explain the maintenance of the interrenal tissue which was necessary for the freshwater adaptation of the killifish. Growth hormone could not be assayed or identified, but cells were present for its production and no elements of the pars intermedia of the hypophysis were observed. The neurohypophysial stalk (infundibulum) with the reconstituted hypophysial fragment contained a great deal of neurosecretory material (Ball, 1965), indicating proper function of the neurosecretory cells of the preoptic nucleus. In the goldfish and midshipman, *Porichthys notatus*, partial or complete hypophysectomy resulted in regeneration of cut axons of the infundibular stalk (Sathyanesan, 1965; Sathyanesan and Gorbman, 1965). A new relationship was established with nearby blood vessels to form a new neurohypophysial-like organ.

In addition to the regenerative capacity of the hypophysis, pituitary glands from a donor animal survive when implanted into the musculature of host *Poecilia formosa*. Subsequent examination of the heterograft demonstrated that it contains growth hormone secreting cells (alpha and beta cells) that secrete teleostean prolactinlike hormone, corticotropin secreting cells (epsilon cells), and thyrotropin secreting cells (delta cells). The cell types that secrete gonadotropins regress and are no longer able to stimulate the ovaries. There was selective retention of some of the functions of the pituitary gland of this fish following ectopic transplantation of a donor pituitary (Ball *et al.*, 1965; Olivereau and Ball, 1966).

V. MESENCEPHALON

A. Anatomy

The mesencephalon of fish can be divided into two general anatomical subdivisions, the optic tectum and tegmentum. In all fish the optic tectum is laminated and is the superior border of the third ventricle. The tegmentum is the inferior border of the third ventricle. Six layers have been described in the optic tectum by Ariëns Kappers et al. (1960), and 10 layers have been described by Ramon y Cajál (1952). In cross section from the pial surface to the third ventricle the zones according to Ariëns Kappers et al. (1960) are as follows:

First is a stratum opticum (layer 1) which contains the primary optic fibers from the medial and lateral optic tract. A second zone (layer 2) of alternate layers of nerve fibers and neurons receives strong afferent connections from tracts such as the ascending spinotectal and bulbartectal tracts. This is followed by a central zone (layer 3) of gray matter which is rich in neurons and interconnections between neurons. This zone appears to be similar in structure to the mammalian cerebral cortex (with primative cell types) and contains the neurons of origin for the larger efferent tracts. The efferent tracts are the tecto-cerebellar, dorsal and ventral tecto-bulbar, tecto-spinal tract, tecto-commissural tract, and connections to the dorsal thalamus. The next zone (layer 4) is a stratum of central white matter which contains the axons of the efferent tracts to lower centers. A white central stratum (layer 5) follows which is comprised of efferent nerve fibers whose nerve cells of origin were in the layer of gray matter. Fascicles of nerve fibers are located in this zone that decussate in the tectal commissure. The ipsilaterally derived nerve fibers of the tectal commissure tend to terminate in the contralateral torus longitudinalis (a continuation of the tectal gray on the superior border of the tegmentum). The next stratum (layer 6) is a periventricular stratum of gray matter. This layer is made up of neurons whose axons enter into the central white matter and are thus some of the efferent outflow from the optic tectum (Ariëns Kappers et al., 1960; Aronson, 1963; Ramon y Cajál, 1952). The superior colliculus (tectum) appears to have the same general structure throughout the vertebrates. The superior colliculus can be thought to be made up of a superficial incoming layer or layers of primary optic fibers, a large zone of intermediate gray substance containing neurons rich in interconnections (resembling the cortex of higher mammals) and a lower efferent fiber layer whose outflow ascends and descends to other neural centers.

The gray zones within the optic tectum of the fish are particularly rich in connections (Ramon y Cajál, 1952). The neurons of this area contain stellate cells which are reminiscent of mammals and other cell types which appear to be unique to the tectum of lower vertebrates. The superficial grouping of nerve cells appears to be primitive fusiform cells. There are also horizontal connecting cells within this plexiform layer interconnecting the various neural elements within the optic tectum. In addition, there are large primitive pyramidal-like cells that possess profusely branched apical dendrites and no basal tuft of dendrites (Ramon y Cajál, 1952). These primitive nerve cells may be one of the cell types which are responsible for the regeneration of the optic tectum (Ramon y Cajál, 1952). The optic tectum of the teleost, because of its rich afferent and efferent connections, appears to be the major center for the integration of visual information with other sensory modalities and the integration center for exteroceptive information ascending and descending from other neural centers.

B. Spontaneous Activity of the Optic Tectum

Electroencephalograms recorded from the optic tectum of codfish, *Gadus callarias*, have indicated the existence of a thalamic and mesencephalic reticular activating system (Enger, 1957). The dominant frequency of both tectum and forebrain in dark-adapted animals was 8–13 Hz. The optic tectum signal had a component that resembled the spindling of mammalian alpha rhythm. Arousal was initiated by general illumination and acoustic stimulation and resulted in a dominant rhythm of 18–32 Hz. The goldfish optic tectum had a dominant rhythm of 7–14 Hz which increased in frequency to 18–24 Hz (arousal) after photic stimulation (Schadé, 1959; Schadé and Weiler, 1959).

C. Electrophysiological Characteristics of the Visual Input

There have been extensive mappings of the retinotectal characteristics of the visual input in cyclostomes, plagiostomes, and actinopterygians. In the lamprey, *Lampetra fluviatilis*, illumination of the eye resulted in visually evoked responses in the optic tectum, tegmentum, medulla oblongata, and spinal cord (Karamian *et al.*, 1966). These responses were elicited by stimulation of the lateral as well as parietal eye. Stimulation of the parietal eye elicited a slow negative wave with a 100–150-msec latency. Stimulation of the lateral eye elicited a series of slow oscillations initiated after 60–90 msec as a negative or negative–positive wave of maximal amplitude. The neural units within the tectum of the lamprey were quite susceptible to fatigue even at low frequencies of stimulation (one light flash per 20–30 sec). Flashes at intervals of 2–5 sec were followed by a marked decrease in the amplitude of the first wave. This demonstrated that the primary wave of the visually evoked response of the lamprey was the discharge of retinal "on units"; subsequent oscillations were related to discharges of retinal "off units." The visual evoked response in the lamprey was comprised of two components, a fast component which appeared to be related to the presynaptic fibers of the retinal ganglionic cells and a slow component which was a negative or a positive–negative wave and evidently represented the dendritic potentials of radially oriented optic tectal units. The occurrence of a single wave in response to light stimulation warranted the assumption that, in contrast to teleost fish which have a bimodal (Konishi, 1960) or trimodal fiber spectrum (Dawson and Bernstein, 1970), the optic nerve of the lamprey consists of nerve fibers with uniform conduction velocity and therefore a unimodal fiber spectrum (Karamian et al., 1966). In addition to this finding, visual evoked responses resulted in discharges in the medulla and the spinal cord, although no responses were recorded from the telencephalon. The response in the medulla and spinal cord appeared to be due to discharges of Müller's fibers. The Müller neurons in the lamprey aggregate and form the tegmental motor nucleus. The response recorded from the Müller fibers in the spinal cord were of the same latency as the response recorded from the optic tectum. Ablation of the optic tectum did not affect the afferent visual input to the Müller cells; therefore, the visual input was derived from another pathway of the primary afferent optic nerve fibers (Karamian et al., 1966).

Evoked visual responses could be recorded from the optic tectum of the plagiostomes *Raja* and *Trygon* following stimulation of the eye by light or electrical stimulation of the optic nerve (Karamian et al., 1966). The latency of the response in the contralateral optic tectum was 50–60 msec. Direct stimulation of the optic nerve resulted in a slow negative–positive wave with a latency of 6.0 msec. Oscillations did not precede this slow optic tectal response. The slow wave of the optic tectal response of these skates showed neither double nor accessory peak which indicated that the skate like the lamprey has a unimodal distribution of nerve fiber diameter in the optic nerve (Karamian et al., 1966).

The visual evoked response in the optic tectum has been studied in the teleosts (goldfish, bass, carp, and bluegill) using perimetry. The stimuli were small spots of light which subtend angles of 0.5°, 1°, 2° up to 30°, as well as small circular disks placed on a background that encompassed the visual field (Fig. 5). The three superficial tectal layers of primary afferent optic fibers terminated on cells which had the same characteristic responses to the visual field as neural units found within the retina (Jacobson and Gaze, 1964; MacNichol et al., 1961; Schwassman and Kruger, 1965; Wagner et al., 1963). The retinotectal projection of retinal units recorded from the upper three zones of the tectum demonstrate tectal neurons with three types of responses (Jacobson and Gaze, 1964). These responses can be classified as units which are "on," "off," or "on–off." Units with on centers can be subdivided according to

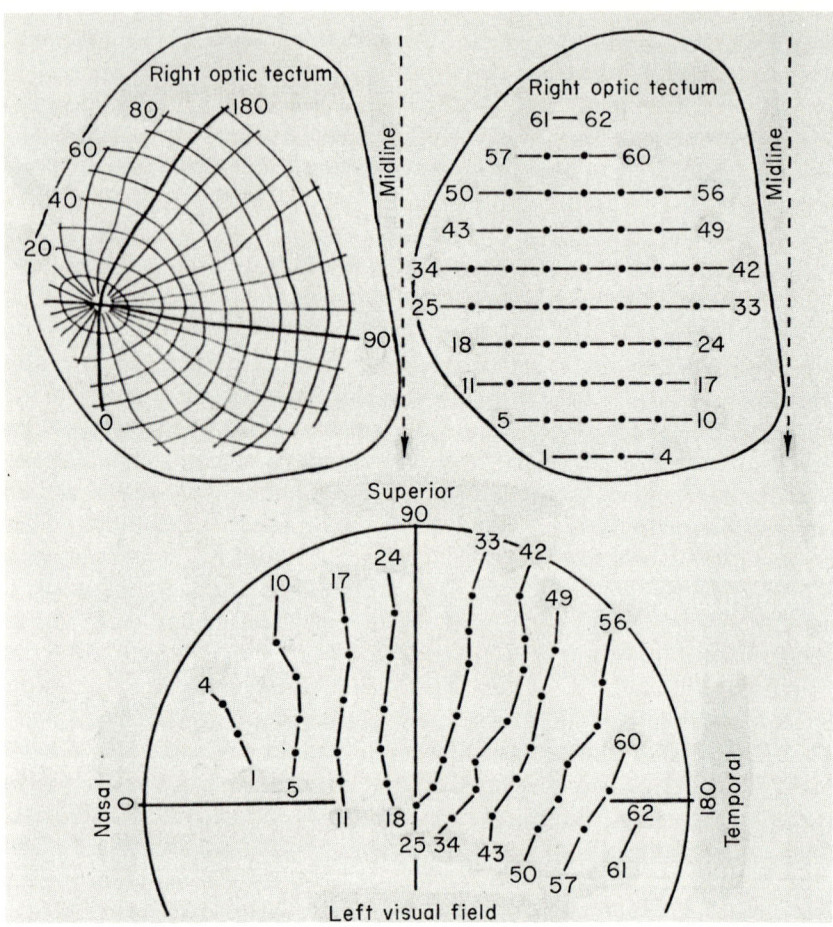

Fig. 5. Projection of the left visual field onto the right optic tectum of the goldfish. The arrows are midline and point rostrally. Upper right: dorsal tectal surface showing rows of numbered electrode positions. Lower: left visual field with optimal stimulus positions corresponding to appropriate electrode positions in upper right of figure. The chart covers 100° from the center of the field. Upper left: map of medians and parallels calculated for visual field on the optic tectum. After Jacobson and Gaze (1964).

response surround: One type of unit responded only to the onset of light; a second unit responded to the onset of light but also induced an inhibitory surround and discharged in response to changes in general background illumination. The layers of the tectum will be A–F, as designated by Jacobson and Gaze (1964). In the most superficial fiber layer (A), stratum opticum of the units of Ariëns Kappers et al. (1960) responded with on and on–off responses with on–off units predominating. There was no orderly sequence of the different types of unit responses in the next two tectal layers [B and C, alternating layers of fibers and cells of Ariëns Kappers et al. (1960)]. After passing through the upper three layers (A, B, and C) of the tectum the electrode entered a zone of sustained on and off responses [layer D, within central gray of Ariëns Kappers et al. (1960)]. Units in this zone discharged actively in the light and some were continuously active in the dark. This was an extremely precise zone since a 20-μ excursion of the tip of the electrode resulted in the loss of units with this response. Recordings made from layer F [periventricular gray of Ariëns Kappers et al. (1960)] from the optic tectum demonstrated units which were on, off, and on–off and were presumably units whose axons comprise portions of the tectal efferent system. The responses of these units were similar to those units which were found in the most superficial layers of the optic tectum. The three types of unit responses were found at various depths within the tectum and were also recorded from nerve fibers in the optic chiasma. On units with inhibitory surrounds were common in the superficial layers (A, B, and C) of the optic tectum and in layer F but only one such unit was found in layer D. The diameter of the receptive field center of the tectal neurons varied in different units when the visual angle of the stimulus subtended from 10° to 30° (Fig. 5). Directionally sensitive units were located when the stimulus was moved parallel to the horizontal meridian of the visual field. Directionally sensitive units could be detected more frequently when stimuli were moved from the temporal to nasal field rather than for movements in the opposite direction (Jacobson and Gaze, 1964; Schwassman and Kruger, 1965).

These electrophysiological results indicate that visual units are better equipped to detect horizontal rather than vertical movement within the visual field. This has been demonstrated behaviorally by Mackintosh and Sutherland (1963). This visual phenomenon has been correlated with rheotactic behavior (orientation of the longitudinal axis of the body in moving water) since the animal is visually better equipped to detect movement in the temporal-nasal direction (Cronly-Dillon, 1961, 1964). On–off units with well-demarcated response zones were the most common type of unit found in all the fiber layers except layer D. These units were also directionally selective in their response. In some units a 2° spot of light moved through the field in a temporal-nasal direction resulted in a discharge, but reversal of this pattern of movement did not result in a response. The receptive field of on–off units increased in size following dark adaptation. On units were recorded from all the fiber layers of the tectum but were found most frequently in layer D in which on–off units or on units with inhibitory surrounds were rarely found. The response of the on unit was a sustained discharge when a spot of light was placed in any part of the receptive field. Diameters of the receptive field ranged from 15° to more than 40°. Off units were recorded from all the fiber layers of the tectum but again were most frequently found in layer D (Jacobson and Gaze, 1964).

The organization of the retinotectal projection within the optic tectum of teleost fish was similar in all species observed (Jacobson and Gaze, 1964; Schwassman and Kruger, 1965; Sutterlin and Prosser, 1968). The only variation appeared to be in the bluegill which departed from the strict linearity of the representation of the visual projection upon the tectum. In the bluegill distances between the receptive field were compressed for a unit discharging to a stimulus below the horizontal meridian (Schwassman and Kruger, 1965).

Retinal unit responses have been recorded from the optic nerve after stimulation of the eye by means of perimetry (Mark and Davidson, 1966). The units in the retina gave the same type of response as the units in the tectum. They were sensitive to small spots of light in a limited visual field or to the movement of small dark or light objects in and out of the receptive field. This, together with the data on the retinal response to color in which the units of the retina (MacNichol et al., 1961) responded like the units in the tectum (Jacobson, 1964a,b), demonstrates that the neurons of the optic tectum of the goldfish do not respond differently from the primary input from the optic nerve. Information recorded from the optic chiasma (Jacobson and Gaze, 1964) and recordings from the optic tectum by other authors show that the response latency is locked to the response time of the primary retinal input.

Computer-averaged signals from the unanesthetized visual pathway exhibit great sensitivity to all electrical events without the sampling bias inherent to unit recording techniques. Electroretinogram (ERG), fast retinal potential (FRP), and tectal evoked response (TER) were recorded at three stimulus luminances, 800, 1000, and 4000 foot-lamberts (ft-L) (Fig. 6). Light stimuli were 50 μsec in duration (Dawson and Bernstein, 1970). All signal component amplitudes of response of the ERG and FRP were directly (logarithmically) related to luminance "brightness of stimulus" as were the three early components of the tectal response. The triple peak tectal signal was suggestive of fibers of three different conduction velocities and thus of a trimodal optic nerve fiber spectrum. In support of this finding, electrical stimulation of the optic nerve produced a triple peaked compound action potential. The amplitude of later tectal signals (following triple peaked response) was not related to luminance of the stimulus since some tectal signals were observed at 800 and 1000 ft-L but were absent or diminished at 4000

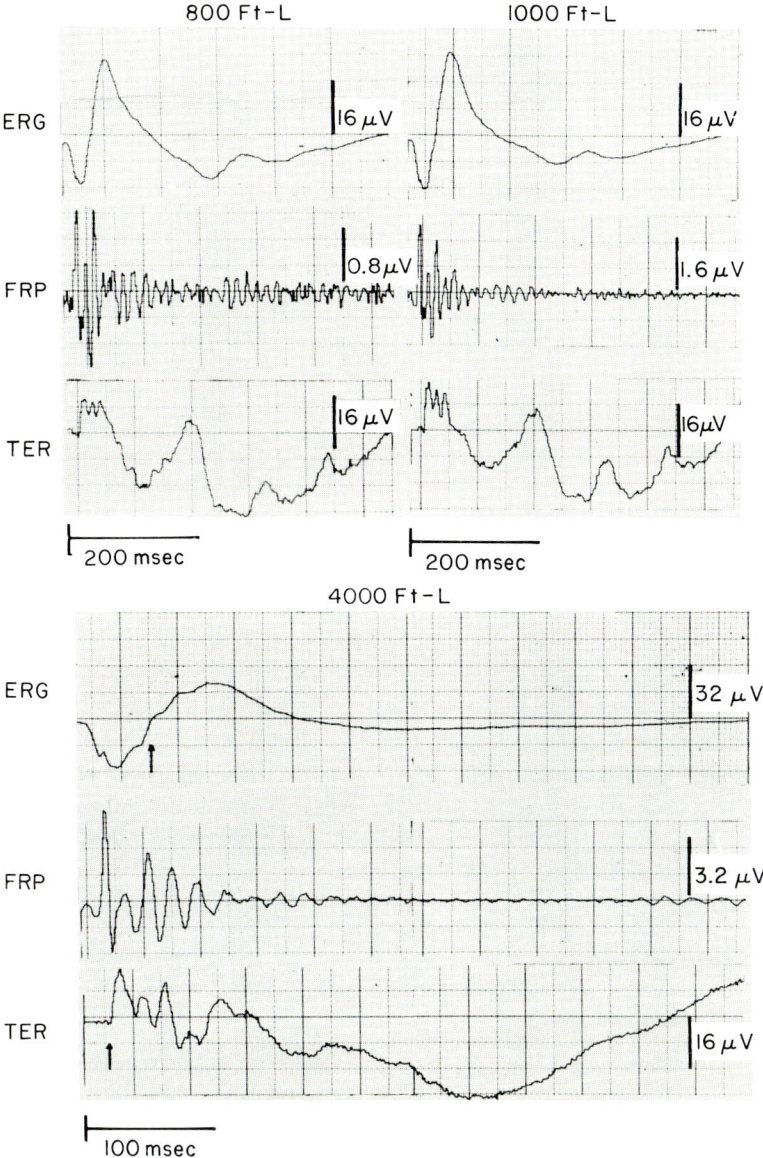

Fig. 6. Signals in the visual pathways of the goldfish. Responses at the eye and tectum were evoked by three light intensities (foot-lamberts). The results of computer averaging are shown for ERG, FRP, and TER. Details are discussed in the text.

ft-L. At this time it is not possible to draw a direct correlation (isomorphism) between retinal and tectal events in the goldfish.

The brief conduction time in the goldfish optic nerve allowed a clear demonstration that the ERG b wave (latency to arrow is 39 msec) occurred after the tectum had been excited (latency to arrow is 16.7 msec) and was not directly involved in information transmission. Relative biological and system noise averages are shown as an indicator of signal validity. The order of presentation from the top is ERG, FRP, and TER (Dawson and Bernstein, 1970).

D. Spectral Sensitivity of Optic Tectal Units

By means of perimetry, goldfish were presented movable spots of colored light (subtending angles of 1° or 2°) which were of 22 different wavelengths (341–779 mμ) and were adjusted to threshold (Jacobson, 1964a,b). Electrophysiological recordings were taken from units in the optic tectum. After dark adaptation the scotopic spectral sensitivity curve was determined for on, off, and on–off units. The scotopic sensitivity curve fitted the absorption spectra of a visual pigment at 533 mμ. This good fit between the scotopic spectral sensitivity curve of all types of units recorded and the absorption spectrum of the visual pigment at 533 mμ demonstrated that the most probable visual pigment in the goldfish rods is porphyropsin. The spectral sensitivity curve (light adapted) corresponded to the absorption spectrum of a visual pigment at 467 mμ which is indicative of the visual pigment cyanopsin. Therefore, it was concluded by these electrophysiological measures that visual pigments porphyropsin and cyanopsin are the two cone visual pigments within the fish retina (Jacobson, 1964a,b). These data agreed with the spectral maxima of two of the three pigments of goldfish cones analyzed by chemical methods (Marks, 1963). On and off units within the optic tectum had different spectral sensitivity functions than could be estimated from the visual pigments (Jacobson, 1964a,b). After light adaptation photic sensitivity curves of tectal units were narrower than the visual pigment absorption curves. The units had maxima at 462, 497–517, 552–584, and 605–651 mμ. There are therefore three classes of these tectal units which can be categorized as red–green units (maxima of 630–651 and 497–517 mμ), red–blue units (maxima of 605–651 and 462 mμ), and yellow–blue units (maxima of 552–605 and 462 mμ). These tectal recordings demonstrate that there may be three types of cones in the goldfish retina, each type containing a different pigment with absorption and sensitivity maxima at about 467, 533, and 620 mμ. However, the spectral sensitivity of red–green, yellow–blue, and red–blue units may result from the interaction between only two different classes of cones, each set of which contains one of the three visual pigments (Jacobson, 1964a,b). These results are in agreement with the retinal electrophysiological and chemical results derived from studies of the retina.

E. Interhemispheric Transfer of Visual Information

The study of interhemispheric transfer of visual information between the hemispheres of the optic tectum has played a major role in the advancement of the knowledge of the function of the optic tectum of fish. This is particularly interesting in respect to the problem of how information in one part of the brain is retrieved and becomes available to other parts of the brain. In general, information presented to one eye and stored in the contralateral tectal hemisphere transfers to the ipsilateral hemisphere thereby enabling the animal to respond appropriately using only the untrained or naive eye. The optic nerve of the fish appears to be completely decussated: the right eye projecting to only the left tectal hemisphere and the left eye projecting to only the right optic tectum (Ariëns Kappers et al., 1960; Papez, 1929). Therefore, the fish appears to be an excellent animal for studies on storage and retrieval of visual information in the central nervous system. Tectal transfer in the fish does not appear to be perfectly developed (Sperry and Clark, 1949). The failure of goldfish in the interocular transfer of visual discrimination in shuttle boxes, however, results from the failure of the transfer of visual motor learning (McCleary, 1954, 1960). If the response measure is cardiac deceleration and does not involve a visual motor response, interocular transfer of color visual information is excellent (Bernstein, 1962). From these studies on interocular transfer of visual input in the fish it appears that under certain conditions there is a significant degree of interocular transfer and that the completeness of the transfer is not a problem of perception but a problem in sensory-motor integration.

The cichlid fish, *Astronotus ocellatus*, were trained monocularly to make a hue discrimination (Arora, 1959; Arora and Sperry, 1961). Following acquisition of the problem, the optic nerve of the trained eye or the tectal projection of the trained eye was transected or extirpated. The untrained or naive eye was then tested, and the fish responded appropriately. It was postulated that such interocular transfers may result from the establishment of a dual memory trace system, one on each side of the brain, or from the retrieval by the untrained side from memory traces established in the subtectal centers on the trained side.

Interocular transfer of a pattern discrimination using various types of discriminanda was complete in goldfish when appropriate controls were carried out (Schulte, 1957; Shapiro, 1965). These controls were for orientational difficulties, specific response biases associated with the monocular vision, restriction of the visual field, and general learning about the experimental situation which was not task specific (Shapiro, 1965).

Although the forebrain commissures were not essential for the transfer of color visual information in goldfish (Bernstein, 1962), the tectal commissure was necessary for the interocular transfer of pattern discrimination in the Oscar, *Astronotus ocellatus* (Mark, 1966). *Astronotus* were trained to respond by jumping out of the water for food reward to different patterns floating on the surface of an aquarium. Animals learned this task binocularly or monocularly (one eye covered). Animals trained monocularly demonstrated interocular transfer. Following training and subsequent transection of the tectal commissure the fish failed to make the interocular transfer of the discrimination. Normal fish also were trained to the patterns with one eye covered and then trained on the reversal problem with the naive eye (trained eye covered). Following acquisition of reversal learning by the naive eye, the eye that was originally trained demonstrated a strong preference for the stimuli on which the naive eye had been trained. Initial learning with one eye then was effectively replacing the original problem through interocular transfer of experience gained through the other eye.

Attempts were made to produce chronic lesions of the posterior commissure in *Astronotus;* unfortunately, the animals did not survive (Mark, 1966). It would be of interest to study the transfer of information in the posterior commissure and the commissure which lies above it, the geniculate commissure. Since the transection of the tectal commissure resulted in the loss of interocular transfer of pattern and the strong suspicion that the posterior commissure was involved with interocular transfer, it is possible that there are two anatomically distinct systems within the mesencephalon of fish for the interocular transfer of visual information (Mark, 1966). Although there was no histological verification of the lesion, interocular transfer of a conditioned avoidance response in goldfish was blocked by posterior commissure transection prior to monocular training (Ingle, 1965a).

Electrophysiological recordings from tectal commissural fibers in goldfish and *Astronotus ocellatus* have not shown any clear evidence that the tectal commissural cells respond to small spots of light or small dark objects moving anywhere within the visual field of either eye (Mark and Davidson, 1966). However, the units whose axons make up the tectal commissure did respond to the general level of illumination. The units were inhibited by light and discharged regularly in the dark. During dark adaptation, the units whose axons were in the tectal commissure were spontaneously active and discharged at a regular rate which varied widely from unit to unit. Sudden increase in the general level of room illumination inhibited the discharge rate and was followed by a rebound acceleration at light-offset. Similar response characteristics have been found during cellular recordings of tectal units (Cronly-Dillon, 1964; Jacobson and Gaze, 1964; Mark and Davidson, 1966).

Every commissural fiber was a slowly adapting unit which gave a prolonged off response (Mark and Davidson, 1966). Units with this characteristic were found most frequently in layer D (Jacobson and Gaze, 1964) of the optic tectum which was located in the zone of central gray matter that contains the cell bodies of origin of the tectal commissure (Ariëns Kappers et al., 1960). There were no on unit responses recorded from commissural axons. The interocular exchange between the two tectal hemispheres is somehow involved in the perceptual mechanism which is related to fine pattern vision in the fish (Mark and Davidson, 1966).

F. The Optic Tectum and Learning

Using crucian carp, *Carassius carassius*, Bianki (1960, 1961, from Bianki and Demina, 1964) found that animals trained to react to light or sound stimuli still respond appropriately after the tectal hemispheres are separated by a longitudinal midline incision. However, the animals were unable to distinguish the positions of the light (spatial orientation) after sectioning of the optic tectum. The animals also appeared to have difficulties in locating the position of the sound stimulus. No similar disturbance in behavior was found after midline incision of the cerebellum. In another series of experiments crucian carp, *Carassius carassius*, and carp, *Cyprinus carpio*, were trained to respond to light or sound by swimming to one side of an aquarium (Bianki and Demina, 1964). Following acquisition of the task one-half of the optic tectum, right or left, was removed. It was found that within the first 3 days postoperatively, removal of the optic tectum had no effect on the ability of the animals to give a positive response to the onset of light. Approximately 21 days postoperatively the animals started to lose their ability to spatially orient the stimuli in the testing apparatus. This removal of the optic tectum had no effect on the animal's ability to make a brightness discrimination of light of differing intensity. Removal of one-half of the cerebellum did not cause spatial orientation difficulty with visual or acoustically oriented problems (Bianki and Demina, 1964). These results are interesting since removal of the optic tectum on one side results in the animal's being blind on that side and untrainable to photic stimulation (Karamyan, 1957).

In addition, it has been found (Arora, 1962) that the engram or memory trace is acquired bilaterally in the fish. Therefore, training one eye and removing the optic tectum contralateral to the trained eye did not affect the ability of an animal to respond appropriately by utilizing the naive untrained eye and the remaining tectal input.

G. Retinotopic Respecification of the Regenerated Optic Nerve

The regeneration of the optic nerve into the optic tectum of the goldfish as well as other species of fish has been demonstrated (Healey, 1957). Recent experiments have been carried out on the goldfish and cichlid *Astronotus ocellatus*. Ten to 12 days after section of the optic nerve, regenerating axons began to reinnervate the tectal hemispheres from the medial and lateral optic tracts (Attardi and Sperry, 1963). In 14–18 days, the regenerating nerve fibers entered the plexiform layer of the optic tectum. This coincided with the onset of the recovery of vision. At this time all the new reinnervating nerve fibers were unmyelinated. During the following weeks the nerve fibers underwent a slow process of maturation and increased in diameter with a concomitant deposition of myelin. The representation of the retina on the optic tectum is upside down and backwards in comparison with the retinal projection (Attardi and Sperry, 1963; Brett, 1957; Healey, 1957). When the dorsal half of the retina was removed and the optic nerve of the same side severed, the regenerating nerve fibers in the optic nerve could originate only from ganglion cells in the ventral half of the retina (Attardi and Sperry, 1963). The regenerating nerve fibers were found to enter their original tract, the medial bundle of the optic tract. With few exceptions the lateral bundle was completely without nerve fibers following regeneration of the transected optic nerve. The nerve fibers were parallel within the medial tract and only entered the plexiform layers in the dorsal tectum. Conversely, if the ventral half of the retina was removed prior to nerve transection nearly all of the regenerated nerve fibers were found to enter only the lateral bundle and to innervate only the ventral half of the tectal hemisphere. In a series of elegant experiments it was shown that peripheral, central, dorsal, and ventral optic nerve fibers

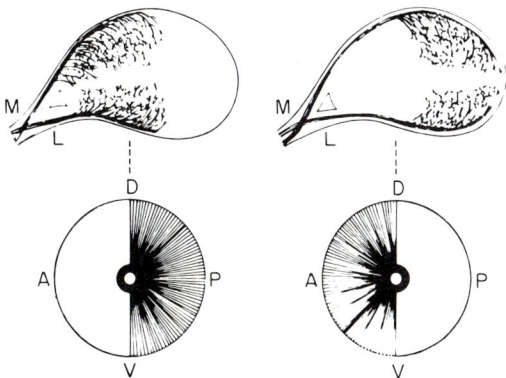

Fig. 7. Schematic representation of the regeneration patterns of the nerve fibers in the optic nerve of the goldfish following removal of the anterior (nasal) or posterior (temporal) hemiretina, respectively. A, anterior; D, dorsal; L, lateral division of optic nerve; M, medial division of optic nerve; P, posterior; and V, ventral. After Attardi and Sperry (1963).

from the retina all regenerate back to their original sites of origin (Fig. 7).

Specific chemical affinities (Sperry, 1955) appear to govern the neuronal retinotopic respecification of synaptic endings from ganglion cells within the retina to the neurons of the tectum. The affinities of regenerating nerve fibers to their original sites of origin are complex and strong. In the fish *Astronotus ocellatus*, cross union of the proximal stump of the medial optic tract with the distal lateral optic tract resulted in restoration of the regenerating nerve fibers to their original pathways upon entering into the tectum. This also occurred following a lateral-medial transposition. Nerve fibers deliberately directed in the wrong channels would not grow into a foreign tract (Arora and Sperry, 1962; Sperry, 1965a).

Optic nerve regeneration has been followed electrophysiologically (Jacobson and Gaze, 1965; Westerman, 1965). Reestablishment of retinotectal projections utilizing point for point mapping following optic nerve regeneration was examined by electrophysiologically mapping projections after operative alteration of the input to the optic tectum or by surgically manipulating the tectum itself. These experiments were carried out on adult goldfish using Attardi and Sperry's anatomical mapping of the optic tectum (1963) and Jacobson and Gaze's electrophysiological mapping of the optic tectum (Jacobson and Gaze, 1964, 1965). The lateral or medial optic tract input into the optic tectum of the goldfish was transected. This procedure deafferented superior-nasal and inferior-temporal projections from the retina to the tectum. The nasal half of the visual field (temporal half of the retina) was removed from the visual input. By using visual evoked responses the remaining area occupied by the intact optic nerve fibers was mapped after sectioning the contralateral optic nerve. The area of the tectum which was deafferented by optic nerve section differed only slightly in different experiments. Normal evoked responses were obtained from the rostral portion of the tectum with no response obtained from the deafferented caudal portion.

If the optic tract was cut and nerve fibers were not permitted to regenerate up either the lateral (nasal) or medial (temporal) divisions of the optic tract, regenerating optic nerve fibers reestablished retinotectal projections which were similar to that shown in animals that had previously undergone cutting of the divisions of the optic tract and immediate recording (Jacobson and Gaze, 1965). This demonstrated a retinotectal reestablishment of the original synaptic patterns upon the neurons within the optic tectum. However, following nasal or temporal hemisections and crushing of the proximal optic nerve, there was significant departure from the normal course of nerve fibers found within the tectum. Such an operative procedure followed by regeneration resulted in the complete absence of sustained on and off responses which were localized in layer D (central gray) of the optic tectum (Jacobson and Gaze, 1964). There was histological verification of this electrophysiological finding since there was an absence or near absence of nerve fibers in layer D of the tectum. It must be emphasized that retinotectal regeneration was nearly perfect and in all animals recovery was excellent except for the recovery of this one layer. In another group of goldfish the medial half of the optic tectum was removed and the left optic nerve crushed (Jacobson and Gaze, 1964). Following regeneration, normal projections were restored to the lateral half of the tectum from the corresponding portion of the retinal visual field. Visual evoked responses could be elicited in the normal retinotectal area of projection in the tectum. In addition, stimulation of the visual field that normally projected to the medial half of the tectum did not result in visual evoked responses in the remaining lateral portion of the optic tectum. These data show that optic nerve fibers destined for the ablated half of the tectum fail to make connections whereas a normal projection to the intact half of the tectum is reestablished.

These electrophysiological findings are direct confirmation of the anatomical study of Attardi and Sperry (1963). In addition, Jacobson and Gaze (1965) have shown that the regenerating optic nerve fibers do not go through as early a phase of spreading out and then respecifying in a maturation process such as is found in amphibia (Gaze *et al.*, 1963, 1965a) but establish original retinotectal connections immediately upon entry into the tectum.

Not only does it appear that regeneration of the nerve into the tectum is specific but excision and reimplantation of a tectal flap with a normal orientation resulted in the restitution of the normal visual projections over the area of the graft (Gaze *et al.*, 1965b). However, restoration of the visually evoked retinal map on the tectal surface was not accompanied by restoration of the normal sequence of visual responses throughout the entire depth of the excised and reimplanted optic tectum.

Following regeneration of the optic nerve, fish can make a hue discrimination (Arora and Sperry, 1963). *Astronotus ocellatus* were trained to jump out of water to two different color stimuli for food reward. Red, blue, green, and yellow were the colored stimuli which were controlled for brightness. Following optic nerve crush and regeneration, previously trained fish were still able to make a hue discrimination. This demonstrates that regeneration of the optic nerve results in retention of perception of the same quality of hue by the animal (Arora and Sperry, 1963).

Pattern vision appears to be restored following optic nerve regeneration since fish respond appropriately in an optokinetic drum (Healey, 1957) and the return of visual responses to moving objects has been demonstrated electrophysiologically (Jacobson and Gaze, 1964); however, critical testing is necessary to establish the degree of functional return of pattern vision. Panels of dots of progressively reduced intervals were used as a measure of visual acuity and visual angle (Weiler, 1966). The average value of visual acuity in the normal eye of *Astronotus* is 5.3 min of visual angle. Since it was found that the mean diameter of unfixed cones in the retina is 2.92 μ a single cone appears to correspond to 5 min of visual angle. Thus, the actual cone dimensions were close to the predicted value and in the upper limits of the visual acuity measured. In all cases there was regeneration of the severed optic nerve and return of vision. The average percentage of the restoration of acuity was 78% of normal. In another group of animals the fish were trained to discriminate between a uniform dot pattern and a pattern of randomly arranged dots of various sizes. The learning curves of these animals did not differ significantly from normals. This was a gross indication that the regeneration and restoration of central connections was uniform and probably distributed evenly over the areas of the tectum. It is interesting that the high degree of recovery found by Weiler (1966), which he suggests under ideal conditions can approach 100% can occur when a major class of neurons within the layers of the tectum (layer D; Jacobson and Gaze, 1965) can possibly be deafferented following optic nerve regeneration.

These studies demonstrate that there is selective regeneration of nerve fibers derived from the retina to the original cells of termination within the optic tectum of fish. The special sensory nerve fibers from the retina entering the tectum appear to grow directly to the area of previous innervation and selectively reinnervate the neurons of original termination. The retinotectal map appears to be reestablished except for the loss of some fibers in a specific area of the optic tectum.

H. Regeneration of the Optic Tectum

The regenerative capacity of the optic tectum in fish has been studied extensively (Kirsche, 1960, 1965b, 1967; Kirsche and Kirsche, 1961; Richter, 1966, 1968). The cells which completely regenerate a totally ablated optic tectum were derived from matrix zones located in the ependymal cell zone (lining the third ventricle) of the optic tectum. There were three matrix zones: a dorsal matrix zone next to the torus longitudinalis, a caudal matrix zone at the caudal pole of the optic tectum consisting of groups of undifferentiated and pluripotential cells, and a basic matrix zone at the base of the tectum. These matrix zones also act in the development of the optic tectum. During embryological development, cells which were differentiated from the matrix zone migrated to the center of the tectum. Thus, the amount of matrix material (germinative zone) which was available to the animal decreased throughout the animal's life. According to Kirsche (1967) these matrix zones continually produced cells within the optic tectum. However, the ability to produce parenchyma was age dependent since as the animal got older, the matrix zones became thinner until the animal reached an age when the matrix no longer produced cells for cell renewal. The cells which were derived were specific for the three matrix zones which have been mentioned earlier. Extensive unilateral lesions within the optic tectum of the crucian carp, *Carassius carassius*, resulted in the complete restitution of the transected fiber tracts and the neuronal and glial components of the optic tectum. However, if the matrix zones were damaged by this operation, the optic tectum, even after 300 days of postoperative recovery, did not show normal cytoarchitectonics. Regeneration did not occur if the rostral portion of the dorsal matrix zone and the caudal matrix zone were removed. Thus, regeneration only occurred in those areas where the matrix zone was left intact. After extensive unilateral lesions of the tectum, and removal of the caudal and dorsal matrix zones, only the basal portions of the tectum would regenerate. Regeneration was initiated at the basal matrix cells. However, regeneration was less extensive because the caudal matrix zone was absent. The caudal matrix zone appeared to be most active in the regeneration of the optic tectum. Regeneration would not occur if all of the matrix zones had been removed by the operation. Richter (1965, 1966, 1968) has found mitoses in the matrix layers of the optic tectum of *Lebistes reticulatus* and *Leucaspius delineatus*. Mitotic activity showed a rapid decrease in zonal activity at the time of sexual maturity (Richter, 1966). This regenerative capacity of the optic tectum of fish was truly remarkable. The fact that areas remain pluripotential throughout the life of the animal and can redifferentiate ependymal cells into neuroblasts and spongioblasts was a unique phenomenon.

I. Axoplasmic Flow

Recent evidence has demonstrated that proteins produced in the perikaryon of neurons were transported by sol–gel changes in the neuron to the bouton terminaux of the axon (Lubinska, 1964). Tritiated leucine or histidine injected into the vitreous chamber of the goldfish or crucian carp eye was incorporated into the perikarya of the ganglion cells of the retina and transported to the axon terminations in the optic tectum (Grafstein, 1967; Rahmann, 1968).

Radioactive leucine was transported along the goldfish ganglion cell axon in a rapid component of 0.4 mm/day and a slow component that arrived at the axon terminals for 23 days after injection (Grafstein, 1967). Tritiated histidine injected into the vitreous body of crucian carp and zebra fish, *Brachydanio rerio,* was also transported biphasically (Rahmann, 1967). There was a rapid interneuronal protein transport of 5.0–7.0 mm/day and a slow component that lasts up to 3 weeks. Three weeks postinjection tritiated histidine was found in cerebrospinal fluid. Deposition of tritiated thymidine in cerebrospinal fluid increased from 20 to 50% of the initial dose between 21 and 46 days.

In zebra fish following intraperitoneal injection of radioactive labeled phosphate in orthophosphate and tritiated histidine, cytidine and uridine could be located in the perikarya of neurons in the body of the cerebellum and the optic tectum (Rahmann, 1965, 1967). These amino acid precursors were transported along the axon after incorporation into macromolecular ribonucleic acid. Rahmann (1965) theorized that all components for maintenance of the axon (mitochondria, neuronal vesicles, etc.) are transported from the perikaryon. However, other studies on transected Mauthner cell axons have shown that a portion of protein synthesis can be derived from axonal and perhaps glial metabolism (Edström, 1964a,b, 1966).

VI. CEREBELLUM

A. Anatomy

In fish, as in most other vertebrates, the cerebellum is the most variable structure within the nervous system. In the Mormyrids the valvula cerebelli is hypertrophied and extends over the dorsal surface of the telencephalon (Aronson, 1963; Nieuwenhuys, 1967a; Nieuwenhuys and Nicholson, 1967). In contrast, in *Bdellostoma*, a myxinid, the cerebellum is represented by a small commissure which consists of eighth nerve and lateral line nerve fibers, the octavo-lateralis system (Bone, 1963; Larsell, 1967; Nieuwenhuys, 1967a).

Larsell (1967) demonstrated that in most groups of fishes the cerebellum is comprised of two fundamental divisions, a cortical basal lobe (vestibulo-lateralis) which receives vestibular and lateral line nerve fibers and a more rostrally situated body of the cerebellum (corpus cerebelli, often convoluted) which receives afferent nerve fibers from the spinal cord, fifth nerve, optic tectum, as well as other systems. The vestibulo-lateralis lobe in the majority of fishes consists of a small intermediate portion with occasionally enlarged lateral auricles (auriculae cerebelli). The auricles are largest in those animals that possess extensive lateral line systems: Chondrichthyes, Polypteriformes, and Crossopterygi (Nieuwenhuys, 1967a). In the actinopterygians dense neural centers situated lateral to the corpus cerebelli, the eminentiae granulares, represent vestibular and lateral line nerve centers. In these fish the cerebellum also possesses a bilateral rostral subtectal projection, known as the valvula cerebelli. The development of this rostral outgrowth is concomitant with the extensive development of the lateral line.

The structure of the cerebellum appears to be generally similar throughout the vertebrate kingdom. It is comprised of an internal stratum albium containing both the afferent axonal inflow from lower and higher centers and the efferent axonal outflow to lower and higher neural centers. The next layer is the stratum granulosum which is an area of granular cells. This layer is followed by the Purkinje cell layer.

The Purkinje cells within this area take various forms in different species of fish. The more evolved fish demonstrate the most complex dendritic patterns of these extraordinary nerve cells (Nieuwenhuys, 1967a; Nieuwenhuys and Nicholson, 1967).

B. Deficit Function

Extensive work on the cerebellum has demonstrated that this portion of the brain is involved in several functions. Earlier work resulted in different deficits from similar cerebellar lesions. After partial or extensive removal of the cerebellum, these ranged from no deficit in equilibrium, swimming, courtship, spawning, or parental behavior to severe impairment in swimming movements and sexual behavior (Aronson, 1963; Healey, 1957). In addition, cerebellar ablation had differential effects on the ability of operated animals to learn to respond to both light and sound stimuli.

In recent studies complete removal of the body of the cerebellum of the dogfish (a selachian) resulted in a decrease in motor activity, changes in sensory functions, and marked reduction in activity of the fish when attempts were made to capture it (Karamyan, 1956). Ablation of the entire cerebellum resulted in complete loss of locomotion in the dogfish; however, recovery of function appeared within 5–8 hr but the recovered pattern of swimming was aberrant. In addition to general locomotor effects, gill movements were disturbed, the animals did not respond to pain stimuli, and the animals flexed toward the point of stimulation. The dogfish appeared disoriented in space and often leaped out of the aquarium. In skates, total removal of the cerebellum did not appear to result in any motor disorders (Karamyan, 1956). This was linked with the fact that removal of the cerebellum did not result in degeneration of layers in the optic tectum. Karamyan's work suggests a difference in extent of sensory or motor disorder after cerebellum ablation in dogfish and skates.

In teleosts (goldfish and carp; Karamyan, 1956) complete removal of the cerebellum resulted in an initial loss of muscle tonus followed by a spontaneous recovery of function within 4–8 hr; however, the operation invariably resulted in death. The fish were hypersensitive to touch or pain and ipsiversive circular motions were often evident. Reaction to auditory and visual stimuli disappeared completely. Removal of only the body of the cerebellum resulted in less severe disorders in carp than in goldfish; however, there appeared to be severe motor and sensory disorders in both species. The histological findings showed degenerative changes in areas of efferent outflow which were in the expected areas for the projections from the cerebellum (Karamyan, 1956).

The rheobase and chronaxie for the erection of the dorsal fins upon stimulation were tested after removal of the entire cerebellum (Karamyan, 1956). After total removal, the rheobase was reduced by more than one-half while the chronaxie increased. In addition, several trophic disorders were found after removal of the teleost cerebellum. These included proliferation of cartilaginous tissue around the mouth and fins, concomitant scale loss on various parts of the body, and marked muscle emaciation (Karamyan, 1956).

Light and sound have been used as conditioned stimuli following cerebellar ablation (Bianki and Demina, 1964; Karamyan, 1956, 1957). Fish were placed in the center of an aquarium with the sources of light and sound above the water on the left or right side of the aquarium. The fish were to choose left side versus right side following stimulus onset. Following conditioning, removal of the right or left portion of the cerebellum (Bianki and Demina, 1964) in the crucian carp, *Carassius carassius*, had no effect on the ability of operated fish to differentiate between acoustic and photic stimuli. Therefore, removal of one-half of the cerebellum has no effect on previously learned conditioned responses to photic and acoustic stimuli. However, loss of the ability to spatially orient acoustic stimuli (which did not affect photic stimuli) was noted upon removal of one-half of the cerebellum (Bianki and Demina, 1964). Following complete extirpation of the cerebellum the fish could not be conditioned to light or to sound stimuli (Karamyan, 1956). However, 30 days after the operation, fish spontaneously recovered the ability to learn differences between acoustic or photic stimuli.

Partial or total removal of the cerebellum resulted in disorders in sensory and motor activity and general tropic disorders (Karamyan, 1956, 1957). Karamyan (1956, 1957) feels that the cerebellum is the visual and auditory integrative center of the fish brain and is the primary organ for the establishment of temporary visual and auditory associations. However, at least some of the visual information is integrated in the telencephalon and optic tectum. More experimental work has to be done on the cerebellum before these types of functions can wholly be ascribed to this neural center.

C. Stimulation of the Cerebellum

The cerebellum of the goldfish, *Carassius auratus*, the sunfish, *Lepomis*, and the catfish, *Ictalurus*, has been stimulated while the animals were free swimming (Clark et al., 1960). Electrodes were chronically implanted in the cerebellum of the animals and were attached to a stimulator via flexible wires. The results of cerebellar stimulation of these teleost fish varied with the strength and duration of stimulus. In general, stimulation was immediately followed by a more-or-less rapid motion which usually involved the turning of the head or tail or both. The direction of turning following stimulation was toward the side contralateral to the electrical stimulus. Stimulation of areas near the midline, especially in the anterior half of the corpus cerebelli resulted in a pattern described as a typical response "stimulus-rebound." The fish turned toward the side stimulated (ipsiversive) only to reverse the direction to the contralateral side following the cessation of stimulation (Clark et al., 1960).

The magnitude of response increased with increasing stimulus amplitude. With mild stimulation there might be gentle bending of the tail or body or both; increased strength of stimulus resulted in increased speeds of reaction. During rapid movement of the body, the fins and eyes were set in positions which were appropriate to the initial response but returned to a mirror image pattern when stimulus-rebound occurred. It was often found that different stimulus thresholds would give two completely opposite responses from the same site of stimulation. Thus, an ipsilateral response at 0.45 mA would result in an ipsiversive turning with an indication of rebound reversal of direction at termination. A stronger stimulus of 0.5 mA to the same area produced turning to the side contralateral to the original effect. In many areas of the cerebellum it appeared that there was a topical response pattern, so that as depth changed with microelectrode advance, subsequent stimulations could reverse the direction of turning. These differences in depth were often only a fraction of a millimeter (Clark et al., 1960). The turning motion of the fish was not always in one plane. In addition to the simple horizontal turning there was often movement with a rolling component and a combination of rolling and turning, which often produced a spiral motion. Upward and downward motions were also produced by stimulation.

Stimulation also resulted in retraction of the barbels of the catfish. In addition, most animals demonstrated coordinated eye movements in the direction of turning or rotation and often retraction or extrusion of the eyeball from the orbit. One case of stimulus evoked nystagmus was observed (Clark et al., 1960). Long aftereffects, like the seizures elicited in mammals, have not been observed in the fish following cerebellar stimulation. However, prolonged effects of other types were elicited from stimulation of other parts of the catfish brain, such as prolonged inability of the fish to orient itself. This was manifested by the animal lying on one side following periods of stimulation (Clark et al., 1960).

The stimulus-rebound phenomena is found in all classes of vertebrates. It would appear then, that the mechanisms of cerebellar function have evolved early. The function of the cerebellum in fish has yet to be completely defined. These preliminary results, however, do indicate that the cerebellum is involved with the maintenance of muscle tonus in the animal, postural reflexes, and integration of visual, acoustic, and perhaps gustatory stimuli.

VII. MEDULLA OBLONGATA

A. Anatomy

The boundary between medulla oblongata and spinal cord in fish is indistinct (Ariëns Kappers et al., 1960; Aronson, 1963; Papez, 1929), and it is not surprising, therefore, to find that the medulla can be easily divided into columns of nerve fibers based on the type of information transmitted. Thus, there are somatic and visceral sensory columns, and somatic and visceral motor columns. The somatic sensory column carries information derived from general sensory, cutaneous, vestibular, lateral line, and trigeminal (Cr.N.V) nerve fibers. These nerve complexes comprise the cutaneous sensory nerves arising from the medulla, are proprioceptive on extroceptive in function, and can be classified according to the type of information carried by the composite nerve fibers in the cranial nerves (Fig. 8). The visceral sensory column carries nerve fibers which are derived from the nerves of chemoreceptors (taste) and nerves arising in the viscera. This column is formed by the combined centers and nerve fibers which are the sensory branches of the facial (Cr.N.VII), glossopharyngeal (Cr.N.IX), and vagus cranial nerves (Cr.N.X). Nerve fibers in this category function in the transmission of introceptive information. There are, in addition, two motor columns: the visceral motor column that carries efferent motor nerve fibers derived from the facial, glossopharyngeal, and vagus nerves, and a somatic motor column that carries efferent nerve fibers to the ocular muscles and muscles of the pharyngeal complex. The visceral motor components are efferent to the glands and musculature of the viscera. These nerve fibers are motor-secretory, visceral motor, and vasomotor in function. Included in this efferent outflow are nerve fibers in the vagus nerve that are visceral

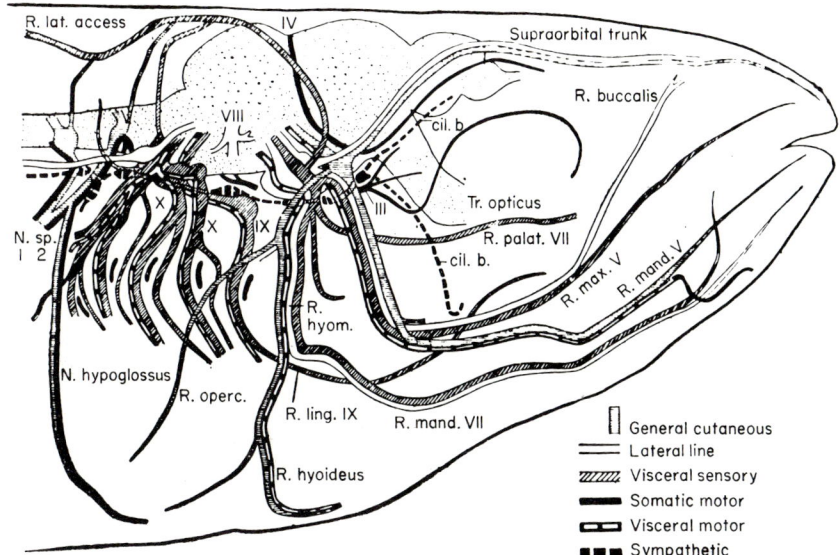

Fig. 8. The functional components of the cranial nerves of the teleost *Menidia*. After Papez (1929), from J. B. Johnston, after reconstruction by Herrick.

motor to the gut. The somatic motor components supply the striated musculature of the body, and they are thus involved in locomotion.

In some Ostariophysi (carp and goldfish) the medulla oblongata is everted for the accommodation of large gustatory lobes. In these animals an elaborate system of taste buds has been developed in the mouth, palatal organ, pharynx, fins, and body wall. In addition to the vagal lobes (Fig. 3), there are other enlargements usually found projecting from the medulla. A prominent facial lobe often is found and is associated with the facial and somatic sensory lobe associated with the trigeminal nerves. A central acoustic lobe is associated with the auditory nerve (CR.N.VIII) and an acoustical-lateral lobe usually is associated with the nerves of the lateral line. The medulla oblongata of fish can contain many extremely interesting cell types. Among these are Müller nerve cells and fibers, Mauthner cells, supramedullary cells, and large reticular cells.

B. Spontaneous Activity of the Medulla Oblongata

The spontaneous activity (EEG) of the medulla has been recorded from the codfish, *Gadus callaris* (Enger, 1957), and the goldfish (Schadé and Weiler, 1959). When the codfish was at rest in the dark, the dominant activity of the medulla was 8–13 Hz with a less-pronounced low frequency activity. After a few seconds of delay, the onset of light induced an increase in the discharges of the dominant 8–13 Hz mode. Evoked responses induced by acoustic stimulation (clicks) were recorded only from the medulla oblongata and never from other parts of the brain. The amplitudes of these potentials reached between 80–110 mV with stimulations of 25 clicks/sec. The evoked response followed up to frequencies of 100–140 clicks/sec. Increasing the frequency to 80 clicks/sec generated 2–3 high spikelike potentials of large amplitude which usually were followed by 2–4 rhythmic waves of 0.5–1.0 sec duration. Cessation of acoustic stimulation was either followed by no response or a spike potential followed by a volley of high frequency waves. Noise in the room during the time of the experiment occasionally could prevent the appearance of 8–13 Hz activity in the cod indicating that acoustic stimulation can result in "arousal."

Electroencephalograms were also obtained from the medulla oblongata of goldfish (Schadé and Weiler, 1959). The wave forms were characterized by slow 0.5–2.0 Hz potentials of low amplitude. In addition, two groups of high frequency waves were superimposed on this slow rhythm resulting in a pattern of 8–11 Hz dominate in the dark, and a pattern of 20–35 Hz observed during light narcosis or during general illumination of the room. Unlike the cod, the electrical activity of the medulla of goldfish was not influenced noticeably by light or noise. However, the recordings from the medulla were made from the gustatory or vagal lobes which could result in the lack of photic or acoustically induced responses (Schadé and Weiler, 1959).

C. Reticulomotor System

The electrical activity of the medullary reticular cells of the dogfish, *Squalus lebruni,* were studied using glass micropipettes (Restieaux and Satchell, 1958). Reticular cells in the medulla of this fish were extremely large, 40–50 μ in diameter and as much as 200–250 μ long. Groups of reticular cells lie on either side of and send their axons into the medial longitudinal fasciculus. The reticular motor nerve fibers descended into the midline of the floor of the medulla to relay information to the motor neurons of the efferent output to the musculature of the trunk and tail. Using intracellular electrodes it was found that the resting membrane potential varied between 60–70 mV in amplitude. Action potentials attained amplitudes of 70–120 mV. The action potential was followed by a post-spike hyperpolarization which followed the falling phase of the action potential. A maximum height of 11 mV was recorded for the 3–7 msec hyperpolarization with a 1–3-msec spike duration. Excitatory postsynaptic potentials were illicited by subliminal afferent volleys. This monosynaptic stimulation resulted in complex synaptic potentials with durations up to 40 msec, indicating the presence of a polysynaptic (interneuron) linkage with a reverberating circuit to account for the prolonged synaptic potential of the reticular cell. Thus, a variety of polysynaptic pathways converge on reticulomotor cells. The polysynaptic input was studied by initiating simultaneous stimulation to the left and right ophthalmic and left and right hyomandibular nerves. The response of the reticulomotor cells to these inputs differed in latency, distinctness in rising phase of the excitatory postsynaptic potential, and amount of late activity induced within the cell by afferent stimulation. Antidromic stimulation of reticulomotor cells could be affected by stimulating the reticulomotor axons within the spinal cord. The conduction velocity of reticulomotor fibers was calculated at 25–66 meters/sec by recording the transmission rates of known lengths of this axon (Restieaux and Satchell, 1958).

The majority of reticulomotor cells produced more than one action potential in response to a single stimulus. The latencies of these responses were usually 7–14 msec, but longer latencies (up to 50 msec) were observed. Many units produced action potentials as much as two or three times to a single stimulus resulting in repetitive discharges and the production of long spike trains. Increasing stimulus strength above threshold increased the number of spikes in a repetitively discharging unit and shortened the latency of the first spike. This indicated that nerve fibers of different thresholds exist in the peripheral nerve being stimulated for the afferent input to the neurons so that increased stimulus strength resulted in increased afferent input to the neurons by the recruitment of axons of higher threshold. This increased afferent activity acting upon internuncial neurons relays information to the reticulomotor cells until the synaptic potential reaches sufficient amplitude to initiate a spike. Action potentials could be evoked by the summation of two independent afferent stimuli neither of which could independently trigger an action potential. The summation of response of the two afferent volleys, either of which could produce a spike independently, resulted in the initiation of a spike earlier in latency than either could produce alone. Units were capable of discharging two or three times to a maximal stimulus applied to a single afferent channel, but discharged only once to a liminal stimulus applied to a single afferent channel. These data demonstrate that the multiple synaptic input can be summed when liminal afferent volleys are presented over separate channels. Again, as could be expected, there was a decrease in latency of response as more afferent channels were activated. Units discharging trains of spikes show the effect of summation of the excitatory influences on the dendritic field.

The receptive fields of the reticulomotor cells were mapped (Restieaux and Satchell, 1958). These afferent fields were grouped into four types of reticulomotor neuron responses: one group of neurons had a receptive field which could be excited by stroking of the skin over any part of the body; a second group had receptive fields which were limited to the region anterior to the posterior margin of the pectoral fins and could also be activated from the stimulation of the ophthalmic and hyomandibular nerves; a small number of units had an extremely restricted field and only responded to stimulation of the ophthalmic nerve or stroking of the surface of the snout; however, the large majority of reticulomotor neurons had receptive fields which were limited to the region anterior to the pectoral fins and responded to stimulation of the ophthalmic and hyomandibular nerves. Afferent nerve fibers were also found to innervate the reticulomotor cells from the optic tectum, since stimulation of the optic tectum resulted in the discharge of these cells with latencies similar to those found from the afferent input derived from stimulation of the ophthalmic nerve (Restieaux and Satchell, 1958).

There was a balance of excitatory and inhibitory afferent input into the reticulomotor system which was partially correlated with asymmetry of the bilateral afferent input. Units were excited and discharged with ipsilateral stimulation of the left ophthalmic nerve but were inhibited by a synchronous contralateral stimulation of the right ophthalmic nerve. The contralateral area of inhibition corresponded to the ipsilateral area of excitation. The integration of afferent sensory information of the body and its resolution into a pattern of reticulomotor discharges appear to be the function of the reticulomotor system within the brain stem of the shark. The reticulomotor nerve fibers distribute information from the brain stem to the motor neuron pool of the trunk and tail. This information is carried in the reticulospinal system within the spinal cord (Restieaux and Satchell, 1958).

D. Taste

Electrophysiological studies have revealed that some fish possess chemoreceptors which respond to the four classic types of taste quality, i.e., sweet, salty, sour, and bitter. However, the stimulation effect of sapid substances on fish chemoreceptors was relative to the species of animal utilized and not to the substance tested (Bardach and Case, 1965;

Bardach et al., 1967; Fujiya and Bardach, 1966; Hidaka and Yokota, 1967; Konishi et al., 1966; Konishi and Zotterman, 1961, 1963; Tateda, 1961, 1964; Teichmann, 1962; Tester, 1963). Recordings of chemoreceptive function have been made from palatine nerves (Cr.N.IX) of ostariophysid fish, nerves from barbels (Cr.N.VII), and nerves (spinal nerve 3) from the modified fin rays of the hake *Urophysis chuss* and sea robin *Prionotus carolinus* (Bardach and Case, 1965). The palatine nerve fibers in Japanese and Swedish carp, *Cyprinus carpio*, were found to respond to salt (sodium chloride solution), acid, sugars, and to a broad gustatory spectrum. Broad spectrum gustatory fibers which comprise the large majority of axons in the palatine nerve responded differently in the Swedish carp and the Japanese carp when the same solution stimulated these chemoreceptors. For instance, an extract of earthworm resulted in a massive response from nerve fibers in the Japanese carp and only a moderate response in the Swedish carp (Konishi and Zotterman, 1963). It is interesting that there was no response of the taste receptors to water, unless animals were previously adapted to isotonic salt solutions (Konishi and Zotterman, 1963). In addition, there were chemoreceptors which were specific for sodium chloride and responded to changes in the ionic environment (Hidaka and Yokota, 1967).

Using Japanese carp, *Cyprinus carpio*, Hidaka and Yokota (1967) recorded the pattern of response from palatine nerves to increasing molarities of sucrose. The response appeared to summate and rise to a maximum level, reaching asymtote with increasing sucrose molarities. This was a typical response when increasing molarities of all the sweet substances were utilized except glycine (sucrose, dextrose, and levulose). Mixtures of glycine and sucrose produced responses which were smaller in magnitude than the sum of the responses produced by the two substances separately. As the concentration of sucrose in the glycine mixture increased, a synergistic effect resulted in larger than normal responses. Mixing glycine with various concentrations of dextrose, sucrose, and levulose yielded comparable results. There appeared to be a competitive inhibition of the receptive sites for sweet substances.

Prestimulation of the chemoreceptors of the carp palatal organ with mercuric chloride resulted in the complete loss of response to subsequent stimulation with dextrose, sucrose, and levulose. The response to glycine was only partially depressed, and the response to sodium chloride usually was not affected. Occasionally, the response to sodium chloride stimulation was augmented by pretreatment with mercuric chloride. These data indicated that there were at least two types of chemoreceptors.

Using carp, Konishi and Zotterman (1961) classified the groups of nerve fiber responses from the glossopharyngeal (Cr.N.IX) nerve fibers innervating the palatal organ into seven groups, one group responding only to acetic acid, but other types responding to two or more different taste solutions (i.e., type 1 responded to acetic acid and sodium chloride, type 3 responded to acetic acid and sucrose, etc.).

In addition to the palatine nerve preparation, preparations of the facial nerve (Cr.N.VII) innervating the lips and barbels of fresh- and saltwater catfish have been used to test the taste responses (Konishi et al., 1966; Tateda, 1961, 1964). Following physiological stimulation of the barbel of the freshwater catfish, *Ameiurus melas*, with sodium chloride solutions, the chemoreceptors responded with a train of summated spikes which continued for the duration of stimulation. In the catfish, potassium chloride produced larger responses in the nerve fibers than sodium chloride (Tateda, 1961). Hydrochloric acid resulted in a large initial burst of impulses which declined rapidly (10–20 sec). Several nerve fibers within the isolated barbel of *Parasilurus asotus* responded to hydrochloric acid alone but not to other taste solutions. Most fibers responded to two or more taste solutions (Tateda, 1964), the majority responding to both acid and salt. Recordings from the facial nerve of sea catfish, *Plotosus anguillaris*, demonstrated a response to hypertonic sodium chloride solution, to quinine and to acid, but not to sugar (Konishi *et al.*, 1966). This insensitivity to sugar solutions was also demonstrated by Bardach and Case (1965) behaviorally (on the sea robin) and physiologically (on the fin preparations from the hake).

There appears then to be differences in the chemoreceptive response patterns to sapid substances by different animals. In addition, the difference between animals which lived in saltwater and animals which lived in freshwater are quite apparent.

Comparisons between the reaction of chemoreceptors to sapid substances in freshwater catfish, *Ictaluris natalis* and *I. nebulosus* (innervation to taste buds, facial nerve only), the chemoreceptors of the tomcod, *Microgadus tomcod* (cranial and spinal innervation to taste buds), and the chemoreceptors of the sea robin *Prionotus carolinus* with modified fins (innervation to taste buds on these fins, spinal nerve 3) were carried out by Bardach *et al.* (1967). The animals were classified according to whether they lived in freshwater or saltwater. Recordings were made from the chemoreceptors of isolated barbels or fin rays. A larger concentration of acetic acid was required for the stimulation of the marine animals than freshwater animals. However, it was pointed out (Bardach *et al.*, 1967) that the presence of acid in seawater can result in the binding of hydrogen ions on the substance upon which the stimulation depends. Therefore, although freshwater fish did seem more sensitive in reality, they may not be, since more acid must be added to the marine medium in order to bring the pH of the stimulating medium to comparable levels in salt- and freshwater. Sodium chloride elicited a more gradual and sustained nerve activity in both barbel preparations than did the acid. In the marine species, increased discharges of taste receptors resulted from an increase in the sodium chloride in the environment. However, lowering of the sodium chloride resulted in an inhibition of the firing pattern in the marine species. When freshwater was applied there was inhibition of discharge rate almost resulting in cessation of discharge. This inhibition was followed by a rebound effect resulting in greater activity upon restimulation with saltwater. In all species of fish studied by Bardach et al. (1967), varying numbers of fibers responded to stimulation with quinine hydrochloride. Less than 10% of all the fibers tested responded to sweet substances, and fibers from the marine sea robin did not respond at all to sugar. This appears to be in keeping with results of Konishi et al. (1966) on the barbel of sea catfish which also did not respond to sugars. As shown by other investigators (Konishi and Zotterman, 1963), Bardach et al. (1967) found that certain taste fibers reacted selectively to different substances. Single nerve fibers responded selectively to acid, salt, sugar, or quinine, respectively. In addition, in the bullhead, catfish, and sea robin, there were fibers that responded selectively to extracts of meat. Homocysteine produced no responses in the sea robin taste receptors, and the bullhead and tomcod tested with the same substance did not initially respond. However, after rinsing the preparation, the tomcod and catfish receptors responded to cysteine. Since homocysteine is only one carbon atom longer than cysteine and will not produce responses in taste receptors, it appears that specific compounds acting on specific response sites are responsible for the reaction of the taste buds. A series of amino acids were tested for their ability to produce signals from chemoreceptors. Of interest is indole which is an electrochemically active substance and leads to strong stimulation of the preparation while indole acetic acid, although related to indole, failed to elicit responses. Recordings made from the olfactory tract of brook trout, *Salvelinus fontinalis,* demonstrated that amino acids perfused over the olfactory organ produced discharges down the olfactory tract. These substances also produce responses to stimulation of chemoreceptors in fish, thus demonstrating olfactory and gustatory chemoreceptor overlap. However, there are certain chemical sensitivities which appear to be specific to one or the other of the chemoreceptive systems. Although fish can both smell and taste certain amino acids, there are some organic substances which can only be tasted or smelled (Bardach et al., 1967). Konishi and Zotterman (1963) found an active lipid within substances which stimulated carp palatal organs. Some of the lipids tested by

Bardach *et al.* (1967) were capable of eliciting responses in chemoreceptors. Therefore, fish may have more taste responses than the four conventional ones formerly found in fresh- and saltwater fish. However, we are dealing here with two systems: a system of free chemoreceptive nerve endings which respond to taste stimulation on the fins of tomcod, and a system of taste buds for chemoreception which are found in the palatal organs or the barbel of carp or catfish. In fact, it has been found that free chemosensitive nerve endings in the sea robin fin do not respond to sugars and that spinally innervated chemoreceptors of sea robins respond to a smaller variety of substances than do the spinally and cranially innervated fins of the tomcod.

It is not yet clear whether or not different sapid substances react on selective sites of the chemoreceptors of the taste end organs of the fish. It is obvious, however, that certain animals such as freshwater carp, respond maximally to certain stimuli. The palatal nerves respond grossly to sweet substances. On the other hand, the sea catfish barbel does not respond at all to sugar substances. It also appears that there is competition for sites on the receptor organ and that the balances and synergistic effects of different substances yield quantitative differences in electrophysiological patterns of discharge.

E. Audition

In general the experiments on audition can be divided into three classes. In one, behavioral end points are used to determine the acoustic threshold of an animal. The second utilizes autonomic unconditioned responses (heart rate) to obtain audiograms. The last approach is an electrophysiological approach to the determination of the central nervous system information used in the integration of acoustic stimuli. The behavioral work has established that the pars inferior (the sacculus and lagena) are the chief organs for sound perception in the teleost and vestibular function resides in the pars superior (utriculus and semicircular canals). However, different parts of labyrinthine function have not been exclusively isolated in the organ so far. In general, members of the Ostariophysi have much better auditory discrimination than nonostariophysid fish. This is most probably because of the morphological connection between the swim bladder and the inner ear through the three pairs of Weberian ossicles (Alexander, 1966; Enger, 1963; Healey, 1957; Kleerekoper and Roggenkamp, 1959; Lowenstein *et al.*, 1968). Fish have perceived some frequencies up to 13,000 Hz, although in most species the upper hearing limit lies around 5000–7000 Hz and is the approximate limit found in ostariophysid fishes. For most nonostariophysid fishes the upper limit for sensitivity is well below 1000 Hz. However, in some species of nonostariophysids there are specializations of the swim bladder which extrude and contact some part of the labyrinth (Sparidae) or there are air-filled cavities around the inner ear (Mormyridae, Anabantidae, and Clupeadae). These animals are sensitive to tone frequencies which can go as high as 1250 and 3000 Hz, respectively. However, the morphological basis of pitch discrimination is not known. The labyrinth has no obvious morphological frequency analyzers within the receptive units. The three organs which would receive sound frequencies are embedded in cavities which are filled with endolymph and completely covered by membranes with areas lined by sensory epithelium. It would appear then that pitch discrimination can be found in the absence of a morphological discriminator only by two possible mechanisms: one, a synchronization between sound frequency and the frequency of impulse discharge, and the other, the use of different sensory units for different sound frequency ranges (Enger, 1963).

The anatomy and ultrastructure of the lamprey labyrinth have been described (Lowenstein et al., 1968). Except for a new type of hair cell and large central labyrinthine ciliated chambers (Lowenstein et al., 1968), it conforms with other fish labyrinths (Young, 1962).

Approximate acoustic thresholds have been obtained in the shark by conditioning to low frequency sounds and using unconditioned cardiac response as the response to the frequency (Wisby et al., 1964). There was usually no heart response to frequencies higher than 500 Hz. The number of sharks responding at each frequency decreased as the frequency increased. Deceleration of heart rate was more pronounced and less variable at frequencies from 7.5 to 40 Hz than at higher frequencies. Sharks can be conditioned to sound (Clark, 1963; Kirtzler and Wood, 1961; Wisby et al., 1964). Using conditioned responses, it was found that the upper detectable frequency limit for the shark was approximately 500 Hz. These data are in accordance with unconditioned heart rate experiments.

The mechanisms of producing sound in marine fishes have been reviewed (Tavolga, 1964) as well as the role of the lateral line in audition (Dijkgraaf, 1963, 1964; Flock, 1965) and will not be covered here. Fish have been trained to a two-way conditioned avoidance response using pure tone as the conditioned stimulus and shock as the unconditioned stimulus. Two marine fish, the squirrelfish, *Holocentrus ascensionis*, and the blue-striped grunt, *Haemulon sciurus*, were utilized by Wodinsky and Tavolga (1964) and Tavolga and Wodinsky (1963, 1965). The lowest thresholds in the squirrelfish were at 800 Hz at pressure levels of approximately −24 dB. At 100 Hz the threshold to response rose to +4 dB, and at 2400 Hz the threshold level was +35.5 dB. In the grunt the threshold at 1100 Hz was about +43 dB and at 600 Hz at −4 dB. At frequencies below 600 Hz there was a high variability in the threshold values obtained. There appeared to be a significant drop in threshold following repeated testing of the animals. This threshold drop was as much as 28 dB at frequencies of 100 Hz. These results were interpreted as being a mechanism for two acoustic detecting systems. That is, there were two sense modalities for receiving pressure waves and displacements which perhaps could be the inner ear and lateral line.

Goldfish have been classically conditioned to sound followed by a food reward (Enger, 1966). Sound pressure thresholds for frequencies 600–700 Hz and lower were dependent upon the distance of the fish from the sound source. There were no significant threshold differences at high levels of stimulation. The lowest thresholds (−42 to −45 dB) were at 1000–1500 Hz which were found to increase to −10 dB for 5000 Hz and to −6 dB (2 meters distance) or −38 dB (0.1 meter distance) for 50 Hz (Enger, 1966).

Enger (1963) used the sculpin, *Cottus scorpius,* a nonostariophysid, for microelectrode recording from the various branches of the auditory nerve. The sound stimuli were usually pure tones of 1–2 sec duration. A wide variety of frequencies were used between 60 and 500 Hz delivered at many intensities. Microelectrodes were used to record single unit activity in the utricular, saccular, and lagenar branches of the auditory nerve.

The spontaneous activity of the units in the three branches of the nerve were of four response types: units which showed the regular activity, units which showed burst activity, units which showed irregular activity, and units that were not active (Fig. 9). More than half of the units responded to the pure tones (Enger, 1963). This response consisted of the initiation of action potentials in units without spontaneous activity and of increased discharge rates in units with regular or irregular spontaneous activity. The units with spontaneous burst activity showed a disruption of this activity into a regular discharge pattern of the same frequency as the stimulus tone used to activate the unit. The action potentials from single units were usually monophasic, positive spikes, but occasionally they were positive–negative spikes with the positive phase having the highest amplitude. The spike recorded from these units were all alike in appearance, had constant amplitudes, bore an all-or-none relationship to stimulus intensity, and disappeared immediately upon a slight movement of the recording electrode. Amplitudes were from less than 1 to 12 mV with durations from 1 to 2 msec. Units of the

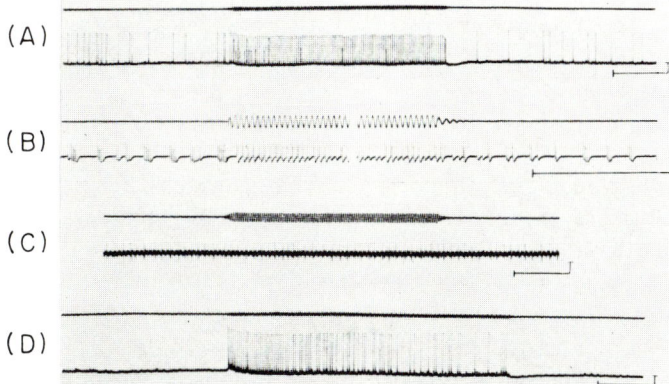

Fig. 9. Spontaneous activity of the four main types of units recorded from cranial nerve VIII (auditory) of the fish, and their responses to a 100 Hz tone, 28 dB, above 1 μb. Vertical bars are 2 mV and the horizontal bars are 250 msec. The top tracing of (A)–(D) is the recorded signal from the hydrophone. Upward deflections are positive voltages. (A) Irregular spontaneous activity averaging 20–30 spikes/sec, lagenar branch; (B) spontaneous burst activity, utricular branch (with 1 sec record interruption); (C) regular spontaneous activity, utricular branch, demonstrating no response; and (D) no spontaneous activity, saccular branch. After Enger (1963).

irregular spontaneous activity type had average discharge rates from 5–80 spikes/sec. These units were found mostly in the utricular and lagenar nerves (approximately one-half) but made up about one-third of the units in the saccular nerve. The burst activity units had trains of discharges in bursts of 2–4 impulses which lasted up to 20 msec with a frequency of 20–70/sec and a total discharge frequency which never went below 50 or exceeded 200 spikes/sec. Units with regular activity had discharge rates of 15 Hz and 60 spikes/sec with a mean discharge frequency of 58 spikes/sec.

Over half of the responses from utricular, saccular, and lagenar units were to pure tone stimulation. The great majority of saccular units responded to sound, whereas only one-third and two-thirds of the utricular and lagenar units did so. Units with spontaneous burst activity never failed to respond to sound. Units with irregular activity were only partially responsive to sound stimulation, and with regular activities responded only to low frequency, high intensity tones if at all. Sound thresholds for a 100 Hz tone demonstrated that there were four types of units, with great variations within the types. Units with irregular spontaneous activity reacted to the entire sound pressure range, but those units which were sound responsive (threshold below 40 dB) had on the average a lower threshold (12 dB) than any other type of unit. Individual units which were not spontaneously active could have low thresholds, but the average value was the same as for the burst activity unit (21 dB). Units with irregular spontaneous activity were invariably not responsive to the 100 Hz tone.

There were two responses of units to pure tone stimulation, the adaptive (habituating) and the nonadaptive. Adaptive units responded initially with an increased rate of discharge which was highest during the initial phase of stimulation and then gradually decreased or habituated. This initial high discharge rate was of the same frequency as the stimulus tone, or a multiple thereof. Adaptation occurred in two steps, a fast adaptation within 200 msec and a slow one within 1 min. This adaptive response was found in all units except those with spontaneous burst activity. The nonadaptive response was found in units with spontaneous burst activity only. The spontaneous response was disrupted by the perception of sound waves, and the unit started to follow the response with little or no adaptation. Units usually gave the same response to each of two successive stimulations which were separated in time by at least 2 sec. However, stimuli of long duration resulted in a definite reduction in response to the second stimulus. The spontaneous activity of the unit always showed a poststimulatory depression (Enger, 1963).

A relationship was found between sound intensity and the response of the units. After a slight initial increase in discharge rate, increased sound pressure resulted in a single spike which followed until further increase of sound pressure resulted in a train of impulses. At frequencies up to 150 Hz the units followed the tone used for stimulation. At 300 Hz or above, neither of the two stimuli elicited more than a few spikes at following rate. There was also a relationship between the sound frequency and the response of the units. As frequency increased the three different types of units were clearly distinguishable (Enger, 1963).

It would appear from these data that ostariophysid fishes have pitch discrimination based upon a synchronization between sound frequency and neuronal discharge frequency. The ability to discriminate low and medium tone frequencies by synchronization of nerve impulses with some frequency is called the "volley theory." This theory seeks to explain pitch discrimination as an interplay between several neurons, each one following with a different phase relation to the sound waves (Enger, 1963). The discharge rate would depend upon both frequency and intensity and a number of neurons should, together, impose on the central nervous system a complete following response. This has been found to be true in Enger's studies (1963). However, frequency discrimination based on the following mechanism does not appear to be extremely elaborate in the fish. There may be other mechanisms involved. Perhaps different types of units are involved which cover different frequency ranges. Both units, however, respond to low frequencies, but only units with certain special spontaneous activity respond to high frequencies. These two types of units could provide information on the frequency of sound.

Saccular microphonics have been recorded from curarized midshipman, *Porichthys notatus* (Cohen and Winn, 1967). Following cautery of the sacculus the microphonic potential virtually disappears. The microphonic appears to arise from the hair cells on the macula of the sacculus as in other vertebrates. The hearing frequency curve for the midshipman based on a 20-μV microphonic response indicated that the ear can detect a frequency range of 30–210 Hz. The low threshold of the ear at 150 Hz corresponded to the fundamental frequency of the sounds produced by these fish. Cohen and Winn (1967) make the point that most of the higher frequencies of fish sound recorded on sonograms apparently (as seen by their data) cannot be utilized as information by the fish ear and are most probably artifacts of the sound-transmitting structures within the fish body.

Saccular microphonics have been utilized in codfish, *Gadus morhua*, and sculpin, *Cottus scorpius*, to determine the animals' ability to perceive sound in the far field, up to 10 meters from the source (Enger and Andersen, 1967). Other studies have been on near field acoustic reception in laboratory studies. Codfish (ostariophysids) displayed a microphonic amplitude potential which was a function of sound pressure only for distances from 0.7 to 10 meters, from an underwater loudspeaker, and for frequencies up to 200 Hz. Microphonic responses were also recorded for frequencies up to 800–1000 Hz. In the sculpin (nonostariophysid) the microphonic potential was dependent upon both the sound pressure and distance. However, no microphonic was recorded after the animal was 1 meter or more from the sound source. It appears that a swim bladder is essential for fish to hear propagated sound waves. The swim bladder generated near-field effects by local water displacements. This local near-field effect in turn is capable of stimulating auditory receptors (Enger and Andersen, 1967).

In a theoretical paper dealing with directional and nondirectional hearing in fish, van Bergeijk (1964) argues that since most fish have only one pressure receptor for far-field acoustic discrimination, they cannot localize the source of sound in the far field. However, near-field acoustic localization requires two displacement receptors which are present in ample numbers in the lateral line. From the data of Enger and Andersen (1967) it would appear that far-field acoustic perception is dependent upon the presence of a swim bladder which results in the production of a propagated near field. Thus, the sensory receptor may not be the essential factor in far field localization and ostariophysid fish may well localize propagated sound waves.

In the herring, *Clupea harengus*, there is a thin duct from the swim bladder which extends into two air-filled bullae close to each labyrinth. This anatomical mechanism apparently acts in the propagation of auditory signals in a manner similar to Weberian ossicles. Extracellular recordings have been made from the acoustic medullary centers of the herring (Enger, 1967). Electrodes were introduced into the medulla through the cerebellum. The first responses were recorded 1.5–2.0 mm below the cerebellar surface and were obtained from 1000 μ ventrad and 700–800 μ in the anterior–posterior direction. This area included the acoustic medullary lobe and the afferent acoustic input. Acoustic stimulation produced increased (excitatory) or decreased (inhibitory) responses over the spontaneous activity of medullary acoustic neurons. Responses were obtained from units to pure tone stimulation for sound frequencies of 30–4000 Hz at moderate sound pressure (35 dB) and up to 8000 Hz for high sound pressures (50–60 dB). Acoustic neurons were classified into two groups according to response to tone stimulation: (1) units responding by excitation over spontaneous rate to frequencies up to 500 Hz, and (2) units responding by inhibition or excitation over spontaneous rate up to 2000 Hz or higher. Although there were increased thresholds for higher frequencies, the most sensitive auditory neurons had thresholds of 20–25 dB at their optimal sound frequencies which ranged from less than 100 to 1200 Hz for different units. Thus, the tentative audiogram for herring is −20 to 25 dB for frequencies of 30–1200 Hz with an increase in threshold to +20 dB for 3000 Hz, +35 dB for 4000 Hz.

VIII. SPINAL CORD

A. Anatomy

The fish spinal cord occupies the entire vertebral canal (Nieuwenhuys, 1964) and in the actinopterygians ends in a specialization, the caudal neurosecretory system (Fridberg and Bern, 1968). The cord consists of a series of segments from which a dorsal and a ventral root arise. In all vertebrates except the petromyzonts the two roots join to form spinal nerves (Ariëns Kappers *et al.*, 1960; Bone, 1963). In the spinal cord there are large dorsal cells (Rohon–Beard cells) present in the larvae of many species of fish. In species of primitive fish the Rohon–Beard cells exist throughout the life of the animal, but in the later life of more advanced animals these cells disappear.

In the cyclostomes the spinal cord is flattened. This may result from the absence of an intramedullary blood supply within the spinal cord. In the cyclostomes the dendrites of the neural elements extend from the ventral gray columns into the dorsal white matter. The dendrites of the spinal neurons tend to run parallel to the long axis of the spinal cord and join the nerve fibers in the unmyelinated white columns. Commissural cells are found in the spinal cord of cyclostomes. The axons of these cells decussate ventral to the central canal and then bifurcate to ascend or descend within the spinal cord. Müller fibers constitute a conspicuous system of descending motor coordinating nerve fibers that can be traced throughout the spinal cord of the lamprey. These nerve fibers do not decussate in the medulla, thus, differing from the well-myelinated, decussated, and collateral bearing Mauthner axons found in the spinal cord of bony fish.

The spinal cord of the Chondrichthyes is more differentiated than that of the cyclostomes. In Chondrichthyes there is an increased number of well-myelinated fiber tracts. In addition, the gray matter is differentiated into a distinct dorsal and ventral horn. Neurons possess extensive dendritic fields and individual neurons send dendrites from gray into white matter. Commissural cell axons and other nerve fibers are numerous in the ventral horn. Commissural cell axons decussate in the midline of the spinal cord in two different commissures which are situated immediately below the central canal.

The spinal cord of the Osteichthyes shows a general resemblance in organization to that of the Chondrichthyes. In the bony fish there is a definite division of the gray matter into dorsal and ventral horns. The dorsal horn is represented as an undivided mass of gray so that the gray substance of the spinal cord is in the shape of an inverted Y. The Rohon–Beard cells disappear in the adult forms of actinopterygians. However, unique to this group of animals one often finds supramedullary neurons of large size which can be considered to be in the cervical spinal cord and medulla oblongata. These giant cells can often be seen with the naked eye and can be as large as 1 mm in diameter. These cells are linked by unique synaptic configurations. The synapses are electrically conducting; therefore, the supramedullary cells are considered to be electrotonically coupled (Bennett et al., 1959a–c, 1967a–d; Hagiwara and Saito, 1959).

In the bony fish, gray matter contains numerous large motor horn cells. These usually occur in two groups: One group is located in the dorsomedial gray and is efferent to the trunk musculature, and one group is located in the ventral gray and is efferent to more specialized areas such as the pectoral fin (Ariëns Kappers et al., 1960). As in the sharks and rays, the spinal cord of bony fish contains many commissural neurons and nerve fibers. Many nerve fibers are connected to the propriospinal system and may ascend to the medulla oblongata, cerebellum, or perhaps the roof of the midbrain (Nieuwenhuys, 1964). The descending pathways in the bony fish consist mainly of vestibulospinal and reticulospinal fibers. One of the conspicuous elements of the spinal cord of bony fish is the large Mauthner fibers which give rise to many short collaterals within the spinal cord. These short collaterals have extensive connections with motor horn cells in the ventral horn.

These features are found to be rather constant in all fish, including some of the most highly evolved bony fish, the Crossopterygi and the Dipnoi. The spinal cord of fish seems to have reached this organization early within the evolution of the animal. One finds the massiveness of the gray matter increases and the connections which are made by the interstitial cells in the spinal cord become more complex as the evolutionary scale is advanced within the fish from the cyclostomes through the Osteichythes (Nieuwenhuys, 1964).

B. Caudal Neurosecretory System and Urophysis

In the actinopterygians the urophysis is a definite neurohemal organ comprised of a high density of blood vessels and the axonal terminations of the neurons of the caudal neurosecretory system. In the teleost the urophysis is a definite organ that is situated ventral to the spinal cord but is extremely variable in shape (Bern and Takasugi, 1962; Fridberg, 1962a,b; Fridberg and Bern, 1968; Hamana, 1962). The neurohemal organ is innervated by a nerve tract which may form an infundibulum or stalk from the neurosecretory cells within the spinal cord. In contrast to actinopterygians the elasmobranch urophysis is not a distinct organ (Bern and Hagadorn, 1959; Fridberg, 1959, 1962b; Hamana, 1962). However, there is an extensive complex of blood vessels ventral to the spinal cord that receives scattered neural input from the neurosecretory neurons that form the caudal neurosecretory system.

The caudal neurosecretory system of fish is composed of neurosecretory neurons (Dahlgren cells) located within the spinal cord (Fridberg and Bern, 1968). In the elasmobranchs the caudal neurosecretory system is extensive with neurosecretory cells found in as much as 22 segments of the caudal spinal cord. These specialized neurons occur in two size classes: One group of cells is small whereas the other is extremely large. The neurosecretory neurons send small unmyelinated axons into the extensive plexus of blood vessels ventral to the spinal cord.

In actinopterygian fish, caudal neurosecretory neurons are ubiquitous and are located in a definite neural center in the distal portion of the spinal cord. Although neurosecretory cells have been found in all fish that have been investigated, the morphology of the caudal neurosecretory system is extremely variable (Fridberg and Bern, 1968). The morphological variation is usually confined to the size and location of the units in the spinal cord. In some small species such as the guppy *Lebistes reticulatus* (Fridberg, 1962a; Sano and Kawamoto, 1959), the neurosecretory cells are extremely small. However, neurosecretory cells of extremely small size appear to be the exception and not the rule (Fridberg and Bern, 1968).

Caudal neurosecretory neurons have unmyelinated axons which in some species gather into an infundibulum or stalk. The infundibulum is reminiscent of the infundibulum of the hypothalamus which leads into the hypophysis. Variation in the location of the nerve fiber tract of the caudal neurosecretory system is a result of the variability in location of the urophysis ventral to the spinal cord. To reach the spinal cord the infundibulum must pierce the leptomenix (primitive meninges) of the spinal cord. Thus, the length and location of the infundibulum is also determined by the shape and location of the urophysis (Fridberg and Bern, 1968). The caudal neurosecretory tract of teleosts usually is free of glial cell processes; however, at the zone of termination within the urophysis one finds ependymal and glial elements interspersed among the neurosecretory fibers (Bern *et al.*, 1965).

The Dahlgren or neurosecretory cells are truly neurosecretory as demonstrated by their ultrastructure. These modified neurons contain elementary neurosecretory granules in the size range of 800–2500 Å within the perikaryon (Afzelius and Fridberg, 1963; Bern and Takasugi, 1962; Enami and Imai, 1958; Fridberg, 1962a; Holmgren and Chapman, 1960; Palay, 1957, 1960; Sano and Knoop, 1959). These elementary neurosecretory granules are found not only in the perikaryon but also have been located along the axons of neurosecretory cells, indicating a migration of neurosecretory material from the perikaryon to the urophysis. The elementary secretory granules are associated with and produced by the Golgi apparatus in the perikaryon of neurosecretory cells in elasmobranchs and teleosts (Bern *et al.*, 1961; Fridberg, 1963; Sano and Knoop, 1959).

The physiology of this extremely interesting system is in doubt. Maetz *et al.* (1963, 1964) using the goldfish, *Carassius auratus*, found that urophysial extracts acted like certain neurohypophysial factors, in particular a prolactinlike hormone (see section on diencephalon). Urophysial extracts affected sodium ion influx across the gills, renal diaresis, and sodium ion excretion. It is also interesting to note that although a prolactinlike hormone was present, urophysial extract was not sufficient to sustain hypophysectomized *Fundulus heteroclitus* in freshwater. Therefore, the prolactinlike influence of the extract was not of sufficient quantity to maintain the animal without supplementation by neurohypophysial function (Pickford et al., 1965).

The urophysis has been implicated in the regulation of buoyancy in the fish (Enami, 1959). However, further attempts to demonstrate this function have been unsuccessful (Fridberg and Bern, 1968; Kawamoto, 1962). It appears then that the caudal neurosecretory system is a neural center within the spinal cord whose discharge to its endorgan (urophysis) results in the release of substances that regulate the osmotic balance in fish.

C. Electrophysiological Properties of the Neurosecretory Cells and Intramedullary Neurons

The discharge patterns of the caudal neurosecretory cells (Dahlgren cells) have been recorded. In general, the response of the cells was different from that of the other spinal neurons in the area. Many species of fish have been utilized such as skates, *Raja* (Bennett and Fox, 1962), the Japanese eel, *Anguilla japonica* (Ishibashi, 1962; Morita et al., 1961), the fluke, *Paralichthys dentatus* (Bennett and Fox, 1962) and *Tilapia mossambica* (Yagi and Bern, 1963, 1965). The results of the former experiments demonstrate an interesting parallel between the function of the caudal neurosecretory and the neurohypophysial system. Comparisons can be made between the influence of osmotic changes owing to sodium ion flux on the discharge rate of neurosecretory cells as well as the influence of higher nervous system centers on the activity of cells within the secretory systems.

Caudal neurosecretory neurons were distinguished from other intramedullary neurons by several characteristics (Bennett and Fox, 1962). Orthodromic subthreshold stimulation of intramedullary neurons were obtained by stimulation of the spinal cord rostral to the unit (stimulation of descending pathways). This stimulation produced graded subthreshold postsynaptic potentials in the neuron. Increased stimulus intensity increased the amplitude of the postsynaptic potential until adequate to produce a spike. Antidromic spikes also can be evoked within intramedullary neurons by stimulating the spinal cord caudal to the unit. This stimulation produced a mixed sequence of discharges which contain both antidromic and synaptic components. The mixed discharge and stimulating effect demonstrated that the units were probably motor neurons. Neurosecretory neurons in skates and teleosts were activated synaptically by stimulation of the rostral spinal cord. However, the presynaptic pathway of these specialized neurons was of high threshold and low conduction velocity (Bennett and Fox, 1962).

The unit spike responses ascribable to the motor neuron and neurosecretory cell were quantitatively different. The unit spike potentials of neurosecretory cells were 4–10 msec in duration. Intracellular recordings from the caudal neurosecetory cell bodies in the Japanese eel (Ishibashi, 1962; Morita *et al.*, 1961) and in the fluke and skate (Bennett and Fox, 1962) demonstrated similar action potentials of long duration. This response was significantly longer than other intramedullary cells and were from 1 to 2 msec in duration. The responses from the neurosecretory cells usually were followed by large and long-lasting undershoots of the resting potential. A spike that was evoked in neurosecretory cells by intracellularly applied depolarizing current was found to be of a much shorter latency with stronger stimuli. The two types of unit spike potentials (from motor cells and neurosecretory cells) were observed in recordings from axons in the caudal neurosecretory system in both *Tilapia* (Yagi and Bern, 1965) and *Cyprinus carpio* (Kinosita *et al.*, 1962, from Yagi and Bern, 1965). These data indicated that both secretory and nonsecretory neurons of the spinal cord innervate the urophysis. It is interesting to note that the size of the neurosecretory neurons appeared to differ but the response characteristics of these types of neurons did not.

On the basis of response to osmotic manipulation of the blood, the neurosecretory units of *Tilapia* (Yagi and Bern, 1963, 1965) were classified into two major and a third minor type. The first major type increased in discharge rate (over spontaneous rate) following intravenous infusion of hypotonic sodium chloride solutions (sodium ion poor). The second major type was activated by infusion of hypertonic solutions of sodium chloride (sodium ion rich). The third or minor type of neurosecretory unit did not respond to osmotic stimuli even though other neurosecretory units which were being recorded simultaneously were discharging. The observed decrease in the electrical activity of some of the units of the neurosecretory system upon the infusion of solutions of different osmolarity indicated the presence of some inhibitory mechanism from other centers within the nervous system that partially controls the neurosecretory cells.

The discharge rate of caudal neurosecretory neurons was influenced by the level of sodium ions of the blood and did not appear to depend on other components of compounds infused into the bloodstream (Yagi and Bern, 1965). Blood osmotic pressure per se was not the stimulus to which the caudal neurosecretory units were responding. In addition, caudal neurosecretory cells that were activated by sodium chloride were generally also activated by sodium sulfate or sodium nitrate. Other neurosecretory units were activated by sucrose and choline chloride which are sodium ion free. In other words, the presence of sodium ions consistently reduced responses in some neurosecretory cells and the absence of sodium ions activated others, thus, producing the two major classes of neurosecretory cells. This response was independent of other ions which were present in the solutions infused (Yagi and Bern, 1965).

In an attempt to locate the center for the control of the caudal neurosecretory system in *Tilapia*, Yagi and Bern (1965) transected the spinal cord. Following spinal cord transection the neurosecretory cells did not respond to physiological stimulation although the local blood supply to the cells was not interrupted. The fact that the neurosecretory cells did not spontaneously generate action potentials after transection of the spinal cord although they retained their excitability suggests that the center regulating the electrical activity of the caudal neurosecretory system is within the brain proper (Yagi and Bern, 1965).

Spinal influences on the caudal neurosecretory system and olfactory influences on the neurohypophysial system may well be similar in mechanism. The pattern of impulses from stimulation of the olfactory organ affects the discharge rate and the excitability of preoptic neurons in the neurohypophysis and partially regulates the release of neurohypophysial hormones. It may well be that a similar sensory modulation of hormonal release is the mechanism by means of which the caudal neurosecretory system acts in osmoregulation.

D. Regeneration of the Spinal Cord

Spinal cord regeneration in larval (Hibbard, 1963; Koppanyi, 1955; Marón, 1959; Niazi, 1963) and adult fish (Clemente, 1964; Healey, 1962; Kirsche, 1965b) was both anatomically and physiologically successful. The regenerative capacity of the spinal cord of fish was unparalleled among the vertebrates. This capacity extended from the simple regrowth of axons across the lesion to reconstitution of the parenchyma into the former neural cytoarchitectonics and complete restitution of new nerve cells and glia.

The relationship between the age and the regenerative capacity of the goldfish spinal cord has been determined. In the goldfish the ability of the animal to reconstitute the spinal cord following complete transection was in relation to the age of the animal (Bernstein, 1964). Young animals (less than 1 year) were able to reconstitute almost 90% of the available axons whereas approximately 60% of the axons were reconstituted in 2- and 3-year-old animals. The ability of the glia to regenerate and reconstitute the diameter of the spinal cord was also related to the age of the animal, the younger animals reconstituting the diameter of the spinal cord almost completely (Bernstein, 1964). In some species of teleosts (Kirsche, 1965b), transection of the spinal cord resulted in the formation of new neurons derived from ependymal cells. The axons of the regenerating neurons grew along glial bridges and aided in the reconstitution of the former zone of the lesion (Kirsche, 1965b).

Not only was the spinal cord reconstituted with the formation of new neurons and glia but the physiological capacity of the spinal cord also appeared to be reestablished after regrowth of the neurites (Bernstein, 1964; Kirsche, 1965b). Therefore, there appears to be respecification of the regenerating axons in the spinal cord which was similar to the regeneration of the optic nerve (Attardi and Sperry, 1963; Sperry, 1963, 1965).

One of the interesting aspects of spinal cord regeneration can be found in the lack of effect of a glial ependymal scar upon the regeneration of the spinal cord of the fish (Bernstein and Bernstein, 1967). Although the glial ependymal scar appears to be a barrier in regeneration of the mammalian spinal cord, regenerating intramedullary axons in fish grew through this alleged mechanical barrier. However, if the spinal cord was arrested (by Teflon) in its growth for 30 days or more, the severed axons no longer retained the capacity to regrow into the area of the lesion (Bernstein and Bernstein, 1967). The mechanism for the cessation of growth has been found to be a special case of contact inhibition with the regrowing neurites forming foreign synaptic contacts just proximal to the glial scar (Bernstein and Bernstein, 1968).

Not only did the spinal cord of teleost fish regenerate after being severed but also one could find the complete reconstitution of areas of spinal cord following ablation. This type of growth pattern was found in the regeneration of the caudal neurosecretory system of *Tilapia* after removal of the caudal peduncle, tail fin, and caudal spinal cord segments (Fridberg et al., 1966; Fridberg and Nishioka, 1966; Imai, 1965). After extirpation of the entire caudal neurosecretory system of *Tilapia*, the entire system was reconstituted (Fridberg et al., 1966). The first elementary neurosecretory granules were found in the perikaryon by the eleventh day of regeneration. These neurosecretory neurons were derived from dedifferentiated and subsequently redifferentiated ependymal cells. Large numbers of these specialized neurons and glial elements were developed and reconstituted the caudal neurosecretory system with almost its original orientation. The reconstitution of the caudal neurosecretory system (Fridberg *et al.,* 1966) demonstrated that the spinal cord of teleost fish can reorganize cytologically and organogenically and thus reconstitute an extirpated area.

Imai (1965) has studied the caudal neurosecretory system of the Japanese eel and has found malformations of this sysem which he considers to be regenerative responses to wounding of the tail area. The new Dahlgren cells took their origin from ependyma. The regenerated neurosecretory area of the spinal cord developed rather rapidly and was fully reconstituted by the end of 5 months, but the urophysis did not regenerate into its original form. The urophysis was found incorporated into the area of the regenerated caudal neurosecretory cells which demonstrated the return of function by the presence of neuroscretory elementary granules. In addition, axons of other nerve fibers which have their cell bodies in the upper portion of the spinal cord were located in the urophysis. Thus, the dual system of innervation was reconstructed (Fridberg *et al.,* 1966).

IX. MYOTYPIC RESPECIFICATION OF REGENERATED NERVES

Studies on myotypic respecification following the heterotrophic regeneration of cranial or peripheral nerves asks the question: Can the specificity of a given muscle alter nerve function in such a manner as to result in return of normal function following regeneration of a foreign innervation? Experimentally this was accomplished by surgical section and cross-union of branches of nerves thus forcing foreign innervation to muscles. This was followed by an assessment of functional recovery. In the Oscar, *Astronotus ocellatus,* the trigeminal nerve (Cr.N.V.) supplies two motor mandibular branches to the muscles of mastication (Arora and Sperry, 1957). Opening the jaw was controlled by the intermandibular muscle and hyoid muscles. The jaw was closed by the abductor or levator mandibulae muscles. Both of these antagonistic muscle complexes were innervated by separate rami of the mandibularis trigemini (motor root Cr.N.V). Transection of the entire motor root of the trigeminal nerve resulted in complete paralysis of the jaw. However, within 16 days, mandibular movement had returned to normal.

The two separate nerve roots (derived from the trigeminal nerve) to the muscles responsible for opening and closing the mandible can be transected and cross-united (Arora and Sperry, 1957). The nerve which normally innervated muscles responsible for closing the mandible now was innervating muscles that were responsible for opening the mandible and vice versa. In four specimens (*Astronotus*), the nerve cross-unions completely regenerated and innervated the muscles with normally opposite function. In these animals, the muscles reacted properly and mandibular movement returned to normal. Electrical stimulation of the cross-united nerves resulted in muscle contraction demonstrating that the foreign nerve had indeed grown into the muscle (Arora and Sperry, 1957). These experiments demonstrated that regenerating motor fibers did establish functional connections with foreign mandibular muscles and that these muscles induced a local muscle specificity so that information derived from the regenerated motor axons resulted in the return of function specific for the muscle (myotypic respecification).

Regeneration of the oculomotor nerve of *Astronotus* resulted in return of normal function following regeneration into extraocular muscles. However, cross-union of the original branches to the different extraocular muscles of the eye in the anglefish, *Pterophyllum scalare*, did not result in normal restitution of function (Arora, 1962). It would appear then that there was no myotypic respecification of the regenerated nerve fibers of this cranial nerve. This also appears to be the case in the regeneration of transposed peripheral nerves to pectoral fins.

Regeneration of spinal axons to peripheral fin musculature has been studied in *Astronotus* (Mark, 1965). In this cichlid fish, there were three groups of antagonistic muscles that moved the pectoral fins. Each of these muscle groups was innervated by a separate nerve bundle derived from the brachial plexus. The muscles were elevators, adductors, or abductors of the pectoral fins. Following transection of the root from the brachial plexus on one side, or partial transection of the root (leaving the nerve fibers to abductor muscles intact), there was an initiation of recovery of function following 7–10 days of paralysis. Recovery was completed in all five animals tested 3 weeks postoperative. Cross-union and regeneration of the innervation of the abductor muscle to the distal stump of the nerve innervating of the adductor musculature (and vice versa) resulted, in the majority of animals, in uncoordinated movements of the fin for up to 10 months postoperatively (Mark, 1965).

Neither respecification of muscle specificity nor central nervous system plasticity of the fish was adequate to compensate for inappropriate nerve muscle connection in the fin musculature. This suggested that the reconstitution of muscular control as seen in the work with the trigeminal nerve primarily depends on the selective reestablishment of peripheral connections of the majority of the muscle fibers reestablishing correct nerve–muscle connections.*

ACKNOWLEDGMENT

The author wishes to dedicate this manuscript to his wife Mary Bernstein. Her love and guidance have been a constant source of inspiration.

REFERENCES

Afzelius, B. J., and Fridberg, G. (1963). The fine structure of the caudal neurosecretory system in Raia batis. Z. Zellforsch. Mikroskop. Anat. 59, 289–308.
Alexander, R. McN. (1966). Physical aspects of swimbladder function. Biol. Rev. 41, 141–176.
Altner, H. (1966). Uber die Aktivitat von Ependym und Glia im Gehirn niederer Wirbeltiere: Sekretorische Phanomene im Hypothalamus von Chimeara monstrosa L. (Holocephali). Z. Zellforsch. Mikroskop. Anat. 73, 10–26.
Ariëns Kappers, C. U., Huber, G. C., and Crosby, E. C. (1960). "The Comparative Anatomy of the Nervous System of Vertebrates, Including Man." Hafner, New York.
Aronson, L. R. (1948). Problems in the behavior and physiology of a species of African mouthbreeding fish (Tilapia macrocephala). Trans. N. Y. Acad. Sci. [2] 2, 33–42.
Aronson, L. R. (1963). The central nervous system of sharks and bony fishes with special reference to sensory and integrative mechanisms. In "Sharks and Survival" (P. W. Gilbert, ed.), pp. 165–241. Heath, Boston, Massachusetts.
Aronson, L. R. (1967). Forebrain function in teleost fishes. Trans. N. Y. Acad. Sci. [2] 29, 390–396.
Aronson, L. R., and Herberman, R. (1960). Persistence of a conditioned response in the cichlid fish, Tilapia macrocephala after forebrain and cerebellar ablations. Anat. Record 138, 332.
Aronson, L. R., and Kaplan, H. (1963). Forebrain function in avoidance conditioning. Am. Zoologist 3, 483–484.
Arora, H. L. (1959). Effects on learning and memory of removal of forebrain, optic lobes and cerebellum in fishes. Ann. Rept. C.I.T. Biol. p. 69.
Arora, H. L. (1962). Role of commissural fibers in interocular transfer of visual discrimination in fishes. Ann. Rept. C.I.T. Biol. p. 88.
Arora, H. L., and Sperry, R. W. (1957). Myotypic respecification of regenerated nerve-fibers in cichlid fishes. J. Embryol. Exptl. Morphol. 5, 256–263.
Arora, H. L., and Sperry, R. W. (1961). Localization of learning and memory in the brain of fishes. Ann. Rept. C.I.T. Biol. p. 129.
Arora, H. L., and Sperry, R. W. (1962). Optic nerve regeneration after surgical cross-union of medial and lateral optic tracts. Am. Zoologist 2, 389.

* Since this chapter was written, many excellent papers have been published. The author would like to mention an excellent symposium ("The Central Nervous System and Fish Behavior," D. Ingle, ed., University of Chicago Press, 1968) which contains contributions by many of the authors quoted in this chapter.

Arora, H. L., and Sperry, R. W. (1963). Color discrimination after optic nerve regeneration in the fish *Astronotus ocellatus*. *Develop. Biol.* **7**, 234–243.
Attardi, D. G., and Sperry, R. W. (1963). Preferential selection of central pathways by regenerating optic fibers. *Exptl. Neurol.* **7**, 46–64.
Bakay, L., and Lee, J. C. (1966). Ultrastructural changes in the edematous central nervous system. III. Edema in shark brain. *Arch. Neurol.* **14**, 644–660.
Ball, J. N. (1965). A regenerated pituitary remnant in a hypophysectomized killifish *Fundulus heteroclitus*): Further evidence for the cellular source of the teleostean prolactin-like hormone. *Gen. Comp. Endocrinol.* **5**, 181–185.
Ball, J. N., Olivereau, M., Slicher, A. M., and Kallman, K. D. (1965). Functional capacity of ectopic pituitary transplants in the teleost *Poecilia formosa* with a comparative discussion on the transplanted pituitary. *Phil. Trans. Roy. Soc. London* **B249**, 66–99.
Bardach, J. E., and Case, J. (1965). Sensory capabilities of the modified fins of squirrel hake (*Urophycis chuss*) and searobin (*Prionotus carolinus* and *P. evolans*). *Copeia* pp. 194–206.
Bardach, J. E., Fujiya, M., and Holl, A. (1967). Investigations of external chemoreceptors of fishes. *In* "Olfaction and Taste II" (T. Hayashi, ed.), pp. 647–665. Pergamon Press, Oxford.
Bennett, M. V. L., and Fox, S. (1962). Electrophysiology of caudal neurosecretory cells in the skate and fluke. *Gen. Comp. Endocrinol.* **2**, 77, 96.
Bennett, M. V. L., Crain, S. M., and Grundfest, H. (1959a). Electrophysiology of supramedullary neurons in *Spheroides maculatus*. I. Orthodromic and antidromic responses. *J. Gen. Physiol.* **43**, 159–188.
Bennett, M. V. L., Crain, S. M., and Grundfest, H. (1959b). Electrophysiology of supramedullary neurons in *Spheroides maculatus*. II. Properties of the electrically excitable membrane. *J. Gen. Physiol.* **43**, 189–219.
Bennett, M. V. L., Crain, S. M., and Grundfest, H. (1959c). Electrophysiology of supramedullary neurons in *Spheroides maculatus*. III. Organization of the supramedullary neurons. *J. Gen. Physiol.* **43**, 221–250.
Bennett, M. V. L., Nakajima, Y., and Pappas, G. D. (1967a). Physiology and ultrastructure of electrotonic junctions. I. Supramedullary neurons. *J. Neurophysiol.* **30**, 161–179.
Bennett, M. V. L., Pappas, G., Aljure, E., and Nakajima, Y. (1967b). Physiology and ultrastructure of electrotonic junctions. II. Spinal and medullary electromotor nuclei in mormyrid fish. *J. Neurophysiol.* **30**, 180–208.
Bennett, M. V. L., Nakajima, Y., and Pappas, G. D. (1967c). Physiology and ultrastructure of electrotonic junctions. III. Giant electromotor neurons of *Malapterurus electricus*. *J. Neurophysiol.* **30**, 209–235.
Bennett, M. V. L., Pappas, G. D., Gimenez, M., and Nakajima, Y. (1967d). Physiology and ultrastructure of electrotonic junctions. IV. Medullary electromotor nuclei in gymnotid fish. *J. Neurophysiol.* **30**, 236–300.
Bennett, M. V. L., Gimenez, M., and Ravitz, M. J. (1968). Synaptically evoked impulse activity of morphologically identified neurosecretory cells. *Anat. Record* **160**, 313–314.
Bern, H. A., and Hagadorn, I. R. (1959). A comment on the elasmobranch caudal neurosecretory system. *In* "Comparative Endocrinology" (A. Gorbman, ed.), pp. 725–727. Wiley, New York.
Bern, H. A., and Takasugi, N. (1962). The caudal neurosecretory system of fishes. *Gen. Comp. Endocrinol.* **2**, 96–110.

Bern, H. A., Nishioka, R. S., and Hagadorn, I. R. (1961). Association of elementary neurosecretory granules with the Golgi complex. *J. Ultrastruct. Res.* **5**, 311–320.

Bern, H. A., Yagi, K., and Nishioka, R. S. (1965). The structure and function of the caudal neurosecretory system of fishes. *Arch. Anat. Microscop. Morphol. Exptl.* **54**, 217–238.

Bernstein, J. J. (1961a). Loss of hue discrimination in forebrain-ablated fish. *Exptl. Neurol.* **3**, 1–17.

Bernstein, J. J. (1961b). Brightness discrimination following forebrain ablation in fish. *Exptl. Neurol.* **3**, 297–306.

Bernstein, J. J. (1962). Role of the telencephalon in color vision in fish. *Exptl. Neurol.* **6**, 173–185.

Bernstein, J. J. (1963). The relation of the telencephalon to the color vision of fish. *Anat. Record* **145**, 207.

Bernstein, J. J. (1964). Relation of spinal cord regeneration to age in adult goldfish. *Exptl. Neurol.* **9**, 161–174.

Bernstein, J. J. (1965). The regenerative capacity of the goldfish telencephalon. *Anat. Record* **151**, 323–324.

Bernstein, J. J. (1967). The regenerative capacity of the telencephalon of the goldfish and rat. *Exptl. Neurol.* **17**, 44–56.

Bernstein, J. J., and Bernstein, M. E. (1967). Effect of glial-ependymal scar and teflon arrest on the regenerative capacity of goldfish spinal cord. *Exptl. Neurol.* **19**, 25–32.

Bernstein, J. J., and Bernstein, M. E. (1968). Contact inhibition: A mechanism of abortive regeneration in the goldfish spinal cord. *Anat. Record* **160**, 315–316.

Bernstein, J. J., and Gelderd, J. B. (1970). Regenerative capacity of long spinal tracts in the goldfish. *Brain Res.* **19**, 21–26.

Bernstein, J. J., and Sadlack, F. J. (1969). The formation of new neurons following telencephalic lesions in goldfish. *Anat. Rec.*, **163**, 154.

Bertolini, B. (1964). Ultrastructure of the spinal cord of the lamprey. *J. Ultrastruct. Res.* **11**, 1–24.

Bertolini, B. (1966). Desmosomi "intracitoplasmatici" nelle cellule di glia. *Atti Accad. Nazl. Lincei, Rend., Classe Sci. Fis., Mat. Nat.* [8] **39**, 367–371.

Bertolini, B., and Mangia, F. (1966). Osservazioni sulla ultrastruttura dell'occhio pineale della lampreda. *Atti Accad. Nazl. Lincei, Rend., Classe Sci. Fis., Mat. Nat.* [8] **41**, 147–153.

Bianki, V. L., and Demina, G. A. (1964). Conditioned reflexes in fish after removal of half of cerebellum or half of optic tectum. *Federation Proc.* **23**, Trans. Suppl., T729–T732.

Bone, Q. (1963). The central nervous system. *In* "Biology of Myxine" (A. Brodal and R. Fänge, eds.), pp. 50–91. Oslo Univ. Press, Oslo.

Boyd, E. S., and Gardner, L. C. (1962). Positive and negative reinforcement from intracranial stimulation of a teleost. *Science* **136**, 648–649.

Brett, J. R. (1957). The sense organs: The eye. *In* "Physiology of Fishes" (M. E. Brown, ed.), pp. 121–154. Academic Press, New York.

Breucker, H., and Horstmann, E. (1965). Elektronenmikroskopische Untersuchungen am Pinaelorgan der Regenbodenforelle (*Salmo irideus*). *Progr. Brain Res.* **10**, 259–269.

Bridges, C. D. B. (1956). The visual pigments of the rainbow trout (*Salmo irideus*). *J. Physiol. (London)* **134**, 620–629.

Clark, E. (1963). Maintenance of sharks in captivity with a report on their instrumental conditioning. *In* "Sharks and Survival" (P. W. Gilbert, ed.), pp. 115–149. Heath, Boston, Massachusetts.

Clark, S. L., Chung, M. Y., Shine, L., and Clark, M. R. (1960). Responses in free swimming fishes to electrical stimulation of the cerebellum. *Am. J. Anat.* **106,** 121–132.

Clemente, C. D. (1964). Regeneration in the vertebrate central nervous system. *Intern. Rev. Neurobiol.* **6,** 257–301.

Cohen, M. J., and Winn, H. E. (1967). Electrophysiological observations on hearing and sound production in the fish, *Porichthys notatus. J. Exptl. Zool.* **165,** 355–370.

Cronly-Dillon, J. R. (1961). Analysis of single units in the optic tract and tectum of goldfish. *Ann. Rept. C.I.T. Biol.* p. 128.

Cronly-Dillon, J. R. (1964). Units sensitive to direction of movement in goldfish optic tectum. *Nature* **203,** 214–215.

Dawson, W., and Bernstein, J. J. (1968). Retinal Na^+ transport blocked by tetrodotoxin: ERG, B-wave obliteration and tectal input. *Psychonom. Bull.* **2.**

Dewsbury, D. A., and Bernstein, J. J. (1969). The role of the telencephalon in the performance of conditioned avoidance responses by goldfish. *Exptl. Neurol.* **23,** 445–456.

Dijkgraaf, S. (1963). The functioning and significance of the lateral-line organs. *Biol. Rev.* **38,** 51–105.

Dijkgraaf, S. (1964). The supposed use of the lateral line as an organ of hearing in fish. *Experientia* **20,** 1–4.

Dixit, V. P. (1967). The nucleus lateralis tuberis in the fresh water teleost, *Clarias batrachus* Linn. *Experientia* **23,** 760.

Dodt, E. (1963). Photosensitivity of the pineal organ in the teleost, *Salmo irideus* (Gibbons). *Experientia* **19,** 642.

Döving, K. B., and Gemne, G. (1965). Electrophysiological and histological properties of the olfactory tract of the burbot (*Lota lota* L.). *J. Neurophysiol.* **28,** 139–153.

Döving, K. B., and Gemne, G. (1966). An electrophysiological study of the efferent olfactory system in the burbot. *J. Neurophysiol.* **29,** 665–674.

Droogleever Fortuyn, J. (1961). Topographical relations in the telencephalon of the sunfish, *Eupomotis gibbosus. J. Comp. Neurol.* **116,** 249–263.

Eakin, R. M. (1963). Lines of evolution of photoreceptors. *In* "General Physiology of Cell Specialization" (D. Mazia and A. Tyler, eds.), pp. 393–425. McGraw-Hill, New York.

Edström, A. (1964a). The ribonucleic acid in the mauthner neuron of the goldfish. *J. Neurochem.* **11,** 309–314.

Edström, A. (1964b). Effect of spinal cord transection on the base composition and content of RNA in the mauthner nerve fiber of the goldfish. *J. Neurochem.* **11,** 557–558.

Edström, A. (1966). Amino acid incorporation in isolated mauthner nerve fiber components. *J. Neurochem.* **13,** 315–321.

Enami, M. (1959). The morphology and functional significance of the caudal neurosecretory system of fishes. *In* "Comparative Endocrinology" (A. Gorbman, ed.), pp. 697–724. Wiley, New York.

Enami, M., and Imai, K. (1958). Studies in neurosecretion. XII. Electron microscopy of the secrete-granules in the caudal neurosecretory system of the eel. *Proc. Japan Acad.* **34,** 164–168.

Enger, P. S. (1957). The electroencephalogram of the codfish (*Gadus callarias*). *Acta Physiol. Scand.* **39**, 55–72.
Enger, P. S. (1963). Single unit activity in the peripheral auditory system of a teleost fish. *Acta Physiol. Scand.* **59**, Suppl. 210, 1–48.
Enger, P. S. (1966). Acoustic threshold in goldfish and its relation to the sound source distance. *Comp. Biochem. Physiol.* **18**, 859–868.
Enger, P. S. (1967). Hearing in herring. *Comp. Biochem. Physiol.* **22**, 527–538.
Enger, P. S., and Andersen, R. (1967). An electrophysiological field study of hearing in fish. *Comp. Biochem. Physiol.* **22**, 517–525.
Flock, A. (1965). Transducing mechanisms in the lateral line canal organ receptors. *Cold Spring Harbor Symp. Quant. Biol.* **30**, 113–145.
Fridberg, G. (1959). A histological evidence of the homology between Dahlgen cells in rays and teleosts. *Acta Zool.* (*Stockholm*) **40**, 101–104.
Fridberg, G. (1962a). Studies on the caudal neurosecretory system in teleosts. *Acta Zool.* (*Stockholm*) **43**, 1–77.
Fridberg, G. (1962b). The caudal neurosecretory system in some elasmobranchs. *Gen. Comp. Endocrinol.* **2**, 249–266.
Fridberg, G. (1963). Electron microscopy of the caudal neurosecretory system in *Leuciscus rutilus* and *Phoxinus phoxinus*. *Acta Zool.* (*Stockholm*) **44**, 245–267.
Fridberg, G., and Bern, H. A. (1968). The urophysis and the caudal neurosecretory system of fishes. *Biol. Rev.* **43**, 175–199.
Fridberg, G., and Nishioka, R. S. (1966). Secretion into the cerebrospinal fluid by caudal neurosecretory neurons. *Science* **152**, 90–91.
Fridberg, G., Nishioka, R. S., Bern, H. A., and Fleming, W. R. (1966). Regeneration of the caudal neurosecretory system in the cichlid teleost *Tilapia mossambica*. *J. Exptl. Zool.* **162**, 311–336.
Fujiya, M., and Bardach, J. E. (1966). A comparison between the external taste sense of marine and fresh water fishes. *Bull. Japan. Soc. Sci. Fisheries* **32**, 45–56.
Gaze, R. M., Jacobson, M., and Székely, G. (1963). The retinotectal projection in *Xenopus* with compound eyes. *J. Physiol.* (*London*) **165**, 484–499.
Gaze, R. M., Jacobson, M., and Székely, G. (1965a). On the formation of connections by compound eyes in *Xenopus*. *J. Physiol.* (*London*)**176**, 409–417.
Gaze, R. M., Jacobson, M., and Sharma, S. C. (1965b). Visual responses from the goldfish brain following excision and reimplantation of the optic tectum. *J. Physiol.* (*London*) **183**, 38–39P.
Grafstein, B. (1967). Transport of protein by goldfish optic nerve fibers. *Science* **157**, 196–198.
Grimm, R. J. (1960). Feeding behavior and electrical stimulation of the brain of *Carassius auratus*. *Science* **131**, 162–163.
Hafeez, M. A., and Ford, P. (1967). Histology and histochemistry of the pineal organ in the sockeye salmon, *Oncorhynchus nerka*, Walbaum. *Can. J. Zool.* **45**, 117–126.
Hagiwara, S., and Saito, N. (1959). Membrane potential change and membrane current in supramedullary nerve cell of puffer. *J. Neurophysiol.* **22**, 204–221.
Hainsworth, F. R., Overmier, J. B., and Snowdon, C. T. (1967). Specific and permanent deficits in instrumental avoidance responding following forebrain ablation in the goldfish. *J. Comp. Physiol. Psychol.* **63**, 111–116.
Hamana, K. (1962). Uber die Neurophysis spinalis caudalis bei Fischen. *J. Kyoto Prefect. Med. Univ.* **71**, 478–490.

Hara, T. J., and Gorbman, A. (1967). Electrophysiological studies of the olfactory system of the goldfish, *Carassius auratus* L. I. Modification of the electrical activity of the olfactory bulb by other central nervous structures. *Comp. Biochem. Physiol.* **21**, 185–200.

Hara, T. J., Ueda, K., and Gorbman, A. (1965). Electroencephalographic studies of homing salmon. *Science* **149**, 884–885.

Hasler, A. D. (1957). Olfactory and gustatory senses of fishes. In "The Physiology of Fishes" (M. E. Brown, ed.), Vol. 2, pp. 187–193. Academic Press, New York.

Hasler, A. D. (1966). "Underwater Guideposts." Univ. of Wisconsin Press, Madison, Wisconsin.

Healey, E. G. (1957). The nervous system. In "The Physiology of Fishes" (M. E. Brown, ed.), Vol. 2, pp. 1–119. Academic Press, New York.

Healey, E. G. (1962). Experimental evidence for regeneration following spinal section in the minnow (*Phoxinus phoxinus*). *Nature* **194**, 395–396.

Herrick, C. J. (1910). The morphology of the forebrain in amphibia and reptilia. *J. Comp. Neurol.* **20**, 413–547.

Hibbard, E. (1963). Regeneration in the severed spinal cord of chordate larvae of *Petromyzon marinus*. *Exptl. Neurol.* **7**, 175–185.

Hidaka, I., and Yokota, S. (1967). Taste receptor stimulation by sweet-tasting substances in the carp. *Japan. J. Physiol.* **17**, 652–666.

Hodgson, E. S., Mathewson, R. F., and Gilbert, P. W. (1967). Electroencephalographic studies of chemoreception in sharks. In "Sharks, Skates and Rays" (P. W. Gilbert, R. F. Mathewson, and D. P. Rall, eds.), pp. 491–501. Johns Hopkins Press, Baltimore, Maryland.

Holmgren, U. (1959). On the structure of the pineal area of teleost fishes. *Goteborgs Veten kaps-Vitterhetssamh. Handl.* **B8**, No. 3, 5–66.

Holmgren, U., and Chapman, C. B. (1960). The fine structure of the urophysis spinalis of the teleost fish *Fundulus heteroclitus* L. *J. Ultrastruct. Res.* **4**, 15–25.

Imai, K. (1965). Malformed caudal neurosecretory system in the eel, *Anguilla japonica*. *Embryologia* **9**, 78–97.

Ingle, D. J. (1965a). The use of the fish in neuropsychology. *Perspectives Biol. Med.* **8**, 241–260.

Ingle, D. J. (1965b). Behavioral effects of forebrain lesions in goldfish. *Proc. 73rd Am. Psychol. Assoc.* pp. 143–144.

Ishibashi, T. (1962). Electrical activity of the caudal neurosecretory cells in the eel, *Anguilla japonica*, with special reference to synaptic transmission. *Gen. Comp. Endocrinol.* **2**, 415–424.

Jacobson, M. (1964a). The spectral sensitivity of single units in the optic tectum of the goldfish. *J. Physiol. (London)* **173**, 28–29P.

Jacobson, M. (1964b). Spectral sensitivity of single units in the optic tectum of the goldfish. *Quart. J. Exptl. Physiol.* **49**, 384–393.

Jacobson, M., and Gaze, R. M. (1964). Types of visual response from single units in the optic tectum and optic nerve of the goldfish. *Quart. J. Exptl. Physiol.* **49**, 199–209.

Jacobson, M., and Gaze, R. M. (1965). Selection of appropriate tectal connections by regenerating optic nerve fibers in adult goldfish. *Exptl. Neurol.* **13**, 418–430.

Jasinski, A., Gorbman, A., and Hara, T. (1967). Activation of the preopticohypophysial neurosecretory system through olfactory afferents in fishes. In "Neurosecretion" (I. Stutinsky, ed.), pp. 106–123. Springer, Berlin.

Kamrin, R. P., and Aronson, L. R. (1954). The effects of forebrain lesions on mating behavior in the male platyfish, *Xiphophorus maculatus*. *Zoologica* **39**, 133–140.

Kandel, E. R. (1964). Electrical properties of hypothalamic neuroendocrine cells. *J. Gen. Physiol.* **47**, 691–717.

Kaplan, H., and Aronson, L. R. (1967). Effect of forebrain ablation on the performance of a conditioned avoidance response in the teleost fish, *Tilapia h. macrocephala*. *Animal Behaviour* **15**, 438–448.

Karamyan, A. I. (1956). Evolution of the function of the cerebellum and cerebral hemispheres. *Fiziol. Zh. SSSR* **25**, No. 3; (*English Transl.*) *U. S. Dept. Comm. Office Tech. Serv., P. B. Rept.* **31**,014, No. TT 61, 1–161 (1962).

Karamyan, A. I. (1957). Some mechanisms of physiologico-morphological evolution of the higher divisions of the central nervous system of vertebrates. *Materialy po Evolyut. Fiziol.* **2**, 86–101; (*English Transl.*) *U. S. Dept. Comm., Office Tech. Serv. P. B. Rept.* **591**,382, No. TT (1959).

Karamian, A. I., Vesselkin, N. P., Beleknova, M. G., and Zagorulko, T. K. (1966). Electrophysiological characteristics of tectal and thalamo-cortical divisions of the visual system in lower vertebrates. *J. Comp. Neurol.* **127**, 559–576.

Kawamoto, N. Y. (1962). Studies on the swimbladder of goldfish (*Carassius auratus* L.) in special reference to the relation between the bladder and the caudal neurosecretory system. *Rept. Fac.-Fisheries, Prefect. Univ. Mie* **4**, 1–6.

Kholodov, Y. A. (1960). Simple and complex food-obtaining conditioned reflexes in normal fish and in fish after removal of the forebrain. *Works Inst. Higher Nervous Activity, Physiol. Ser.* (*English Transl.*) **5**, 194–201.

Kirsche, W. (1960). Zur Frage der Regeneration des Mittelhirns der Teleostei. *Verhandl. Anat. Ges.* (*Jena*) **56**, 259–270.

Kirsche, W. (1965a). Die Wirkung von Hirnembryonalextrakt auf das Vorderhirn von *Amblystoma mexicanum* nach einseitiger Resektion des rechten Vorderhirnpoles. *J. Hirnforsch.* **7**, 463–480.

Kirsche, W. (1965b). Regenerative Vorgange im Gehirn und Rückenmark. *Ergeb. Anat. Entwicklungsgeschichte* **38**, 143–194.

Kirsche, W. (1967). Uber postembryonale Matrixzonen im Gehirn verschiedener Vertebraten und deren Beziehung zur Hirnbauplanlehre. *Z. Mikroskop. Anat. Forsch.* **77**, 313–406.

Kirsche, W., and Kirsche, K. (1961). Experimentelle Untersuchungen zur Frage der Regeneration und Funktion des Tectum Opticum von *Carassius carassius* L. *Z. Mikroskop. Anat. Forsch.* **67**, 140–182.

Kirtzler, H., and Wood, L. (1961). Provisional audiogram for the shark, *Carcharhinus leucas*. *Science* **133**, 1480–1482.

Klatzo, I. (1967). Cellular morphology of the lemon shark brain. *In* "Sharks, Skates and Rays" (P. W. Gilbert, R. F. Mathewson, and D. P. Rall, eds.), pp. 341–360. Johns Hopkins Press, Baltimore, Maryland.

Kleerekoper, H. (1967). Some aspects of olfaction in fishes, with special reference to orientation. *Am. Zoologist* **7**, 385–395.

Kleerekoper, H., and Roggenkamp, P. A. (1959). An experimental study on the effect of the swimbladder on the hearing sensitivity in *Ameiurus nebulosus* (Lesueur). *Can. J. Zool.* **37**, 1–8.

Konishi, J. (1960). Electric response of visual center to optic nerve stimulation in fish. *Japan. J. Physiol.* **10**, 28–41.

Konishi, J., and Zotterman, Y. (1961). Taste functions in the carp. An electrophysiological study on gustatory fibers. *Acta Physiol. Scand.* **52**, 150–161.

Konishi, J., and Zotterman, Y. (1963). Taste functions in fish. *In* "Olfaction and Taste" (Y. Zotterman, ed.), pp. 215–233. Macmillan, New York.

Konishi, J., Uchida, M., and Mori, Y. (1966). Gustatory fibers in the sea catfish. *Japan. J. Physiol.* **16**, 194–204.

Koppanyi, T. (1955). Regeneration in the central nervous system of fishes. *In* "Regeneration in the Central Nervous System" (W. F. Windle, ed.), pp. 3–19. Thomas, Springfield, Illinois.

Kruger, L., and Maxwell, D. S. (1966). The fine structure of ependymal processes in the teleost optic tectum. *Am. J. Anat.* **119**, 479–498.

Kruger, L., and Maxwell, D. S. (1967). Comparative fine structure of vertebrate neuroglia: Teleosts and reptiles. *J. Comp. Neurol.* **129**, 115–142.

Larsell, O. (1967). *In* "The Comparative Anatomy and Histology of the Cerebellum from Myxinids through Birds" (J. Jansen, ed.). Univ. of Minnesota Press, Minneapolis, Minnesota.

Lederis, K. (1964). Fine structure and hormone content of the hypothalamo-neurohypophysial system of rainbow trout (*Salmo irideus*) exposed to sea water. *Gen. Comp. Endocrinol.* **4**, 638–661.

Long, D. M., Bodenheimer, T. S., Hartmann, J. F., and Klatzo, I. (1968). Ultrastructural features of the shark brain. *Am. J. Anat.* **122**, 209–236.

Lowenstein, O., Osborne, M. P., and Thornhill, R. A. (1968). The anatomy and ultrastructure of the labyrinth of the lamprey (*Lampetra fluviatilis* L.). *Proc. Roy. Soc.* **B170**, 113–134.

Lubinska, L. (1964). Axoplasmic streaming in regenerating and in normal nerve fibers. *Prog. Brain Res.* **13**, 1–71.

McCleary, R. A. (1954). Neural implications of interocular transfer in the goldfish. *Am. Psychologist* **9**, 423.

McCleary, R. A. (1960). Type of response as a factor in interocular transfer in the fish. *J. Comp. Physiol. Psychol.* **53**, 311–321.

McCleary, R. A., and Bernstein, J. J. (1959). A unique method for control of brightness cues in study of color vision in fish. *Physiol. Zool.* **32**, 284–292.

Mackintosh, J., and Sutherland, N. S. (1963). Visual discrimination by the goldfish: The orientation of rectangles. *Animal Behaviour* **11**, 135–141.

MacNichol, E. F., Jr., Wolbarsht, M. L., and Wagner, H. G. (1961). Electrophysiological evidence for a mechanism of color vision in the goldfish. *In* "Light and Life" (W. D. McElroy and B. Glass, eds.), pp. 795–816. Johns Hopkins Press, Baltimore, Maryland.

Maetz, J. (1963). Physiological aspects of neurohypophyseal function in fishes with some reference to the amphibia. *Symp. Zool. Soc. London* **9**, 107–140.

Maetz, J., and Julien, M. (1961). Action of neurohypophyseal hormone on the sodium fluxes of a freshwater teleost. *Nature* **189**, 152–153.

Maetz, J., Bourguet, J., and Lahlou, B. (1963). Action de l'urophypophyse sur les échanges de sodium (étudiés à l'aide du ^{24}Na) et sur l'excrétion urinaire du téléostéen *Carassius auratus*. *J. Physiol.* (*Paris*) **55**, 159–160.

Maetz, J., Bourguet, J., and Lahlou, B. (1964). Urophyse et osmorégulation chez *Carassius auratus*. *Gen. Comp. Endocrinol.* **4**, 401–414.

Mark, R. F. (1965). Fin movement after regeneration of neuromuscular connections: An investigation of myotypic specificity. *Exptl. Neurol.* **12**, 292–302.

Mark, R. F. (1966). The tectal commissure and interocular transfer of pattern discrimination in cichlid fish. *Exptl. Neurol.* **16**, 215–225.

Mark, R. F., and Davidson, T. M. (1966). Unit responses from commissural fibers of the optic lobes of fish. *Science* **152**, 797–799.
Marks, W. B. (1963). Difference spectra of the visual pigments in single goldfish cones. Ph.D. Dissertation, Johns Hopkins University, Baltimore, Maryland.
Marón, K. (1959). Regeneration capacity of the spinal cord in *Lampetra fluviatilis* larvae. *Folia Biol. (Warsaw)* **7**, 179–189.
Marón, K. (1963). Endbrain regeneration in *Lebistes reticulatus*. *Folia Biol. (Warsaw)* **11**, 1–10.
Meier, A. H., and Fleming, W. R. (1962). The effects of pitocin and pitressin on water and sodium movements in the euryhaline killi fish *Fundulus kansae*. *Comp. Biochem. Physiol.* **6**, 215.
Morita, H., Ishibashi, T., and Yamashita, S. (1961). Synaptic transmission in neurosecretory cells. *Nature* **191**, 183.
Morita, Y. (1966). Entladungsmuster pinealer Neurone der Regenbogenforelle (*Salmo irideus*) bei Belichtung des Zwischenhirns. *Arch. Ges. Physiol.* **289**, 155–167.
Mugnaini, E. (1964). The ultrastructure of the cerebral hemispheres in *Myxine glutinosa* (L.). *Proc. 3rd Reg. Conf. (Eur.) Electron Microscopy Prague, 1964* pp. 277–278. Publ. House Czecho. Acad. Sci., Prague.
Mugnaini, E. (1967). On the occurrence of filamentous rodlets in neurons and glia cells of *Myxine glutinosa* (L.). *Sarsia*, **29**, 221–232.
Mugnaini, E., and Walberg, F. (1965). The fine structure of the capillaries and their surroundings in the cerebral hemispheres of *Myxine glutinosa* (L.) *Z. Zellforsch. Mikroskop. Anat.* **66**, 333–351.
Nakajima, Y., Pappas, G. D., and Bennett, M. V. L. (1965). The fine structure of the supramedullary neurons of the puffer with special reference to endocellular and pericellular capillaries. *Am. J. Anat.* **116**, 471–477.
Niazi, I. A. (1963). The histology of tail regeneration in the ammocoetes. *Can. J. Zool.* **41**, 125–145.
Nieuwenhuys, R. (1959). The structure of the telencephalon of the teleost *Gasterosteus aculeatus*. *Koninkl. Ned. Akad. Wetenschap. Proc.* **C62**, 341–362.
Nieuwenhuys, R. (1960). Some observations on the structure of the forebrain of bony fishes. *In* "Structure and Function of the Cerebral Cortex" (D. B. Tower and J. B. Schadé, eds.), pp. 144–149. Elsevier, Amsterdam.
Nieuwenhuys, R. (1962a). Some aspects of the comparative anatomy of the forebrain. *Arch. Neerl. Zool.* **14**, 568–601.
Nieuwenhuys, R. (1962b). The forebrain in some groups of fishes. *Anat. Record* **142**, 262.
Nieuwenhuys, R. (1962c). The morphogenesis and the general structure of the actinopterygian forebrain. *Acta Morphol. Neerl.-Scand.* **5**, 65–78.
Nieuwenhuys, R. (1962d). Trends in the evolution of the actinopterygian forebrain. *J. Morphol.* **111**, 69–88.
Nieuwenhuys, R. (1963). The comparative anatomy of the actinopterygian forebrain. *J. Hirnforsch.* **6**, 171–192.
Nieuwenhuys, R. (1964). Comparative anatomy of the spinal cord. *Progr. Brain Res.* **11**, 1–57.
Nieuwenhuys, R. (1966). The interpretation of the cell masses in the teleostean forebrain. *In* "Evolution of the Forebrain, Phylogenesis and Ontogenesis of the Forebrain" (R. Hassler and H. Stephan, eds.), pp. 32–39. Thieme, Stuttgart.

Nieuwenhuys, R. (1967a). Comparative anatomy of the cerebellum. *Progr. Brain Res.* **25**, 1–93.

Nieuwenhuys, R. (1967b). Comparative anatomy of olfactory centres and tracts. *Progr. Brain Res.* **23**, 1–64.

Nieuwenhuys, R., and Bodenheimer, T. S. (1966). The diencephalon of the primitive bony fish *Polypterus* in the light of the problem of homology. *J. Morphol.* **118**, 415–450.

Nieuwenhuys, R., and Nicholson, C. (1967). Cerebellum of mormyrids. *Nature* **215**, 764–765.

Oksche, A., and Kirschstein, H. (1967). Ultrastruktur der Sinneszellen im Pinealorgan von *Phoxinus laevis*, L. *Z. Zellforsch. Mikroskop. Anat.* **78**, 151–166.

Olivereau, M., and Ball, J. N. (1966). Histological study of functional ecoptic pituitary transplants in a teleost fish (*Poecilia formosa*). *Proc. Roy. Soc.* **B164**, 106–129.

Oztan, N. (1966). The fine structure of the adenohypophysis of *Zoarces viviparus* L. *Z. Zellforsch. Mikroskop. Anat.* **69**, 699–718.

Oztan, N. (1967). Occurrence of large hyaline bodies in the preoptic nerve cells of *Zoarces viviparus* L. *Z. Zellforsch. Mikroskop. Anat.* **83**, 53–57.

Palay, S. L. (1957). The fine structure of the neurohypophysis. *In* "Ultrastructure and Cellular Chemistry of Neural Tissue" (H. Waelsch, ed.), pp. 31–49. Harper (Hoeber), New York.

Palay, S. L. (1960). The fine structure of secretory neurons in the preoptic nucleus of the goldfish (*Carassius auratus*). *Anat. Record* **138**, 417–443.

Palay, S. L., McGee-Russell, S. M., Gordon, M. S., and Grillo, M. A. (1962). Fixation of neural tissues for electron microscopy by perfusion with solutions of osmium tetroxide. *J. Cell. Biol.* **12**, 385–410.

Papez, J. W. (1929). "Comparative Neurology." Crowell, New York.

Pappas, G. (1966). Electron microscopy of neuronal junctions involved in transmission in the central nervous system. *In* "Nerve as a Tissue" (K. Rodahl and B. Issekutz, eds.), pp. 49–87. Harper, New York.

Pflugfelder, O. (1953). Wirkungen der Epiphysektomie auf die Postembryonalentwicklung von *Lebistes reticulatus* Peters. *Arch. Entwicklungsmech. Organ.* **146**, 115–136.

Pflugfelder, O. (1954). Wirkungen partieller Zerstörungen der Parietalregion von *Lebistes reticulatus* Peters. *Arch. Entwicklungsmech. Organ.* **147**, 42–60.

Pflugfelder, O. (1964). Wirkungen lokaler Hirnläsionen auf Hypophyse und Thyreoidea von *Carassius gibelio auratus*, Bloch. *Arch. Entwicklungsmech. Organ.* **155**, 535–548.

Pickford, G. E., and Atz, J. W. (1957). "The Physiology of the Pituitary Gland of Fishes." N.Y. Zool. Soc., New York.

Pickford, G. E., Robertson, E. E., and Sawyer, W. H. (1965). Hypophysectomy, replacement therapy, and the tolerance of the euryhaline killifish, *Fundulus heteroclitus*, to hypotonic media. *Gen. Comp. Endocrinol.* **5**, 160–180.

Potter, D. D., and Loewenstein, W. R. (1955). Electrical activity of neurosecretory cells. *Am. J. Physiol.* **183**, 652.

Rahmann, H. (1965). Zum Stofftransport im Zentralnervensystem der Vertebraten. Autoradiographische Untersuchungen mit P-32-Orthophosphat, H-3-Histidin, H-3-Cytidin und H-3-Uridin an Mausen und Fischen. *Z. Zellforsch. Mikroskop. Anat.* **66**, 878–890.

Rahmann, H. (1967). Autoradiographische Untersuchungen zum RNS-Stoffwechsel im Tectum opticum von *Brachydanio rerio* Ham. Buch. (Cyprinidae, Pisces). *Histochemie* **11**, 205–215.

Rahmann, H. (1968). Syntheoseort und Ferntransport von Proteinen im Fischhirn. *Z. Zellforsch. Mikroskop. Anat.* **86**, 214–237.

Ramon y Cajál, S. (1952). "Histologie du système nerveux de l'homme et des vertèbres." Consejo Superior de Investigaciones Cientificas, Madrid.

Rasquin, P. (1958). Studies in the control of pigment cells and light reactions in recent teleost fishes. *Bull. Am. Museum Nat. Hist.* **115**, 1–68.

Restieaux, N. J., and Satchell, G. H. (1958). A unitary study of the reticulomotor system of the dogfish, *Squalus lebruni* (Vaillant). *J. Comp. Neurol.* **109**, 391–416.

Richter, W. (1965). Regeneration im Tectum opticum bei *Leucaspius delineatus* (Heckel, 1843). *Z. Mikroskop. Anat. Forsch.* **74**, 46–68.

Richter, W. (1966). Mitotische Aktivitat in den Matrixzonen des Tectum opticum von juvenilen und adulten *Lebistes reticulatus* (Peters, 1859). *J. Hirnforsch.* **8**, 195–206.

Richter, W. (1968). Regeneration im Tectum opticum bei adulten *Lebistes reticulatus* (Peters, 1859) (Poecilidae, Cyprinodontes, Teleostei). *J. Hirnforsch.* **10**, 173–186.

Robertson, J. D. (1963). The occurrence of a subunit pattern in the unit membranes of club endings in mauthner cell synapses in goldfish brains. *J. Cell Biol.* **19**, 201–221.

Robertson, J. D., Bodenheimer, T. S., and Stage, D. E. (1963). The ultrastructure of mauthner cell synapses and nodes in goldfish brains. *J. Cell Biol.* **19**, 159–199.

Ruch, T. C. (1958). Motor systems. *In* "Handbook of Experimental Psychology" (S. S. Stevens, ed.), pp. 154–208. Wiley, New York.

Rüdeberg, C. (1966). Electron microscopical observations on the pineal organ of the teleosts *Mugil auratus* (Risso) and *Uranoscopus scaber* (Linne). *Pubbl. Staz. Zool. Napoli* **35**, 47–60.

Rüdeberg, C. (1968a). Receptor cells in the pineal organ of the dogfish *Scyliorhinus canicula* Linne. *Z. Zellforsch. Mikroskop. Anat.* **85**, 521–526.

Rüdeberg, C. (1968b). Structure of the pineal organ of the sardine, *Sardina pilchardus sardina* (Risso) and some further remarks on the pineal organ of *Mugil* spp. *Z. Zellforsch. Mikroskop. Anat.* **84**, 219–237.

Sano, Y., and Kawamoto, M. (1959). Entwicklungsgeschichtliche Beobachtungen an der Neurophysis spinalis caudalis von *Lebistes reticulatus* Peters. *Z. Zellforsch. Mikroskop. Anat.* **51**, 56–64.

Sano, Y., and Knoop, A. (1959). Elektronenmikroskopische Untersuchungen am kaudalen neurosekretorischen System von *Tinca vulgaris*. *Z. Zellforsch. Mikroskop. Anat.* **49**, 464–492.

Sathyanesan, A. G. (1965). The reorganization of the hypophysial stalk following hypophysectomy in the teleost *Porichthys notatus* Girard. *Z. Zellforsch. Mikroskop. Anat.* **67**, 734–739.

Sathyanesan, A. G., and Gorbman, A. (1965). Typical and atypical regeneration and overgrowth of hypothalamo-hypophysial neurosecretory tract after partial or complete hypophysectomy in the goldfish. *Gen. Comp. Endocrinol.* **5**, 456–463.

Sawyer, W. H., Musnick, R. A., and van Dyke, H. B. (1960). Antidiuretic hormone. *Circulation Res.* **21**, 1027.

Schadé, J. P. (1959). Bilateral synchrony and arousal in EEG of fish. *Electroencephalog. Clin. Neurophysiol.* **11**, 613–614.
Schadé, J. P., and Weiler, I. J. (1959). Electroencephalographic patterns of the goldfish (*Carassius auratus* L.). *J. Exptl. Biol.* **36**, 435–452.
Schnitzlein, H. N. (1962). The habenula and dorsal thalamus of some teleosts. *J. Comp. Neurol.* **118**, 225–268.
Schnitzlein, H. N. (1964). Correlation of habit and structure in the fish brain. *Am. Zoologist* **4**, 21–32.
Schulte, A. (1957). Transfer und Transpositionsversuche mit monokular Dressierten Fischen. *Z. Vergleich. Physiol.* **39**, 432–476.
Schultz, R., Berkowitz, E. C., and Pease, D. C. (1956). The electron microscopy of the lamprey spinal cord. *J. Morphol.* **98**, 251–273.
Schwassmann, H. O., and Kruger, L. (1965). Organization of the visual projection upon the optic tectum of some freshwater fish. *J. Comp. Neurol.* **124**, 113–126.
Segaar, J. (1956). Brain and instinct with *Gasterosteus aculeatus. Koninkl. Ned. Akad. Wetenschap. Proc.* **C59**, 738–749.
Segaar, J. (1961). Telencephalon and behaviour in *Gasterosteus aculcatus. Behaviour* **18**, 256–287.
Segaar, J. (1965). Behavioural aspects of degeneration and regeneration in fish brain: A comparison with higher vertebrates. *Progr. Brain Res.* **14**, 143–231.
Segaar, J., and Nieuwenhuys, R. (1963). New etho-physiological experiments with male *Gasterosteus aculeatus* with anatomical comment. *Animal Behaviour* **11**, 331–344.
Shapiro, S. M. (1965). Interocular transfer of pattern-discrimination in the goldfish. *Am. J. Psychol.* **78**, 21–38.
Sperry, R. W. (1955). Functional regeneration in the optic system. *In* "Regeneration in the Central Nervous System" (W. F. Windle, ed.), pp. 66–76. Thomas, Springfield, Illinois.
Sperry, R. W. (1963). Chemoaffinity in the orderly growth of nerve fiber patterns and connections. *Proc. Natl. Acad. Sci. U. S.* **50**, 703–710.
Sperry, R. W. (1965a). Embryogenesis of behavioral nerve nets. *In* "Organogenesis" (R. L. Dehaan and H. Ursprung, eds.), pp. 161–186. Holt, New York.
Sperry, R. W. (1965b). Selective communication in nerve nets: Impulse specificity vs. connection specificity. *Neurosci. Res. Program Bull.* **3**, 37.
Sperry, R. W., and Clark, E. (1949). Interocular transfer of visual discrimination habits in a teleost fish. *Physiol. Zool.* **22**, 372–378.
Sutterlin, A. M., and Prosser, C. L. (1968). Properties of optic nerve fibers and tectal cells in goldfish optic tectum. *Federation Proc.* **27**, 517.
Tateda, H. (1961). Response of catfish barbels to taste stimuli. *Nature* **192**, 343–344.
Tateda, H. (1964). The taste response of the isolated barbel of the catfish. *Comp. Biochem. Physiol.* **11**, 367–378.
Tavolga, W. N. (1964). Sonic characteristics and mechanisms in marine fishes. *In* "Marine Bio-Acoustics" (W. N. Tavolga, ed.), pp. 195–211. Pergamon Press, Oxford.
Tavolga, W. N., and Wodinsky, J. (1963). Auditory capacities in fishes: Pure tone thresholds in nine species of marine teleosts. *Bull. Am. Museum Nat. Hist.* **126**, 179–239.
Tavolga, W. N., and Wodinsky, J. (1965). Auditory capacities in fishes: Threshold

variability in the blue-striped grunt, *Haemulon sciurus*. *Animal Behaviour* **13**, 301–311.

Teichmann, H. (1962). Die Chemorezeption der Fische. *Ergeb. Biol.* **25**, 177–205.

Tester, A. L. (1963). Olfaction, gustation and the common chemical sense in sharks. *In* "Sharks and Survival" (P. W. Gilbert, ed.), pp. 255–282. Heath, Boston, Massachusetts.

Ueda, K., Hara, T. J., and Gorbman, A. (1967). Electroencephalographic studies on olfactory discrimination in adult spawning salmon. *Comp. Biochem. Physiol.* **21**, 133–143.

van Bergeijk, W. A. (1964). Directional and nondirectional hearing in fish. *In* "Marine Bio-Acoustics" (W. N. Tavolga, ed.), pp. 281–299. Pergamon Press, Oxford.

Wagner, H. G., MacNichol, E. F., Jr., and Wolbarsht, M. L. (1963). Functional basis for "on"-center and "off"-center receptive fields in the retina. *J. Opt. Soc. Am.* **53**, 66–70.

Warren, J. M. (1960). Reversal learning by paradise fish (*Macropodus opercularis*). *J. Comp. Physiol. Psychol.* **53**, 376–378.

Warren, J. M. (1961). The effect of telencephalic injuries on learning by paradise fish, *Macropodus opercularis*. *J. Comp. Physiol. Psychol.* **54**, 130–132.

Weiler, I. J. (1966). Restoration of visual acuity after optic nerve section and regeneration, in *Astronotus ocellatus*. *Exptl. Neurol.* **15**, 377–386.

Westerman, R. A. (1965). Specificity in regeneration of optic and olfactory pathways in teleost fish. *In* "Studies in Physiology" (D. R. Curtis and A. K. McIntyre, eds.), pp. 263–269. Springer, Berlin.

Westerman, R. A., and von Baumgarten, R. (1964). Regeneration of olfactory paths in carp (*Cyprinus carpio* L.). *Experientia* **20**, 519–521.

Wilson, J. A. F., and Westerman, R. A. (1967). The fine structure of the olfactory mucosa and nerve in the teleost *Carassius auratus* L. *Z. Zellforsch. Mikroskop. Anat.* **83**, 196–206.

Wisby, W. J., Richard, J. D., and Nelson, D. R. (1964). Sound perception in elasmobranches. *In* "Marine Bio-Acoustics" (W. N. Tavolga, ed.), pp. 255–268. Pergamon Press, Oxford.

Wodinsky, J., and Tavolga, W. N. (1964). Sound detection in teleost fishes. *In* "Marine Bio-Acoustics" (W. N. Tavolga, ed.), pp. 269–280. Pergamon Press, Oxford.

Yagi, K., and Bern, H. A. (1963). Electrophysiologic indications of the osmoregulatory role of the teleost urophysis. *Science* **142**, 491–493.

Yagi, K., and Bern, H. A. (1965). Electrophysiologic analysis of the response of the caudal neurosecretory system of *Tilapia mossambica* to osmotic manipulations. *Gen. Comp. Endocrinol.* **5**, 509–526.

Young, J. Z. (1962). "The Life of Vertebrates." Oxford Univ. Press, London and New York.

2

THE PINEAL ORGAN

JAMES CLARKE FENWICK

I. Introduction	91
II. The Pineal Dilemma	91
III. Ontogeny and Structure of the Pineal Organ	92
A. Ontogeny	92
B. Structure	93
C. Innervation	95
IV. Physiology of the Pineal Body	96
A. The Pineal Body as a Sensory Organ	97
B. The Pineal Body as a Secretory Organ	100
C. The Pineal Body as a Sensory *cum* Secretory Organ	103
V. Synopsis	103
References	104

I. INTRODUCTION

Despite the vast number of publications on the pineal body of fishes, no distinct function has been unequivocally demonstrated. It is still impossible to present a unified concept of pineal function; rather an attempt will be made to review some of the evidence for the various current hypotheses concerning its function. Several monographs on the pineal body are available: Studnička (1905), Tilney and Warren (1919), Bargmann (1943), and Kitay and Altschule (1954); in addition, the recently published proceedings of an international round-table conference on its structure and function, edited by Ariëns Kappers and Schadé (1965), contains many facts pertinent to the pineal body of fishes.

II. THE PINEAL DILEMMA

The most persistent question concerning the pineal gland of fishes is whether it is the vestigial remnant of an ancestral third eye or whether it is a specialized functional organ. Its location on the top of the head orients the pineal organ directly toward the primary source of light and thereby effectively eliminates the pineal body as a receptor of a visual field. But this orientation renders the pineal well suited for detecting changes in light intensity or for acting as a dosimeter of incident radiation. In other words, the pineal of fishes probably never was, in the true sense, an eye. Nevertheless, numerous investigators have shown that the pineal organ of the fishes does have a light receptive function. The problem then is not the presence or absence of a function, but rather, a question of the precise function or functions.

Although several of the current hypotheses are not mutually exclusive, they can, for present purposes, be divided into two categories—one emphasizing primarily a sensory role and the other a secretory function. Of the sensory theories, the following are the most credible: (1) The pineal body is a photosensory structure, (2) it acts as a baro- or chemoreceptor for the cerebrospinal fluid (CSF), or, (3) it performs the function of a mediator in the olfactory induced response to exohormones. Conversely, (4) the pineal may be primarily a gland of external secretion related to the chemical composition of the CSF or the metabolism of the brain tissue, or (5) it may be a gland of internal secretion with an endocrine function.

It is not proposed here to review all of the literature concerning these various hypotheses but to provide a framework into which the accumulated knowledge of pineal function in the fishes can be channeled.

III. ONTOGENY AND STRUCTURE OF THE PINEAL ORGAN

The description of pineal ontogeny, morphology, and histology given here may be amplified from detailed accounts and bibliographies by Hill (1891, 1894), Terry (1910), Tilney and Warren (1919), Holmgren (1958b, 1959a,b), Oksche (1965), Van de Kamer (1965), and Rüdeberg (1966).

A. Ontogeny

Much of the polemic surrounding the pineal organ of the fishes originates from the prevailing ignorance of its embryological origin. This has given rise to inconsistent nomenclature and a failure to distinguish clearly the pineal body from other epithalamic outgrowths.

In vertebrates, two approximately middorsal evaginations develop from the roof of the diencephalon. The rostral evagination represents the anlage of the parapineal body (Hill, 1891, 1894; Friedrich-Freksa, 1932; U. Holmgren, 1959a) which is homologous with the parietal eye of certain reptiles (U. Holmgren, 1965). The pineal organ develops from the more posterior of the diencephalic evaginations (Rasquin, 1958). Therefore, the parapineal and pineal organs are two independent structures and are not homologous (Tilney and Warren, 1919). The paraphysis, which has occasionally been confused with the parapineal (Rasquin, 1958), is morphologically part of the telencephalon and is not part of the pineal area.

In fishes, the parapineal anlage usually shows some early ontogenetic development (U. Holmgren, 1959a) but is generally absent or rudimentary in adults (Kelly, 1962; Van de Kamer, 1965), including the primitive ganoid *Amia calva* (Kingsbury, 1897) and the dipnoan *Protopterus annectens* (U. Holmgren, 1959b). However, a parapineal body resembling a parietal eye of lizards was described in at least one adult teleost (Steyn and Webb, 1960). In other groups of fishes, a well-developed parapineal body with distinguishable photosensory-like cells is found only among the nonmyxinoid cyclostomes (Oksche, 1965). However, the same author notes that the problem of the parapineal organ is not resolved and should be completely reexamined.

B. Structure

1. Topography and General Morphology

The pineal body, *sensu stricto*, is a conspicuous structure in most adult fishes (U. Holmgren, 1959a); it is, however, rudimentary in *Syngnathus acus* and *Hippocampus spinosa* (Studnička, 1896) and absent in *Torpedo ocellata*, *T. marmorata* (Selachii), and the adult *Myxine glutinosa* (Oksche, 1965). Despite contradictory opinions, a pineal organ is present in the tuna *Thynnus thynnus* (U. Holmgren, 1958a).

In basic form, the fish pineal organ consists of a hollow, variously invaginated and well-vascularized structure lying dorsal to the diencephalon to which it is connected by a hollow narrow stalk. The pineal body consists of an expanded distal vesicular area containing a central lumen which connects with the third ventricle via the narrow proximal stalk. Posteriomedially the stalk connects with the subcommissural organ and the commissura posterior and makes contact with the commissura habenulae on its anterior side before joining the dorsal sac epithelium which comes to lie directly ventral to the distal vesicle of the pineal body. With advancing age, the epithelium of the distal vesicle undergoes a gradual infolding with the result that the central lumen, although retaining its connection with the third ventricle, becomes divided into narrow sinuous spaces. Many of the blood vessels which surround the pineal body in embryonic and larval stages, are, with the increased development of the animal, carried into the infoldings of the pineal epithelium and come to lie in the intralobular septa. This centripetal invasion by the blood vessels results in the extensive vascularization characteristic of the adult pineal. Generally, the distal vesicle is closely apposed to the overlying cranial roof but is not, in teleosts, associated with a pineal foramen in the skull (U. Holmgren, 1959b; Hafeez and Ford, 1967). However, as in many deep-sea fishes, the pineal body often lies beneath a more or less depigmented and translucent area of the cranial roof (U. Holmgren, 1959b). In other fish, the pineal body is located in a depression in the roof of the skull or may be associated with a very complex structure of skull bones which may allow the passage of light directly to the pineal body (Rivas, 1953). Some authors, in questioning a photosensory role of the pineal body, have attached special significance to the fact that some fish possess thick skull bones and have no special structure to allow light to pass though to the pineal region. It should be noted, however, that measurable quantities of light have been demonstrated to penetrate the skulls of sheep, dogs, rabbits, and rats (Ganong *et al.*, 1963).

2. HISTOLOGY

Many of the references on the pineal histology are found in papers concerned with the secretory activity of this organ and will be discussed in a later section (Section IV, B, 1 and 2). Histologically, the pineal body of fishes most closely resembles a sensory structure. The structure of the vesicular epithelium resembles the retina except that the pineal "retina" is of a reversed type with the sensory cells in the inner retinal layer; thus, they are turned in the direction from which the light enters (Ariëns Kappers, 1965). In cyclostomes the dorsal wall of the pineal forms a lenslike pellucida, but a lenslike differentiation is never found among other fishes where the pineal seems to have acquired some characteristics of a secretory organ (N. Holmgren, 1918a,b; U. Holmgren, 1958b; Altner, 1965; Hafeez and Ford, 1967). Despite this histological progression from an eyelike structure in some fish to a glandular organ in others, there is now general agreement that the pineal body of fish is largely composed of three cell types: sensory cells, supporting cells, and ganglion or nerve cells (Rasquin, 1958; U. Holmgren, 1959b; Rüdeberg, 1968b).

Most of the pineal cells are sensory cells (U. Holmgren, 1959a). These are most abundant in the vesicular area on the side of the lumen (Hafeez and Ford, 1967) but occur throughout the pineal epithelium. These cells are large, ovoid, or club-shaped with well-defined nuclei surrounded by clear cytoplasm containing many mitochondria and capped with a membranous structure—an outer segment (Rüdeberg, 1968b). Cytoplasmic outgrowths arising from the sensory cells and extending into the pineal lumen have also been described (Friedrich-Freksa, 1932; Van de Kamer, 1955; Holmgren, 1959a; Hafeez and Ford, 1967). On the pole to the cytoplasmic outgrowths, the sensory cells give off axons (or the neuraxes of Hafeez and Ford, 1967), which, together with the dendrites from the ganglion cells (see below), form a network of axons and dendrites within the pineal organ (U. Holmgren, 1959a).

Supporting cells of various types show a marked affinity for nuclear stains and lack the typical structure of the sensory cells (U. Holmgren, 1959a; Hafeez and Ford, 1967). It is most likely that these cells are of an ependymal glia type.

Ganglion or nerve cells together with the dendrites and Nissl substance indicative of their nervous nature, have been described in the pineal of some but not all fishes examined (U. Holmgren, 1959a; Morita, 1966; Hafeez and Ford, 1967; Rüdeberg, 1968b). Although few in number compared with the other cell types, ganglion cells may be readily distinguished by their large size and by the presence of two or three well-stained nucleoli (U. Holmgren, 1959a). Recently, Rüdeberg (1968b) has reported that within the pineal of the sardine, *Sardina pilchardus sardina*, the ganglion cells are limited to three small groups. If this localization of ganglion cells proves to be the case in other fishes, it may account for the failure of some workers to find pineal ganglion or nerve cells.

C. Innervation

Apart from the sensory cells proper, afferent nerve fibers connecting the pineal body with other parts of the brain have been described (Hill, 1894; Studnička, 1905; Ariëns Kappers, 1965; Rüdeberg, 1968b). These nerves, which appear to represent the axons of ganglion cells (Rüdeberg, 1968b), arise from the extensive nerve net formed within the pineal by the dendrites of the ganglion cells (U. Holmgren, 1959a) and neuraxis of the primary sensory cells (Hafeez and Ford, 1967). In the stalk region, these fibers form variable numbers of nerve bundles (Ariëns

Kappers, 1965) before reaching the roof of the diencephalon. Efferent fibers originating within the central nervous system and ending on pineal cells have rarely been described (Van de Kamer, 1955; Hafeez and Ford, 1967). These efferent fibers are probably the same type of aberrant commissural fibers described in mammals (Ariëns Kappers, 1965).

Aside from a well-documented nervous connection between the pineal body and the posterior commissure (Holt, 1891; Hill, 1894; Studnička, 1905; Ariëns Kappers, 1964; Rüdeberg, 1968b) relatively little is known of the central terminations of the pineal nerve (tractus epiphyseos of Ariëns Kappers, 1965). Possible connections have been suggested with the superior commissure (Holt, 1891), and with the right habenular nucleus (Ariens Kappers, 1964) as well as other habenular structures (Holt, 1891; N. Holmgren, 1918b). In one case only, fibers have been described as running to the optic tectum (N. Holmgren, 1918c). Some workers have suggested a link between the pineal body and the subcommissural organ (Dendy, 1907; U. Holmgren, 1959a), but this is apparently not present in all fishes. Le Gros Clarke (1932) believes that the pineal organ of cyclostomes is heavily supplied with olfactory input channels and this possibility should be investigated. Notwithstanding the paucity of information concerning the exact site(s) of the pineal nerve terminations, connections already described could permit a link between the pineal body and the efferent centers of the brain through which the pineal body could influence various integrative afferent centers (Ariëns Kappers, 1965). Further information on the innervation of the fish pineal gland together with the phylogenetic implications and earlier bibliographies may be consulted in the excellent papers of U. Holmgren (1959a) and Ariëns Kappers (1965).

IV. PHYSIOLOGY OF THE PINEAL BODY

Although neurophysiological studies have proved that the pineal organ of *Salmo gairdneiri irideus* is responsive to illumination (Dodt, 1963; Morita, 1966) there is only scanty literature on the actual physiology of the pineal organ in fishes. Active cellular metabolism has been indicated by at least two authors (Palayer, 1958; U. Holmgren, 1959b) who studied the uptake of radioactive phosphorus by the fish pineal.

Recently, interest has been directed toward the possible presence of melatonin and other tryptophan derivatives within the fish pineal (Quay, 1965; Oksche and Kirschstein, 1967; Oguri *et al.*, 1968; Rüdeberg, 1968b). Quay (1965) localized the enzyme hydroxyindole-*O*-methyl transferase (HIOMT) which is responsible for the formation of melatonin within the pineal of *Salmo irideus*. Working with the same species, Oguri *et al.* (1968) found that a melatonin precursor, 5-hydroxytryptophan is taken up by the pineal organ in greater quantities than it is by other parts of the brain studied; they infer from this finding that the pineal of fish may also be concerned with melatonin synthesis. These studies have now been substantiated by Fenwick (1970a) who isolated melatonin from the pineal organs of the Pacific salmon *Oncorhyncus tshawytscha*. Thus, the pineal organ of fish is responsive to light, has a high cellular metabolism, and shows an active tryptophan metabolism which is capable of producing the mammalian hormone metatonin.

A. The Pineal Body as a Sensory Organ

There remains little doubt that the pineal body of some, if not all fish, functions as a sensory organ. Krabbe (1916) and Walter (1923) and more recently Van de Kamer (1952) as well as Hafeez and Ford (1967) have suggested that the fish pineal is associated with the detection of the pressure or chemical composition of the CSF. Evidence favoring this hypothesis includes the reported open connection between the pineal lumen and the ventricles of the brain, the changing shape of the pineal body of *Esox lucius* in relation to the pressure of the CSF, and the cytoplasmic extensions of the primary sensory cells toward the lumen of the body thus suggesting a functional relationship with the CSF. Hafeez and Ford (1967) go on to suggest that the similarities between the pineal body and the subcommissural organ with regard to the chemical nature of their secretions and the common direction of release into the ventricles support the view of a CSF–pineal relationship. There have been no reported studies of the effect of pinealectomy on either the pressure or composition of the CSF.

Although the hypothesis that the pineal gland is associated with the limbic system or visceral brain (Relkin, 1966) suffers from the same paucity of experimental evidence, some suggestive findings have been reported. Both Le Gros Clarke (1932) and Boon (1938) noted a close anatomical relationship between the pineal body and the olfactory system of fishes; the latter has been shown to have at least some connection with hypothalamic nuclei responsible for pituitary hormone release (Jasinski *et al.*, 1966; Peter and Gorbman, 1968). Moreover, Hoffman and Reiter (1965) reported that olfactory stimuli are capable of interfering with the primary inhibitory effects which the active pineal gland has on the reproductive system of male hamsters. Such studies have not yet been extended to fishes.

In contrast to the scanty evidence for these pineal functions, many publications suggest a photosensory role as the primary sensory function of the pineal organ. The evidence, based on a number of fishes, comes from direct neurophysiological studies as well as indirect evidence from histology or ultrastructure, color changes, and behavioral tests. Only the more pertinent evidence will be reviewed here.

By light (Studnička, 1905; Dendy, 1907; N. Holmgren, 1920; Adam, 1956, 1957; U. Holmgren, 1959a; Eakin, 1963; Oksche, 1965; Hafeez and Ford, 1967; Rüdeberg, 1968a,b) as well as electron microscopic investigation (Kelly, 1962; Breuker and Horstmann, 1965; Oksche and Vaupel von Harnak, 1965; Oksche and Kirchstein, 1967; Rüdeberg, 1966), it has been clearly demonstrated that morphologically the neurosensory cells present in the epithelium of the pineal body are very similar to the ciliary type of photosensory cells present in the retina of the lateral eyes (Ariëns Kappers, 1965) and possess characteristics which are indisputably conelike (Breuker and Horstmann, 1965; Oksche and Kirchstein, 1967). It has been suggested by De la Motte (1963) that the light receptive role of the pineal body in *Phoxinus* is based on porphyropsin, although Thines and Kahling (1957) had previously reported a different visual pigment in *Anoptichthys*. More directly, Grunewald-Lowenstein (1956) observed histological and histochemical changes in the pineal organ of *Astanax* following prolonged exposure to continuous darkness or illumination. Similar changes, however, were not found in young sockeye salmon held under different light regimes (Hafeez and Ford, 1967). Recently, Dodt (1963) and Morita (1966) have demonstrated the photosensory potential of the fish pineal by electrophysiologically detecting alterations in nervous activity in the pineal of *Salmo irideus* during and after illumination of the pineal even though this species lacks any specialized tissue overlying the pineal organ. This finding is important when attempting to define a light receptive role for the pineal in those fishes which possess apparently light impermeable skulls. It questions the theory that such a skull precludes a photosensory role.

A light receptive role of the pineal organ is also indicated by light-induced alterations of pigment distribution within the chromatophores. Although several studies suggested that the pineal had no effect on pigment distribution (von Frisch, 1911a,b; Scharrer, 1927; Wykes, 1938), von Frisch did find that some part of the diencephalon was involved in chromatophore responses. The pineal was more directly implicated by Young (1935) who found that pineal extirpation abolished the marked diurnal rhythm of color change in ammocoetes and distributed the rhythm in adult *Lampetra*. Subsequently, Breder and Rasquin (1950), Hoar (1955), and Schonherr (1955) reported some degree of pigmentary dispersion after pineal occlusion or destruction—a finding similar to that reported by Young (1935) in pinealectomized ammocoetes. These results are supported by the observation that administration of beef pineal extracts in embryonic and larval *Fundulus* caused marked pallor (Wyman, 1924; Fain and Hadley, 1966) although similar results could not be obtained in adult *Fundulus* (Fain and Hadley, 1966) or adult *Phoxinus* (Hewer, 1926). The difference in responsiveness between the young and adults was presumed by Fain and Hadley (1966) to result from the acquisition, in the adults, of nervous innervation mediated by catecholamines and a loss of sensitivity to the pineal extracts. Whether the pallor following pineal extract administration is the result of pituitary inhibition or a direct action on the pigment cells is not known. A direct action is, however, suggested by the ability of the pineal extract to cause pigment concentration *in vitro* (unpublished).

Photosensitivity of the pineal body has been indirectly demonstrated in several species of fish by the use of behavioral tests. Breder and Rasquin (1947), Hoar (1955), and Fenwick (1970b) report that phototaxis is abolished following pinealectomy. Further, Breder and Rasquin (1947) and Fenwick (1970b) found that phototaxis, whether positive or negative, depended on the presence of intact optic cysts or intact eyes; they concluded that although the sign of phototaxis was governed by the pineal organ, the phenomenon itself depended on the presence of lateral photic receptors. Hoar (1955), however, working with young sockeye salmon smolts, *Oncorhynchus nerka*, found that the negative phototaxis of otherwise intact animals is not disturbed following damage to the pineal region; this difference may result from species or age differences of the animals or differences in the intensity of light employed in the experiments. The pineal organ of fishes becomes increasingly invaginated with age (Hoar, 1955) and undergoes a decrease in sensory cell number with continued development (Hafeez and Ford, 1967) together with a marked degeneration of nervous elements (Ariëns Kappers, 1965). Thus, the pineal body of fishes may change from a primary photosensory structure in the young fish to a secondary photosensory structure in the older fish where it can no longer autonomously produce a phototactic response. On the other hand, Hoar (1955) performed his experiments outside where the light intensities were much greater than those employed in other studies (Breder and Rasquin, 1947; Fenwick, 1970b). Previously, von Frisch (1911a) and Young (1935) suggested a general light sensitivity of the diencephalic roof; thus, since the smolts used by Hoar (1955) had thin skulls, the phototaxis demonstrated in "pinealectomized" fish could have resulted from the effect of high light intensity on the brain itself.

Recent studies by Fenwick (1970b) have shown that intact goldfish in a light gradient were unevenly distributed and spent most of their time in the darker half of the tank. Conversely, pinealectomized, bilaterally enucleated, or pinealectomized plus bilaterally enucleated goldfish were distributed uniformly throughout the gradient. From this evidence it was concluded that the phototactic response of goldfish depends upon the presence of the pineal organ as well as the eyes. Furthermore, in a conditioning situation, pinealectomized animals with intact vision showed significantly more responses to the conditioned stimulus than did the controls when the conditioned stimulus was light, but not when the conditioned stimulus was sound. Blind goldfish, with or without an intact pineal, could not be effectively conditioned to light although they did become conditioned to sound. From this data it was concluded that although the pineal organ of goldfish is a photosensory organ, in the absence of the eyes, the photic information received by the pineal cannot be translated into a directional response or be used as a sensory mechanism for the initiation of active behavior. It appears, therefore, that the photosensory role of the goldfish pineal organ is to modulate the response elicited by photic information received by the eyes.

Although these studies do not provide conclusive evidence of the mode of action or the exact role of the pineal body in light reception, they do demonstrate the importance of the pineal body in phototactic behavior and suggest that the pineal organ and the eyes function as a unit in phototaxis.

B. The Pineal Body as a Secretory Organ

1. EXTERNAL SECRETION

In addition to its photosensory role, many investigators have shown that fish pineal produces an apocrine (Grunewald-Lowenstein, 1956; Hafeez and Ford, 1967) secretion (Tretjakoff, 1915; N. Holmgren, 1918a,b; Friedrich-Freksa, 1932; U. Holmgren, 1958b; Rasquin, 1958; Altner, 1965) which contains glycogen (Grunewald-Lowenstein, 1956; Rasquin, 1958; U. Holmgren, 1959b). The secretion is generally considered to enter the CSF (Friedrich-Freksa, 1932; Hafeez and Ford, 1967). Although Van de Kamer (1955) suggests that the secretory droplets may be experimental artifacts, the high cellular metabolism demonstrated by the cytological studies of Palayer (1958) and U. Holmgren

(1959a) and the presence of a well-developed Golgi complex in the supporting cells and the sensory cells (Rüdeberg, 1968b) indicate secretory activity.

2. INTERNAL SECRETION

There is also the possibility of an internal secretory or endocrine role. Although both Friedrich-Freksa (1932) and U. Holmgren (1959a) suggest that some pineal secretion is taken up by the blood vessels in teleosts, Hafeez and Ford (1967) did not find any secretory granules within the nerve fibers or in the proximity of the blood vessels of sockeye salmon, *Oncorhynchus nerka*. However, since the demonstration of secretory granules depends upon the nature of the secretion and on the specificity of the techniques employed, an internal secretory role of the pineal organ was not definitely ruled out and few investigators have applied techniques suitable for detecting the one known mammalian pineal hormone melatonin (Axelrod et al., 1960).

Melatonin (Fenwick, 1970a) and its precursor serotonin, (Fenwick, 1970a; Owman and Rüdeberg, cited in Rüdeberg, 1968b) together with the enzyme HIOMT (Quay, 1965) necessary for the formation of melatonin (Axelrod et al., 1960), have now been demonstrated in fish pineal organs. Also, Oguri et al. (1968) report that the pineal takes up more ^{14}C-5-hydroxytryptophan than any other tissues studied. Since this substance is a precursor of serotonin, an active tryptophan metabolism within the fish pineal organ is indicated. Thus, an endocrine role of the fish pineal, possibly based on the hormone melatonin, is not improbable.

Despite earlier reports by Krockert (1936a,b) that the growth rate of young *Lebistes* could be decreased by feeding them desiccated bull pineal glands, Pflugfelder (1953, 1954, 1956a) clearly implicated the pineal as an endocrine organ of fishes and suggested a pineal–pituitary relationship in teleosts. His observations on pinealectomized guppies indicated hypertrophy of the adenohypophysis, hyperthyroidism, decreased growth rate, a slight acceleration in the appearance of secondary sex characteristics in young males, an increase in the activity of the interrenal cells, and a disturbed calcium metabolism resulting in spinal curvature. The thyroidal effects could be partially offset by the injection of epiphysan or thyroxin (Pflugfelder, 1956b). Pflugfelder (1964) has since reported alterations in the pars distalis and thyroid in pinealectomized goldfish. U. Holmgren (1959b) reported similar findings with regard to skeletal abnormalities and was able to reduce these by administering beef pineal injections. Further, the latter investigator reported a decreased radiocalcium uptake in pinealectomized fish and an increased uptake in those fish receiving injections of pineal extract. These findings, however, could not be confirmed by Weisbart and Fenwick (1966) who report an undisturbed blood calcium level in pinealectomized goldfish.

Few other studies have examined the endocrine nature of the fish pineal, and these have been conflicting. Contrary to the report of thyroidal hypertrophy by Pflugfelder (1964), Pang (1967) found a decreased thyroidal cell height in pinealectomized *Fundulus* and Rasquin (1958), U. Holmgren (1959b), Fenwick (1970c), and Peter (1968) report no alteration in this tissue following pinealectomy. Further, neither Peter (1968) nor Fenwick (1970c) could find any apparent increase in the interrenal cells as previously reported by Pflugfelder (1953). While the significance of these findings is not clear, the possibility that the pineal body of mammals is related to electrolyte balance (Farrell, 1960) has never been wholly repudiated and should provide the impetus for further investigation in this area.

Although a pineal–gonadal relationship seems well established in the mammals (Kitay and Altschule, 1954), little research has been carried out on the possibility of such an axis in the fishes. Krockert (1936a,b) was able to delay the appearance of secondary sex characteristics in young *Lebistes* by feeding desiccated bull pineal glands. In the same species, Pflugfelder (1954) reports a slight acceleration in the sexual development of males following pinealectomy. In contrast to Pflugfelders results, Schonherr (1955) and Rasquin (1958) reported unsuccessful attempts to hasten reproductive maturation by pinealectomy. Pang (1967) described a delay in the appearance of nuptial coloration in pinealectomized *Fundulus,* but noted that the controls had smaller gonad sizes. It is not clear whether this means that his controls underwent gonadal atrophy or that his pinealectomized fish showed gonadal hypertrophy. Recently, Peter (1968) and Fenwick (1970c), both working with goldfish, have reported contradictory (although not totally incompatible) results. Peter (1968), in one experiment of 9-months duration, could find no signficant differences between the gonosomatic indices (GSI) (gonad weight/body weight $\times$ 100 = GSI) of pinealectomized, sham-pinealectomized, and unoperated control goldfish. However, when experiments were carried out at different stages of the yearly reproductive cycle, Fenwick (1970c) found that although pinealectomy had no effect on the size of the gonads during most of the year, the operation, when carried out just prior to the onset of the final maturation phase preceding the normal spawning period, did result in a highly significant increase in the GSI relative to that found in the groups with intact pineal. The time at which pinealectomy was found to have this effect corresponded with that period of the reproductive cycle during which an increasing day length had its greatest stimulatory effect on the reproductive system (Fenwick, 1970c). Further, the increase in gonad size in goldfish subjected to increasing periods of daily light exposure could be prevented by the daily injection of melatonin (Fenwick, 1970a).

These results may help to explain the earlier contradictory reports. In the first place it seems probable that the pineal gland may not function to a similar extent, or even in an identical manner, during different periods of the life cycle. As seen previously, the invagination of the pineal gland increases with age (Hoar, 1955) and there is a decrease in sensory cell number (Hafeez and Ford, 1967) together with a marked degeneration of nervous structures (Ariëns Kappers, 1965). Further, the pineal gland becomes less accessible to light with increase in body size; even the effect of pinealectomy on pigment distribution differed in young and older lampreys (Young, 1935). Second, if the endocrine function of the pineal organ is dependent upon melatonin as suggested by Fenwick (1970a,c), and if the level of melatonin is depressed by exposure to long periods of bright light as reported in the mammals (Wurtman and Axelrod, 1965a), then the experimental design must be considered before interpreting the results of pinealectomy. If such controls are not carefully enforced, the intensity and duration of the experimental photoperiod may be sufficient to lower the melatonin level to a point where even animals with a pineal are effectively "physiologically pinealectomized."

C. The Pineal Body as a Sensory *cum* Secretory Organ

The apparent disagreement between workers supporting a sensory function as opposed to those who argue for a secretory function does not seem justified. Rather, it seems likely that both functions are linked and that the pineal organ should be viewed neither as an entirely sensory structure nor as a wholly secretory gland; both sensory and secretory functions seem likely. This hypothesis is supported by recent observations made on the chelonian epiphysis. Vivien and Roels (1967) noted that certain cells within the chelonian epiphysis could be seen to alternate between characteristics of reception and secretion; it is not difficult to develop a similar theory for the pineal organ of the fishes.

V. SYNOPSIS

The pineal body of fishes should be reexamined in the light of recent knowledge concerning the physiology of the pineal in higher vertebrates.

The presence of photosensory cells with central connections, the extensive vascularity, high metabolic activity, evidence for both external and internal secretory activity, the pineal involvement in phototaxis, and the diverse effects of pinealectomy on the pigmentary response and the endocrine system together with the evidence for an active tryptophan metabolism favor the view that the fish pineal body, far from being an evolutionary vestige, is an important functional organ. While its small size and simplicity do not indicate it is a visual receptor, it is sufficiently sensory in appearance to suggest a role as a dosimeter of variations in incident radiation. The pineal's location and its sensory *cum* secretory activity suggest that it is well adapted to play a part in the mediation of light influences on the fish pituitary gland; furthermore, it could probably control the pressure and/or composition of the cerebrospinal fluid. As pointed out by Roth (1964), the phylogenetic trend toward a more glandular structure in the higher vertebrates may reflect only a change in the way the information reaches the pineal rather than a completely unrelated change in function.

REFERENCES

Adam, H. (1956). Ventrikel und die Mikroskopische Struktur seiner Wände bei *Lampetra (Petromvzon) fluviatilis* L. und *Myxine glutinosa* L., nebst einigen Bemerkungen über das Infundibularorgan von *Branchiostom* (Amphioxus) *lanceolatum* Pall. In "Progress in Neurobiology" (J. Ariëns Kappers, ed.), p. 146. Elsevier, Amsterdam.

Adam, H. (1957). Beitrag zur Kenntnis der Hirnventrikel und des Ependyms bei Cyclostmen. *Anat. Anz.* 103, 173–188.

Altner, H. (1965). Histologische und Histochemische Untersuchungen an der Epiphyse von Haisen. *Progr. Brain Res.* 10, 145–171.

Ariëns Kappers, J. (1964). Survey of the innervation of the pineal organ in vertebrates. *Am. Zoologist* 4, 47–57.

Ariëns Kappers, J. (1965). Survey of the innervation of the epiphysis cerebri and the accessory pineal organs of vertebrates. *Progr. Brain Res.* 10, 87–153.

Ariëns Kappers, J., and Schadé, J. P., eds. (1965). "Structure and Function of the Epiphysis Cerebri." Elsevier, Amsterdam.

Axelrod, J., MacLean, P. D., Albers, R. W., and Weissback, H. W. (1960). Regional distribution of methyl transferase enzymes in the nervous system and glandular tissues. *In* "Regional Neurochemistry" (S. S. Kety and J. Elkes, eds.), pp. 307–311. Pergamon Press, Oxford.

Bargmann, W. (1943). Die Epiphysis Cerebri. *In* "Handbuch der mikroskopischen Anatomie des Menschen" (W. von Möllendorf, ed.), Vol. 4, Part 4, pp. 310–502. Springer, Berlin.

Boon, A. A. (1938). Comparative anatomy and physiopathology of the autonomic hypothalamic centers. *Acta Psychiat. Neurol. Scand.* Suppl. 18, 1.

Breder, C. M., and Rasquin, P. (1947). Comparative studies in the light sensitivity of blind characins from a series of Mexican caves. *Am. Museum Novitates* 89, 325–351.

Breder, C. M., and Rasquin, P. (1950). A preliminary report on the role of the pineal organ in the control of pigment cells and light reactions in recent teleost fishes. *Science* **111**, 10–12.
Breuker, H., and Horstmann, E. (1965). Elektronemikroskopische Untersuchungen am Pinealorgan der Rogenbogenforelle (*Salmo irideus*). *Progr. Brain Res.* **10**, 259–269.
Byrne, J. (1968). The effect of photoperiod and temperature on the daily pattern of locomotor activity in juvenile sockeye salmon (*Oncorhyncus nerka* Walbaum). Ph.D. Thesis, Dept. of Zoology, University of British Columbia, Vancouver, Canada.
Chaston, I. (1968). Influence of light on activity of brown trout (*Salmo trutta*). *J. Fisheries Res. Board Can.* **25**, 1285–1289.
De la Motte, I. (1963). Untersuchungen zur vergleichenden Physiologie der Lichtempfindlichkeit geblendeter Fische. *Naturwissenschaften* **9**, 1–3.
Dendy, A. (1907). On the parietal sense-organs and associated structures in the New Zealand lamprey (*Geotria australis*). *Quart. J. Microscop. Sci.* **51**, 1–29.
Dodt, E. (1963). Photosensitivity of the pineal organ in the teleost, *Salmo irideus* (Gibbons). *Experientia* **19**, 642–643.
Eakin, R. M. (1963). Lines of evolution of photoreceptors. *In* "General Physiology of Cell Specialization" (D. Mazia and A. Tyler, eds.), pp. 393–425. McGraw-Hill, New York.
Eigenmann, C. H. (1909). Cave vertebrates of America, a study in degenerative evolution. *Carnegie Inst. Wash. Publ.* **104**, 1–241.
Fain, W. B., and Hadley, M. E. (1966). In vitro response of melanophores of *Fundulus heteroclitus* to melatonin, adrenalin and noradrenalin. *Am. Zoologist* **6**, 596–597.
Farrell, G. (1960). Adrenoglomerulotropin. *Circulation* **21**, 1009–1015.
Fenwick, J. C. (1970a). Demonstration and effect of melatonin in fish. *Gen. Comp. Endocr.* **14**, 86–97.
Fenwick, J. C. (1970b). Effects of pinealectomy and bilateral enucleation on the phototactic response and on the conditioned response to light of the goldfish. *Carassius auratus. Can. J. Zool.* **48**, 175–182.
Fenwick, J. C. (1970c). The pineal organ; Photoperiod and reproductive cycles in the goldfish. *J. Endocrinol.* **46**, 101–111.
Friedrich-Freksa, H. (1932). Entwicklung. Bau und Bedeutung der Parietalgegend bei Teleostiern. *Z. Wiss. Zool. Abt.* **A141**, 52–142.
Ganong, W. F., Shepherd, M. D., Wall, J. R., Van Brunt, E. E., and Clegg, M. T. (1963). Penetration of light into the brain of mammals. *Endocrinology* **72**, 962–963.
Grunewald-Lowenstein, M. (1956). Influence of light and darkness on the pineal body in *Astyanax mexicanus* (Filippi). *Zoologica* **41**, 119–128.
Hafeez, M. A., and Ford, P. (1967). Histology and Histochemistry of the Pineal Organ in the Sockeye Salmon, *Oncorhyncus nerka*, Walbaum. *Can. J. Zool.* **45**, 117–126.
Hewer, H. R. (1926). Studies in the color changes of fishes. 1. The action of certain endocrine secretions in the minnow. *J. Exptl. Biol.* **3**, 123–140.
Hill, C. (1891). Development of the epiphysis in *Corogonus albus. J. Morphol.* **5**, 503–510.
Hill, C. (1894). The epiphysis of teleosts and *Amia. J. Morphol.* **9**, 237–266.
Hoar, W. S. (1955). Phototactic and pigmentary responses of the sockeye salmon

smolts following injury to the pineal organ. *J. Fisheries Res. Board Can.* **12**, 178–185.

Hoffman, R. A., and Reiter, R. J. (1965). Influence of compensatory mechanisms and pineal gland on dark induced gonadal atrophy in male hamsters. *Nature* **207**, 658.

Holmgren, N. (1918a). Zum Bau der Epiphyse von *Squalus acanthias*. *Arkiv Zool.* II/**23**, 1–28.

Holmgren, N. (1918b). Uber die Epiphysennerven von *Clupea sparattus* und *harangus*. *Arkiv Zool.* II/**25**, 1–5.

Holmgren, N. (1918c). Zur Frage der Epiphysen-Innervation bei Teleostiern. *Folia Neuro-Biol.* (*Leipzig*) **11**, 1–15.

Holmgren, N. (1920). Zoo Anatomie und Histologie des Uondevund Zwischemhirus des Knochenfische. *Acta Zool.* (*Stockholm*) **1**, 137–315.

Holmgren, U. (1958a). On the pineal organ of the tuna, *Thynnus thynnus*. *Bull. Museum Comp. Zool. Harvard Coll.* **100**, 1–8.

Holmgren, U. (1958b). Secretory material in the pineal body as shown by aldehyde-fuchsin following performic acid oxidation. *Stain Technol.* **33**, 148–149.

Holmgren, U. (1959a). On the structure of the pineal area of teleost fishes with special reference to a few deep sea fishes. *Goteborgs Vetenskaps-Vitterhetssamh. Handl.* **B8**, 1–66.

Holmgren, U. (1959b). Studies on the pineal gland. Ph.D. Thesis, Harvard University, Cambridge, Massachusetts.

Holmgren, U. (1965). On the ontogeny of the pineal and para pineal organs in teleost fishes. *Progr. Brain Res.* **10**, 172–182.

Holt, E. W. (1891). Observations on the development of the teleostean brain with special reference to the brain of *Clupea harengus*. *Zool. Jahrb., Abt. Anat. Ontog. Tiere* **4**, 482–504.

Janzen, W. (1933). Untersuchungen ubev Grosshirnfunktion en des Goldfisches (*Carussius arratus*). *Zool. Jahrb., Abt. Allgem. Zool. Physiol. Tiere* **52**, 591–628.

Jasinski, A., Gorbman, A., and Hara, T. J. (1966). Rate of movement and redistribution of stainable neurosecretory granules in hypothalamic neurones. *Science* **154**, 776.

Kelly, D. E. (1962). Pineal organs: Photoreception, secretion, and development. *Am. Scientist* **50**, 597–625.

Kingsbury, B. F. (1897). The encephalic evaginations of ganoids. *J. Comp. Neurol.* **7**, 37–44.

Kitay, J., and Altschule, M. D. (1954). "The Pineal Gland." Harvard University Press, Cambridge. Massachusetts.

Krabbe, K. H. (1916). Histologische und embryologische Untersuchungen uber die Zirbeldruse des Menschen. *Anat. Anz.* **54**, 187–319.

Krockert, G. (1936a). Die Wirkung der Verfutterung von Schilddrusen und Zirebeldrusen-sustanz an *Lebistes reticulatus* (Zahnkarpfen). *Z. Ges. Exptl. Med.* **98**, 214–220.

Krockert, G. (1936b). Entwicklungsanderungen bei der Fischart Cichliden durch Verabreichung von Schilddruse, Ziebelund thymusdruse. *Z. Ges. Exptl. Med.* **99**, 451–455.

Le Gros Clarke, W. E. (1932). The structure and connections of the thalamus. *Brain* **55**, 406–427.

Morita, Y. (1966). Entladungsmuster pinealer Neurone der Regenbogenforelle

(*Salmo irideus*) bei Belichtung des Zwischenhirns. *Arch. Ges. Physiol.* **289**, 155–167.
Oguri, M., Omura, Y., and Hibiya, T. (1968). Uptake of ¹⁴C-labelled 5-hydroxy tryptamine into the pineal organ of rainbow trout. *Bull. Japan. Soc. Sci. Fisheries* **34**, 687–690.
Oksche, A. (1965). Survey of the development and comparative morphology of the pineal organ. *Progr. Brain Res.* **10**, 3–29.
Oksche, A., and Kirschstein, H. (1967). Die ultrastruktur der Sinneszellen in Pinealorgan von *Phoxinus laevis*. *Z. Zellforsch. Mikroskop. Anat.* **78**, 151–166.
Oksche, A., and Vaupel von Harnak, M. (1965). Vergleichende Elektronenmikroskopische Studien am Pinealorgan *Progr. Brain Res.* **10**, 237–258.
Palayer, P. (1958). Fixation de phosphore radioactif dans différentes parties du cerveau, notamment dans l'epiphyse, et dans quelques tissus chez la truite arc-en-ciel (*Salmon gairdnerii* R.). *Compt. Rend. Soc. Biol.* **152**, 305–308.
Pang, P. K. T. (1965). Light sensitivity of the pineal in blinded *Fundulus heteroclitus*. *Am. Zoologist* **5**, 682.
Pang, P. K. T. (1967). The effect of pinealectomy on the adult male killifish, *Fundulus heteroclitus*. *Am. Zoologist* **7**, 715.
Peter, R. E. (1968). Failure to detect an effect of pinealectomy in goldfish. *Gen. Comp. Endocrinol.* **10**, 443–449.
Peter, R. E., and Gorbman, A. (1968). Some afferent pathways to the preoptic nucleus of the goldfish. *Neuroendocrinology (N.Y.)* **3**, 229–237.
Pflugfelder, O. (1953). Wirkungen partieller Zerstorungen der Parietalregion von wicklung von *Lebistes reticulatus* Peters. *Arch. Entwicklungsmech. Organ.* **146**, 115–136.
Pflugfelder, O. (1954). Wirkungen partieller Zerstorungen der Parietalregion von *Lebistes reticulatus*. *Arch Entwicklungsmech. Organ.* **147**, 42–60.
Pflugfelder, O. (1956a). Wirkungen von Epiphysan und Thyroxin auf die Schilddruse epiphusektomierter *Lebistes reticulatus* (Peters). *Arch. Entwicklungsmech. Organ.* **148**, 463–473.
Pflugfelder, O. (1956b). Physiologie der epiphyse. *Zool. Anz.* **20**, 53–75.
Pflugfelder, O. (1964). Wirkungen lokaler Hirnlasionen auf Hypophyse und Thyreoidea von *Carassius gibelio auratus*, Bloch. *Arch. Entwicklungsmech. Organ.* **155**, 535–548.
Quay, W. B. (1965). Retinal and Pineal Hydroxy-O-Methyl Transferase activity in Vertebrates. *Life Sci.* **4**, 983–991.
Rasquin, P. (1958). Studies in the control of pigment cells and light responses in recent Teleost Fishes. *Am. Museum Novitates* **115**, 8–68.
Relkin, R. (1966). The pineal gland. *New Engl. J. Med.* **274**, 944–950.
Rivas, L. R. (1953). The pineal apparatus of tunas and related scombrid fishes as a possible light receptor controlling phototactic movements. *Bull. Marine Sci. Gulf Caribbean* **3**, 168–180.
Roth, W. D. (1964). Comments on Ariëns Kappers review and observations on pineal activity. *Am. Zoologist* **4**, 53–57.
Rüdeberg, C. (1966). Electron microscopical observations on the pineal organ of the teleosts *Mugil auratus* (Risso) and *Uranoscopus scaber* (Linne). *Pubbl. Staz. Zool. Napoli* **35**, 47–60.
Rüdeberg, C. (1968a). Receptor cells in the pineal organ of the dogfish *Scyliorhinus canicula* (Linne). *Histochemie* **85**, 521–526.
Rüdeberg, C. (1968b). Structure of the pineal organ of the sardine, *Sardina pil-*

chardus (Risso) and some further remarks on the pineal organ of the *Mugil* spp. *Histochemie* **84**, 219–237.

Scharrer, E. (1927). Die Lichtempfindlichkeit blinder Elritzen (Untersuchingen uber das Zwischenhirn der Fische). *Z. Vergleich. Physiol.* **7**, 1–38.

Schonherr, J. (1955). Uber die Abhangigkeit der Instinkthandlugen vom Vordenhirn und Zwischenhirn (Epiphyse) bei *Gasterosteus aculeatus* L. *Zool. Jahrb., Abt. Allgem. Zool. Physiol. Tiere* **65**, 357–386.

Steyn, W., and Webb, M. (1960). The pineal complex in the fish *Labeo umbratus*. *Anat. Record* **136**, 79–85.

Studnička, F. K. (1896). Beitrage zur Anatomie und Entwicklungsgeschichte des Vorderhirns der Kranioten. *Stizber. Bohmischen Ges. Wiss.* Sect. 2, No. 15, 1–32.

Studnička, F. K. (1905). Die Parietalorgane. *In* "Lehrbuch der vergleichenden mikroskopischen Anatomie der Wirbeltiere" (A. Oppel, ed.). Springer, Vienna.

Terry, R. J. (1910). The morphology of the pineal region in teleosts. *J. Morphol.* **21**, 321–358.

Thines, G., and Kahling, J. (1957). Untersuchungen uber die Farbempfindlichkeit des Hohlenfisches *Anoptichthys jordani* Hubbs and Innes (Characidae). *Z. Biol.* **109**, 150–160.

Tilney, F., and Warren, W. F. (1919). "The Morphology and Evolutionary Significance of the Pineal Body," Am. Anat. Mem. No. 9. Wistar Inst., Philadelphia, Pennsylvania.

Tretjakoff, D. (1915). Die Parietalorgane von *Petromyzon fluviatilus*. *Z. Wiss. Zool., Abt. A* **113**, 1–112.

Van de Kamer, J. C. (1952). On the function of the pineal organ. *Arch. Neerl. Zool.* **9**, 565–566.

Van de Kamer, J. C. (1955). The pineal organ in fish and amphibian. *In* "Progress in Neurobiology" (J. Ariëns Kappers, ed.), pp. 113–120. Elsevier, Amsterdam.

Van de Kamer, J. C. (1965). Histological structure and cytology of the pineal complex in fishes, amphibia, and reptiles. *Progr. Brain Res.* **10**, 30–48.

Vivien, J. H., and Roels, B. (1967). Ultrastructure of the Chelonian epiphysis. *Compt. Rend. Soc. Biol.* **264**, 1743–1754.

von Frisch, K. (1911a). Das Parietalorgan der Fische als functionierendes Organ. *Sitzber. Ges. Morphol. Physiol. Munich* **27**, 16–18.

von Frisch, K. (1911b). Beitrage zur Physiologie der Pigmentzellen in der Fischaut. *Arch. Ges. Physiol.* **138**, 319–387.

Walter, F. K. (1923). Weitere Untersuchungen zur Pathologie und Physiologie der Zirbeldruse. *Z. Ges. Neurol. Psychiat.* **83**, 411–463.

Weisbart, M., and Fenwick, J. C. (1966). Effect of pinealectomy on osmotic and ionic regulation in the goldfish, *Carassius auratus*. *Am. Zoologist* **6**, 257.

Wurtman, R. J., and Axelrod, J. (1965a). The pineal gland. *Sci. Am.* **213**, 50–64.

Wurtman, R. J., and Axelrod, J. (1965b). The formation, metabolism, and physiologic effects of melatonin in mammals. *Progr. Brain Res.* **10**, 520–529.

Wykes, U. (1938). The control of photopigmentary responses in eyeless catfish. *J. Exptl. Biol.* **15**, 363–370.

Wyman, L. C. (1924). The reactions of the melanophores of embryonic and larval *Fundulus* to certain chemical agents. *J. Exptl. Biol.* **40**, 161–180.

Young, J. Z. (1935). The photoreceptors of lampreys. *J. Exptl. Biol.* **12**, 254–270.

3

AUTONOMIC NERVOUS SYSTEMS

GRAEME CAMPBELL

I. Introduction	109
II. Cyclostomes	110
A. Cranial Autonomic Nerves	110
B. Spinal Autonomic Nerves	112
III. Gnathostomatous Fish	114
A. Cranial Autonomic Nerves	114
B. Spinal Autonomic Nerves	122
C. Summary	128
References	128

I. INTRODUCTION

The current revival of interest in the general physiology of fishes has emphasized our ignorance of autonomic functions in these animals. Since Nicol reviewed the anatomy and function of the autonomic nervous system of fish in 1952, there have been many new investigations; these have often been concerned with details of the innervation of systems already examined in some detail, namely, the heart, the gastrointestinal tract musculature, and chromatophores. It is rather surprising that there has not been greater interest in the innervation of, for example, the vasculature, the internal genitalia, and the digestive glands. It is possible that this will be the last of the old-style reviews of this subject, for new techniques, including the now widespread use of delicate mechanoelectric transducer recording systems and the introduction of the histochemical technique for visualizing tissue catecholamines, are already starting to revise our opinions on the autonomic nervous system of fishes.

II. CYCLOSTOMES

A. Cranial Autonomic Nerves

The oculomotor nerve is absent in hagfish, in which the lateral eyes are degenerate. The eyes of lampreys are also small, but oculomotor nerves are found. Anatomical evidence for an autonomic component of the oculomotor in lampreys is controversial, and there would appear to be no function for such a component since the eyes are probably completely devoid of intrinsic musculature.

An autonomic component of the facial nerve has been claimed for *Myxine*, largely on the basis of brainstem analysis. Lindström (1949) has suggested that the autonomic fibers innervate the lingual artery, but he does not exclude the possibility that the fibers are afferent. No physiological investigations of this nerve have been made.

In both hagfish and lampreys, the two vagus nerve trunks unite to form a cardiac plexus, lying on the gut wall, containing numerous nerve cell bodies and a few ganglionic aggregations. From this plexus, autonomic fibers are distributed to the region of the heart, major veins, gall bladder, and, via an unpaired nerve trunk, to the entire length of the intestine (see Nicol, 1952; Johnels, 1956, 1957; Peters, 1963; Fänge *et al.*, 1963). The intestinal wall contains a nervous plexus with numerous cell bodies (Brandt, 1922), some of which may be sensory neurons (Milokhin, 1958, 1959a,b). The neurons become sparse in the posterior region of the gut (Fänge *et al.*, 1963), although the fiber plexus persists. Many of these enteric neurons are probably innervated by vagal fibers.

Of the vagal innervation systems, that of the heart has been studied in greatest detail. Greene (1902) first reported that the heart of the hagfish *Polistotrema* (=*Bdellostoma*) *stouti* was not affected by stimulation of the vagus nerve, or indeed any other nerve. This was confirmed for the same species by Carlson (1904) and for another hagfish, *Myxine glutinosa*, by Augustinsson *et al.* (1956). In agreement with this observation, no nerve fibers were found during an electron microscopic investigation of *Myxine* heart (Hoffmeister *et al.*, 1961). Normal hearts of both species are virtually unaffected by acetylcholine and by catecholamines (Östlund, 1954; Fänge and Östlund, 1954; Chapman *et al.*, 1963). In contrast, the lamprey heart is innervated by the vagus nerves. The main response to stimulation of the medulla oblongata is an acceleration of the heart followed by a period of deceleration after periods of intense stimulation (*Lampetra*, Zwaardemaker, 1924; Augustinsson *et al.*, 1956), although a period of inhibition preceding the acceleration has been reported (*Ichthyomyzon*, Carlson, 1906). The heart of *Lampetra* does not contain adrenergic nerve fibers demonstrable by fluorescence histochemistry (see Falck *et al.*, 1966), and catecholamines are variously stated to cause a slight depression (Otorii, 1953) or an acceleration and augmentation (Augustinsson *et al.*, 1956; Falck *et al.*, 1966) of the cardiac beat. It would therefore seem that the vagal innervation is cholinergic, since acetylcholine accelerates the heartbeat, at the same time depressing the force of contractions. Similar actions are exerted by other choline esters, including succinylcholine, and nicotine, but not by muscarine; the response to acetylcholine is inhibited by the nicotinic blocking drugs hexamethonium and tubocurarine but is unaffected by atropine (Otorii, 1953; Augustinsson *et al.*, 1956; Falck *et al.*, 1966). The cholinergic receptors are therefore of the nicotinic type. Because of this, Augustinsson *et al.* (1956) suggested that acetylcholine was acting on ganglion cells in the heart. It seems a simpler view that the lamprey heart muscle has a closer affinity to the skeletal muscle than to the cardiac muscle of higher vertebrates (Itina, 1959) and that acetylcholine is acting directly on nicotinic receptors in the heart muscle itself. In this case, one would expect treatment with curare to prevent vagal transmission to the heart, but Zwaardemaker (1924) observed cardiac responses to medulla oblongata stimulation in lampreys immobilized with curare. Perhaps the cardiac cholinergic receptors are relatively resistant to blockade by curare. The vagus does not appear to innervate the heart in larval lampreys (Carlson, 1904, 1906).

Stimulation of either vagal trunk in *Myxine* causes a contraction of the strongly muscular gallbladder (Fänge and Johnels, 1958). This pathway involves a ganglionic synapse in the hepatic plexus since topical application of nicotine to the plexus abolishes the responses. The terminal neurons are probably cholinergic, for acetylcholine causes a contraction of the gallbladder, whereas epinephrine appears to relax it.

The vagal innervation of the intestine is poorly understood. The muscularis of the gut wall is poorly developed, but it seems that all regions of the intestine are contracted by acetylcholine and relaxed by catecholamines (Fänge, 1948; von Euler and Östlund, 1957; Johnels and Östlund, 1958). In addition, the isolated intestine is reported to undergo peristaltic movements (Olcott, 1931; Johnels and Östlund, 1958). It is therefore clear that the gut musculature is functional. However, Fänge (1948) and Fänge and Johnels (1958) could not detect any intestinal response to vagal stimulation in *Myxine*. Previously, Patterson and Fair (1933) had found that stimulation of the vagus in *Bdellostoma* caused a very slight relaxation of the gut, but their records were obtained *in situ*, and in view of the thinness of the gut wall the response could have been artifactual. On the other hand, Fänge and Johnels (1958) pointed out that they had allowed the animal to warm up to room temperature (18°C), and they suggested that the vagus might be ineffective at this high temperature (see also Burnstock, 1958a). The only independent support for an inhibitory vagal innervation of the gut is the observation that nicotine causes a relaxation of previously contracted *Myxine* intestine *in vitro* (Fänge, 1948), which may indicate the presence of inhibitory neurons in the gut wall. It has been postulated elsewhere (Campbell and Burnstock, 1968) that the vagus nerve is primitively inhibitory to the vertebrate gastrointestinal tract and that the inhibitory nerve fibers are of a nonadrenergic type recently demonstrated in the mammalian gastric vagus (Martinson, 1965; Campbell, 1966).

The possibility has been raised that the vagus nerves contain adrenergic nerve fibers (Fänge *et al.*, 1963) since the vagus of *Myxine* contains considerable concentrations of catecholamines, predominantly norepinephrine (von Euler and Fänge, 1961). Adrenergic fibers have not been found in the cranial autonomic outflow of any higher vertebrate, but this does not necessarily mean that they will be absent in cyclostomes. Alternatively, sympathetic adrenergic fibers could enter the vagi from the spinal outflow, as claimed by Marcus (1910), since the vagi run extremely close to the mixed ventral rami of the anterior spinal nerves (see Peters, 1963). Finally, the catecholamines could be stored in chromaffin cells; the vagus contains many ganglion cells, and one must remember that the "ganglion cells" found in cyclostome hearts by Augustinsson *et al.* (1956) were later shown to be specialized chromaffin cells (Johnels and Palmgren, 1960; Bloom *et al.*, 1961; Hoffmeister *et al.*, 1961). A fluorescence histochemical study of this problem would be most welcome.

B. Spinal Autonomic Nerves

There is still much confusion about the distribution of autonomic nerve fibers from the spinal cord. Some of this confusion arises from the fact that many peripheral nerve cell bodies have been observed in cyclostomes, but it is not always clear which of these neurons are autonomic and which sensory. In addition, there are reports of fibers leaving the spinal cord of *Lampetra* in both dorsal and ventral spinal nerves and running directly, without an intervening ganglionic synapse, to visceral structures (Johnels, 1956). Whether all or any of these fibers are autonomic, i.e., efferent, is unknown and will probably remain so until physiological observations have been made.

There is, however, general agreement that there are neither organized sympathetic chains nor strictly segmental sympathetic ganglia in cyclostomes. There are collections of ganglion cells along the cardinal veins in the abdominal cavity of lampreys which may represent sympathetic ganglia. In *Lampetra*, visceral branches of the spinal nerves run directly to the region of blood vessels, the kidneys, and the gonads. In posterior regions there is a spinal outflow to the cloaca, rectum, and ureters, with many ganglion cells in the pathway (Johnels, 1956). A similar segmental visceral outflow, possibly vasomotor, has been demonstrated in *Myxine* (Fig. 1; Fänge *et al.*, 1963), but in this animal the outflow to the posterior gut appears to be mainly somatic motor, supplying the striated muscles of the cloacal sphincter, with a sensory component. It is highly likely that there is no spinal autonomic innervation of the gut in hagfish but that the smooth muscles of at least the posterior intestine of

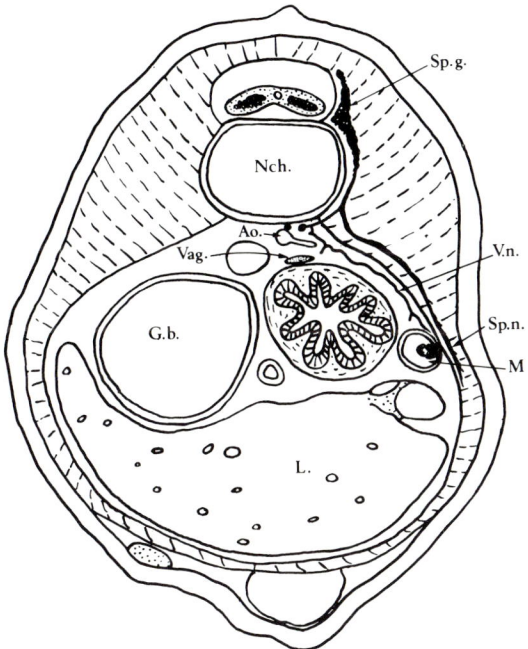

Fig. 1. Spinal autonomic fibers in *Myxine*. Diagrammatic illustration of the course of the visceral (spinal sympathetic) fibers as seen in a projection of a transverse section. Longitudinal fibers below the notochord indicated by two dots. Ao., dorsal aorta, at the beginning of a segmental artery; G.b., gallbladder; L., liver; M, mesonephros; Nch., notochord; Sp.g., spinal ganglion; Sp.n., spinal nerve; Vag., vagus nerve; V.n., visceral nerve. Redrawn with slight modification from Fig. 6 of Fänge *et al.* (1963).

lampreys is innervated from the spinal cord. Support for a vasomotor function of the sympathetic has come from the observation of Leont'eva (1966), who showed histochemically that there is an adrenergic innervation of blood vessels in cyclostomes.

Both lampreys and hagfish possess a rich subcutaneous nerve plexus containing uni- and bipolar nerve cells, a feature unique among the vertebrates. The ganglion cells are innervated by spinal nerve fibers and are almost certainly autonomic (see Bone, 1963). A number of functions could be subserved by the fibers. Bone (1963) has found that the plexus is particularly concentrated at the orifices of the slime glands in *Myxine*, associated with bundles of smooth muscle which form sphincters for the glands. In addition, the plexus is continuous about the walls of the gland and may regulate the activity of smooth muscle cells around the gland. Bone has also noted that cell bodies in the plexus often lie beside the small blood vessels of the subcutis, and he has suggested that the plexus may control blood flow in the subcutaneous blood sinuses. One further function may be proposed for the subcutaneous plexus, that of chromatophore control. Color changes in most cyclostomes appear to be extremely slow if they occur at all (see Young, 1935). However, Wild (1903) has reported that *Petromyzon marinus* shows dramatic and rapid color changes, indicating a nervous control of the chromatophores in this species, presumably via the plexus.

III. GNATHOSTOMATOUS FISH

Physiological observations on the autonomic nervous system of gnathostome fishes have been virtually restricted to selachians and teleosts. Although there are conspicuous anatomical differences between the autonomic systems of these two groups, especially in the arrangement of the spinal outflow, it seems likely that there are no gross physiological differences.

A. Cranial Autonomic Nerves

Autonomic fibers leave the brainstem of selachians in the oculomotor, facial, glossopharyngeal, and vagus nerves. In teleosts, the outflow is restricted to the oculomotor and vagus nerves. There have been no physiological investigations of the autonomic functions mediated by the selachian VIIth and IXth nerves, but it has been suggested that both outflows provide vasomotor fibers to the pharyngeal region. Presumably this function has been taken over completely by the vagus in teleosts.

1. Oculomotor Nerve

Oculomotor nerves are absent from forms with reduced eyes, but where present they supply preganglionic fibers to the ciliary ganglion, from which postganglionic fibers pass to the eyeball (Fig. 2). Stimulation of the oculomotor nerves causes pupillary dilatation in the teleost *Uranoscopus scaber* (Young, 1931a) and in one selachian, *Scyllium*, although this was not seen in two others (*Mustelus, Trygon*; Young, 1933a). This response is in marked contrast to the pupillary constriction

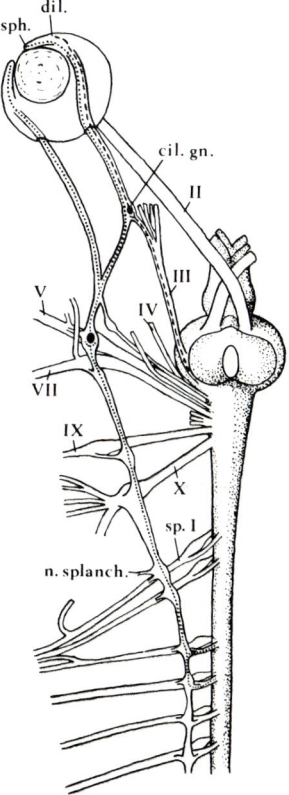

Fig. 2. Diagram of a ventral view of the anterior sympathetic system in *Uranoscopus*, showing the sympathetic and parasympathetic innervation of the iris. cil. gn., ciliary ganglion; dil., pupil dilator muscle; n. splanch., anterior splanchnic nerve; sph., pupil sphincter muscle; sp. 1, first spinal nerve; cranial nerves numbered II, III, IV, V, VII, IX, and X. (. . .) Sympathetic pathways and (- - -) parasympathetic pathways; positions of ganglionic synapses shown by filled circles. Redrawn, slightly modified, from Fig. 1 of Young (1931a).

caused by oculomotor nerve stimulation in mammals. There is no direct evidence to show whether the response represents an inhibition of activity in the sphincter or an excitation of the dilator muscle. In *Uranoscopus*, both epinephrine and acetylcholine can cause both dilatation and constriction of the pupil; in the selachians studied both epinephrine and acetylcholine caused predominantly dilator reactions, and it has been found that acetylcholine causes contraction of the isolated iris dilator muscle of *Scyllium* (Young, 1931a, 1933a,b). Atropine inhibits constrictions of *Uranoscopus* pupil caused both by sympathetic nerve stimulation and by acetylcholine (Young, 1931a), but the effects of blocking drugs on responses to oculomotor nerve stimulation have not been tested. The most simple, but not necessarily the correct interpretation of these limited observations, is that both the sympathetic and the oculomotor are cholinergic and that the sympathetic provides an excitatory innervation to the sphincter (in teleosts only) while oculomotor stimulation excites the dilator muscle.

2. Vagus Nerve

The posttrematic rami of the vagus nerve in teleosts (as well as of the glossopharyngeal nerve in selachians) almost certainly innervate the branchial vascular bed. Direct evidence of efferent innervation is wanting, but elevations of branchial vascular resistance are known to occur during, for instance, periods of anoxia (*Squalus acanthias*, Satchell, 1962; *Salmo gairdneri*, Holeton and Randall, 1967). Determination of whether this phenomenon is nerve mediated is made difficult by indications that the gill vasculature can constrict in direct response to the anoxic conditions (Satchell, 1962). It is well established that catecholamines dilate and acetylcholine constricts the branchial vasculature in teleosts (Krawkow, 1913; Keys and Bateman, 1932; Östlund and Fänge, 1962) and in dipnoans (Johansen *et al.*, 1968), although no vasomotor responses could be obtained from the isolated gills of the selachian *Squalus acanthias* (Östlund and Fänge, 1962). Similar vasoconstrictor actions of acetylcholine have been observed in the amphibian pulmonary vascular bed (Voigt, 1939; Brecht, 1947), a system derived from branchial arch vasculature. It is then reasonable to suggest that the vagi provide cholinergic vasoconstrictor fibers to the gills, if only because of the analogy with the established cholinergic vasoconstriction mediated by the vagi in amphibian lungs (Luckhardt and Carlson, 1921).

The visceral vagal rami distribute efferent fibers to the heart, the stomach, and usually no more than the most oral region of the intestine, the swim bladder in teleosts, and possibly to accessory digestive organs and to blood vessels in this region. The innervation of the heart has been reviewed by Nicol (1952) and more recently by Johansen and Martin (1965). (See also Chapter 4 by Randall, this volume.) The evidence for a cholinergic vagal supply to the heart, causing a reduction in the rate of beating, is so overwhelmingly strong that no further comment is needed. The consensus has been that the vagi do not contain cardiac augmentor or accelerator fibers, whether of sympathetic or parasympathetic origin (but see Section III, B), and it is well established that there are no distinct sympathetic nerves innervating the heart directly. Jullien and Ripplinger (1957) found that after degenerative sections made high in the vagal trunk stimulation of the vagal cardiac branches still caused negative chronotropic and inotropic responses in teleosts, indicating that the ganglionic synapses in the vagal pathway occur within the vagal trunk itself. They also observed that the heart rate was slow and that arrhythmias were common in bivagotomized tench and that the heart returned to a fast, regular beat after atropine treatment. They suggested that the vagal postganglionic cell bodies are tonically active, holding the heart muscle under an inhibitory drive, and that the activity of the neurons is normally inhibited by fibers from the central nervous system running in the vagi, a suggestion which deserves further consideration.

A further observation made by Jullien and Ripplinger is also worth reconsidering. In a series of papers around 1950 (summarized in Jullien and Ripplinger, 1957), they provided evidence for a unique system of noncholinergic "negative tonotropic" nerve fibers in the vagal supply to the teleost heart. The crux of their argument is that in nonbeating hearts, e.g., after treatment with acetylcholine, stimulation of the vagus causes elongation of the heart (Fig. 3a); on the other hand, in hearts treated with atropine, stimulation of the vagi causes elongation of the heart when all signs of negative inotropic and chronotropic responses have vanished (Fig. 3b). Since their recordings have been made by attaching the heart to a kymograph lever *in situ*, leaving the heart attached to the body by the large veins alone, one might suggest that the "tonotropic" responses represent inhibition of venous smooth muscle tone. This suggestion is striking enough but is perhaps more acceptable than the alternative that the cardiac muscle cells are in a condition of tonus which is not related to cardiac action potential activity.

The vagal innervation of the gut of teleosts and selachians has been reviewed by Nicol (1952), Barrington (1957), and Campbell and Burnstock (1968). All reviewers agree that stimulation of the vagi causes contraction of the stomach, but not of the intestine, in those animals possess-

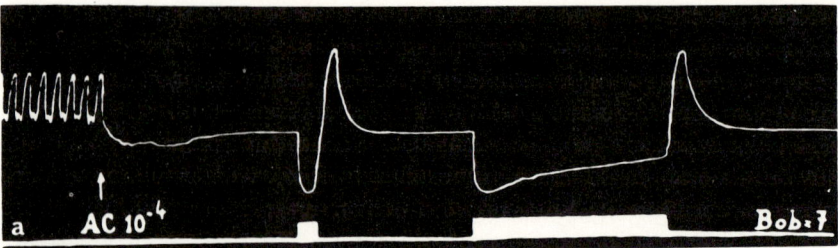
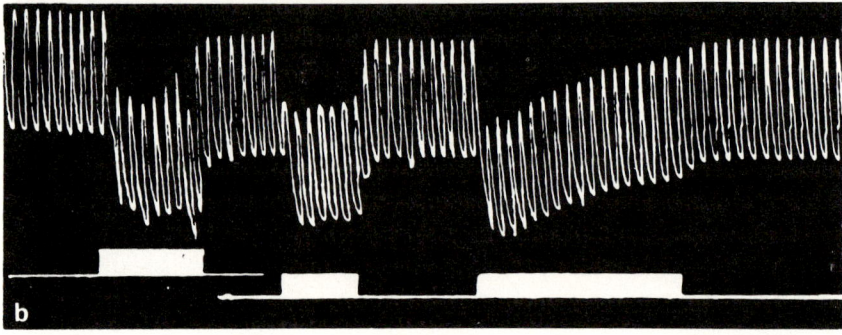

Fig. 3. "Negative tonotropic" vagal innervation of the heart of *Tinca*. Heart left *in situ*, bulbus arteriosus cut. (a) Application of acetylcholine (Ac, 10^{-4} g/ml) stops the heart in diastole. Stimulation of the vagus nerve (white markers) causes an apparent further elongation of the heart. (b) Heart treated with atropine. Stimulation of the vagus (white markers) has no negative inotropic or chronotropic effects, but a negative "tonotropic" effect persists. From Jullien and Ripplinger (1957).

ing a stomach. When there is no stomach, the vagi may cause contraction of part or all of the intestine. However, cautious interpretation of this result is needed for two reasons. First, part or all of the responses observed might result from stimulation of sympathetic nerve fibers running with the vagus, at least in teleosts. Young (1931b) described connections between the cranial sympathetic chain and the vagus nerve in *Uranoscopus* (Fig. 2), and Fänge (1953) has already raised this question with respect to the innervation of the swim bladder. It is pertinent to raise this possibility since all of the excitatory nerve fibers in the vagosympathetic trunk supplying the gastric musculature of an amphibian, *Bufo marinus*, are of sympathetic origin, the vagus supplying only nonadrenergic inhibitory fibers to the stomach (Campbell, 1969). The presence of these nonadrenergic inhibitory fibers in the gastric vagi of both mammals (Martinson, 1965; Campbell, 1966) and amphibians provides the second reason for caution, for it is not yet clear whether such fibers exist in fish. Campbell and Burnstock (1968) have pointed out that many of the excitatory responses to vagal stimulation recorded by earlier workers on both selachian and teleostean gut do not start until some time after the period of stimulation of the nerves is over (see Fig. 4). In this respect the contractions appear more like the "rebound" or recovery contractions which follow responses to inhibitory nerve stimulation in mammalian gut than like the virtually immediate primary contractions caused by stimulation of the excitatory innervation. In other words, there is still considerable doubt as to whether the contractions seen in fish stomach preparations following vagus nerve stimulation are mediated by excitatory or inhibitory nerves. The absence of records of actual inhibition could be easily explained under the conditions of experimentation for the nerves have been stimulated for relatively short periods, the spontaneous contractions occur at a very low rate, if at all, and the musculature has little or no tonus. Pharmacological investigations have been of little help in answering this question so far. For instance, vagal excitation of the stomach is not inhibited by atropine in the brown trout (Burnstock, 1958b) or in *Raja* (Babkin *et al.*, 1935), whereas vagal excitation of the smooth muscle of the intestine of the tench is prevented by atropine (Méhés and Wolsky, 1932); in all three preparations, atropine prevents the excitatory action of acetylcholine. Only further experimentation can settle this question.

There have been no further investigations of neural control of gastric, hepatic, or pancreatic secretion in any fish since the subject was reviewed by Barrington in 1957. One can only agree with Barrington that there is no evidence for such a nervous control, while remarking that there have been too few investigations to allow any conclusion to be reached.

The processes causing filling and emptying of the swim bladder of teleosts are complex (see chapter by Steen, this volume). At least three processes seem to be implicated in determining whether a net secretion or absorption of gases occurs. First, the relative amounts of secretory and absorptive epithelium exposed to the lumen of the swim bladder can be varied by differential contraction of the muscularis mucosae or by variable closure of the oval (Fänge, 1953). Second, the rates of blood perfusion of the secretory epithelium, and therefore of the retia mirabilia, and of the resorptive epithelium are probably independently controlled. Third, the activity of the secretory cells themselves, which appear to act by adding metabolites to the blood leaving the secretory epithelium (see Kuhn *et al.*, 1963), is probably under nervous control.

Bohr (1894) first showed that following section of the vagi, experi-

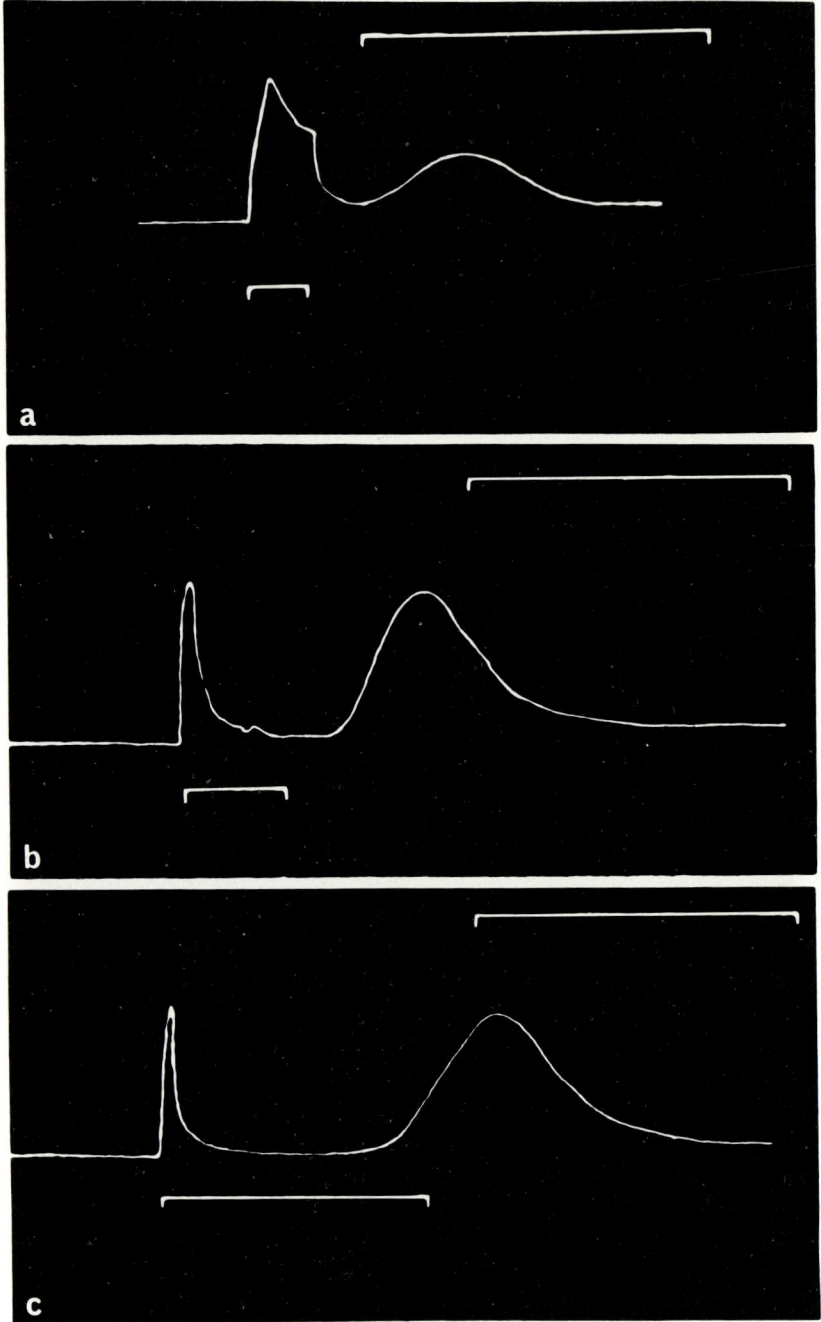

mentally deflated swim bladders were no longer refilled. This observation shows that the vagi contain nerve fibers which are in some way responsible for secretion into the swim bladder. However, Fänge (1953) was unable to show any filling responses to vagal stimulation. In fact, the only clear response to vagal stimulation that he observed was an increase in the relative area of resorptive epithelium exposed to the enclosed gases brought about in a number of ways by contraction of the muscularis mucosae, and a vasodilation in the resorptive epithelium. This response would favor emptying of the swim bladder, and it probably masked completely those effects of vagal stimulation favoring filling. Fänge showed that these "emptying" responses of the muscularis mucosae and of the vasculature were probably mediated by adrenergic nerves which, he suggested, contracted the muscularis mucosae of the secretory portion but relaxed that of the resorptive portion of the swim bladder and dilated the blood vessels in the resorptive region. In view of their adrenergic nature, Fänge postulated that these fibers were sympathetic in origin, joining the vagus intracranially. In the absence of conflicting evidence, this is still the safest interpretation. Fänge's pharmacological arguments that the nerves are adrenergic have received support from studies using the histochemical technique for demonstrating tissue monoamines. Fahlén et al. (1965), using this technique, have shown that there is an adrenergic (norepinephrine) innervation of blood vessels throughout the swim bladder and of the muscularis mucosae only in the secretory region in cod and trout; Fahlén (1965) found that the muscularis of only the postulated resorptive region in *Argentina* was adrenergically innervated. Many or all of these adrenergic neurons have their cell bodies in or near the swim bladder (Fahlén et al., 1965), a fact which would explain why vagotomy does not alter catecholamine levels in this organ (von Euler and Fänge, 1961).

It is highly likely that the vagal nerve fibers which cause secretion are cholinergic, for Fänge (1953) has shown that atropine treatment, like vagotomy, inhibits gas secretion into a deflated swim bladder. The problem is which structure or structures are being affected by the

---

Fig. 4. Effects of vagal stimulation on the gut of *Tinca*. During stimulation of the vagus (white bars below traces), the striated muscle contracts. The smooth muscle contraction occurs only after a latent period of about 20 sec in (a) (10-sec period of stimulation), 30 sec in (b) (20-sec period of stimulation), and 40 sec in (c) (stimulated for 50 sec). The increase in latency with increasing durations of stimulation indicates an inhibitory effect of vagal stimulation and suggests that the smooth muscle contractions may be "rebound" in nature. Time markers, 1 min. From R. Mahn (1898), Untersuchungen über das physiologische Verhalten des Schleiendarms, *Arch. Ges. Physiol.* **72**, 273–304. Springer, Berlin.

cholinergic fibers. Fänge has suggested that the vagi provide cholinergic fibers to contract the muscularis mucosae of resorptive regions, a response which would tend to cause filling by decreasing the area of resorptive epithelium exposed. He reached this conclusion because he found that isolated preparations of the tissue were contracted by acetylcholine and relaxed by atropine. However, he did not observe muscular contractions of this type in response to vagal stimulation, nor did atropine treatment affect the position of the swim bladder diaphragm in *Ctenolabrus*. Fänge has provided evidence for vagal cholinergic fibers causing vasodilatation in the secretory portion of the swim bladder. He noted that during secretion there was a marked vasodilatation in the gas gland, whereas after vagotomy, and especially after atropine treatment, there was considerable vasoconstriction in the gland. But Fänge rightly raised the still unresolved question of whether the vasodilatation occurring during activity is a primary effect caused by specifically vasodilator nerves or a secondary effect caused by the release of metabolites from cholinergically activated secretory cells.

B. Spinal Autonomic Nerves

The sympathetic nervous system of selachians consists of a series of paravertebral ganglia, one or more occurring per spinal nerve, linked together at most by a loose plexus of nerve bundles. There is no compact sympathetic chain. The ganglia are connected to the spinal nerves by white rami communicantes. Young (1933c) claims that gray rami are absent, i.e., that there is no entry of postganglionic fibers into the spinal nerves to reach dermal structures, but there is some evidence for sympathetic nervous control of dermal melanophores. The ganglia extend no further caudally than the posterior end of the mesonephros in adults. There are no cranial sympathetic ganglia nor is there any semblance of a sympathetic chain entering the head region; however, this does not preclude a sympathetic innervation of cranial structures via perivascular nerve plexuses. There are no prevertebral ganglia, and fibers from the paravertebral ganglia are collected into anterior, middle, and posterior splanchnic nerves extending to the alimentary canal via plexuses on the arterial supply. Other fibers run directly to the genital and urinary ducts.

In teleosts, the sympathetic system has a well-defined structure resembling that found in tetrapods. Paravertebral sympathetic ganglia, usually two per spinal segment, are linked together regularly to form two sympathetic chains which fuse to a greater or lesser extent in different species. The sympathetic chains are connected to spinal nerves by both gray and white rami communicantes. The chains extend caudally to the end of the tail and also extend rostrally into the cranium where connections are usually made with the IIIrd, Vth, VIIth, IXth, and Xth cranial nerves (Fig. 2). The gastrointestinal tract is mainly innervated by an anterior splanchnic nerve, leaving the right sympathetic chain, but a posterior splanchnic nerve has been found in trout (Burnstock, 1958b). Genital and vesical nerves leave the posterior regions of the chains, supplying the urinary and genital ducts via periarterial nerve plexuses.

A sympathetic innervation of the iris occurs in teleosts, but it has never been observed in selachians. The fibers involved leave the spinal cord in the anterior spinal nerves and run rostrally up the chain to form synapses in the sympathetic ganglion in the trigeminal nerve (*Uranoscopus*; Young, 1931a; Fig. 2). Stimulation of the sympathetic chain causes pupilloconstriction. The response is prevented by treatment with atropine, indicating that the nerves are cholinergic (Young, 1931a, 1933b). The sympathetic also appears to provide a motor innervation to the striated extraocular muscles in teleosts because sympathetic section causes abnormal protrusion of the eyeball (Young, 1931a). There is no evidence concerning the transmitter substance mediating this effect, but it has been found that sympathetic adrenergic fibers can cause contraction of the superior rectus muscle in a mammal (Eakins and Katz, 1967).

It has long been denied that there is a sympathetic innervation of the heart in fish (see Johansen and Martin, 1965; Randall, 1968). However, fluorescence histochemical studies of the heart of teleosts have now shown that in spite of one report to the contrary (von Mecklenburg, 1966, cited in Falck *et al.*, 1966) there is an adrenergic innervation of the cardiac chambers (Govyrin and Leont'eva, 1965; Gannon and Burnstock, 1969; Fig. 5). Nerve fibers containing small granular vesicles, typical of adrenergic nerves in other animals, have been seen during an electron microscopic investigation of the trout heart (Yamauchi and Burnstock, 1968). The adrenergic nerve fibers appear to come both from the vagi and along the coronary arteries and occur most densely in the sinus venosus and auricle. Gannon and Burnstock (1969) have confirmed Kulaev's report (1957) that stimulation of the vagi can cause cardioacceleration in teleosts. They have also shown that this response is prevented by treatment with the adrenergic neuron-blocking drugs, bretylium and guanethidine, or with an epinephrine-blocking drug, propranolol, a congener of which inhibits cardiac responses to catecholamines in plaice (Falck *et al.*, 1966). As in the case of the vagal adrenergic innervation of the swim bladder, it is probably safest to assume

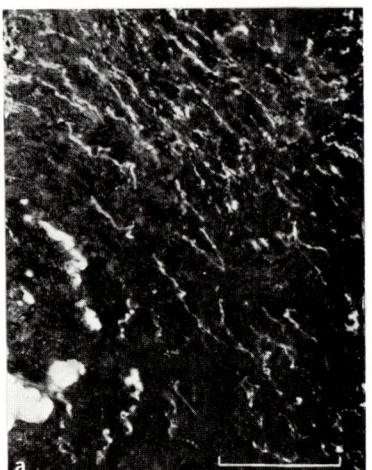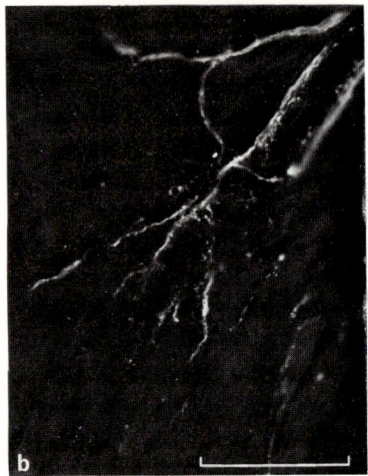

Fig. 5. Adrenergic nerve fibers in the heart of *Salmo*. Material treated with paraformaldehyde vapor at 80°C for 1 hr. Paraffin sections 10 μ thick. (a) Oblique section of wall of sinus venosus to show dense innervation of muscle. Note large nerve bundles in lower left corner. Sinus pretreated *in vitro* with α-methyl-norepinephrine (10^{-5} g/ml) to enhance the normally weak specific fluorescence. (b) Section of dense layer of ventricle wall to show adrenergic plexus investing a coronary artery (upper right) and nerve bundles passing out into the ventricular muscle. Tissue not pretreated. Calibration in (a) and (b), 100 μ. From Gannon and Burnstock (1969).

that these adrenergic cardioaccelerator nerve fibers enter the vagi from the sympathetic chain via the intracranial connections. In selachians, considering the absence of sympathetic nerves running directly to the heart and the absence of intracranial connections between the sympathetic and the vagus, it seems unlikely that a sympathetic adrenergic innervation of the heart will be found.

In both selachians and teleosts we are temptingly close to being able to say that there is sympathetic nervous control of the vascular system. On the one hand are reports that catecholamines injected into the circulation cause pressor responses in selachians (MacKay, 1931; Lutz and Wyman, 1932; Burger and Bradley, 1951) and teleosts (Mott, 1951; Randall and Stevens, 1967). On the other hand are the many reports of dense nervous plexuses investing blood vessels and, recently, a number of histochemical studies which show that there are adrenergic nerve fibers in some of these plexuses, at least in teleosts (Fahlen *et al.*, 1965; Leont'eva, 1966; Baumgarten, 1967; Read and Burnstock, 1968; Kirby and Burnstock, 1969; Gannon and Burnstock, 1969). Kirby and Burnstock (1969) have shown that adrenergic and cholinergic nerves, both mediating contraction of isolated spiral strips, occur in the ventral aorta of trout and eels. And yet there is still no report of a direct effect of autonomic nerve stimulation on the resistance of any vascular bed in fishes. What little evidence there is goes against belief in a sympathetic vasomotor tone. Wilber and Sudak (1960) found that there were no compensatory cardiovascular responses to hemorrhage in *Mustelus* and *Squalus*, while Burger and Bradley (1951) showed that a ganglion-blocking drug, tetraethylammonium, and an epinephrine-blocking drug, dibenamine, did not cause any alteration of ventral or dorsal aortic pressure in *Squalus*. Randall and Stevens (1967) found that the epinephrine-blocker, phenoxybenzamine, caused a fall of only 1–2 mm Hg in dorsal aortic pressure in coho and sockeye salmon. At the moment it seems likely that if there is a sympathetic vasoconstrictor innervation of systemic vascular beds, it is used to regulate blood flow to individual organs rather than to sustain a certain level of blood pressure. Any generalized control of systemic vascular resistance is likely to be mediated by circulating epinephrine from chromaffin cells (Randall and Stevens, 1967).

The sympathetic innervation of the gastrointestinal tract has been reviewed recently (Campbell and Burnstock, 1968). Interpretation of the available results is again made difficult by the possibility that the observed excitatory responses to sympathetic nerve stimulation are in reality rebound contractions caused by the stimulation of inhibitory nerves. However, these authors concluded that a number of workers had shown true primary contractions mediated by sympathetic nerves. Other excitatory responses could be interpreted as rebound contractions, and there is one clear example of an inhibitory response to posterior splanchnic nerve stimulation in trout (Burnstock, 1958b). In summary, these authors suggested that the sympathetic supply to all regions of the gut in both selachians and teleosts contains a mixture of excitatory and inhibitory nerve fibers. One may suggest that the excitatory nerve fibers are cholinergic, for at least some of the excitatory responses are abolished by atropine treatment (Burnstock, 1958b). The inhibitory nerve fibers may be adrenergic since catecholamines cause relaxation of the gut of a number of species (see Campbell and Burnstock, 1968) and an adrenergic innervation of gastrointestinal muscle has been demonstrated histochemically in trout, eel, and tench (Baumgarten, 1967; Read and Burnstock, 1968), but direct evidence is lacking.

In addition to the adrenergic, presumably sympathetic, nerves reaching the swim bladder in the vagus trunk of teleosts (see Section III, A, 2), there is a direct sympathetic innervation by a branch of the splanchnic nerve. Fänge (1953) showed that if this branch is cut there is a slight rise in the oxygen content of the swim bladder gases. Fänge suggested that this might occur because of a loss of sympathetic vasoconstrictor tone in the gas gland. However, Fahlen et al. (1965) have shown that adrenergic nerve terminals, deriving from cell bodies in the coeliac sympathetic ganglion, form pericellular networks about the adrenergic neurons of the gas gland ganglion in cod. This raises the possibility that by sectioning the splanchnic nerve one removes an inhibitory influence on synaptic transmission in the gas gland ganglion, thus enhancing a secretory activity being mediated by that neural pathway.

A few passing observations have been made on the sympathetic innervation of the urinogenital system. Stimulation of the sympathetic ganglia in selachians causes contractions of the oviducts, as does the application of acetylcholine (Bottazzi, 1902; Young, 1933c). Similar responses of the ovaries occur in the teleosts *Lophius* and *Uranoscopus* (Young, 1936). In addition, stimulation of the sympathetic innervation of the urinary bladder in these teleosts causes a contraction and the expulsion of urine, a response also mimicked by acetylcholine. Young found that the excitatory responses of the bladder to nerve stimulation were reduced by atropine, suggesting that the nerves are cholinergic.

The number of investigations of nervous control of dermal chromatophores in fish is so great that no detailed discussion will be given here (see Parker, 1948; Fingerman, 1963; chapter by Fujii, Volume III). The main conclusion reached has been that the melanophores in many if not all teleosts are under the control of sympathetic adrenergic nerve fibers, stimulation of which causes concentration of the melanophore pigment. A comparable sympathetic adrenergic innervation of dermal photophores has been postulated for the teleost *Porichthys* (Nicol, 1957). Although it has been suggested that there are cholinergic sympathetic fibers which mediate pigment dispersal in melanophores, it seems to this reviewer that there is no definitive evidence for their existence. In selachians, color changes are generally slow or absent, but evidence has been produced for melanophore-concentrating nerves in *Mustelus* (e.g., Parker, 1935). In a few teleosts, catecholamines cause dispersal rather than the normal concentration of melanophore pigments (e.g., Enami, 1955), and it is not clear how this will affect the chromatophore innervation. There is very little information concerning the innervation of other chromatophore types (xanthophores and erythrophores), although it is well known that they may, for example, disperse under conditions which induce melanophore concentration (e.g., Hewer, 1927). Virtually nothing is known of the control of internal melanophores. Perhaps the most striking feature of chromatophore control in fish is that the animals can not only match the shade of the background but can also match its pattern, an ability which is highly developed in various flatfish. The presence of such responses indicates that the sympathetic system can be used to pick out single or, at most, small groups of chromatophores as required, showing a fineness of control which is not usually attributed to the sympathetic system.

There is no clear evidence for the existence of a "spinal parasympathetic" autonomic outflow in fish. But Friedman (1935) has shown that stimulation of posterior spinal nerves in skates causes erection of the claspers by both muscular contraction and vasodilatation and elicits se-

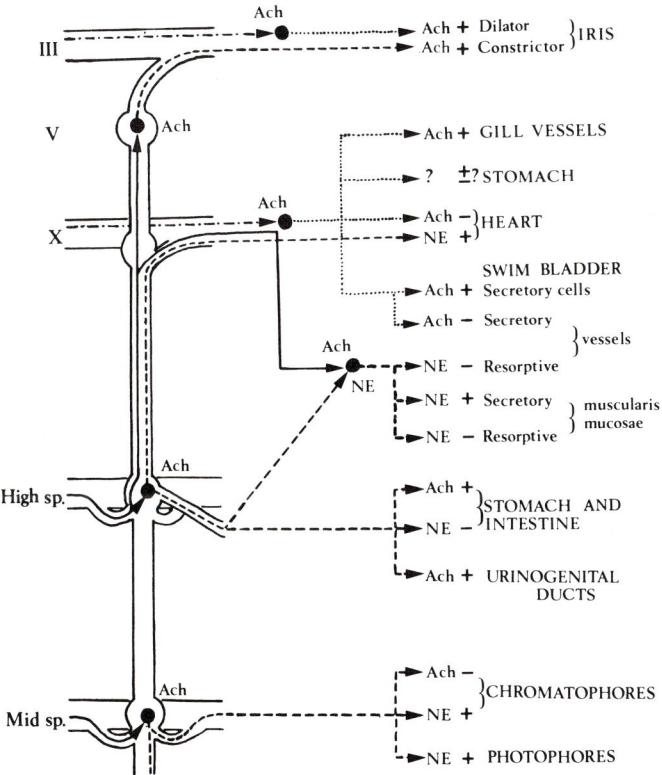

Fig. 6. Diagram to illustrate some principles of the arrangement of the autonomic nervous system in teleosts, showing the autonomic outflow in the oculomotor (III), vagus (X), anterior spinal (high sp.), and midspinal (mid sp.) nerves. Nerve types shown as (——) preganglionic sympathetic, (- - -) postganglionic sympathetic, (— · —) preganglionic parasympathetic, (· · ·) postganglionic parasympathetic. The filled circles represent ganglionic synapses. Transmitter substances shown as Ach (acetylcholine) and NE (norepinephrine). Effects of the nerves shown as + (excitation of muscle and gland, concentration of chromatophores) and −.

cretion from the clasper gland. As Nicol (1952) has remarked, these phenomena resemble effects of stimulation of the sacral parasympathetic system in mammals.

C. Summary

Rather than give a verbal summary, a diagram is presented indicating some of the features of the autonomic system in gnathostome fishes (Fig. 6). The diagram represents an interpretation of some of the data for teleosts which have been discussed in the text. A similar diagram for a selachian would differ mainly in the lack of a contribution of sympathetic fibers to the cranial nerves and the presence of an autonomic outflow in the VIIth and IXth cranial nerves. As will be seen from the text, the evidence for a number of features illustrated in the diagram is scant.

REFERENCES

Augustinsson, K. B., Fänge, R., Johnels, A., and Östlund, E. (1956). Histological, physiological and biochemical studies on the heart of two cyclostomes, hagfish (*Myxine*) and lamprey (*Lampetra*). *J. Physiol.* (*London*) **131**, 257–276.

Babkin, B. P., Friedman, M. H. F., and MacKay-Sawyer, M. E. (1935). Vagal and sympathetic innervation of the stomach in the skate. *J. Biol. Board Can.* **1**, 239–250.

Barrington, E. J. W. (1957). The alimentary canal and digestion. In "The Physiology of Fishes" (M. E. Brown, ed.), Vol. 1, pp. 109–161. Academic Press, New York.

Baumgarten, H. G. (1967). Vorkommen und Verteilung adrenerger Nervenfasern im Darm der Schleie (*Tinca vulgaris* Cuv.). *Z. Zellforsch. Mikroskop. Anat.* **76**, 248–259.

Bloom, G., Östlund, E., von Euler, U. S., Lishajko, F., Ritzén, M., and Adams-Ray, J. (1961). Studies on catecholamine-containing granules of specific cells in cyclostome hearts. *Acta Physiol. Scand.* **53**, Suppl. 185, 1–34.

Bohr, C. (1894). The influence of section of the vagus nerve on the disengagement of gases in the airbladder of fishes. *J. Physiol.* (*London*) **15**, 494–500.

Bone, Q. (1963). Some observations upon the peripheral nervous system of the hagfish, *Myxine glutinosa J. Marine Biol. Assoc. U. K.* **43**, 31–47.

Bottazzi, F. (1902). Untersuchungen über das viscerale Nervensystem der Selachier. *Z. Biol.* **43**, 372–442.

Brandt, W. (1922). Das Darmnervensystem von *Myxine glutinosa*. *Z. Anat. Entwicklungsgeschichte* **65**, 284–292.

Brecht, K. (1947). Über die wirkung elektrischer reizung des Vagosympathicus auf die glatte Muskulatur der Froschlunge und ihre Beeinflussung durch Ionen bei künstlicher Durchströmung. *Arch. Ges. Physiol.* **249**, 94–111.

Burger, J. W., and Bradley, S. E. (1951). The general form of the circulation in the dogfish (*Squalus acanthias*). *J. Cellular Comp. Physiol.* **37**, 389–402.

Burnstock, G. (1958a). Reversible inactivation of nervous activity in a fish gut. *J. Physiol.* (*London*) **141**, 33–45.

Burnstock, G. (1958b). The effect of drugs on spontaneous motility and on response to stimulation of the extrinsic nerves of the gut of a teleostean fish. *Brit. J. Pharmacol.* **13**, 216–226.

Campbell, G. (1966). The inhibitory nerve fibers in the vagal supply to the guinea-pig stomach. *J. Physiol. (London)* **185**, 600–612.

Campbell, G. (1969). The autonomic innervation of the stomach of a toad (*Bufo marinus*). *Comp. Biochem. Physiol.* **31**, 693–706.

Campbell, G., and Burnstock, G. (1968). Comparative physiology of gastrointestinal motility. *In* "Handbook of Physiology" (Am. Physiol. Soc., J. Field, ed.), Sect. 6, Vol. IV, pp. 2213–2266. Williams & Wilkins, Baltimore, Maryland.

Carlson, A. J. (1904). Contributions to the physiology of the heart of the California hagfish (*Bdellostoma dombeyi*). *Z. Allgem. Physiol.* **4**, 259–288.

Carlson, A. J. (1906). The presence of cardio-regulative nerves in the lampreys. *Am. J. Physiol.* **16**, 230–232.

Chapman, C. B., Jensen, D., and Wildenthal, K. (1963). On the circulatory control mechanisms in the Pacific hagfish. *Circulation Res.* **12**, 427–440.

Eakins, K. E., and Katz, R. L. (1967). The effects of sympathetic stimulation and epinephrine on the superior rectus muscle of the cat. *J. Pharmacol. Exptl. Therap.* **157**, 524–531.

Enami, M. (1955). Melanophore contracting hormone (MCH) of possible hypothalamic origin in the catfish, *Parasilurus*. *Science* **121**, 36–37.

Fahlén, G. (1965). Histology of the posterior chamber of the swimbladder of *Argentina*. *Nature* **207**, 94–95.

Fahlén, G., Falck, B., and Rosengren, E. (1965). Monoamines in the swimbladder of *Gadus callarias* and *Salmo irideus*. *Acta Physiol. Scand.* **64**, 119–126.

Falck, B., von Mecklenburg, C., Myhrberg, H., and Persson, H. (1966). Studies on adrenergic and cholinergic receptors in the isolated hearts of *Lampetra fluviatilis*. (Cyclostomata) and *Pleuronectes platessa* (Teleostei). *Acta Physiol. Scand.* **68**, 64–71.

Fänge, R. (1948). Effects of drugs on the intestine of a vertebrate without sympathetic nervous system. *Arkiv. Zool.* [1] **40**, A1–A9.

Fänge, R. (1953). The mechanisms of gas transport in the euphysoclist swimbladder. *Acta Physiol. Scand.* **30,**, Suppl. 110, 1–133.

Fänge, R. (1962). Pharmacology of poikilothermic vertebrates and invertebrates. *Pharmacol. Rev.* **14**, 281–316.

Fänge, R., and Johnels, A. G. (1958). An autonomic nerve plexus control of the gall bladder in *Myxine*. *Acta Zool. (Stockholm)* **39**, 1–8.

Fänge, R., and Östlund, E. (1954). The effects of adrenaline, noradrenaline, tyramine and other drugs on the isolated hearts from marine vertebrates and a cephalopod (*Eledone cirrosa*). *Acta Zool. (Stockholm)* **35**, 289–305.

Fänge, R., Johnels, A. G., and Enger, P. S. (1963). The autonomic nervous system. *In* "Biology of Myxine" (A. Brodal and R. Fänge, eds.), pp. 124–136. Oslo Univ. Press, Oslo.

Fingerman, M. (1963). "The Control of Chromatophores." Pergamon Press, Oxford.

Friedman, M. H. F. (1935). The function of the claspers and clasper-glands in the skate. *J. Biol. Board Can.* **1**, 261–268.

Gannon, B. J., and Burnstock, G. (1969). Excitatory adrenergic innervation of the fish heart. *Comp. Biochem. Physiol.* **29**, 765–773.

Govyrin, V. A., and Leont'eva, G. R. (1965). Distribution of catecholamines in

vertebrate myocardium. *Zh. Evolyutsionnoi Biokhim. i Fiziol.* **1**, 38–44 (abstract only seen).
Greene, C. W. (1902). Contributions to the physiology of the California hagfish, *Polistotrema stouti*. II. The absence of regulative nerves for the systemic heart. *Am. J. Physiol.* **6**, 318–324.
Hewer, H. R. (1927). Studies on colour changes of fish. II. An analysis of the colour patterns of the dab. *Phil. Trans. Roy. Soc. London* B **215**, 177–187.
Hoffmeister, H., Lickfield, K., Ruska, H., and Ryback, B. (1961). Sécrétions granulaires dans le coeur branchial de *Myxine glutinosa* L. *Z. Zellforsch. Mikroskop. Anat.* **55**, 810–817.
Holeton, G. F., and Randall, D. J. (1967). Changes in blood pressure in the rainbow trout during hypoxia. *J. Exptl. Biol.* **46**, 297–305.
Itina, N. A. (1959). "Functional Properties of Neuro-Muscular Arrangements in Lower Vertebrates." Akad. Nauk. S. S. R., Leningrad (cited in Fänge, 1962).
Johansen, K., and Martin, A. W. (1965). Comparative aspects of cardiovascular function in vertebrates. *In* "Handbook of Physiology" (Am. Physiol. Soc., J. Field, ed.), Sect. 2, Vol. III, pp. 2583–2614. Williams & Wilkins, Baltimore, Maryland.
Johansen, K., Lenfant, C., and Hanson, D. (1968). Cardiovascular dynamics in the lungfishes. *Z. Vergleich. Physiol.* **59**, 157–186.
Johnels, A. G. (1956). On the peripheral autonomic nervous system of the trunk region of *Lampetra planeri*. *Acta Zool.* (Stockholm) **37**, 251–286.
Johnels, A. G. (1957). On the cardiac plexus of the vagus intestinalis nerves in *Myxine*. *Acta Zool.* (Stockholm) **38**, 283–288.
Johnels, A. G., and Östlund, E. (1958). Anatomical and physiological studies on the enteron of *Lampetra fluviatilis* (L.). *Acta Zool.* (Stockholm) **39**, 9–12.
Johnels, A. G., and Palmgren, A. (1960). "Chromaffin" cells in the heart of *Myxine glutinosa*. *Acta Zool.* (Stockholm) **41**, 313–314.
Jullien, A., and Ripplinger, J. (1957). Physiologie du coeur des poissons et de son innervation extrinsèque. *Ann. Sci. Univ. Besancon, Zool. Physiol.* [2] **9**, 35–92.
Keys, A., and Bateman, J. B. (1932). Branchial response to adrenaline and pitressin in the eel. *Biol. Bull.* **63**, 327–336.
Kirby, S., and Burnstock, G. (1969). Comparative pharmacological studies of isolated spiral strips of large arteries from lower vertebrates. *Comp. Biochem. Physiol.* **28**, 307–319.
Krawkow, N. P. (1913). Über die Wirkung von Giften auf die Gefässe isolierter Fischkiemen. *Arch. Ges. Physiol.* **151**, 583–603.
Kuhn, W., Ramel, A., Kuhn, H. J., and Marti, E. (1963). The filling mechanism of the swimbladder. Generation of high gas pressures through hairpin countercurrent multiplication. *Experientia* **19**, 497–511.
Kulaev, B. S. (1957). Nervous regulation of the rhythm of cardiac activity in fish. I. The effect of stimulation and section of the cardiac branches of the vagus on the cardiac rhythm. *Bull. Exptl. Biol. Med.* (USSR) (English Transl.) **44**, 771–774.
Leont'eva, G. R. (1966). Distribution of catecholamines in blood-vessel walls of cyclostomes, fishes, amphibians and reptiles. *Zh. Evolyutsionnoi Biokhim. i Fiziol.* **2**, 31–36 (abstract only seen).
Lindström, T. (1949). On the cranial nerves of the cyclostomes, with special reference to the N. trigeminus. *Acta Zool.* (Stockholm) **30**, 315–458.
Luckhardt, A. B., and Carlson, A. J. (1921). Studies on the visceral sensory nervous

system. VIII. On the presence of vasomotor fibres in the vagus nerve to the pulmonary vessels of the amphibian and the reptilian lung. *Am. J. Physiol.* **56**, 72–112.
Lutz, B. R., and Wyman, L. C. (1932). The effect of adrenaline on the blood pressure of the elasmobranch, *Squalus acanthias*. *Biol. Bull.* **62**, 17–22.
MacKay, M. E. (1931). The action of some hormones and hormone-like substances on the circulation in the skate. *Contrib. Can. Biol.* **7**, 17–29.
Mahn, R. (1898). Untersuchungen über das physiologische Verhalten des Schleiendarms. *Arch. Ges. Physiol.* **72**, 273–304.
Marcus, H. (1910). Über den Sympathicus. *Sitzbere Ges. Morphol. Physiol. Munich* **25**, 119–131.
Martinson, J. (1965). Vagal relaxation of the stomach. Experimental re-investigation of the concept of the transmission mechanism. *Acta Physiol. Scand.* **64**, 453–462.
Méhés, J., and Wolsky, A. (1932). Untersuchungen an der quergestreiften Muskulatur des Darmes der Schleie (*Tinca vulgaris*). *Arb. Ung. Biol. Forsch.-Inst.* **5**, 139–154.
Milokhin, A. A. (1958). Über eigene rezeptorische Neuronen des vegetativen Nervensystems. *Z. Mikroskop.-Anat. Forsch.* **63**, 497–517.
Milokhin, A. A. (1959a). Die afferente Innervation des Verdauungstraktes bei einigen niederen Wirbeltieren. I. Die afferente Innervation des Verdauungstraktes bei dem kaspischen Neunauge (*Caspiomyzon wagneri* Kessl.). *Z. Mikroskop.-Anat. Forsch.* **65**, 402–412.
Milokhin, A. A. (1959b). Über synaptische Verbindungen im Darmplexus der Cyclostomen. *Z. Mikroskop.-Anat. Forsch.* **66**, 45–52.
Mott, J. C. (1951). Some factors affecting the blood circulation in the common eel (*Anguilla anguilla*). *J. Physiol. (London)* **114**, 387–398.
Nicol, J. A. C. (1952). Autonomic nervous systems in lower chordates. *Biol. Rev.* **27**, 1–49.
Nicol, J. A. C. (1957). Observations on photophores and luminescence in the teleost *Porichthys*. *Quart. J. Microscop. Sci.* **98**, 179–188.
Olcott, C. (1931). Peristalsis in *Myxine*. *Directors Rept. Mt. Desert Island, Maine* pp. 43–45.
Östlund, E. (1954). The distribution of catecholamines in lower animals and their effect on the heart. *Acta Physiol. Scand.* **31**, Suppl. 112, 1–67.
Östlund, E., and Fänge, R. (1962). Vasodilation by adrenaline and noradrenaline and the effects of some other substances on perfused fish gills. *Comp. Biochem. Physiol.* **5**, 307–309.
Otorii, T. (1953). Pharmacology of the heart of *Entostephnus japonicus*. *Acta Med. Biol. (Niigata)* **1**, 51–59.
Parker, G. H. (1935). The electric stimulation of the chromatophore nerve-fibers in the dogfish. *Biol. Bull.* **68**, 1–3.
Parker, G. H. (1948). "Animal Colour Changes and their Neurohumours." Cambridge Univ. Press, London and New York.
Patterson, T. L., and Fair, E. (1933). The action of the vagus on the stomach-intestine of the hagfish. Comparative studies. VIII. *J. Cellular Comp. Physiol.* **3**, 113–119.
Peters, A. (1963). The peripheral nervous system. *In* "Biology of Myxine" (A. Brodal and R. Fänge, ed.), pp. 92–123. Oslo Univ. Press, Oslo.
Randall, D. J. (1968). Functional morphology of the heart in fishes. *Am. Zoologist* **8**, 179–189.

Randall, D. J., and Stevens, E. D. (1967). The role of adrenergic receptors in cardiovascular changes associated with exercise in salmon. *Comp. Biochem. Physiol.* **21**, 415–424.

Read, J. B., and Burnstock, G. (1968). Comparative histochemical studies of adrenergic nerves in the enteric plexuses of vertebrate large intestine. *Comp. Biochem. Physiol.* **27**, 505–517.

Satchell, G. H. (1962). Intrinsic vasomotion in the dogfish gill. *J. Exptl. Biol.* **39**, 503–512.

Voigt, R. (1939). Ein Verfahren zur Beobachtung vasomotorischer Wirkungen an der Froschlunge und einige damit erzielte Ergebnisse. *Arch. Exptl. Pathol. Pharmakol.* **192**, 179–188.

von Euler, U. S., and Fänge, R. (1961). Catecholamines in nerves and organs of *Myxine glutinosa*, *Squalus acanthias* and *Gadus callarias*. *J. Gen. Comp. Endocrinol.* **1**, 191–194.

von Euler, U. S., and Östlund, E. (1957). Effects of certain biologically occurring substances on the isolated intestine of fish. *Acta Physiol. Scand.* **38**, 364–372.

Wilber, C. G., and Sudak, F. N. (1960). Circulatory response of elasmobranchs to hemorrhage. *Anat. Record* **138**, 389.

Wild, G. (1903). Einige Mitteilungen über Fische und Fischerei in Heilbronn. Quoted in *Bronn's Klassen* **6**, (1), 665, (1924).

Yamauchi, A., and Burnstock, G. (1968). An electron microscopic study on the innervation of the trout heart. *J. Comp. Neurol.* **132**, 567–588.

Young, J. Z. (1931a). The pupillary mechanisms of the teleostean fish *Uranoscopus scaber*. *Proc. Roy. Soc.* **B107**, 464–485.

Young, J. Z. (1931b). On the autonomic nervous system of the teleostean fish *Uranoscopus scaber*. *Quart. J. Microscop. Sci.* **74**, 491–535.

Young, J. Z. (1933a). Comparative studies on the physiology of the iris. I. Selachians. *Proc. Roy. Soc.* **B112**, 228–241.

Young, J. Z. (1933b). Comparative studies on the physiology of the iris. II. *Uranoscopus* and *Lophius*. *Proc. Roy. Soc.* **B112**, 242–249.

Young, J. Z. (1933c). The autonomic nervous system of selachians. *Quart. J. Microscop. Sci.* **75**, 571–624.

Young, J. Z. (1935). The photoreceptors of lampreys. II. The functions of the pineal complex. *J. Exptl. Biol.* **12**, 254–270.

Young, J. Z. (1936). The innervation and reaction to drugs of the viscera of teleostean fish. *Proc. Roy. Soc.* **B120**, 303–318.

Zwaardemaker, H. (1924). Action du nerf vague et radio-activité. *Arch. Neerl. Physiol.* **9**, 213–228.

4

THE CIRCULATORY SYSTEM

D. J. RANDALL

I. Introduction 133
II. Blood Pressures in Arteries and Veins 135
III. The Heart 138
    A. Anatomy 138
    B. Electrical Properties 141
    C. Mechanical Properties 143
    D. Regulation of Cardiac Activity 149
IV. General Properties of the Cardiovascular System 158
    A. Cardiac Output 158
    B. Blood Volume 159
    C. Blood Distribution 160
    D. Skeletal Muscle Circulation 162
V. The Effects of Some Substances on the Circulatory System . . 165
    A. Acetylcholine 165
    B. Atropine 166
    C. Catecholamines 166
    D. Reserpine 167
    E. Eptatretin 167
    F. Histamine 167
    G. 5-Hydroxytryptamine (Serotonin) 167
    H. Renin and Angiotensin II 168
    I. Oxytocin and Vasopressin 168
    J. Urophysial Extracts 168
References 168

I. INTRODUCTION

Mott (1957) reviewed the literature on the cardiovascular system of fishes and noted the fragmentary nature of available information. Since Mott's review many of the techniques of mammalian circulatory physiology have been applied to fishes, and much more information is now available. There are, however, still some areas in which there is a marked paucity of data; for example, very little is known about the hemodynamics of blood flow in arteries and veins in fish.

Recently, Robb (1965) has reviewed the anatomy of the cardiovascular system and Johansen and Martin (1965) have discussed some comparative aspects of cardiovascular function in vertebrates. Randall (1968) and Johansen and Hanson (1968) have discussed the functional morphology of fish and dipnoan hearts, respectively. Hunn (1967) has compiled a bibliography on the blood chemistry of fishes.

Fishes, including the Agnatha and Dipnoi, as well as the Chondrichthyes, Osteichthyes, Choanichthyes, and teleosts, represent a diverse group including over 20,000 species, extending in space over a wide variety of environments, and in time over widely divergent evolutionary pathways. The requirements of an aquatic environment enforces limitations on various systems and maintains morphological and physiological similarities between the groups. These similarities, however, may be superficial, and there is a danger of assuming conformity in function simply because of a general structural resemblance.

There is an apparent conformity of structure in the circulatory systems of fishes, with the exception of cyclostomes and lungfishes, blood pumped by the heart passes through the gill and systemic circulation before returning to the heart. Variations in the general scheme are, first, the large blood volume, low blood pressure, and the development of accessory hearts in the hagfish, and, second, the presence of a pulmonary circulation and a partial separation of oxygenated and deoxygenated blood in the heart of lungfish, associated with the transition from an aquatic to an aerial environment. A number of air-breathing fish have a well-developed circulation to respiratory surfaces other than the gills. The blood supply to these accessory respiratory organs is in parallel with the systemic circulation but in series with the gill circulation.

Of the small number of species of fish used in studies of the circulation, only a few have been investigated extensively. These include hagfish, dogfish, skate, ratfish, Port Jackson shark, trout, salmon, carp, tench, cod, lingcod, eel, and lungfish. The data have been collected using a variety of techniques under a variety of conditions. In some instances the fish were not intact and were either out of water, restrained, or anesthetized; in other experiments the recordings were obtained from intact, unanesthetized, and relatively unrestrained fish. The circulatory system is of necessity very responsive to changes in the environment and often adjusts rapidly to an experimental situation. The use of telemetry and other techniques has permitted experimentation on the cardiovascular system with relatively little interaction between the animal and the experimental procedure. However, the type of experimental procedure used can affect the results obtained. Differences in results reported in the literature may, therefore, result from the techniques used or represent interspecific or intraspecific variations in function. Thus, in discussing the cardiovascular system of fishes, one must be aware, first that generalizations rest on studies of only a small number of fish, and, second, that variations in results may reflect either species differences or differences in the techniques used.

II. BLOOD PRESSURES IN ARTERIES AND VEINS

The branchial or systemic heart in fish is usually the main propulsive organ for circulating the blood. Contractions of the heart serve to convert chemical energy into mechanical energy in the form of pressure and flow. The dissipation of energy as blood flows through the vascular network results in a fall in blood pressure. The hydraulic resistance to flow is inversely proportional to the fourth power of the radius of a blood vessel; therefore, the major portion of the drop in pressure occurs across the smallest blood vessels.

In fish, blood from the heart passes first through the ventral aorta into the gill circulation and then via the dorsal aorta into the general body circulation. In the majority of fish, the capillary bed of the gills represents a considerable proportion of the total vascular resistance to flow, and it is in series with all other capillary networks. In the lungfish, blood passes from the heart through the branchial vessels into the systemic and pulmonary circulations. The pulmonary vascular bed is in series with the gill circulation but in parallel with the systemic circulation (Johansen and Hanson, 1968). In the lungfish *Neoceratodus forsteri* there are capillaries in all gill arches, but in other species of lungfish, *Protopterus aethiopicus* and *Lepidosiren paradoxa*, there are no capillaries in the first and second gill arches, and branchial vessels connect the ventral and dorsal aortae directly.

In all fish, except the hagfish, the highest blood pressures have been recorded from the ventricle during systole. The resistance to flow between ventricle and ventral aorta is low; therefore, when the intervening valves are open, pressures in the ventricle and ventral aorta do not differ from each other by more than 1 mm Hg. Arterial pressures reported by earlier workers (see Mott, 1957) are sometimes unreliable because the fish were either restrained, out of water, ventral side up, anesthetized, or not intact. Johansen (1962) recorded ventral aortic pressures of 30 mm Hg in the cod, *Gadus morhua;* his fish were restrained and ventral side up, but not anesthetized. Robertson *et al.* (1966) measured pressures of 82/50 mm Hg in the ventral aorta and 44/37 mm Hg in the dorsal aorta of the spring salmon. There was no water flow over the gills when records were taken, initiating a reflex bradycardia as indicated by the low heart rates in these fish. Blood pressures have been recorded in unrestrained trout (Randall *et al.*, 1965; Holeton and Randall, 1967a; Stevens and Randall, 1967a), salmon (Smith *et al.*, 1967; Davis, 1968), and carp (Garey, 1967). Hanson (1967) has recorded blood pressure in a number of unrestrained Chondrichthyes (skate, dogfish, and ratfish). Arterial and intraventicular pressures have been recorded in immobilized elasmobranchs by Sudak (1965a,b), Satchell (1961), and Satchell and Jones (1967). All values were in the range of 30–70 mm Hg but show variability within a species as well as between species. In general, arterial blood pressures appear to be lower in elasmobranchs than in teleosts.

During diastole, ventral aortic pressure falls, valves in the conus or bulbus close, and pressure declines as the blood leaves the aorta. The pulse pressure in the ventral aorta of fishes is between 10 and 30 mm Hg, increasing to values as high as 40 mm Hg during hypoxia (Holeton and Randall, 1967a).

The first appreciable drop in blood pressure occurs across the gills. Hanson (1967) recorded dorsal aortic diastolic pressures which were 15% in skate *Raja binoculata,* 20% in dogfish *Squalus suckleyi,* and 26% in ratfish *Hydrolagus colliei,* of the ventral aortic blood pressure. The respective reductions in systolic pressure were 25, 25, and 40% of the ventral aortic pressure. The recorded pressure drop in trout, *Salmo gairdneri* (Holeton and Randall, 1967a; Stevens and Randall, 1967a), carp, *Cyprinus carpio* (Garey, 1967), and lungfish (Johansen and Hanson, 1968) was between 40 and 50% of the respective ventral aortic pressure. Hypoxia probably increases the resistance to flow through the gills in the trout (Holeton and Randall, 1967a) and dogfish, *Squalus acanthias* (Satchell, 1961), whereas exercise (Stevens and Randall, 1967a,b) and catecholamines (Burger and Bradley, 1951) cause a decrease in resistance to blood flow through the gills.

The dorsal aorta of fishes is a long tube, running the length of the body, which distributes blood to the systemic circulation. It is much less elastic than the ventral aorta. Blood pressure and flow in the dorsal aorta of the cod, *Gadus morhua,* is pulsatile (Shelton *et al.*, 1969) even though blood flows through the capillaries of the gills before entering this vessel. Pulse pressures in the dorsal aorta are usually less than 50% of those in the ventral aorta.

The total peripheral resistance to blood flow probably decreases during swimming in the trout (Stevens and Randall, 1967a,b) but increases during hypoxia (Holeton and Randall, 1967a). The pulmonary circulation in the lungfish has a lower resistance to blood flow than the systemic circulation (Johansen and Hanson, 1968). When the lungfish, *Protopterus aethiopicus*, breathes, blood flow to the lungs increases, whereas during motor activity, systemic blood flow increases.

Venous pressures in fish range from slightly below atmospheric to about 10 mm Hg. Contractions of the heart within a noncompliant pericardium in elasmobranchs and lungfish create subatmospheric pressures in the venous system near the heart (Satchell and Jones, 1967; Johansen and Hanson, 1968), producing aspiratory forces which increase venous return to the heart. Thus blood flow in veins near the heart in these animals is determined by the pumping action of the heart and skeletal muscles and aspiratory forces resulting from contractions of the heart within a noncompliant pericardium. The relative importance of these factors will be determined by the size of the vascular resistance to flow between the heart and the vein in question and the magnitude of the subatmospheric pressure created in the venous system by contractions of the heart. Flow in the vena cava of the lungfish, *Protopterus aethiopicus*, is largely determined by aspiratory forces when the animal is resting and by the pumping activity of skeletal muscle during exercise. Flow in the pulmonary vein of lungfish is largely dependent in the pumping action of the heart, but aspiratory forces do produce small increases in blood flow (Johansen and Hanson, 1968).

Pressures in the cardiovascular system of hagfish, are very low and are maintained by the action of the branchial and several accessory hearts (Johansen, 1960; Chapman *et al.*, 1963; Jensen, 1965). The hearts work independently of each other and circulate blood to a specific area of the body. The branchial heart pumps blood from the body and liver into the ventral aorta. The portal heart pumps blood from the gut and anterior cardinal vein to the liver. The action of skeletal muscles on a cartilaginous plate in the caudal region propels blood forward from large subcutaneous sinuses. This cartilaginous plate, the associated skeletal muscles, and a pair of lateral sacs with valves to regulate the direction of blood flow constitute the caudal heart. The activity of the skeletal muscles of this heart is initiated by impulses from the central nervous system. Both the branchial and portal hearts are aneural (Chapman *et al.*, 1963). Systolic arterial pressures are of the order of only a few mm Hg, and are determined by the interaction of the activity of the various hearts. Chapman *et al.* (1963), working on the Pacific hagfish, *Eptatretus stoutii*, noted that the branchial heart often stops and occasionally the portal hearts contract while the branchial heart is inactive. Johansen (1960) recorded large oscillations in the dorsal aortic blood pressure in *Myxine glutinosa* that were associated with contractions of the gill musculature and suggested that the gills take an active part in the propulsion of blood. Chapman *et al.* (1963) did not observe any effect of gill contractions on blood flow in the Pacific hagfish.

The velocity of the pressure pulse through different parts of the circulatory system has not been measured in fish. Nothing is known about the reflection of pressure waves from arborizations in the circulation or the contribution of reflected waves to the pulse pressure at any point in the circulation.

III. THE HEART

A. Anatomy

The systemic or branchial heart of fishes is the main propulsive organ for circulating the blood. It consists of four chambers in series (Fig. 1). Venous blood enters the sinus venosus from the liver and the ducts of Cuvier, passes into the atrium, and is then pumped into the ventricle and finally into the ventral aorta, via either a conus (Chondrichthyes and Dipnoi) or bulbus (teleosts) arteriosus. All chambers with the exception of the bulbus are contractile, and a unidirectional flow through the heart is maintained by valves at the sinoatrial and atrioventricular junctions and at the junction of the ventricle and bulbus or conus. The contractile conus wall consists of a thin layer of cardiac muscle covering a fibrous elastic sheath, and it may have up to seven transverse rows of valves along its length. The bulbus, elastic and noncontractile, has only a single pair of valves guarding the exit of the ventricle.

The heart of the cyclostome is similar to that of other fish except that there is only a single duct of Cuvier in the adult (paired in other fish), and the conus is poorly developed. The exit of the ventricle is guarded by a pair of pocket valves, but there are no other valves in the conus. In *Petromyzon*, however, there are two longitudinal ridges similar to those found in the embryonic conus of other forms.

The heart of the lungfish (Johansen and Hanson, 1968) is somewhat different from the typical fish pattern described above. The heart lies behind the pectoral girdle; the sinus and atrium dorsal to the ventricle. The atrioventricular valves are absent and are replaced by an atrioventricular plug. There is a partial subdivision of the heart by the

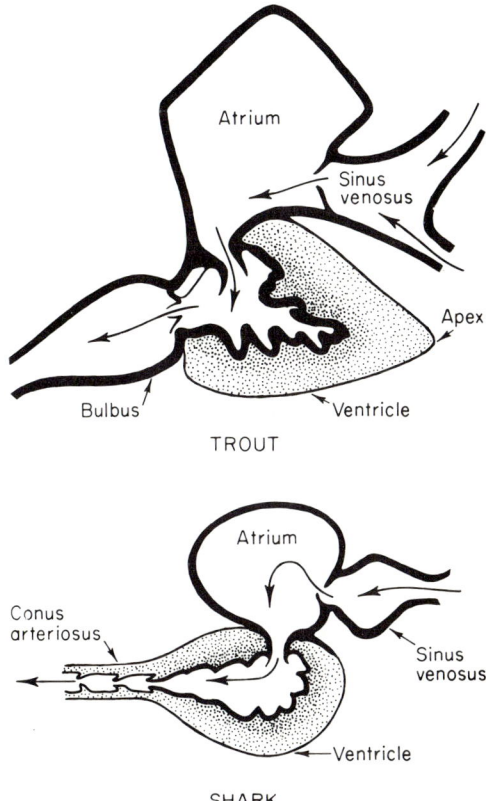

Fig. 1. Diagrams of the heart in a teleost (trout) and an elasmobranch (shark). From D. J. Randall (1968), *Am. Zoologist* 8, 179–189.

incomplete atrial and ventricular septa and the spiral folds in the conus arteriosus (bulbus chordis). The ventral aorta is either very short or nonexistent, and, as in amphibia, the branchial arteries arise from the anterior end of the conus arteriosus. The degree of subdivision of the heart varies between different genera of lungfish and is correlated with the prevalence of air breathing. *Neoceratodus*, the Australian lungfish, breathes air only when the oxygen content of the water is low, the African lungfish, *Protopterus aethiopicus*, and the South American lungfish, *Lepidosiren paradoxa*, are predominantly air breathers and spend several months each year on land. Division of the heart is least complete in *Neoceratodus* and best developed in *Lepidosiren* (Johansen and Hanson, 1968). The pulmonary vein and sinus venosus represent separate pathways for venous blood returning to the heart. Blood from the lungs is directed toward the left side of the atrium, whereas blood from the systemic circulation is directed into the right side. The partial separation of right and left atrium and ventricle aids the selective passage of blood from the lung into the left side of the ventricle. There are no conal valves but there are two folds or ridges, one large the other small, arising from opposite sides of the internal surface of the conus arteriosus (bulbus cordis). The free edges of the folds almost touch and act as a barrier which partially divides the conus into two outflow channels. The conus twists and turns, and in the anterior region the two ridges fuse to form a horizontal septum that completely divides the lumen into a dorsal and ventral channel. Blood from the left side of the ventricle is preferentially directed into the ventral channel and subsequently into the systemic circulation, whereas blood from the right side of the ventricle preferentially enters the dorsal channel which directs blood into the two posterior pairs of branchial arches, one pair of which directs blood to the pulmonary arteries (Johansen and Hanson, 1968).

The heart in all fishes, with the exception of that of the hagfish, has a coronary blood supply. The origin of the coronary supply is variable in lungfish; *Neoceratodus,* as in teleosts and elasmobranchs, has a coronary vessel derived from the anterior hypobranchial system. In *Lepidosiren* the coronaries arise from the second afferent artery (Foxon, 1955). This transition from an efferent to an afferent origin for the coronary supply is associated with the absence of a capillary network in the second gill arch. In addition, this gill arch presumably contains oxygenated blood returning via the left side of the heart from the pulmonary circulation.

The hagfish branchial heart has no coronary supply, and blood perfusing the heart is almost totally deoxygenated (Chapman *et al.,* 1963). The hagfish heart is only sporadically active and the pressures developed are small; hence, the work done by the branchial heart is also small compared with that done by other vertebrate hearts. It is possible that heart work in the hagfish is limited by the absence of a coronary supply. The extensive development of accessory hearts in hagfish may represent an attempt to spread the load of circulating the blood in the absence of a heart capable of prolonged activity at high work rates. However, the hagfish branchial heart is capable of producing systolic pressures of the order of 20–30 mm Hg. The heart obeys Starling's law and when venous pressures are increased peak intraventricular pressures rise to 30 mm Hg, and the work done by the heart increases. Such increases in heart work, if large, may be anaerobic. The hagfish "usually lies virtually immobile in lightless, frigid marine canyons for very long periods of time" (Chapman *et al.,* 1963, p. 430). When food appears it swings into violent and effective action. Nothing is known of the changes in blood flow that may occur during these periods of activity.

The heart of fish is composed of typical vertebrate cardiac muscle fibers (Yamauchi and Burnstock, 1968). The reported fiber diameters are smaller than those of mammals and may account for the relatively small number of intracellular potentials recorded from fish hearts. Jensen (1965) reported atrial and ventricular fiber diameters in the hagfish heart as 6.1 and 7.1 μ, respectively. These are about half the size of atrial and ventricular fibers in the dog.

B. Electrical Properties

Action potentials and electrocardiograms recorded from fish hearts are similar to those recorded from other vertebrate hearts (Fig. 2). Jensen (1965) recorded cardiac resting potentials of between 40 and 60 mV in the hagfish portal heart and in atrial and ventricular fibers in the branchial heart of three species of hagfish, an elasmobranch, and two marine teleosts. Other investigators have recorded resting potentials of about 70 mV in atrial strips of the skate heart (Seyama and Irisawa, 1967), in the goldfish atrium and ventricle (Kuriyama et al., 1960), and in the trout ventricle (Fig. 2). Jaeger (1965) recorded somewhat larger resting potentials from the atrium and ventricle of the roach, *Leuciscus rutilus* L.

Initiation of the heartbeat usually occurs in the sinoatrial node (Mott, 1957), but the actual site and extent of the pacemaker region varies from fish to fish. Jensen (1965) was able to record pacemaker potentials from a large number of fibers in the branchial (systemic) and portal heart of the hagfish. Those fibers having a pacemaker potential were situated in both the atrium and the ventricle of the branchial heart. McWilliam (1885) and Kisch (1948) have shown that in fish many regions of the heart are capable of pacemaker activity.

The conduction velocities of the wave of excitation spreading out from the pacemaker region are slower than in mammalian hearts (Randall, 1968). The rate of spread of excitation varies in different parts of the heart, and there are fast conducting pathways in the trout ventricle, *Salmo gairdneri*, that transmit the wave of excitation rapidly to the apex and cause it to contract before other parts of the ventricle closer to the atrium and bulbus are excited (see Fig. 1).

The electrocardiogram has been recorded from a number of fish (for references, see Randall, 1968; Mott, 1957) and consists of a P wave followed by a QRS complex and a T wave (Fig. 2). Oets (1950) recorded a V wave which preceded the P wave and was associated with a contraction of the sinus venosus in the eel. Chapman et al. (1963) recorded a typical electrocardiogram from the hagfish; however, an-

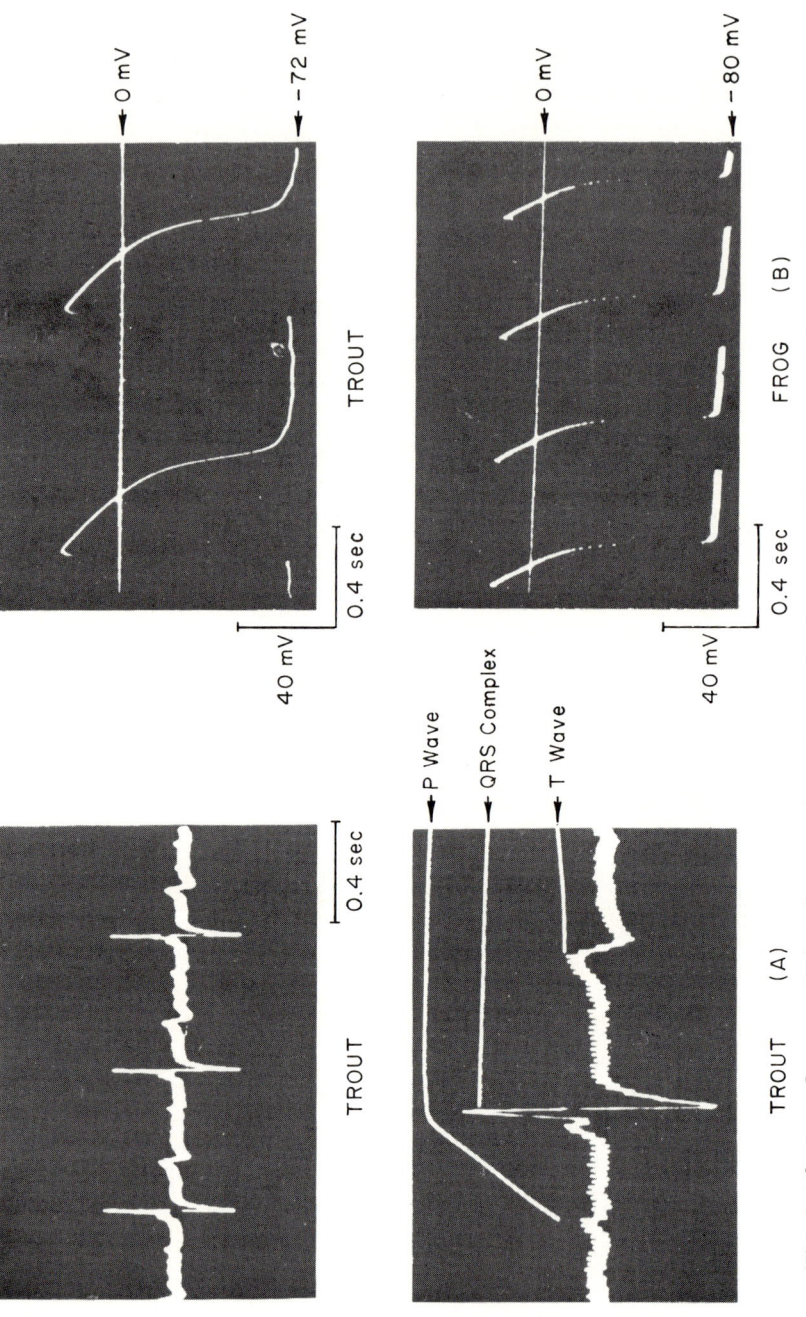

Fig. 2. Electrocardiograms (A) and ventricular resting and action potentials (B) recorded from the heart of the rainbow trout, *Salmo gairdneri*, and the frog, *Rana pipiens* (Bennion, 1968). From D. J. Randall (1968), *Am. Zoologist* **8**, 179–189.

other diphasic deflection resulting from activity of the portal heart appeared on the record. As might be expected it had no phase relationship with the electrocardiogram of the systemic (branchial) heart. Tebēcis (1967) recorded a wave complex (termed the "Bd complex") in the electrocardiogram of the Port Jackson shark. The Bd complex normally occurs between the QRS complex and the T wave and is associated with the depolarization of the conus. The exact position of the Bd complex in the ST segment is variable and may even coincide with the T wave. In a few preparations, Tebēcis recorded a Br complex between the T wave of one cardiac cycle and the P wave of the next cycle. The Br complex is associated with repolarization of the conus. Tebēcis (1967) calculated the conduction velocity of the conal wave of excitation from the time delay between two conal electrograms recorded from different points along the length of the conus and found that the velocity of conduction was very slow, of the order of 2–4 cm/sec. The conduction velocity of the excitation wave passing over the surface of the ventricle of the Port Jackson shark is much faster, in the range of 40–100 cm/sec.

C. Mechanical Properties

There are differences in function between the teleost and elasmobranch heart. The elasmobranch heart is contained in a noncompliant pericardium and has a contractile conus. The teleost heart is within a thin and compliant pericardium and has a noncontractile but very elastic bulbus.

In teleosts, the rate of atrial filling is determined by venous pressure. The sinus venosus appears to play little or no role in actively moving blood through the heart, but it forms part of an extensive venous reservoir serving the atrium. The volume of the atrium prior to systole is adequate to fill the ventricle, and the atrioventricular valves do not open until atrial systole (Randall, 1968). Therefore, contractions of the atrium in the presence of sinoatrial valves to prevent the reflux of blood into the sinus venosus serve to fill the ventricle, and, unlike the situation in mammals, there is no direct inflow of blood into the ventricle from the venous system during ventricular diastole.

Contractions of the ventricle result in large increases in intraventricular pressure (Fig. 3), the atrioventricular valves close and, in teleosts at least, the ventricle contracts isovolumetrically until pressures rise above that in the bulbus, when the valves at the exit of the ventricle open and blood leaves the ventricle. Pressures in the ventricle, bulbus,

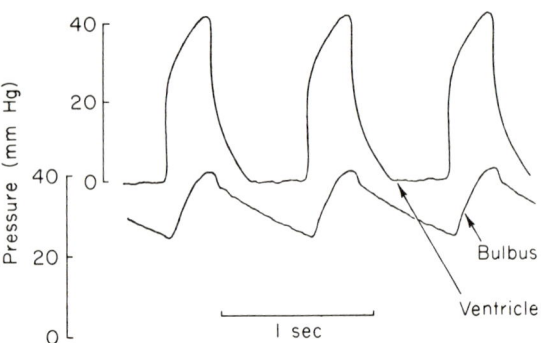

Fig. 3. Pressures recorded from the ventricle and bulbus of the lingcod, *Ophiodon elongatus* (Stevens *et al.*, 1969).

and ventral aorta rise simultaneously and do not differ by more than 1 mm Hg, indicating a low resistance to flow between the ventricle and ventral aorta (Fig. 3). Intraventricular pressure falls as the ventricle relaxes; the exit valves close and relaxation is isovolumetric until atrial systole opens the atrioventricular valves and fills the ventricle once more.

Blood flow in the ventral aorta and pressures in the ventricle and bulbus have been recorded in the lingcod, *Ophidon elongatus* (Fig. 4). Flow in the ventral aorta can be related to three phases of the cardiac cycle. There is a rapid increase in ventral aortic blood flow as the ventricle contracts, a rapid decrease in flow as the ventricle relaxes, and, finally, a slow decline in the rate of flow as the volume of the bulbus decreases. Closure of the valves between the ventricle and bulbus occurs at the transition from a rapid to a slow decline in ventral aortic blood flow rate. Flow in the ventral aorta only falls to zero if the heart rate is very low (Fig. 5). Usually the minimum flow rate of blood is between 9 and 12 ml/min in a 2–3-kg lingcod, *Ophiodon elongatus*. In Fig. 5A the mean blood flow is 15 ml/min, the period when blood flow is due to the elastic rebound of the bulbus is 44% of each cardiac cycle, and mean blood flow in the ventral aorta during this period is 10 ml/min. Blood flow due to the elastic rebound of the bulbus therefore represents about 29% of the total cardiac output. At lower rates the proportion of blood flowing during this period is undoubtedly larger. Thus the bulbus plays a significant role in maintaining blood flow in the ventral aorta during ventricular diastole. If the bulbus were absent, and the ventricle connected directly to a short ventral aorta leading to the afferent branchial arteries, pressures developed by the ventricle would need to be much larger than those in the presence of a bulbus if cardiac output

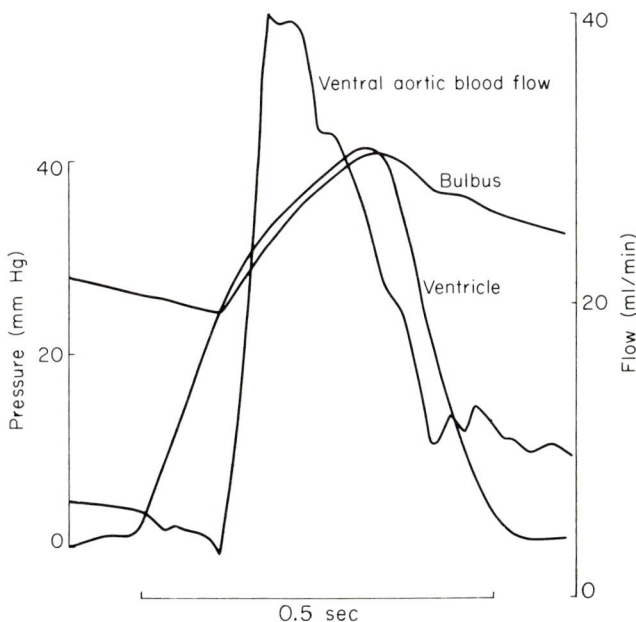

Fig. 4. Pressures in the ventricle and bulbus and blood flow in the ventral aorta replotted and superimposed upon one another to illustrate the relationships of the pressure and flow in the ventricle, bulbus, and ventral aorta of the lingcod, *Ophiodon elongatus* (Stevens *et al.*, 1969).

were to be maintained at the same level. The presence of a bulbus therefore dampens the oscillations in pressure and flow imposed upon the ventral aorta by contractions of the ventricle and also diminishes the work done by the ventricle in maintaining a given cardiac output.

In elasmobranchs, contractions of the heart within a noncompliant pericardium produce subatmospheric pressures within the pericardial cavity. These subatmospheric intrapericardial pressures have been recorded in lungfish (Johansen and Hanson, 1968), as well as in a number of elasmobranchs (Sudak, 1965a,b; Hanson, 1967; Johansen and Martin, 1965). The reduction in intrapericardial pressure produces aspiratory forces which increase venous return to the heart. This increase in venous return presumably augments cardiac output, and Hanson (1967) found that opening the pericardial cavity of the ratfish, *Hydrolagus colliei*, decreased cardiac output. However, Satchell and Jones (1967) were unable to detect any change in cardiac output when the pericardium of the Port Jackson shark, *Heterodontus portus jacksoni*, was opened.

The magnitude of the intrapericardial subatmospheric pressure depends on the rigidity of the pericardium and the magnitude and rate

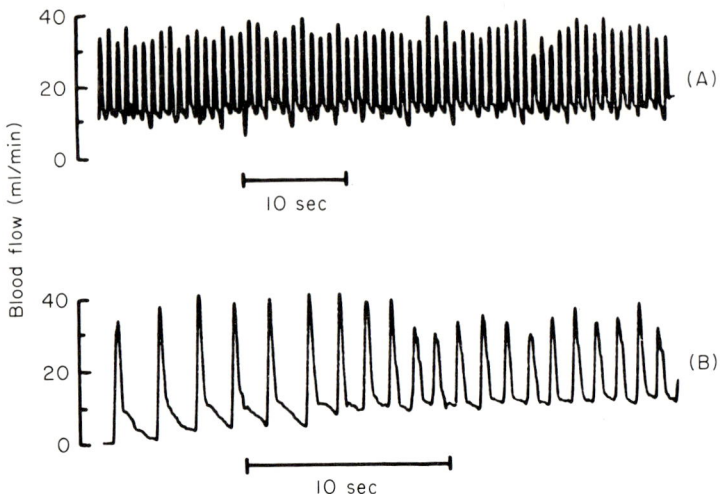

Fig. 5. The effect of changing heart rate on the stroke volume and velocity of blood flow in the ventral aorta of a 2.7-kg lingcod, *Ophiodon elongatus*. At low heart rates the blood flow falls to zero during diastole. At high heart rates the elastic rebound of the bulbus maintains blood flow in the ventral aorta during diastole. (A) Fish immobile in tank, heart rate normal. (B) Changes in blood flow associated with recovery from bradycardia induced by a short burst of muscular activity. From D. J. Randall (1968), *Am. Zoologist* **8**, 179–189.

of change in volume of the heart. The respiratory movements have also been shown to exert an effect on the magnitude of the intrapericardial subatmospheric pressure (Sudak, 1965b). Ventricular ejection leads to a rapid decrease in the volume of blood contained within the heart; hence, the major reduction in intrapericardial pressure occurs during ventricular systole. Conal systole causes a less marked reduction in intrapericardial pressure (Sudak, 1965a). The aspiratory forces produced by contractions of the ventricle and conus within a noncompliant pericardium increases venous return to the heart and augments atrial filling.

There are several sets of valves in the conus of the elasmobranch heart. Only the most distal valves are closed during ventricular diastole, maintaining the pressure difference between the ventral aorta and the conus and ventricle. The valves in the conus nearer the ventricle are patent, and the ventricular and conal chambers are interconnected. Pressures rise in both the ventricular and conal chambers during atrial systole, indicating that contractions of the atrium move blood into the conus as well as the ventricle (Hanson, 1967). As in the teleost, ventricular filling during diastole is solely dependent upon atrial systole.

As the proximal conal valves are open prior to ventricular systole and the ventricular lumen and conal chamber are interconnected, blood moves out of the ventricle into the conus at the onset of ventricular systole. Thus there can be no truly isovolumetric phase to ventricular systole in elasmobranchs.

The number of valves in the conus varies in different fish. The Port Jackson shark has three sets of valves with three semilunar valves in each set. Satchell and Jones (1967) refer to the valves as lower (proximal), middle, and upper (distal) conal valves. The lower valves are those near the junction of the conus and ventricle. The middle valves divide the conus into upper and lower conal chambers. A diagram by Hanson (1967) shows four rows of conal valves in the dogfish, *Squalus acanthias*, whereas a drawing by Sudak (1965a) of a freshly obtained heart from a smooth dogfish, *Mustelus canis*, shows only proximal and distal conal valves.

During ventricular diastole the proximal (lower and middle) conal valves of the Port Jackson shark are open (Satchell and Jones, 1967), the ventricular and conal chambers are interconnected, and the pressure difference between ventricle and conus and ventral aorta is maintained by the closed distal (upper) conal valves. The proximal and middle valves are not large in the Port Jackson shark, and when the conus is relaxed their free edges do not meet and the valves are incompetent. Conal systole narrows the conal lumen, and under these conditions the valves are competent. The conus swells at the onset of ventricular systole as blood moves from the ventricle into the conus. Pressures rise simultaneously in the ventricular and conal chambers and eventually exceed that in the ventral aorta; the distal (upper) conal valves open and blood flows into the ventral aorta. Conal systole begins before the end of ventricular systole. The contraction starts at the junction of the ventricle and conus and passes forward toward the ventral aorta. Conal systole aids the closure of first the proximal (lower) and then the middle conal valves as the ventricle relaxes. As the conus relaxes the proximal (lower) and middle valves become incompetent, there is a small backflow, and the distal (upper) conal valve closes. This valve, unlike the proximal and middle valves, is competent in the absence of conal systole; hence, it remains closed after the conus relaxes. Satchell and Jones (1967) recorded blood flow in the ventral aorta of the Port Jackson shark and found that flow occurs only during the period between the opening of the distal conal valves and the closing of the proximal conal valves. During this period the changes in flow and pressure in the ventral aorta can be ascribed to ventricular systole. The period between closure of the proximal and distal valves corresponds to conal systole, and during this period, Satchell and Jones did not record any blood flow in the ventral aorta. There is hardly any backflow in the ventral aorta as the valves close, but if conal systole is prevented, the proximal (lower) and middle valves are incompetent throughout the cardiac cycle and there is a large backflow of blood in the ventral aorta as the distal valves close. The conus is within a rigid pericardium, and the subatmospheric intrapericardial pressures tend to dilate the conus and increase the incompetence of the conal valves. If the intrapericardial pressure is experimentally reduced, the backflow in the ventral aorta increases as the distal valve closes. Thus, Satchell and Jones (1967) suggest that conal systole plays no part in ejecting blood into the ventral aorta but maintains the competence of the proximal and middle conal valves for a short period of time subsequent to the ventricular ejection of blood. The suggestions made by Satchell and Jones (1967) are not mutually exclusive. Conal systole could play a part in ejecting blood into the ventral aorta and also maintain the competence of the conal valves. Sudak (1965a), Johansen et al. (1966), and Hanson (1967) have reported data that either indicate or demonstrate enhanced flow in the ventral aorta that can be ascribed to conal systole. Thus, in most elasmobranchs studied, conal systole maintains flow in the ventral aorta during ventricular relaxation.

A series of conal valves are required because the conduction velocity of the wave of excitation over the conus is relatively slow (Tebēcis, 1967). The number of sets of valves present is presumably related to the length of the conus and the velocity of conduction. A series of valves is required if proximal portions of the conus relax, and the valves become patent, while systole is still occurring in more distal portions of the conus. Thus a series of valves permits the progressive contraction and relaxation of the conus and reduces any backflow resulting from conal systole.

Work done by the contracting ventricle and the efficiency of contraction have not been calculated for any fish heart. Work done is force times distance moved, which, in terms of the heart, is pressure times volume ejected. Pressure and flow continually change during ventricular systole. Work is done only during periods when blood is ejected from the heart. A measure of ventricular stroke work can be obtained by plotting ventricular pressure against ventricular volume (Fig. 6). The area of the enclosed loop is a measure of the stroke work of the ventricle (see Rushmer, 1961). Although ventricular pressures have been measured, there are no reported data in the literature of associated volume changes in the ventricle during systole, or indeed in any other part of the heart. In Fig. 6, changes in ventricular volume were calculated from increments of ventral aortic flow. This is not an accurate assessment of volume change in the ventricle because, during systole, the volume of

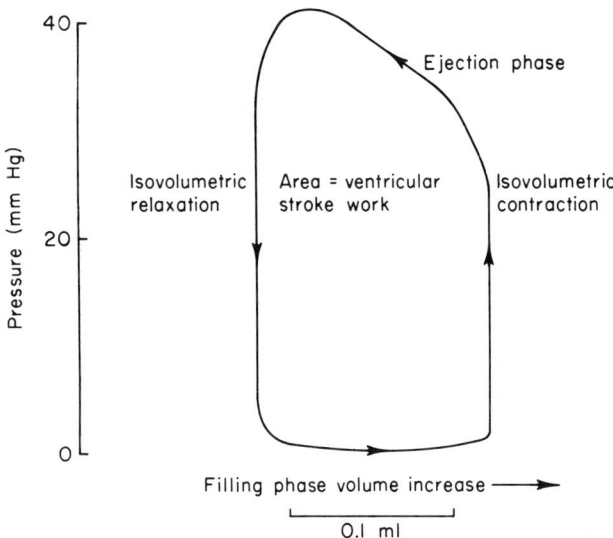

Fig. 6. Pressure and volume relationships in the ventricle of the lingcod, *Ophiodon elongatus*. From data reported in Fig. 4. Volume changes in the ventricle were calculated from increments of blood flow in the ventral aorta (see text).

the bulbus is increased, and the volume of blood flowing into the ventral aorta is less than that leaving the ventricle. During ventricular diastole, when the valves between the bulbus and ventricle close, the volume of the bulbus decreases and the elastic rebound of the bulbus maintains flow in the ventral aorta. Thus the estimates of ventricular volume changes reported in Fig. 6 are on the low side.

The construction of pressure–volume loops for the ventricle and bulbus or conus, under a variety of conditions, would enable one to assess the influence of the bulbus or conus on the pattern of blood flow in the ventral aorta.

D. Regulation of Cardiac Activity

1. Nervous Regulation

All fish hearts, with the exception of that of the hagfish, are innervated by a branch of the vagus nerve (Mott, 1957; Randall, 1968). Stimulation of the vagus in teleosts and elasmobranchs causes a marked bradycardia which can be blocked by atropine. The application of acetylcholine has been shown to slow the heart (Johansen *et al.*, 1966), and it has been assumed that the cardiac vagus contains cholinergic fibers (Randall, 1966). Some investigators, however, have been able to produce cardioacceleration by vagal stimulation (see chapter by Campbell, this volume), and recently adrenergic fibers have also been found to innervate the trout heart (Yamauchi and Burnstock, 1968; Gannon and Burnstock, 1969).

The lamprey heart is innervated by a branch of the vagus nerve. This nerve has not been stimulated directly, but stimulation of the medulla oblongata causes cardioacceleration (Augustinsson *et al.*, 1956). The effects of nervous stimulation are similar to those produced by the application of acetylcholine on the heart. Acetylcholine has a positive chronotropic (increased rate of contraction) but a negative inotropic (decreased force of contraction) effect on the lamprey heart (Falck *et al.*, 1966). The effects of acetylcholine cannot be explained in terms of the release of catecholamines from stores within the heart or of the action of these compounds on β-adrenergic receptors known to be present in the lamprey heart (Falck *et al.*, 1966). The effects of acetylcholine are not blocked by atropine, but they are blocked by curare. The lamprey heart, like most other fish, appears to receive a cholinergic nervous supply; the receptor sites, however, are different. The lamprey cholinergic receptors are blocked by curare, whereas those in teleosts and elasmobranchs are blocked by atropine.

A rise in dorsal aortic blood pressure in teleosts causes a bradycardia (Mott, 1957) which can be blocked by atropine (Randall and Stevens, 1967). Irving *et al.* (1935) have shown that increases in blood pressure increase the level of afferent activity in nerves innervating the branchial arches of the dogfish. Recently, Laurent (1967) has recorded chemoreceptor and baroreceptor discharges from nerves innervating the pseudobranch in teleosts. Dorsal aortic pressure varies little in the intact animal, and the pseudobranch may play a role in regulating heart rate, via the level of vagal tone, to maintain a constant dorsal aortic pressure.

The level of vagal tone to the heart varies between species and is altered, within a single species to a variable extent, by a large number of parameters. Bradycardia has been observed in response to light flashes, mechanical vibrations, salinity changes, atmospheric pressure changes, anoxia, removal from water, and touch (see Randall, 1968, for references). The exact nature of the response varies from fish to fish, some fish slow their heart in response to almost any stimulus, whereas others are more selective in terms of the stimuli required to evoke bradycardia. Most of the above parameters probably slow the heart by increasing the level of vagal tone. The functional significance of many of the

4. THE CIRCULATORY SYSTEM

heart rate changes is not clear. Usually only rate changes have been measured. A decrease in heart rate does not necessarily indicate a decreased cardiac output. In the trout, bradycardia which develops during hypoxia is offset by an increase in stroke volume in such a way that only minor changes in cardiac output are observed.

The level of vagal tone to the heart oscillates in phase with the breathing cycle in many fish and tends to inhibit the heart as the mouth opens (Randall, 1966). Satchell (1961) has suggested that this sinus arrhythmia serves to correlate maximum flows of blood and water at the respiratory surface.

Swimming in freshwater teleosts is associated with a decrease in vagal tone and an increase in heart rate. The magnitude of the increase is related to the level of vagal tone prior to activity. Large heart rate changes indicate a high level of vagal tone before exercise, and *vice versa*. When Chondrichthyes or some marine teleosts are forced to swim, exercise is associated with an increase in vagal tone and a bradycardia, which disappears if the activity is prolonged (Randall, 1968). The bradycardia may not occur during spontaneous swimming.

2. ANEURAL REGULATION

Starling's law of the heart is a basic aneural cardiac control mechanism. This law states that the energy of contraction is a function of the initial length of the muscle fiber. The greater the end diastolic volume, the greater will be the force of contraction and therefore the stroke volume. The venous input pressure to the heart usually determines end diastolic volume.

Catecholamines are known to increase both the rate (positive chronotropic effect) and myocardial contractility (positive inotropic effect) of mammalian hearts. The increased myocardial contractility results in a larger stroke volume being ejected from an unchanged end diastolic volume, i.e., there is greater systolic emptying of the heart. Both the positive inotropic and chronotropic responses result from the action of catecholamines on β-adrenergic receptor sites in the heart. Catecholamines, therefore, alter the nature of the Starling relationship between end diastolic volume and stroke volume by their action on β-adrenergic receptors in the heart.

The fish heart appears to obey Starling's law. The *in situ* aneural hagfish heart responds to increased filling by increasing its force of contraction (Chapman *et al.*, 1963). The isolated *in vitro* trout heart, *Salmo gairdneri*, also obeys Sartling's law (Bennion, 1968). As the input pressure is raised both stroke volume and apparent stroke work (calculated by

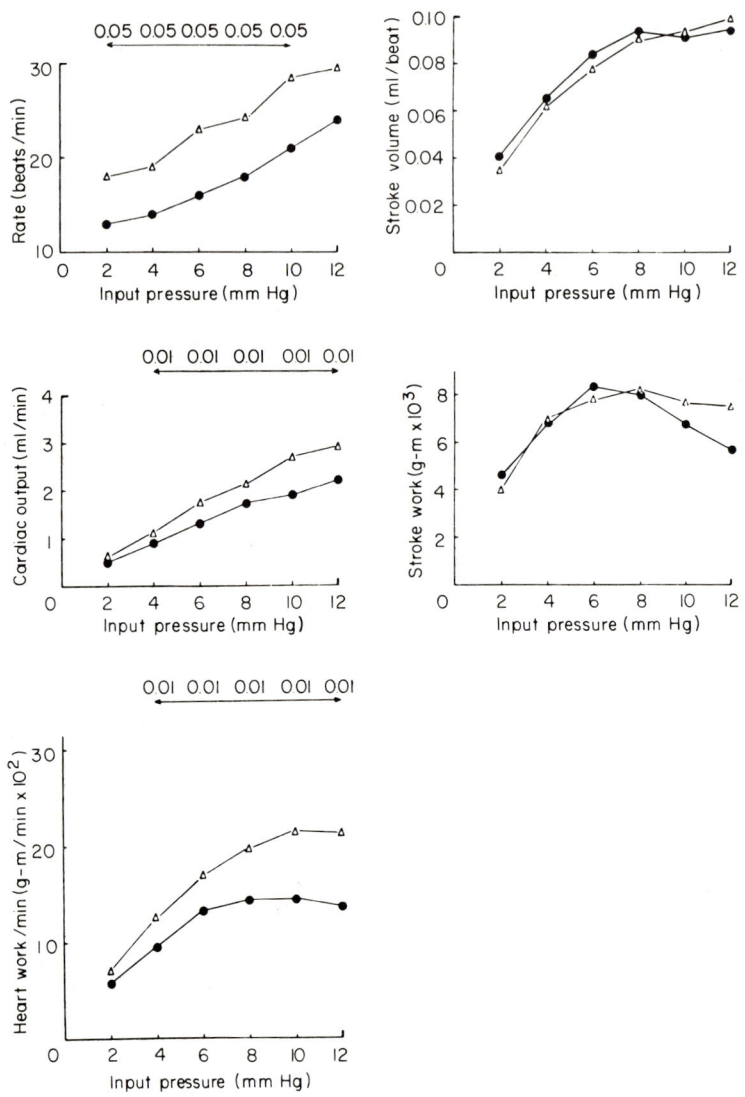

Fig. 7. Heart function curves showing the effect of increasing the input pressure on five cardiac variables at two epinephrine concentrations at 6°C: (●) 0.01 µg/ml epinephrine in saline; (△) 0.1 µg/ml epinephrine in saline. The experiments were carried out on the *in vitro* trout heart, *Salmo gairdneri*. The arrows above the graphs indicate the range over which the pairs of values are significantly different. Each point on each graph is a mean of seven values. From Bennion (1968).

multiplying stroke volume by the difference between the mean output and input pressures) increase (Figs. 7 and 8). Increasing input pressures also cause the rate to increase in the isolated heart. In all probability, this results from a direct effect of stretch on the pacemaker fibers causing changes in ionic conductance. Both the increase in rate and stroke volume contribute to a large increase in cardiac output as the input pressure is raised.

A rise in temperature increases heart rate both in the isolated preparation and in the intact animal. In the *in vitro* heart, the rate increases are offset by decreases in stroke volume as temperature increases, and cardiac output remains constant. Decreases in temperature, therefore, appear to have a positive inotropic effect on the heart.

Epinephrine alters the relationship between end diastolic volume and stroke volume in the isolated trout heart (Figs. 7 and 8). In all cases, epinephrine increases cardiac output and apparent heart work (calculated by multiplying cardiac output by the difference between mean output and input pressures to the heart; in all of these experiments the heart had to pump saline through a fixed resistance to a height above the input level). The effect of epinephrine on stroke volume and heart rate is temperature dependent. At low temperature (6°C), increased epinephrine levels raise heart rate but have little effect on stroke volume (Fig. 7), whereas at high temperatures (15°C), stroke volume is increased but heart rate is decreased (Fig. 8).

In the *in vitro* trout heart, stroke volume and heart rate are affected by the input pressure, temperature, and epinephrine. At low temperatures (6°C), the positive inotropic effect of a decrease in temperature masks the positive inotropic effect of epinephrine, and only rate increases are observed when epinephrine levels are raised. At high temperatures (15°C), however, the increased inotropic effect of epinephrine is seen and stroke volume increases when epinephrine levels are raised. At this temperature (Fig. 8) the initial rate is high. Systolic emptying is increased so greatly when epinephrine levels are raised that a longer interval is needed to fill the heart again. Heart rate appears to be sensitive to stretch of the heart as well as to temperature and epinephrine; hence, as the heart requires more time to fill at 15°C when epinephrine levels are increased, the stretch component determining rate is reduced and heart rate falls (Bennion, 1968). Thus epinephrine exerts both a chronotropic and inotropic effect on the *in vitro* trout heart; the positive chronotropic response predominates at low temperatures, whereas the positive inotropic response predominates at high temperatures. Epinephrine is known to have several metabolic effects on the mammalian heart. It promotes glycogen breakdown by increasing the level of cyclic AMP, which leads to an increase

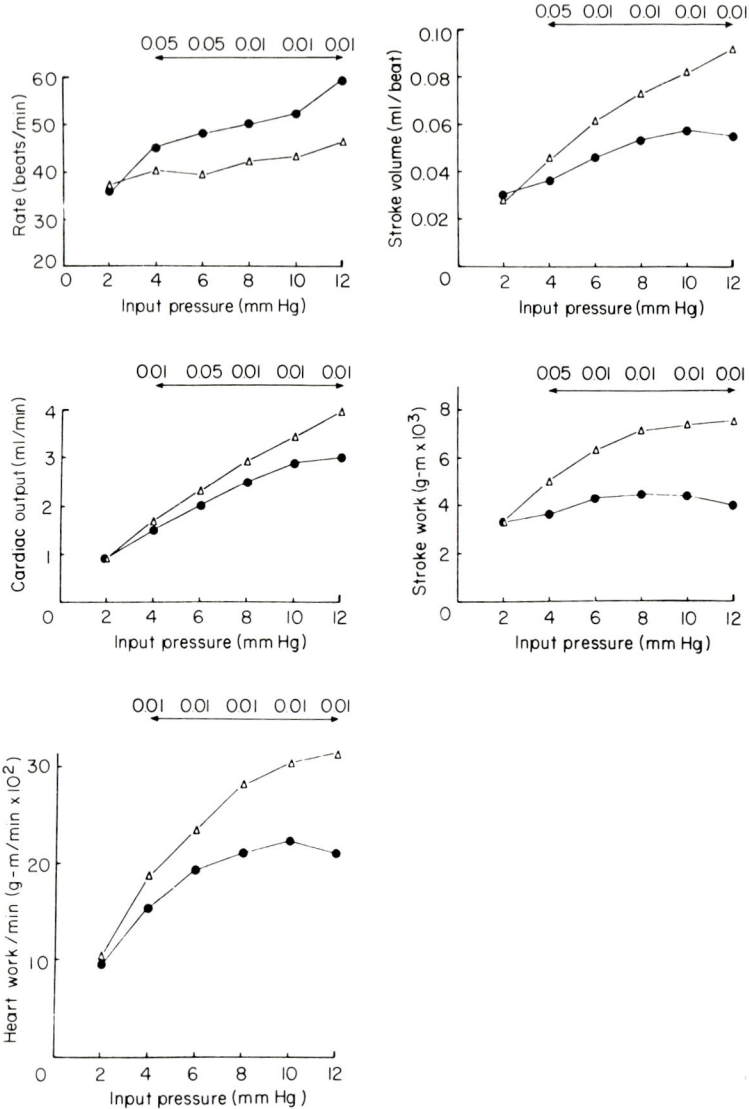

Fig. 8. Heart function curves showing the effect of increasing the input pressure on five cardiac variables at two epinephrine concentrations at 15°C: (●) 0.01 µg/ml epinephrine in saline; (△) 0.1 µg/ml epinephrine in saline. The experiments were carried out on the *in vitro* trout heart, *Salmo gairdneri*. The arrows above the graphs indicate the range over which the pairs of values are significantly different. The probability that the difference results from chance alone for each pressure level is given above each arrow. Each point on each graph is a mean of ten values. From Bennion (1968).

in the active form of the enzyme glycogen phosphorylase. Cyclic AMP is known to sensitize actomyosin toward changes in calcium concentration (Mommaerts et al., 1963). Thus cyclic AMP plays a role in both the inotropic and glycogenolytic effects of epinephrine. Both the glycogenolytic and inotropic effects are blocked by β-adrenergic blocking agents in mammals (Sutherland and Rall, 1960). The β-adrenergic blocking agent, Inderal, blocks the response to epinephrine in the isolated trout heart and causes a general myocardial depression similar to that seen in mammals (Nickerson, 1964). The α-adrenergic blocking agent, phenoxybenzamine, had no effect on the response of the isolated trout heart (Bennion, 1968).

High levels of catecholamines have been found in cyclostome hearts (Bloom et al., 1961). Much lower levels of adrenaline and noradrenaline have been demonstrated in the hearts of other fishes (von Euler and Fänge, 1961), and adrenaline usually represents more than 50% of the total catecholamines present. Catecholamines have been shown to increase the rate and force of contraction of both teleost and elasmobranch hearts (Östlund, 1954; Falck et al., 1966), and β-adrenergic receptors have been demonstrated in the isolated plaice and trout heart. Surprisingly, catecholamines have little effect on the hagfish heart; however, Chapman et al. (1963) did find that adrenaline restored loss of vigor caused by the application of reserpine in the hagfish heart.

In the intact animal the activity of the heart is determined by the sum total of many factors that influence the heart. Johansen (1962) observed an increase in stroke volume when venous return was experimentally increased in the cod, *Gadus morhua;* no rate changes were observed. Labat et al. (1961) observed an increase in heart rate in the catfish when the temperature was increased or if saline was injected into the hepatic vein, increasing venous return to the heart. Laffont and Labat (1966) found that below 8°C epinephrine injections caused a decrease in heart rate. The bradycardia recorded by Laffont and Labat may have been the result of an increase in vagal tone, resulting from a change in blood pressure caused by the action of epinephrine on the circulatory system (Randall and Stevens, 1967).

When trout swim there are large changes in cardiac output which are the result of small increases in heart rate and large increases in stroke volume (Stevens and Randall, 1967b); also the total peripheral resistance to flow decreases. There is no vagal tone to the heart of trout during rest or exercise as long as the fish is in well-aerated water; therefore, any changes that occur during activity cannot be explained in terms of a decrease in vagal tone to the heart. The level of circulating catecholamines increases during exercise in trout (Nakano and Tomlinson, 1967). There are β-adrenergic receptors in the heart (Bennion,

1968) and adrenergic receptors in the gills and elsewhere in the circulation (Randall and Stevens, 1967). The effect of epinephrine is to decrease the resistance to flow in the gill circulation. Thus the changes in cardiac output occurring during exercise may be caused, at least in part, by the action of catecholamines on β receptors in the heart and on other adrenergic receptors in the peripheral circulation. One would suspect, from data on the mammalian circulatory system, that other factors, such as the production of metabolites, play a role in decreasing peripheral resistance during exercise in fish.

Exercise in the lingcod, *Ophiodon elongatus*, is accompanied by stroke volume and cardiac output increases only when the fish is atropinized (Stevens *et al.*, 1969). Exercise in the nonatropinized fish is accompanied by an increase in vagal tone and a bradycardia, which decreases cardiac output. Possibly this is a response to the stimulus which provokes exercise rather than to exercise itself. Lingcod usually remain motionless at the bottom of the tank and must be prodded or startled to produce short bursts of swimming. An increase in vagal tone may not be a physiological response to spontaneous swimming in this animal. Hanson (1967) found that spontaneous activity in a number of Chondrichthyes was accompanied by an increase in cardiac output; if, however, the animal was disturbed and then swam away an initial bradycardia was observed.

An increase in temperature acts directly on the heart, increasing the intrinsic rate of the pacemaker cells. In the intact lingcod, *Ophiodon elongatus*, stroke volume remains constant over a wide range of temperatures, and a rise in temperature causes an increase in cardiac output owing to an increase in heart rate (Fig. 9). The *in vitro* heart of the trout behaves in a different way (see p. 153); a rise in temperature increases heart rate but decreases stroke volume, and cardiac output remains constant over a wide range of temperatures. In the intact animal, temperature changes affect the whole animal, whereas in the *in vitro* preparation only the response of the heart is recorded. Many factors, including the magnitude of the peripheral resistance to blood flow, can affect stroke volume. There are some indications that the peripheral resistance to blood flow decreases as temperature increases (Davis, 1968), and this would tend to maintain stroke volume in the face of a negative inotropic effect on the heart caused by a temperature increase.

The balance of evidence indicates that changes in cardiac output in fish are associated with changes in stroke volume with some adjustment of heart rate (Randall, 1968). The rate changes are mediated via cholinergic nerve innervating the heart, changes in temperature, and changes in the level catecholamines. Stroke volume changes occur in response

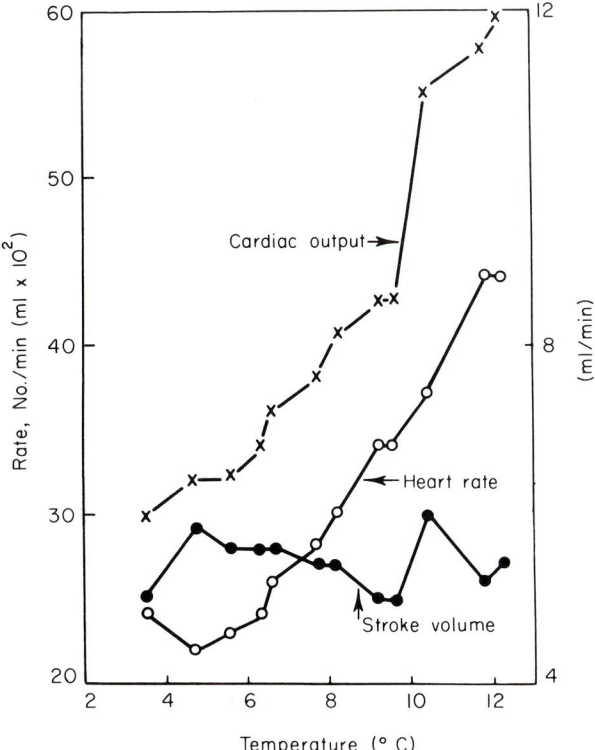

Fig. 9. The effect of temperature on heart rate, stroke volume, and cardiac output in the lingcod, *Ophiodon elongatus* (Stevens et al., 1969). From D. J. Randall (1968), Am. Zoologist **8**, 179–189.

to increased levels of catecholamines and increases in venous return to the heart.

3. The Regulation of Venous Return

An increase in venous return will increase end diastolic volume and increase stroke volume as described by Starling's law. An increase in venous return can be achieved in a number of ways. A decrease in total peripheral resistance will increase the *vis a tergo* effect of the heart on venous return. In Chondrichthyes, increased cardiac activity will lower the negative pressure within the rigid pericardium and raise the aspiratory forces moving venous blood back to the heart. Mobilization of blood from storage organs and venous reservoirs will also aid venous return. The pumping activity of skeletal muscle in the presence of suitably placed valves is another powerful mechanism for increasing venous return during exercise.

Hagfish possess several accessory hearts, namely, the portal, cardinal, and caudal hearts (Chapman *et al.*, 1963). The action of these hearts plus the pumping action of contractions of the gill musculature (Johansen, 1960) aid the circulation of the blood and regulate venous return to the systemic (branchial) heart.

The effectiveness of skeletal muscle contractions in increasing venous return is dependent on the presence of valves directing blood flow toward the heart. Valves have been demonstrated in the segmental arteries and veins in the caudal regions of the Port Jackson shark, *Heterodontus portus jacksoni*. These valves are at the entrance and exit of the myotomal muscles and are so arranged that the contracting muscles pump blood from the arterial to the venous side of the circulation, thus increasing venous return to the heart (Satchell, 1965). Venous valves have been found in teleosts (Dornesco and Santa, 1963). Blood flow in the dorsal aorta of some teleosts may be augmented by the presence of a stiff ligament (de Kock and Symmons, 1959) extending into and partially dividing the lumen of the dorsal aorta. The ligament extends from the basioccipital to the caudal region and undulations of the body, pressing the walls of the dorsal aorta against this stiff ligament, may propel blood in a caudal direction.

Mobilization of blood from storage organs or venous reservoirs may aid venous return and increase cardiac output. Exercise in trout is accompanied by a decrease in splenic volume (Stevens, 1968). Hepatic vein sphincters in elasmobranchs regulate the volume and transit time for blood in the liver (Johansen and Hanson, 1967), and their relaxation would permit the mobilization of blood stored in the liver at times of increased demand on cardiac output. The hepatic vein sphincters relax in response to the topical application of adrenaline and contract in response to acetylcholine. Lowering the dorsal fin in the ratfish decreases the amount of blood in the dorsal fin sinus and may be used to increase venous return in this fish (Hanson, 1967).

IV. GENERAL PROPERTIES OF THE CARDIOVASCULAR SYSTEM

A. Cardiac Output

Three methods have been used to determine cardiac output in fish. These are the application of the Fick principle (Mott, 1957; Holeton and Randall, 1967b; Stevens and Randall, 1967b), the dye dilution method (Murdaugh *et al.*, 1965), and the direct measurement of the velocity of blood flow in the ventral aorta using either an electromagnetic (Satchell and Jones, 1967) or a Doppler ultrasonic blood flowmeter system (Johansen *et al.*, 1966).

Recorded values of cardiac output in a number of elasmobranchs are in the region of 25 ml/kg/min. The fish investigated include dogfish, *Squalas acanthias* (Murdaugh *et al.*, 1965; Robin *et al.*, 1964) and *Scyliorhinus stellaris* (Piiper and Schumann, 1967). Hanson (1967) reported cardiac outputs for the dogfish, *Squalus suckleyi*, of between 9 and 23 ml/kg/min. The cardiac output of the ratfish, *Hydrolagus colliei*, is about 21 ml/kg/min. Cardiac output per kilogram falls with increasing weight in the skate, *Raja binoculata*.

By comparison, measured values for cardiac output in teleosts, are variable, ranging from 5 to 100 ml/kg/min (Goldstein *et al.*, 1964; Johansen, 1962; Garey, 1967; Holeton and Randall, 1967b; Stevens and Randall, 1967b; Randall, 1968). Most values, however, fall within the range of 15–30 ml/kg/min.

Peak ejection velocities of blood flow in the ventral aorta of the Californian Horn shark (*Heterodontus francisci*), skate (*Raja binoculata*), and dogfish (*Squalus suckleyi*) are between 8 and 20 cm/sec (Johansen *et al.*, 1966; Hanson, 1967) which occur 120 msec after the aortic valves open in the Horn shark and 450 msec in the skate; acceleration of the blood reaches a maximum of 120 cm/sec^2 in the Horn shark.

B. Blood Volume

The blood volume of a fish consists of that portion of the extracellular space contained within the cardiovascular system, plus the volume of erythrocytes, leukocytes, and platelets in the blood. It consists of a plasma volume and the volume of cells within the blood, which when expressed as a percentage of the total blood volume is the hematocrit. Both plasma volume and hematocrit vary considerably in fish.

Mott (1957) has reviewed the earlier literature on blood volume measurements in fish. Holmes and Donaldson (1969) have recently completed an exhaustive review of the subject which will only be discussed briefly here. Thorson (1958, 1959, 1961, 1962) has estimated blood volume from measurements of plasma volume and hematocrit in a large number of fish. Agnatha have very large blood volumes in excess of any other fish. The blood volume of Chondrichthyes is about 6.6% body weight, whereas that of Chondrostei, Holostei, and teleosts (both marine and freshwater species) is approximately 3% body weight. Thorson (1961) pointed out that the evolutionary trend in fish appeared to be toward smaller blood volumes, the significance of which is not clear. Schiffman and Fromm (1959), Conte et al. (1963), and Ronald et al. (1964) also reported blood volumes of approximately 3% body weight in trout and cod. Most investigators have used methods developed for determining blood volume in mammals. Many of these methods depend on the injection of a dye and the measurement of its subsequent dilution in the blood. It is important that the dye be evenly distributed throughout the vascular compartment before blood volume is determined. Most investigators, in accord with mammalian practices, have allowed 30–40 min for equilibration of the dye. Smith (1966) found that Evans blue (T-1824) was not evenly distributed in the blood of salmon until at least an hour had elapsed, and reported blood volumes of 6% body weight for a number of salmonids, approximately twice that previously reported for any other teleost. This very slow mixing of the dye indicates the possibility of a slow turnover of blood in some parts of the vascular system.

Violent bursts of swimming result in small increases in weight in the trout. The concentrations of hemoglobin and plasma proteins oscillate in phase with one another, indicating changes in the blood water concentration during exercise. There is an initial hemodilution followed by a hemoconcentration as exercise continues (Stevens, 1968). The weight gain and initial hemodilution could be the result of a net influx of water into the blood across the gills, followed by an osmotic shift of water into the muscles, causing the hemoconcentration. Exercise, or any form of disturbance, increases blood catecholamine levels in salmonids (Nakano and Tomlinson, 1967; Randall and Stevens, 1967), which in turn alter the pattern of blood flow through the gills and increase the rate of ion, water, and gas exchange (Baumgarten et al., 1969). Artificially raising the blood adrenaline levels in trout results in the fish gaining weight. Thus any form of disturbance, which causes a rise in blood catecholamine levels in freshwater fish, may cause a hemodilution resulting from a net influx of water across the gills.

C. Blood Distribution

Circulation time can be calculated by dividing blood volume by cardiac output per minute. If cardiac output is 25 ml/kg/min and blood volume is 5% body weight, the calculated circulation time will be 2 min. Mott (1957) reported that blood took 5.6 ± 1.9 sec to pass through the gills and appeared in the visceral veins 20 sec after leaving the heart. The observed circulation time is much more rapid than the calculated value. The estimated time however is only an average value, and the circulation time through different pathways must vary.

Stevens (1968) found that the blood was unevenly distributed to different parts of the body of the rainbow trout, *Salmo gairdneri* (Table I). White myotomal muscle, constituting 66% of the total weight of the fish, contains only 15.8% of the total blood volume (assuming a blood volume of 5% body weight). Red myotomal muscle contains between two and three times as much blood per unit weight as white muscle, which is in approximate agreement with the ratio of the number of capillaries in the two types of muscle. Combined, the red and white myotomal muscle contains about 20% of the blood and represents two-thirds of the body weight. The remaining one-third of the body therefore contains 80% of the total blood volume. If one assumes the blood volume of the trout studied by Stevens to be 5% body weight, 4 ml of blood are contained in 33% of the body weight in a 100-g fish. The blood volume in parts of the body other than the myotomes is therefore about 12% tissue weight.

The apparent distribution of blood does not change as a result of exercise in the trout, except for a slight decrease in spleen blood volume and a rather surprising increase in gut blood volume (Stevens, 1968). Small changes in the diameter of vessels cause large changes in vascular

Table I

The Distribution of Blood to Various Tissues in Rainbow Trout, *Salmo gairdneri*, Assuming a Blood Volume of 5% Body Weight[a]

Tissue	A (Body weight, %)	B (Blood volume in tissue, %)	A/B
White muscle	66	15.8	0.24
Red muscle	1.0	6	6.0
Heart	0.2	2	10.0
Gills	3.9	7.6	19.0
Gut	5.1	2.4	0.47
Liver	1.4	4.0	2.9
Spleen	0.3	1.4	4.7
Blood in arteries, veins, heart, and kidney	3.0	60.0	20.0
Remainder	19.1	0.9	0.005
Total	100.0	100.0	

[a] After Stevens, 1968.

resistance and blood flow but only small changes in the volume of blood in the vessels. Stevens measured only the instantaneous blood volume in various tissues. The procedure he adopted did not enable him to detect changes in blood flow. Very little is known about the relative changes in flow to various organs during exercise in fish.

D. Skeletal Muscle Circulation

1. Exercise

There are two types of muscle fiber in the myotomes of fishes which can be recognized by their color. Red or dark fibers form thin, lateral superficial sheets just under the skin, white fibers make up the rest of the underlying muscle mass. The two types of muscle fiber form discrete motor systems, receiving separate nervous innervation and having different enzymic distributions, mitochondrial content, and fiber diameter. White muscle is used in burst swimming of short duration, and energy is supplied by anaerobic glycolysis (Bone, 1966). During burst swimming there is a rapid depletion of muscle glycogen and a large lactate production. The lactate diffuses slowly into the blood and up to 12 hr postexercise recovery is necessary before the low preexercise lactate levels are restored (Stevens and Black, 1966). White muscle fatigues rapidly, and the effects of burst activity on glycogen depletion and lactate production are cumulative during the recovery period.

White myotomal muscle, constituting the major portion of the body, is only used in burst activity of short duration. Fish, therefore, drag two-thirds of their body around simply to effect escape reactions and various other burst responses. The fish is streamlined and neutrally bouyant and the cost of moving this volume of muscle is probably small, especially if the maintainance and circulation of the muscle is minimized. The sparse vascularization limits the oxygen supply and during activity the muscle must operate anaerobically. The poor vascularization will also contribute to the slow release of lactate from the muscle and so minimize the effects of a massive lactate production on the rest of the body.

Red myotomal muscle is used in sustained swimming and is almost impossible to fatigue (Bone, 1966). There are no changes in red muscle glycogen content during activity, and energy is supplied by the oxidation of fats. The blood volume and number of capillaries per unit weight are three times that in white muscle (Stevens, 1968). Scombroids and certain elasmobranchs, in addition to the superficial red muscles, have a deeper lying red portion of the myotome, correlated with an ability to sustain high cruising speeds and muscle temperatures that are above ambient.

Nothing is known about the regulation of blood flow to the red and white muscle fibers during exercise. Contractions of skeletal muscle probably create a *vis a tergo* force driving blood back to the heart. Satchell (1965) demonstrated that an isolated trunk preparation of the Port Jackson shark (*Heterodontus portusjacksoni*) was able to pump and increase the flow of a perfusate during flection of the trunk. Arteries in the postpelvic region of the trunk arising from the dorsal aorta have valves which prevent the reflux of blood into the dorsal aorta during muscle contraction. Veins entering the caudal vein in the postpelvic region have valves which prevent reflux of blood into the segmental vessels. The caudal vein is encased in bone, and the contracting muscles squeeze all vessels in that region except the caudal vein propelling blood from the arterial to venous side of the circulation. Flow through the trunk preparation was somewhat higher after electrical stimulation than before, indicating that some other factors also play a role in increasing muscle blood flow during exercise.

2. Temperature Regulation

The muscles of fish are cooled by the blood, loss of heat through the general body surface is insignificant (Carey and Teal, 1969a). Metabolic heat, which warms the blood, is quickly lost at the gill surface, where blood comes into close proximity to the water. Water has a high specific heat, and thermal diffusion is ten times more rapid than gaseous diffusion; thus, thermal equilibrium between blood and water must occur very rapidly in the gills. Tuna and lamnid sharks, however, can maintain muscle temperatures well above ambient. This is particularly true of the bluefin tuna, *Thunnus thynnus*, which can control its muscle temperature so that the warmest part of the muscle mass varies only 6°C over a 20°C range of water temperatures (Carey and Teal, 1969a). The red muscle, used in sustained swimming, is the warmest part of the whole muscle mass (Fig. 10). This ability to thermoregulate is related to the presence of a rete mirabile formed by arteries and veins supplying the red muscles (Carey and Teal, 1966) and to the encapsulation of the red muscle by the white muscle. The retia of bluefin tuna are formed from arteries and veins arising from a pair of cutaneous vessels. The segmental arteries arising from the dorsal aorta are of secondary importance in the bluefin tuna, unlike skipjack and yellowfin tuna, which have central as well as lateral retia. The retia act as countercurrent heat

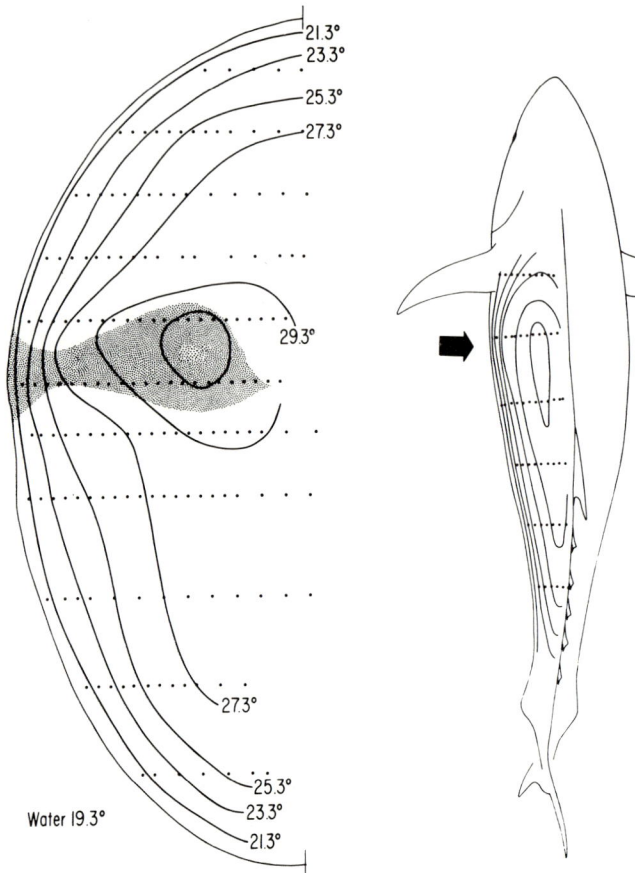

Fig. 10. Temperature distribution in cross section and frontal section of a bluefin tuna. Temperatures were measured with thermistor probes at positions indicated by the dots. Heavy curves are isotherms plotted at 2°C intervals. Dark (red) muscle is indicated by stippling. From F. G. Carey and J. M. Teal (1969a), *Comp. Biochem. Physiol.* **28**, 205–213.

exchangers in such a way that venous blood warms arterial blood entering the muscle and reduces heat loss. The retia, therefore, form a thermal barrier retaining metabolic heat in the tissues and preventing its loss in the gills (Carey and Teal, 1966).

The size of the rete must be regulated to conserve heat but allow the free passage of gases in and out of the muscle. The red muscles operate aerobically, and the design of the rete must take advantage of the differences between thermal and gaseous diffusion. Because thermal diffusion is rapid, a relatively small rete is required to allow thermal equilibration of arterial and venous blood; gases diffuse much more slowly, and so there is probably only an insignificant exchange of gases across this rete. Thermal and gaseous diffusion coefficients across the rete, however, have not been measured.

The ability to thermoregulate varies. Large bluefin tuna appear to have a greater capacity to thermoregulate than small fish. Skipjack and yellowfin tuna can maintain an elevated muscle temperature, but it is only a fixed value above ambient over a wide range of temperatures. Large bluefin tuna can maintain a red muscle temperature of between 26° and 32°C, while the ambient temperature varies between 6° and 30°C. At low water temperatures the muscle is 20°C above ambient, whereas at water temperatures of 30°C the muscle is only 2°C above ambient. This implies regulation of muscle temperature, which could be achieved by altering the functional size of the rete and/or by altering blood flow through the rete.

Lamnid sharks have a similar rete structure to bluefin tuna (Carey and Teal, 1969b). Both groups can thermoregulate and are able to swim swiftly, at a sustained speed, as a result of the extra power available from warm muscles. The bluefin tuna exists over a wide range of temperatures and migrates long distances across several temperature zones. Carey and Teal (1969a) report an instance of a bluefin tuna traveling from the Bahamas to Norway, a distance of 4200 nautical miles, in less than 50 days. This is an excellent testimonial of the ability of warm red muscles to maintain sustained swimming at high speed.

V. THE EFFECTS OF SOME SUBSTANCES ON THE CIRCULATORY SYSTEM

When injected into, or otherwise applied to, the circulatory system of fish, a large number of compounds are known to cause changes in blood pressure and flow. The author has attempted to list the effects of a few compounds on the cardiovascular system of fish; some of these responses have already been discussed in preceding sections.

A. Acetylcholine

Acetylcholine increases the vascular resistance to flow through the gills of teleosts (Keys and Bateman, 1932) directing blood away from the secondary lamellae (Steen and Kruysse, 1964). The resistance to flow in the systemic circulation of elasmobranchs is decreased (Hanson, 1967). Acetylcholine has a negative chronotropic effect on teleost and elasmobranch hearts but acts as a cardiac stimulant in the lamprey (Falck et al., 1966). Methacholine, succinylcholine, pilocarpine, neostigmine, and d-tubocurarine all have similar but less marked effects than acetylcholine, causing bradycardia and fall in blood pressure in the eel (Chan, 1967).

B. Atropine

Atropinization increases heart rate in both elasmobranchs (Johansen et al., 1966) and teleosts (Randall, 1966), blocking the effect of efferent activity in cholinergic fibers innervating the heart. Atropine injected into the vascular system increases heart rate in both the resting and active dogfish causing an increase in ventral aortic systolic blood pressure but a decrease in peak ejection velocity and stroke volume of the heart (Johansen et al., 1966).

C. Catecholamines

Both adrenaline and noradrenaline increase ventral and dorsal aortic blood pressure and may either increase, decrease, or have no effect on heart rate when injected into the intact fish (Hanson, 1967; Randall and Stevens, 1967). A decrease in heart rate is the result of reflex vagal inhibition on the heart, probably resulting from a rise in blood pressure, and can be blocked by atropine. The direct effect of catecholamines on the heart is stimulatory. Adrenaline has a positive chronotropic and inotropic effect on the lamprey, teleost, and elasmobranch heart (Johansen et al., 1966; Falck et al., 1966; Bennion, 1968); but it has little or no effect on the hagfish heart (Bloom et al., 1961). Adrenaline and noradrenaline dilate the vessels in the teleost and elasmobranch gill (Keys and Bateman, 1932; Östlund and Fänge, 1962) decrease the resistance to flow through the gills, and increase blood flow in the secondary lamellae (Steen and Kruysse, 1964). Phenoxybenzamine, an α-adrenergic receptor blocking agent, diminishes the effect of catecholamines on the circulation. Isoprenaline, a β-adrenergic stimulant, decreases the resistance to blood flow through the gills (Randall, 1967). Chan (1967) suggested that isoprenaline dilated vessels in the systemic circulation of the eel. Hanson (1967) concluded that catecholamines increase systemic resistance to flow in elasmobranchs. Keys and Bateman (1932) found that adrenaline increased the resistance to flow through the systemic vessels in the eel's tail.

Tyramine and ephedrine, substances that cause the release of catecholamines from bound sites in mammals, caused a bradycardia and a transient fall, followed by a rise, in blood pressure in the eel, similar to that produced by noradrenaline (Chan, 1967). Cocain, which prevents the release and reduces the storage of noradrenaline in mammalian adrenergic nerves, causes a bradycardia and hypertension when injected into the eel. The bradycardia can be abolished by atropine.

D. Reserpine

This substance depletes the stores and limits the uptake of catecholamines in mammalian tissues. Injections in the eel cause the chromatophores to dilate and the swim bladder to swell, changes consistent with catecholamine depletion (Chan, 1967). Reserpine depletes the catecholamine stores in both the lamprey and hagfish heart (Bloom et al., 1961) and causes a loss in vigor in the hagfish heart which can be restored by adrenaline (Chapman et al., 1963).

E. Eptatretin

Eptatretin is an unstable amine or amide but, although it has similar effects, it is not a catecholamine. It has been found only in the hagfish heart, but has a marked positive chronotropic and inotropic effect on the heart of many vertebrates, including the hagfish (Jensen, 1963).

F. Histamine

Histamine causes vasoconstriction in the perfused gills of pike and a fall in blood pressure in the ventral aorta of the eel (see Mott, 1957). Chan (1967) found that histamine had little or no effect on blood pressure in the dorsal and ventral aorta of the eel. Chan used injected doses of 200 μg, some 100 times greater than that used by Mott.

G. 5-Hydroxytryptamine (Serotonin)

Chan (1967) found that injections of 25 μg of 5-hydroxytryptamine caused a slight increase, followed by a decline, in arterial blood pressure in the eel. No changes in heart rate or pulse pressure were recorded.

H. Renin and Angiotensin II

Angiotensin II, formed by the action of renin on angiotensin I, is a potent vasoconstrictor in mammals. Angiotension II, injected into eels, causes a marked increase in both dorsal and ventral aortic blood pressure, but it has no effect on heart rate or pulse pressure. The corpuscles of Stannius and kidney extracts, possible sources of renin in fish, also cause a rise in blood pressure (Chan, 1967).

I. Oxytocin and Vasopressin

The neurohypophysial hormones oxytocin and vasopressin cause a prolonged and marked rise in ventral aortic blood pressure, but they have little effect on the dorsal aortic blood pressure in the eel. The magnitude of the response varies with different hormones, the order of potency is 4-serine 8-isoleucine oxytocin > oxytocin > 8-arginine oxytocin > arginine vasopressin > lysine vasopressin, with lysine vasopressin having almost no effect on arterial blood pressure (Chan, 1967). The action of these hormones on the circulation appears to be related to interaction between the hormones, catecholamines, and α-adrenergic receptor sites.

J. Urophysial Extracts

Mugil urophysial extracts increase ventral and dorsal aortic blood pressure when injected into the eel. The eel has a caudal lymph heart, consisting of two chambers, an atrium, and a ventricle. The lymph heart collects fluid from the lateral cutaneous and hemal lymphatic vessels and pumps it into the caudal vein. Extracts from the eel urophysis increase the rate of beat of the caudal lymph heart (Chan, 1967).

ACKNOWLEDGMENTS

This manuscript was prepared while the author was a John Simon Guggenheim Memorial Fellow in the Zoology Department, University of Bristol. The author would like to thank Professor G. M. Hughes for his hospitality during this period.

REFERENCES

Augustinsson, K-B., Fänge, R., Johnels, A. G., and Östlund, E. (1956). Histological physiological and biochemical studies on the heart of two cyclostomes, Hagfish (*Myxine*) and lamprey (*Lampetra*). *J. Physiol.* (*London*) 131, 256–276.
Baumgarten, D., Randall, D. J., and Malyusz, M. (1969). Gas exchange versus ion exchange across the gills of fishes. In preparation.

Bennion, G. R. (1968). The control of the function of the heart in teleost fish. M.Sc. Thesis, University of British Columbia, Vancouver, B.C., Canada.

Bennion, G. R. (1968). Unpublished data.

Bloom, G., Östlund, E., von Euler, U. S., Lishajko, F., Ritzén, M., and Adams-Ray, J. (1961). Studies on the catecholamine-containing granules of specific cells in cyclostome hearts. *Acta Physiol. Scand.* 53, Suppl. 185, 1–34.

Bone, Q. (1966). On the function of the two types of myotomal muscle fibre in elasmobranch fish. *J. Marine Biol. Assoc. U.K.* 46, 321–349.

Burger, J. W., and Bradley, S. E. (1951). The general form of the circulation in the dogfish, *Squalus acanthias*. *J. Cellular Comp. Physiol.* 37, 389–402.

Carey, F. G., and Teal, J. M. (1966). Heat conservation in tuna fish muscle. *Proc. Natl. Acad. Sci. U.S.* 56, 1464–1469.

Carey, F. G., and Teal, J. M. (1969a). Regulation of body temperature by the bluefin tuna. *Comp. Biochem. Physiol.* 28, 205–213.

Carey, F. G., and Teal, J. M. (1969b). Mako and Porbeagle: Warm-bodied sharks. *Comp. Biochem. Physiol.* 28, 199–204.

Chan, D. K. O. (1967). Hormonal and haemodynamic factors in the control of water and electrolyte fluxes in the European eel *Anguilla anguilla* L. Ph.D. Thesis, University of Sheffield, Sheffield, England.

Chapman, C. B., Jensen, D., and Wildenthal, K. (1963). On circulatory control mechanisms in the Pacific hagfish. *Circulation Res.* 12, 427–440.

Conte, F. P., Wagner, H. H., and Harris, T. (1963). Measurements of the blood volume of the fish, *Salmo gairdneri gairdneri*. *Am. J. Physiol.* 205, 533–540.

Davis, J. C. (1968). The influence of temperature and activity on certain cardiovascular and respiratory parameters in adult sockeye salmon. M.Sc. Thesis, University of British Columbia.

de Kock, L. L., and Symmons, S. (1959). A ligament in the dorsal aorta of certain fishes. *Nature* 184, 194.

Dornesco, G. J., and Santa, V. (1963). Les structure des aortes et des vaisseaux sanguins de la carpe (*Cyprinus carpio* L.). *Anat. Anz.* 113, 136–145.

Falck, B., von Mecklenburg, C., Myhrberg, H., and Persson, H. (1966). Studies on adrenergic and cholinergic receptors in the isolated hearts of *Lampetra fluviatilis* (Cyclostomata) and *Pleuronectes platessa* (Teleostei). *Acta Physiol. Scand.* 68, 64–71.

Foxon, G. E. H. (1955). Problems of the double circulation in vertebrates. *Biol. Rev.* 30, 196–228.

Gannon, B. J., and Burnstock, G. (1969). Excitatory adrenergic innervation of the fish heart. *Comp. Biochem. Physiol.* 29, 765–773.

Garey, W. F. (1967). Gas exchange, cardiac output and blood pressure in free swimming carp (*Cyprinus carpio*). Ph.D. Dissertation, University of New York, Buffalo, New York.

Goldstein, L., Forster, R. P., and Fanelli, E. M., Jr. (1964). Gill blood flow and ammonia excretion in the marine teleost *Myoxocephalus scorpius*. *Comp. Biochem. Physiol.* 12, 489–499.

Hanson, D. (1967). Cardiovascular dynamics and aspects of gas exchange in chondrichthyes. Ph.D. Thesis. University of Washington, Seattle, Washington.

Holeton, G. F., and Randall, D. J. (1967a). Changes in blood pressure in the rainbow trout during hypoxia. *J. Exptl. Biol.* 46, 297–305.

Holeton, G. F., and Randall, D. J. (1967b). The effect of hypoxia upon the partial pressure of gases in the blood and water afferent and efferent to the gills of rainbow trout. *J. Exptl. Biol.* **46**, 317–327.

Holmes, W. N., and Donaldson, E. M. (1969). The body compartments and the distribution of electrolytes. In "Fish Physiology" (W. S. Hoar and D. J. Randall, eds.), Vol. 1, pp. 1–89. Academic Press, New York.

Hunn, J. B. (1967). "Bibliography on the Blood Chemistry of Fishes," pp. 1–32. Bur. Sport Fisheries and Wildlife, U.S. Govt. Printing Office, Washington, D.C.

Irving, L., Solandt, D. T., and Solandt, O. M. (1935). Nerve impulses from branchial pressure receptors in the dogfish. *J. Physiol. (London)* **84**, 187–190.

Jaeger, R. (1965). Aktionsportentiale der Myokardfasern des Fischherzens. *Naturwissenschaften* **52**, 482–483.

Jensen, D. (1963). Eptatretin: A potent cardioactive agent from the branchial heart of the Pacific hagfish *Eptatretus stoutii*. *Comp. Biochem. Physiol.* **10**, 129–151.

Jensen, D. (1965). The aneural heart of the hagfish. *Ann. N.Y. Acad. Sci.* **127**, 443–458.

Johansen, K. (1960). Circulation in the hagfish, *Myxine glutinosa* L. *Biol. Bull.* **118**, 289–295.

Johansen, K. (1962). Cardiac output and pulsatile aortic flow in the teleost *Gadus morhua*. *Comp. Biochem. Physiol.* **7**, 169–174.

Johansen, K., and Hanson, D. (1967). Hepatic vein sphincters in elasmobranchs and their significance in controlling hepatic blood flow. *J. Exptl. Biol.* **46**, 195–203.

Johansen, K., and Hanson, A. (1968). Functional anatomy of the hearts of lungfishes and amphibians. *Am. Zoologist* **8**, 191–210.

Johansen, K., and Martin, A. W. (1965). Comparative aspects of cardiovascular function in vertebrates. In "Handbook of Physiology" (Am. Physiol. Soc., J. Field, ed.), Sect. 2, Vol. III, pp. 2583–2614. Williams & Wilkins, Baltimore, Maryland.

Johansen, K., Franklin, D. L., and Van Citters, R. L. (1966). Aortic blood flow in free-swimming elasmobranchs. *Comp. Biochem. Physiol.* **19**, 151–160.

Keys, A., and Bateman, J. B. (1932). Branchial responses to adrenaline and pitressin in the eel. *Biol. Bull.* **63**, 327–336.

Kisch, B. (1948). Electrographic investigations of the heart of fish. *Exptl. Med. Surg.* **6**, 31–62.

Kuriyama, H. A., Goto, M., Maeno, T., Abe, Y., and Ozaki, S. (1960). Comparative studies on transmembrane potentials and electrical characteristics of cardiac muscle. In "Electrical Activity in Single Cells" (Y. Katsuki, ed.), pp. 243–260. Igakushoin, Hango, Tokyo.

Labat, R., Raynaud, P., and Serfaty, A. (1961). Réactions cardiaques et variations de masse sanguine chez les Téléostéens. *Comp. Biochem. Physiol.* **4**, 75–80.

Laffont, J., and Labat, R. (1966). Action de l'adrenaline sur la fréquence cardiague de la Carpe commune: Effet de la temperature du milieu sur l'intensité de la reaction. *J. Physiol. (Paris)* **58**, 351–355.

Laurent, P. (1967). La pseudobranchi des Téléostéens: Preuves électrophysiologiques de ses fonctions chemoréceptrice et baroréceptrice. *Compt. Rend.* **264**, 1879–1882.

McWilliam, J. A. (1885). On the structure and rhythm of the heart in fishes, with especial references to the heart of the eel. *J. Physiol. (London)* **6**, 192–245.

Mommaerts, W. F. H. M., Uchida, K., and Seraydarian, K. (1963). Cyclic adenosine, 31–51, phosphate as a regulator of the contractility of actomyosin. *Federation Proc.* **22**, 351.

Mott, J. C. (1957). The cardiovascular system. *In* "The Physiology of Fishes" (M. E. Brown, ed.), Vol. 1, pp. 81–105. Academic Press, New York.
Murdaugh, H. V., Robin, E. D., Millen, J. E., and Drewry, W. F. (1965). Cardiac output determinations by the dye dilution method in *Squalus acanthias*. *Am. J. Physiol.* 209, 723–726.
Nakano, T., and Tomlinson, N. (1967). Catecholamine and carbohydrate concentrations in rainbow trout (*Salmo gairdneri*) in relation to physical disturbance. *J. Fisheries Res. Board Can.* 24, 1701–1715.
Nickerson, M. (1964). Adrenergic regulation of cardiac performance. *Circulation Res.* 14, 15, Suppl. 2, 130–138.
Oets, J. (1950). Electrocardiograms of fishes. *Physiol. Comparata Oecol.* 2, 181–186.
Östlund, E. (1954). The distribution of catecholamines in lower animals and their effects on the heart. *Acta Physiol. Scand.* 31, Suppl. 112, 1–67.
Östlund, E., and Fänge, R. (1962). Vasodilation by adrenaline and noradrenaline, and the effects of some other substances on perfused fish gills. *Comp. Biochem. Physiol.* 5, 307–309.
Piiper, J., and Schumann, D. (1967). Efficiency of O_2 exchange in the gills of the dogfish, *Scyliorhinus stellaris Resp. Physiol.* 2, 135–148.
Randall, D. J. (1966). The nervous control of cardiac activity in the tench (*Tinca tinca*) and the goldfish (*Carassius auratus*). *Physiol. Zool.* 34, 185–192.
Randall, D. J. (1967). Unpublished observations.
Randall, D. J. (1968). Functional morphology of the heart in fishes. *Am. Zoologist* 8, 179–189.
Randall, D. J., and Stevens, E. Don (1967). The role of adrenergic receptors in cardiovascular changes associated with exercise in salmon. *Comp. Biochem. Physiol.* 21, 415–424.
Randall, D. J., Smith, L. S., and Brett, J. R. (1965). Dorsal aortic blood pressure recorded from rainbow trout (*Salmo gairdneri*). *Can. J. Zool.* 43, 863–877.
Robb, J. S. (1965). "Comparative Basic Cardiology." Grune & Stratton, New York.
Robertson, O. H., Krupp, M. A., Thompson, N., Thomas, S. F., and Hane, S. (1966). Blood pressure and heart weight in immature and spawning Pacific salmon. *Am. J. Physiol.* 256, 957–964.
Robin, E. D., Murdaugh, H. V., and Millen, J. E. (1964). Gill gas exchange in dogfish shark. *Federation Proc.* 23, Part 1, 469.
Ronald, K., Macnab, H. C., Stewart, J. E., and Beaton, B. (1964). Blood properties of aquatic vertebrates. I. Total blood volume of the Atlantic cod, *Gadus morhua*. L. *Can. J. Zool.* 42, 1127–1132.
Rushmer, R. F. (1961). "Cardiovascular Dynamics." Saunders, Philadelphia, Pennsylvania.
Satchell, G. H. (1960). The reflex co-ordination of the heart beat with respiration in the dogfish. *J. Exptl. Biol.* 37, 719–731.
Satchell, G. H. (1961). The responses of the dogfish to anoxia. *J. Exptl. Biol.* 38, 531–543.
Satchell, G. H. (1962). Intrinsic vasomotion in the dogfish gill. *J. Exptl. Biol.* 39, 503–512.
Satchell, G. H. (1965). Blood flow through the caudal vein of elasmobranch fish. *Australian J. Sci.* 27, 241–242.
Satchell, G. H., and Jones, M. P. (1967). The function of the conus arteriosus in the Port Jackson shark, *Heterodontus portusjacksoni*. *J. Exptl. Biol.* 46, 373–382.

Schiffman, R. H., and Fromm, P. O. (1959). Measurement of some physiological parameters in rainbow trout (*Salmo gairdneri*). *Can. J. Zool.* 37, 25–32.

Seyama, I., and Irisawa, H. (1967). The effect of high sodium concentration on the action potential of the skate heart. *J. Gen. Physiol.* 50, 505–517.

Shelton, G., Jones, D. R., and Randall, D. J. (1969). In preparation.

Smith, L. S. (1966). Blood volumes of three salmonids. *J. Fisheries Res. Board Can.* 23, 1439–1446.

Smith, L. S., Brett, J. R., and Davis, J. C. (1967). Cardiovascular dynamics in swimming adult sockeye salmon. *J. Fisheries Res. Board Can.* 24, 1775–1790.

Steen, J. B., and Kruysse, A. (1964). The respiratory function of teleostean gills. *Comp. Biochem. Physiol.* 12, 127–142.

Stevens, E. Don (1968). The effect of exercise on the distribution of blood to various organs in rainbow trout. *Comp. Biochem. Physiol.* 25, 615–625.

Stevens, E. Don, Bennion, G. R., Randall, D. J., and Shelton, G. (1969). The effect of temperature, catecholamines and exercise on blood flow and pressure in intact, unrestrained lingcod. In preparation.

Stevens, E. Don, and Black, E. C. (1966). The effect of intermittent exercise on carbohydrate metabolism in rainbow trout, *Salmo gairdneri*. *J. Fisheries Res. Board Can.* 23, 471–485.

Stevens, E. Don, and Randall, D. J. (1967a). Changes in blood pressure, heart rate and breathing rate during moderate swimming activity in rainbow trout. *J. Exptl. Biol.* 46, 307–315.

Stevens, E. Don, and Randall, D. J. (1967b). Changes in gas concentrations in blood and water during moderate swimming activity in rainbow trout. *J. Exptl. Biol.* 46, 329–337.

Sudak, F. N. (1965a). Intrapericardial and intracardiac pressures and the events of cardiac cycle in *Mustelus canis* (Mitchell). *Comp. Biochem. Physiol.* 15, 199–215.

Sudak, F. N. (1965b). Some factors contributing to the development of subatmospheric pressure in the heart chambers and pericardial cavity of *Mustelus canis* (Mitchell). *Comp. Biochem. Physiol.* 15, 199–215.

Sutherland, E. W., and Rall, J. W. (1960). The relation of adenosine, 3′,5′-phosphate and phosphorylase to the action of catecholamines and other hormones. *Pharmacol. Rev.* 12, 265–299.

Tebēcis, A. K. (1967). A study of electrograms recorded from the conus arteriosus of an elasmobranch heart. *Australian J. Biol. Sci.* 20, 843–846.

Thorson, T. B. (1958). Measurement of fluid compartments of four species of marine Chondrichthyes. *Physiol. Zool.* 31, 16–23.

Thorson, T. B. (1959). Partitioning of body water in sea lamprey. *Science* 130, 99–100.

Thorson, T. B. (1961). The partitioning of body water in Osteichthyes: Phylogenetic and ecological implications in Aquatic vertebrates. *Biol. Bull.* 120, 238–254.

Thorson, T. B. (1962). Partitioning of body fluids in the lake Nicaragua shark and three marine sharks. *Science* 138, 688–690.

von Euler, U. S., and Fänge, R. (1961). Catecholamines in nerves and organs of *Myxine glutinosa*, *Squalus acanthias*, and *Gadus callaris*. *Gen. Comp. Endocrinol.* 1, 191–194.

Yamauchi, A., and Burnstock, G. (1968). An electromicroscopic study on the innervation of the trout heart. *J. Comp. Neurol.* 132, 567–588.

5

ACID–BASE BALANCE

C. ALBERS

I. Introduction 173
II. Basic Concepts of Physical Chemistry 174
  A. The Dissociation of Water and the Definition of pH . . 174
  B. Dissociation of Weak Acids 175
  C. Carbonic Acid 177
  D. Buffer Action and Its Mathematical Description . . . 181
  E. Effects of Ionic Strength and Temperature 183
III. The Transport of CO_2 in the Blood 188
  A. The CO_2 Combining Curve of the Blood 188
  B. The pH of the Blood as Related to CO_2 196
IV. The Intracellular pH 203
V. Controlling Mechanisms of the Acid–Base Balance . . . 204
References 205

I. INTRODUCTION

This chapter deals with the chemical reactions and the physiological mechanisms affecting the concentration of hydrogen ions in the various fluid compartments of the body. Since acids are substances capable of delivering hydrogen ions, the concentration of hydrogen ions [H^+] depends primarily on the amount of the various acids present in the body fluids. The most important of these is carbonic acid which is derived from carbon dioxide (CO_2), one of the major end products of metabolism. Hence the equilibrium equations of carbonic acid and its reactions with the so-called buffer substances form the chemical basis of CO_2 transport and acid–base balance. The concentration of CO_2 and the chemical composition of the body fluids are therefore the principal factors governing acid–base balance. The former is controlled by ventilation, the latter by the action of various excretory mechanisms.

In the term "acid–base balance," acid stands for the sum of all anions except the hydroxyl ion and base stands for the sum of all cations except the hydrogen ion. This "medical" definition of acids and bases does not seem to be in accord with the proper chemical definition. This disagreement, however, turns out to be formal rather than substantial, as pointed out by Siggaard-Andersen (1965). The medical definition was adopted several decades ago by van Slyke and his group who contributed most to our basic knowledge of the acid–base balance in mammals. Because much more is known about the acid–base balance in warm-blooded animals, some basic facts about the acid–base balance in mammals will be included in this chapter. The question of whether or not these facts are pertinent to the acid–base balance of fish has to be left open in many cases. It is the hope of the author that his description of the acid–base balance will prompt other investigators to fill in the gaps in this really fascinating field of comparative physiology and biochemistry.

II. BASIC CONCEPTS OF PHYSICAL CHEMISTRY

A. The Dissociation of Water and the Definition of pH

Water molecules are dissociated into hydrogen ions and hydroxyl ions to a very small extent. The ionic product of water

$$K_W = [H^+][OH^-] \tag{1}$$

is of the order of 10^{-14}, depending on the temperature and the ionic strength of the solution. The hydrogen ion is normally present in the hydrated form H_3O, but for the sake of simplicity we shall stick throughout this chapter to the more convenient form of denoting the hydrogen ion as H^+. Since in pure water hydrogen ions equal hydroxyl ions the concentration of H^+ is $\frac{1}{2} K_W$ or about 10^{-7} mole/liter. If $[H^+]$ is changed by adding acids or bases, the concentration of hydroxyl ions $[OH^-]$ changes inversely according to Eq. (1). For instance, if we have a 0.1 N HCl solution, $[H^+] = 10^{-1}$ and $[OH^-] = 10^{-13}$ (for $K_W = 10^{-14}$).

The classic laws of thermodynamics describing the reactions of ions are valid only for ideal, infinitely diluted solutions. Very dilute solutions may be regarded as nearly ideal solutions. In all practical cases, however, the thermodynamic laws are valid only if activities rather than concentrations are used. The activity is obtained by multiplying the concentration by the so-called activity coefficient f. For hydrogen ions we have $a_{H^+} = [H^+] f_{H^+}$. Although for practical purposes the distinction between the concentration and the activity of an ion is important, we shall neglect the difference between [H⁺] and a_{H^+} in the following derivations. This is also justified because all methods for the determination of [H⁺] actually give information about a_{H^+}.

Sometimes it is more convenient to use logarithmic units. A logarithmic scale is also justified on a physicochemical basis, since the chemical potential or the energy associated with the activity of an ion is related to the logarithm of the activity. This leads to the concept of pH, which was introduced first by Sorensen (1909) as the negative logarithm of the concentration of hydrogen ions. The definition of pH commonly accepted now is

$$\text{pH} = -\log a_{H^+} \qquad (2)$$

From this definition it is easy to see that an increase in [H⁺] (or a_{H^+}) is denoted by a decrease in pH and vice versa. The pH scale covers a large concentration range, one unit being equivalent to a tenfold change in [H⁺]. When the pH scale is used, the actual changes in [H⁺] are often underrated. Actually, a change of 0.3 pH units does not look too impressive, but it really means a doubling of [H⁺]. One should not forget this fact when speaking about the relative constancy of arterial pH for instance.

B. Dissociation of Weak Acids

Let us consider an acid AH which dissociates into hydrogen ions and the anion A⁻. According to the law of mass action we may write

$$\frac{[H^+][A^-]}{[AH]} = K \qquad (3)$$

If the equilibrium constant K is very large, the concentration of [AH] is negligible and the dissociation may be regarded as virtually complete. This is the case with strong acids like HCl or HNO_3. If K is very small, only a fraction of the acid is dissociated and the acid is a weak acid. Most organic acids belong to this group. As a thermodynamic constant K depends on the temperature but not on the ionic strength. If we convert to activities in Eq. (3), assuming that $a_{AH} = [AH]$ and $a_{A^-} = [A^-] f_{A^-}$, then

$$\frac{a_{H^+}[A^-]}{[AH]} \cdot f_{A^-} = K = K' \cdot f_{A^-} \qquad (4)$$

K' now depends on the temperature and on the ionic strength. Its negative logarithm is written pK'. If we take the logarithms on both sides of Eq. (4) and multiply by minus one, we obtain the important equation

$$\text{pH} = \text{p}K' + \log \frac{[\text{A}^-]}{[\text{AH}]} \qquad (5)$$

which is known as the Henderson-Hasselbalch equation.

For a given concentration [AH] we obtain from Eq. (4) $[\text{H}^+][\text{A}^-]$ = const, which can be interpreted in a similar manner as Eq. (1). Therefore, if $[\text{H}^+]$ increases $[\text{A}^-]$ must decrease. If we add a strong acid to a weak acid, hydrogen ions recombine with the anion A^- to form the undissociated acid AH. Thus, the dissociation of AH is dependent on $[\text{H}^+]$. If h is the fractional extent to which a weak acid is dissociated, then

$$h = \frac{[\text{A}^-]}{[\text{A}^-] + [\text{AH}]} \qquad (6)$$

Since we have from Eq. (3)

$$[\text{AH}] = \frac{[\text{H}^+] \cdot [\text{A}^-]}{K}$$

we obtain

$$h = \frac{K}{K + [\text{H}^+]} \qquad (7\text{a})$$

or in the logarithmic form

$$h = 1/(1 + 10^{\text{p}K-\text{pH}}) \qquad (7\text{b})$$

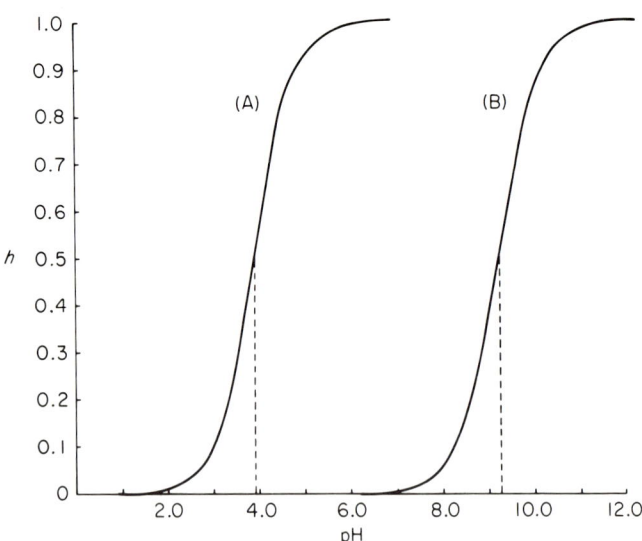

Fig. 1. Fractional dissociation h of lactic acid (A) and of boric acid (B) as a function of pH.

5. ACID–BASE BALANCE

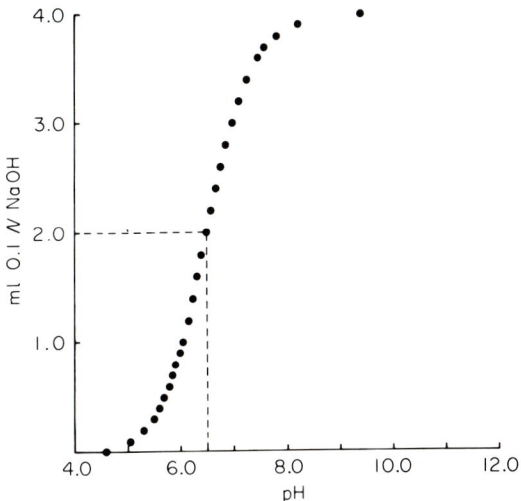

Fig. 2. Titration curve of 4 ml of 0.1 N DMO at 19.6°C (Albers, 1966).

Therefore, if [H⁺] is very large with respect to K, h approaches zero. If [H⁺] is very small, h approaches unity. If [H⁺] = K, h is 0.5, that means exactly one-half of the weak acid is dissociated when pH equals pK'. A plot of h vs. pH according to Eq. (7b) is shown in Fig. 1 for lactic acid (pK' = 3.9 at 25°C) and for boric acid (pK' = 9.2 at 25°C). Exactly the same curve is obtained if we titrate a weak acid with a strong base and plot the fractional extent of neutralization vs. pH. We may call this curve the standardized titration curve, whereas the ordinary titration curve is obtained by plotting the amount of base consumed (e.g., in milliliter of 0.1 N NaOH) versus pH. This results in a sigmoid-shaped curve too. It is customary to determine the pK' value of an acid by reading the pH of half neutralization from a titration curve. Figure 2 shows the titration of the weak acid DMO (dimethyloxazoledinedione), which is used for the indirect determination of the intracellular pH (see below). From Fig. 2 we read for half neutralization pH = 6.48; pK' of DMO is therefore 6.48 at that temperature.

C. Carbonic Acid

The amount of gaseous CO_2 physically dissolved in water or electrolyte solutions is proportional to the partial pressure of CO_2 according to Henry's law

$$[CO_2] = Sp_{CO_2} \qquad (8)$$

where S is the solubility coefficient in millimoles per liter per torr and p_{CO_2} the partial pressure of CO_2 in torr. Tables of S commonly refer to the sum of physically dissolved CO_2 and carbonic acid according to Eq. (9a). If a liquid has been equilibrated with a gas mixture containing CO_2 until the number of CO_2 molecules escaping from the liquid into the gas phase equals the number of CO_2 molecules entering the liquid from the gas phase, the net exchange between the two phases is zero. For this state of equilibrium it is said that the liquid phase has the same partial pressure of CO_2 as has the gas phase, whether or not the liquid is still in contact with the gas phase. Even if a liquid has never been in contact with a gas phase, gaseous components derived from chemical reactions exert a partial pressure which is linked to the amount of dissolved gas by the solubility coefficient S by Eq. (8). It is possible to calculate the partial pressure of CO_2 from analytical data as well as to measure it directly with membrane-covered glass electrodes.

CO_2 dissolved in water reacts to form carbonic acid:

$$CO_2 + H_2O \rightleftharpoons H_2CO_3 \tag{9a}$$

The carbonic acid as a dibasic acid dissociates into bicarbonate ions and carbonate ions

$$H_2CO_3 \rightleftharpoons H^+ + HCO_3^- \tag{9b}$$
$$HCO_3^- \rightleftharpoons H^+ + CO_3^{2-} \tag{9c}$$

From the law of mass action we have

$$[CO_2] = L[H_2CO_3] \tag{10a}$$

$$\frac{[H^+][HCO_3^-]}{[H_2CO_3]} = K_1 \tag{10b}$$

$$\frac{[H^+][CO_3^{2-}]}{[HCO_3^-]} = K_2 \tag{10c}$$

The equilibrium constant L is very large, less than 0.5% of the dissolved CO_2 being transformed into H_2CO_3. K_1 is the true dissociation constant of carbonic acid and about 20 times greater than that of acetic acid. But because of the large value of L in Eq. (10a), dissolved CO_2 acts as a much weaker acid. This may be seen if we combine Eqs. (10a) and (10b):

$$[H^+] = \frac{K_1}{L} \cdot \frac{[CO_2]}{[HCO_3^-]} \tag{11}$$

$K_1' = K_1/L$ is the so-called apparent first dissociation constant of carbonic acid and in the order of 4×10^{-7} at 20°C. Correspondingly, pK_1' of carbonic acid is about 6.4 at 20°C.

5. ACID–BASE BALANCE

At the high body temperature of mammals the concentration of carbonate ions is practically zero. In fish, however, a small fraction of the chemically bound CO_2 is present as carbonate. Therefore, we also have to consider reactions (9c) and (10c). If we denote the total concentration of CO_2 as C_T, we have

$$C_T = Sp_{CO_2} + [HCO_3^-] + [CO_3^{2-}]$$

Solving Eqs. (10b) and (10c) for $[HCO_3^-]$ and $[CO_3^{2-}]$, we finally arrive at the rather complex equation

$$\text{pH} = pK_1' + \log \frac{C_T - Sp_{CO_2}}{Sp_{CO_2}(1 + K_2/[H^+])} \tag{12a}$$

which obviously cannot be solved for pH because $[H^+]$ appears in the denominator on the right side of the equation. After rearranging Eq. (12a) into the form

$$\text{pH} = pK_1' - \log\left(1 + \frac{K_2}{[H^+]}\right) + \log\left(\frac{C_T}{Sp_{CO_2}} - 1\right) \tag{12b}$$

and if

$$pK_1'' = pK_1' - \log\left(1 + \frac{K_2}{[H^+]}\right) \tag{13}$$

then we return to the familiar form of the Henderson-Hasselbalch equation

$$\text{pH} = pK_1'' + \log\left(\frac{C_T}{Sp_{CO_2}} - 1\right) \tag{12c}$$

which is equivalent to Eq. (5) and where pK_1'' now appears to depend not only on the temperature but also on the pH. The first reason for the interaction between pK_1'' and pH is the incorporation of the second dissociation constant K_2 of carbonic acid according to Eq. (13). A second reason is the formation of the ion $NaCO_3^-$ from the reaction $Na^+ + CO_3^{2-} = NaCO_3^-$ (Siggaard-Andersen, 1965).

It is possible to determine the chemically bound CO_2 $[HCO_3^-] + [CO_3^{2-}]$ by titration. If this quantity is introduced in the derivations above, another pK'' is obtained. Likewise some authors prefer the use of the activity of carbonic acid

$$a_{H_2CO_3} = a_0 a_{H_2O} p_{CO_2}$$

where p_{CO_2} is the partial pressure of CO_2, a_0 the solubility of CO_2 in pure water, and a_{H_2O} the activity of water obtained from the freezing point depression. This leads to another definition of pK_1' and hence pK_1''. There

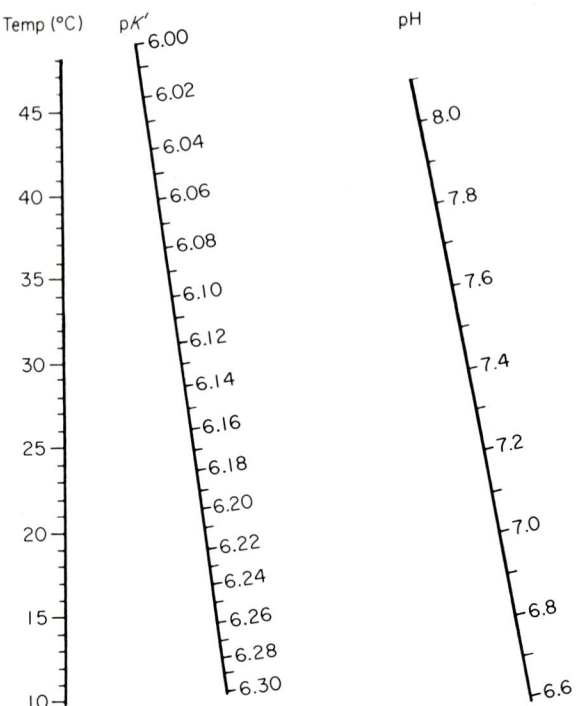

Fig. 3. Operational pK' of carbonic acid as a function of pH and temperature. From Severinghaus et al. (1956). Reprinted by permission of J. Appl. Physiol.

are several ways to define pK_1', depending on the use of concentration or activity of one or more members participating in the equilibrium equation. Therefore, when using tabulated values of pK_1', the basic assumptions underlying such tables must not be neglected. With these reservations in mind we shall simply write pK_1' rather than pK_1'' in the following paragraphs.

For practical purposes, pK_1' is obtained by simultaneous determination of C_T and pH in samples of plasma equilibrated at a known p_{CO_2}. From the analytical data pK_1' can be calculated. Figure 3 shows a line chart for reading pK_1' as a function of pH and temperature. This line chart is valid for mammalian plasma and may also be used for most teleost fish, because ionic strength is similiar in both cases (see Table I). In Fig. 4 pK_1' values of dogfish plasma are shown which differ from the values of mammalian plasma owing to the higher ionic strength of elasmobranch blood. The reader should keep in mind that pK_1' defined here is not a constant nor does it have a thermodynamic meaning. It is nothing but an operational figure serving to fit three measurable quanti-

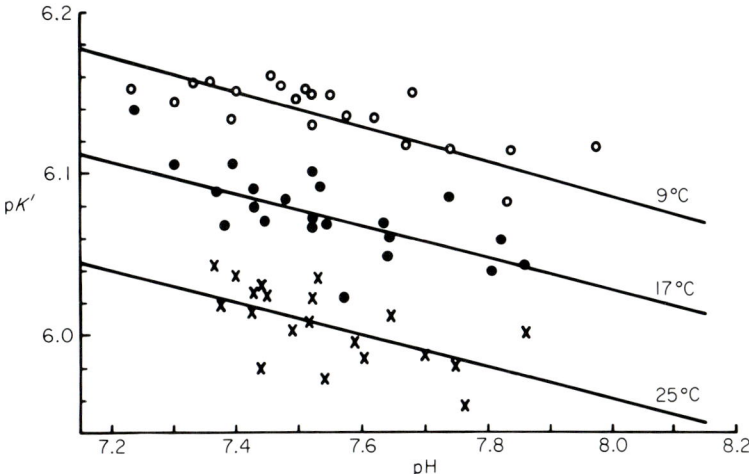

Fig. 4. Operational pK' of carbonic acid in dogfish plasma as a function of pH and temperature. $S_{CO_2} = 73\%$ CO_2 solubility in pure water. From Albers and Pleschka (1967). Reprinted by permission of North Holland Publishing Co.

ties (C_T, pH, and p_{CO_2}) into an equation. This equation has proved to be most useful and its importance is beyond any doubt. These simplifications lead Homer W. Smith (1956) to call this equation "a most useful monument to human laziness."

D. Buffer Action and Its Mathematical Description

Close inspection of the titration curves (Figs. 1 and 2) reveals an important relationship. Both the top and the bottom part of the curves indicate large changes in pH if small amounts of the base are added. On the other hand, in the neighborhood of the inflection point even large amounts of base result in only relatively small changes in pH. At the inflection point half of the acid is neutralized and present as salt. Thus, such a mixture of a weak acid and its salt with a strong base tends to maintain its pH when other acids or bases are added and is called, therefore, a buffer solution. As seen from Figs. 1 and 2 the buffer action is most effective if salt and acid are present in equal amounts. The buffer action can be described quantitatively by the slope of the curve, dB/dpH, where dB is an infinitesimal amount of base added to the solution. If we start with a pure acid AH at a concentration C_P and add B moles of the base per liter, it follows $[A^-] = B$ and $[AH] = C_P - B$. Substitution of these values into Eq. (5) and logarithmic differentiation finally gives

$$\frac{dB}{dpH} = 2.3 C_P \frac{K[H^+]}{(K + [H^+])^2} \tag{14}$$

which apparently has its maximum, if $[H^+] = K$ and $dB/dpH = 0.575\ C_P$. This formula was derived by van Slyke, who called dB/dpH the buffer capacity of a solution. Obviously the buffer capacity depends on (1) the concentration C_P of the buffer substance, (2) the dissociation constant K of the buffer substance, and (3) the concentration of hydrogen ions. Buffer systems where $C_P = [A^-] + [AH]$ is constant are called "homogeneous systems." Biological examples are tissues and blood passing through tissues. The buffer substances in the blood are the plasma proteins and especially hemoglobin in the red cells. Because of their ampholytic dissociation, plasma proteins and hemoglobin can be considered to be much weaker acids than carbonic acid. They are able therefore to buffer the hydrogen ions arising from the carbonic acid. In the tissues, proteins are less important for buffering which is effected chiefly by anorganic and organic phosphate compounds. A homogeneous buffer system is completely described by Eq. (14). If a narrow range of $[H^+]$ is considered, dB/dpH is nearly constant and the relationship between B and pH can be approximated by a straight line. This fact is used for the determination of the so-called buffer lines of true plasma (see below).

Quite another type of buffering is realized in systems where $[AH]$ is kept constant rather than C_P. In the case of a bicarbonate buffer $[AH]$ is kept constant if the partial pressure of CO_2, and hence the product Sp_{CO_2}, remains unchanged. Since addition of an acid to a bicarbonate buffer would primarily increase the concentration of carbonic acid [Eq. (9b)] and of CO_2 [Eq. (9a)], there must be another system linked to the buffer which takes up the excess of CO_2. An example of such a buffer is seawater which is in equilibrium with the p_{CO_2} of the atmosphere. If an acid is added to seawater, CO_2 escapes into the atmosphere; conversely, if a base is added, CO_2 is taken up from the atmosphere until p_{CO_2} is restored to its initial value. Since we have two systems linked to each other, such buffering is said to occur in a heterogeneous system. The quantitative behavior of such a heterogeneous system is obtained from Eq. (12c) where the denominator is kept constant while the numerator is changed. Two curves corresponding to $Sp_{CO_2} = 0.01$ and 0.1 mmole/liter and $pK' = 6.4$ are shown in Fig. 5. In contrast to the sigmoid-shaped curves of Figs. 1 and 2, we obtain curves with a slope and thus with a buffer capacity increasing continuously with the pH. For $Sp_{CO_2} = $ const the buffer capacity is found by differentiation of Eq. (12c) to be $dB/dpH = 2.3\ [HCO_3^-]$. The absolute value of pH depends on p_{CO_2}, a tenfold change in p_{CO_2} results in a change of pH by

5. ACID–BASE BALANCE

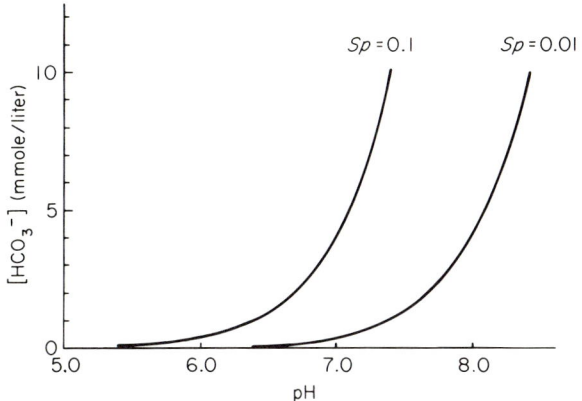

Fig. 5. Relationship between bicarbonate concentration [HCO_3^-] and pH for a heterogeneous buffering system. Physically dissolved CO_2 0.1 and 0.01 mmole/liter, respectively.

one unit. The maintenance of a constant p_{CO_2} therefore is the major factor controlling the pH in such a heterogeneous system. The pH of the blood passing through the capillaries of the lungs is chiefly regulated by the proper adjustment of the ventilation to yield an almost constant arterial CO_2 tension.

We should keep in mind that the basic mechanisms of buffering in a homogeneous (or closed) system and in a heterogeneous (or open) system are fundamentally different. In a homogeneous system the buffering is a purely chemical process, depending on the nature and the concentration of the participating buffer substances. In a heterogeneous system the buffering depends on physical and physiological mechanisms involved in the maintenance of a constant CO_2 tension. As we have seen, acid–base regulation takes advantage of both types of buffering. Within the tissue acids entering the blood are buffered by chemical processes according to a homogeneous system, whereas in the lungs or gills the blood, being in proximity to the environment, shows the typical properties of a heterogeneous buffer system.

E. Effects of Ionic Strength and Temperature

1. Effect of Ionic Strength

All biological fluids contain salts in an appreciable concentration, resulting in an ionic interaction which has to be taken into account. When investigating the acid–base balance, the ionic interaction is especially important for polyvalent ions. The total ionic interaction can

Table I

Ionic Composition of Fish Plasma and Calculated Values of the Ionic Strength μ
(all concentrations in mmole/kg H_2O^a)

Species	Na	K	Ca	Mg	Cl	SO_4	HCO_3^b	HPO_4	μ	Reference
Myxine	558	9.6	12.5	38.8	576	13.3	2	?	0.70	Robertson (1954)
Lampetra fluviatilis	125	3.3	4.0	4.4	100	5.6	6.7	?	0.15	Robertson (1954)
Raia stabuluforis	277	6.5	5.1	3.0	252	1	8.1	2.0	0.29	Smith (1929)
Raia erinacea	270	8.6	12.9	4.1	275	(4)	29.6	?	0.33	Hartman et al. (1941)
Narcine	144	7.5	17.2	3.2	171	(4)	13.3	?	0.22	Pereira and Sawaya (1957)
Rhinobatus	154	13.7	7.8	2.1	155	(4)	24.5	?	0.20	Pereira and Sawaya (1957)
Cyprinus carpio	130	6.3	2.9	1.4	127	(4)	(10)	?	0.15	Field et al. (1943)
Coregonus	141	3.8	5.3	3.4	117	4.6	10.6	?	0.16	Robertson (1954)
Muraena	212	2.0	7.7	4.9	188	11.4	8.0	?	0.25	Robertson (1954)
Thunnus	190	26.8	(4)	(3)	181	(4)	41.8	?	0.24	Becker et al. (1958)

^{a} Water content of plasma was assumed to be 930 g/liter if not stated otherwise in the references. Assumed values in parentheses.
^{b} Bicarbonate values if not given in the reference are calculated as the difference between the sum of anions and the sum of cations.

be expressed as the ionic strength μ, which is half the sum of the products $c_i z_i^2$, where c_i is the concentration of the ion species i and z_i is the corresponding valency. The ionic strength of human plasma is 0.167 at 38°C. Table I shows the ionic composition of plasma of various freshwater and marine fish and the ionic strength calculated from these data. Some freshwater fish such as *Cyprinus carpio* and *Lampetra fluviatilis* have about the same ionic strength as mammals. In many species the ionic strength is higher, reaching about 0.3 for *Raia* and 0.7 for *Myxine*. In solutions having such an ionic strength the law of mass action can be applied only in a much more sophisticated form. Since the quantitative approach to the ionic interaction is beyond the scope of this chapter, the reader is referred to any good textbook on physical chemistry for details. Briefly summarized, an increase in ionic strength has two consequences: (1) it decreases the solubility of gases and (2) it increases the equilibrium constants for most reactions. The latter becomes evident if we recall from Eq. (4) that the operational constant K' is related to the thermodynamic constant K by the activity coefficient $K' = K/f$. Since f is lowered by an increase in ionic strength, K' must increase (or pK' decrease) together with μ. Examples are seen in Table II where K_1' of boric acid and of carbonic acid as well as K_2' of carbonic acid are listed as a function of temperature and chlorinity of seawater. Values for the solubility S of CO_2 are also given. Obviously at any given

Table II

Effect of Temperature and Chlorinity on the Dissociation Constants of Boric Acid and Carbonic Acid and the Solubility of CO_2[a]

t (°C)	Chlorinity (‰)	Boric acid $K' \times 10^9$	Carbonic acid $K_1' \times 10^6$	Carbonic acid $K_2' \times 10^9$	S_{CO_2} (mmole/liter atm)
0	0	0.40	0.26	0.023	77.0
	15	1.10	0.58	0.43	67.4
	20	1.29	0.63	0.57	64.0
10	0	0.50	0.34	0.032	53.6
	15	1.35	0.74	0.60	47.2
	20	1.58	0.80	0.80	45.2
20	0	0.60	0.40	0.042	39.4
	15	1.62	0.89	0.76	35.1
	20	1.95	0.97	1.02	33.7
30	0	0.72	0.45	0.051	29.9
	15	1.95	1.01	0.93	27.0
	20	2.29	1.10	1.26	26.0

[a] From Harvey (1966). Reprinted by permission of Cambridge University Press.

temperature S is highest for pure water (chlorinity $= 0$) and decreases if the salt content and hence the ionic strength increase. The dissociation constants show the opposite effect.

In the same way the ionic strength increases the ionic product of water K_W. If we pass from concentrations to activities we obtain from Eq. (1)

$$K_W = \frac{a_{H^+} \cdot a_{OH^-}}{a_{H_2O}} \tag{15}$$

where $a_{H^+} \cdot a_{OH^-}$ is called the thermodynamic dissociation product. Since K_W and a_{H_2O} are affected by the ionic strength in an opposite direction, the effect of the ionic strength on the thermodynamic dissociation product is small. It seems reasonable, therefore, to derive the definition of neutrality from the thermodynamic dissociation product rather than from the ionic product K_W.

2. Effect of Temperature

There is no constant in the whole realm of physical chemistry which does not depend on temperature. As a general rule the dissociation of a molecule into ions is favored by an increase in temperature. As a consequence most values of K' increase together with the temperature. This can be seen from Table II which also shows that the effect of temperature on the second dissociation constant for carbonic acid is even greater than that on the first.

Of particular interest is the dissociation of water. Since with increasing temperature K_W increases, the pH indicating neutrality decreases. It is only at 25°C that pure water has a pH of 7.0, whereas at 5°C it has a pH of 7.36 and at 50°C it has a pH of 6.65. That is to say, the "meaning" of pH depends on the temperature. A pH of 7.0 "means" a neutral solution at 25°C, a slightly acid solution at 5°C, and a slightly alkaline solution at 50°C. To avoid the difficulty of interpreting a pH value, Winterstein (1954) suggested the use of the ratio $[OH^-]/[H^+]$ which of course is unity for a neutral solution, smaller than unity for an acid solution, and larger than unity for an alkaline solution. For similar reasons Rahn (1967) introduced the term "relative alkalinity" which is defined by $[H^+]_N/[H^+]$, where $[H^+]_N$ is the hydrogen ion concentration at neutrality and $[H^+]$ the actual hydrogen ion concentration. Since $[H^+]_N/[H^+] = \sqrt{K_W}/[H^+]$ and $[OH^-]/[H^+] = K_W/[H^+]^2$, the relative alkalinity of Rahn is the square root of Winterstein's term $[OH^-]/[H^+]$. A line chart for both terms as a function of temperature and pH is shown in Fig. 6.

5. ACID–BASE BALANCE

In a similar way temperature affects the dissociation constants of the substances involved in the buffering processes. This has two consequences: (1) if a blood sample is cooled anaerobically, its pH will increase by roughly 0.015 per each degree centigrade; (2) if blood is equilibrated with CO_2, not only the physically dissolved CO_2 but also the chemically bound CO_2 increases if the temperature is lowered (see Fig. 12). This may be explained as follows: If the blood is cooled, the

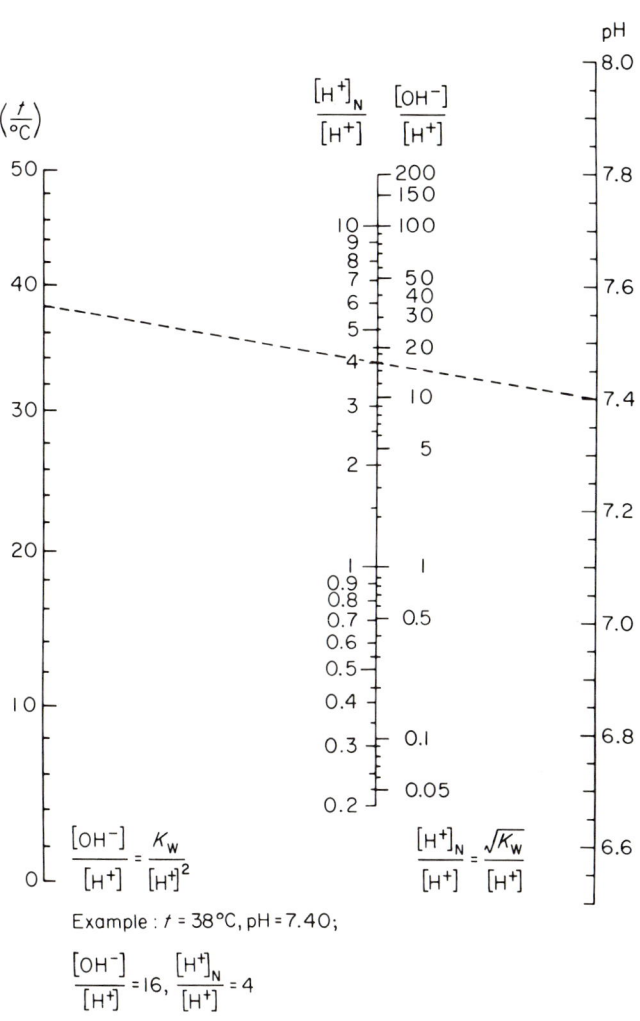

Fig. 6. Relative alkalinity and $[OH^-]/[H^+]$ as a function of pH and temperature. Adopted from Albers (1962).

K' of the buffer substances decreases more than the K' of carbonic acid. Therefore, hydrogen ions recombine with the buffer to increase the undissociated moiety and more cations are available to form the dissociated salts of the carbonic acid: The chemically bound CO_2 increases. In addition, the physically dissolved CO_2 is increased, too, because the solubility of gases varies inversely with the temperature (Table II). Since the effect of temperature on the physically dissolved CO_2 is more pronounced than that on the chemically bound CO_2, the denominator in Eq. (12c) decreases more than does the numerator. Therefore, the pH must increase.

The effect of temperature on the dissociation constants of carbonic acid can be seen in Table II. From the data in Table II it is obvious that pK_2' is affected more than pK_1' especially for salt solutions. Although these effects are important for the physical chemistry of freshwater and seawater, they are of minor importance for the temperature effect on the acid–base balance of the blood when compared with the effects of temperature on the solubility of CO_2 and on the dissociation constants of the buffer substances.

III. THE TRANSPORT OF CO_2 IN THE BLOOD

A. The CO_2 Combining Curve of the Blood

The curve relating the total CO_2 content of the blood to the CO_2 tension is commonly called the CO_2 dissociation curve of the blood. Since total CO_2 comprises physically dissolved CO_2 as well as chemically bound CO_2, the term "CO_2 combining curve" seems to be more appropriate. Figure 7 shows the CO_2 combining curves of oxygenated and deoxygenated blood of the salmon and the dogfish. In the region of low CO_2 tensions, the curves increase steeply but then gradually flatten out. In the trout the curves of deoxygenated blood show a higher CO_2 content than those of oxygenated blood, whereas no such effect could be demonstrated in the dogfish. As shown by Henderson (1932) for mammalian blood and by Ferguson et al. (1938) and Albers and Pleschka (1967) for fish blood, such curves yield straight lines in a limited range of CO_2 tensions when plotted in a double logarithmic system. This is very convenient for practical purposes. Figure 8 shows examples of CO_2 combining curves of the blood from various fishes plotted this way. Obviously, the CO_2 combining power of the blood, as indicated by these parameters, varies considerably. There is some correlation between the

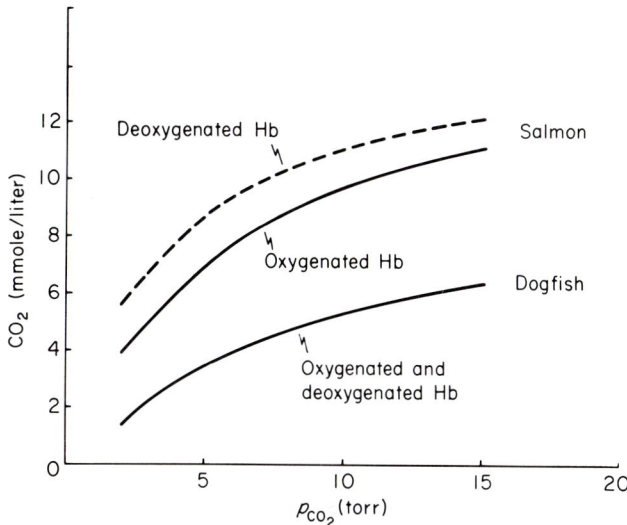

Fig. 7. CO_2 combining curves of the Altantic salmon, *Salmo salar*, and the landlocked salmon, *Salmo salar sebago*, at 5°C and of the dogfish, *Mustelus canis*, at 22°C. Data from Black *et al.* (1966b,c) and from Ferguson *et al.* (1938).

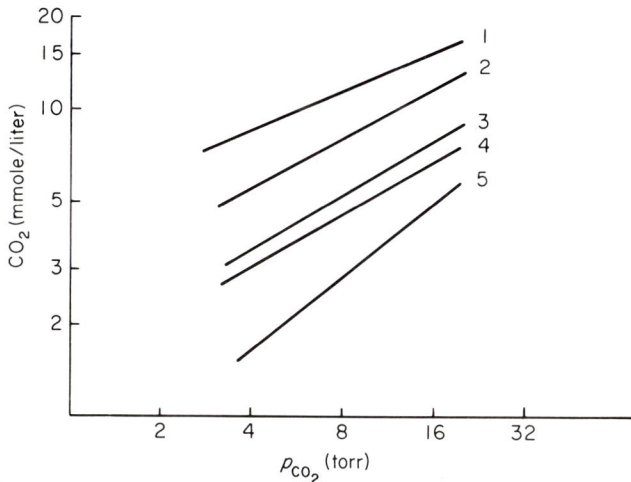

Fig. 8. CO_2 combining curves of various fishes plotted in a double log system. 1, *Cyprinus carpio* (15°C) (Ferguson and Black, 1941); 2, *Raia oscillata* (10°C) (Dill *et al.*, 1932); 3, *Opanus tau* (20°C) (Root, 1931); 4, *Mustelus canis* (22°C) (Ferguson *et al.*, 1938); 5, *Pterodorus* (28°C) (Willmer, 1934, quoted by Fry, 1957).

hemoglobin content and the CO_2 combining power. The reason for this as well as for the effects of oxygenation and deoxygenation will be discussed below.

It seems necessary to stress the narrow range of CO_2 tensions encountered under physiological conditions. Generally the arterial p_{CO_2} is about 1–2 torr only and the venous p_{CO_2} rarely exceeds 10 torr. *In vitro* investigations of the CO_2 transport should concentrate largely on this range. Equilibrium of fish blood with CO_2 tensions as high as 70 torr as done in the early papers in this field does not seem to provide useful information.

1. Physically Dissolved CO_2

If the partial pressure of CO_2 is known, the amount of physically dissolved CO_2 can be calculated with the aid of the solubility coefficient S [see Eq. (8)]. Whereas values of S are well established for mammalian blood, there are almost no data reported on fish blood. In the case of marine fish some authors simply use the solubility of CO_2 in the surrounding seawater. Since the ionic composition and the ionic strength of fish plasma displays a fairly large variability, it is highly desirable to obtain reliable data for S especially if the Henderson-Hasselbalch equation is to be used. If approximations of S are applied, the results of such calculations are jeopardized by substantial errors.

Reaction (9a)

$$CO_2 + H_2O \rightleftharpoons H_2CO_3$$

as a molecular reaction proceeds very slowly. The time needed for full equilibrium is about 200 times longer than the time spent by the blood in the gills or in the lungs. The red cells of all vertebrates, however,

Table III

Concentration of Carbonic Anhydrase in Red Blood Cells of Various Species[a]

Species	Carbonic anhydrase (eu/g)
Squalus acanthias	40
Raia ocellata	32
Ameiurus nebulosus	300
Lophius piscatorius	300
Perca fluviatilis	2400
Dog	1400
Rat	2400
Cat	2000

[a] From Maren (1967).

Table IV
Effect of Blocking Carbonic Anhydrase with Diamox on p_{CO_2}, pH, and CO_2 Content of Blood, Cerebrospinal Fluid, and Aqueous Humor

Specimen		p_{CO_2} (torr)	pH	C_{CO_2} (mmoles/liter)
	Squalus acanthias[a]			
Venous blood	Control	9	7.56	7.96
	Diamox	16	7.47	12.1
Cerebrospinal fluid	Control	8	7.66	8.57
	Diamox	18	7.56	17.1

Specimen		p_{CO_2} (torr)	pH	C_{CO_2} (mmoles/kg H_2O)
	Salvelinus namaycush[b]			
Venous blood	Control	4.4	7.61	6.65
	Diamox	9.3	7.38	8.96
Aqueous humor	Control	4.2	7.65	6.40
	Diamox	8.4	7.22	4.96
Swim bladder	Control	4.6	—	—
	Diamox	10.2	—	—

[a] From Maren (1962).
[b] From Hoffert and Fromm (1966) and Hoffert (1966).

contain the enzyme carbonic anhydrase which speeds up the formation and splitting of carbonic acid. The concentration of the enzyme in the red cells is correlated to the metabolic activity of each species. Although the concentration of the enzyme in fish blood is lower in most cases than that in mammals (Table III) there is evidence that the concentration of the enzyme is always in excess of the physiological demands (Maren, 1962; Larimer and Schmidt-Nielsen, 1960). If the enzyme is blocked by sulfanilamide a respiratory acidosis ensues with an elevation of the arterial CO_2 tension and a decrease in pH (Table IV). The increase of p_{CO_2} is not only seen in the blood but also in the swim bladder, cerebrospinal fluid, and aqueous humor.

2. Chemically Bound CO_2

After carbonic acid has formed, the following reactions take place:

$$H_2CO_3 + BPr \rightleftharpoons BHCO_3 + HPr$$
$$H_2CO_3 + BO_2Hb \rightleftharpoons BHCO_3 + HO_2Hb$$
$$H_2CO_3 + BHb \rightleftharpoons BHCO_3 + HHb$$

where B denotes a cation bound either to plasma proteins (Pr) or to oxygenated or deoxygenated hemoglobin (O_2Hb and Hb, respectively). Since carbonic anhydrase is not present in the plasma, the first reaction is of minor importance. The other two reactions take place within the red cell. From the law of mass action it is apparent that the higher the concentration of hemoglobin, the higher the concentration of bicarbonate; hence, the CO_2 combining power of the blood. At the low body temperatures of most fish a small fraction of the bicarbonate further dissociates into carbonate ions and hydrogen ions. Because of the presence of ion pairs like $NaCO_3^-$ (Siggaard-Andersen, 1965) it is impossible at the present time to obtain quantitative information about the carbonate concentration from the few data available. The cations for the formation of bicarbonate are made available by the ampholytic dissociation of the protein component of the hemoglobin. The functional group involved is the imidazole, the dissociation of which is strongly affected by the oxygenation of the hemoglobin (see below).

Another chemical reaction between CO_2 and hemoglobin depends on some free α-amino groups, which at the alkaline side of the isoelectric point form so-called carbamino hemoglobin according to the equation

$$HbNH_2 + CO_2 \rightleftharpoons HbNCOO^- + H^+$$

Since an increase in the concentration of CO_2 on the left side of the equation is always associated with an increase in the concentration of H^+ at the right side of the equation, the equilibrium of the reaction is hardly affected by the CO_2 tension. It is, however, strongly dependent on the oxygenation of the hemoglobin. Although in mammals the carbamino hemoglobin is only a small fraction of the absolute amount of the total CO_2, it plays an important role in the changes of the CO_2 concentration associated with the delivery of CO_2 from the tissues to the capillary blood. This may be seen from the following example: In human beings arterial blood contains an average 22 mmoles/liter of total CO_2, which is made up of 1.25 mmoles/liter of physically dissolved CO_2 (or 5.7% of the total CO_2), 19.65 mmoles/liter of bicarbonate (or 89.3% of the total), and 1.1 mmoles/liter of carbamino hemoglobin (or 5.0% of the total CO_2). When this blood passes through a capillary it takes up 2.2 mmoles/liter of CO_2, the venous blood having 24.2 mmoles/liter of total CO_2. The arteriovenous difference of 2.2 mmoles/liter is partitioned into 0.18 mmoles/liter of physically dissolved CO_2 (or 8.2% of the arteriovenous difference), 1.39 mmoles/liter of bicarbonate (or 63% of the arteriovenous difference), and 0.63 mmoles/liter of carbamino hemoglobin (or 28.8% of the arteriovenous difference). The changes in carbamino hemoglobin therefore account for almost one-third of the total change in CO_2 when blood passes from an artery to a vein (Roughton, 1964).

Whether or not carbamino compounds are formed in fish blood is not known. Indirect evidence seems to rule out the presence of carbamino hemoglobin in the dogfish, *Mustelus* (Ferguson *et al.*, 1938), and in the carp (Ferguson and Black, 1941), whereas in the trout some caramino hemoglobin may be present (Ferguson and Black).

From the distribution of CO_2 between cells and plasma in the dogfish, Ferguson *et al.* (1938) conclude that substances other than hemoglobin might be involved in the buffering of CO_2 within the red cell. Apart from the so-called Y-bound CO_2 (an association of protein and bicarbonate), they point out that not yet identified substances in the nuclei may participate in the buffering process. This seems to be an interesting phenomenon which also could be of importance for the buffering in the tissues. During heavy exercise it is claimed that creatinine diffuses out of the muscles and increases the buffering capacity of the blood in the trout (Black *et al.*, 1959).

3. Interaction between Red Blood Cells and Plasma

Mammalian red blood cells are practically impermeable to cations but are freely permeable to anions. Because of the negative charges of the nondiffusing protein inside the red blood cell, a Donnan equilibrium exists across the cell membrane. The distribution of the ions is characterized by the ratio r according to

$$r = \frac{[Cl^-]_i}{[Cl^-]_e} = \frac{[HCO_3^-]_i}{[HCO_3^-]_e} = \frac{[A^-]_i}{[A^-]_e} = \frac{[H^+]_e}{[H^+]_i} \tag{15a}$$

where $[A^-]$ is the concentration of all monovalent anions except bicarbonate. The subscripts refer to the red cells (i) and plasma (e). The concentrations have to be expressed per kilogram of cell water and of plasma water, respectively, not per liter of plasma or red cells. As a corollary of the Donnan equilibrium, the increase of bicarbonate within the red cell leads to a redistribution of anions. Freshly formed bicarbonate diffuses out of the cells into the plasma in exchange for chloride, until Eq. (15a) is fulfilled.

Since the bicarbonate formed increases the number of osmotic particles in the cell water, there is also a redistribution of water, which enters the cells and causes a small but measurable increase in cell volume.

The two predominant consequences of these processes in mammalian blood are:

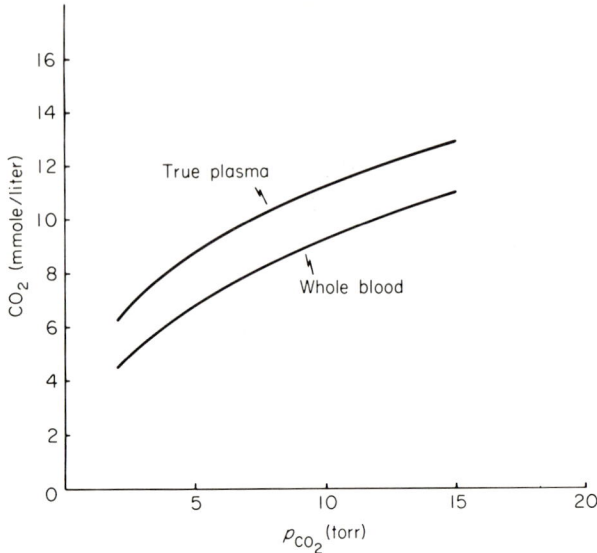

Fig. 9. CO_2 combining curve of true plasma and of whole blood in the rainbow trout, *Salmo gairdneri* Richardson. Data from Ferguson and Black (1941).

(1) From the total CO_2 in whole blood one-third only is found in the red cells and two-thirds in the plasma although the volume of red cells in man is about 45% of the blood. The CO_2 content of the plasma therefore exceeds that of whole blood (see Fig. 9).

(2) When blood passing through a capillary takes up CO_2, almost all bicarbonate is formed within the red cells but only one-half stays in the cells and the other half is exchanged for chloride.

There are almost no data available on the Donnan equilibrium of fish erythrocytes. In the dogfish, Ferguson *et al.* (1938) determined the Donnan ratio for chloride to be 0.49–0.61, whereas for CO_2 the ratio was 1.03–1.97. These authors stress the experimental difficulty of such determinations, especially in fish blood. They state, ". . . it cannot be concluded from these data that there is no shift in water or chloride when CO_2 is added to dogfish blood." Albers *et al.* (1969), however, demonstrated in the dogfish, *Scyliorhinus canicula*, a decrease in the chloride ratio when the CO_2 tension was elevated. According to Ferguson and Black (1941), in the rainbow trout there is a higher Donnan ratio for CO_2 than for chloride, whereas in the carp both ratios are of the same order of magnitude. With increasing CO_2 tensions the red cells of the trout display an unusual increase in volume together with an appreciable decrease in plasma chloride. It is assumed that acids other than carbonic acid are buffered within the red cells and are also exchanged for chloride (Ferguson and Black, 1941).

4. TRUE PLASMA VERSUS SEPARATED PLASMA

Because of the interaction between cells and plasma it is necessary to introduce the distinction between true plasma and separated plasma. If red cells are spun down, the remaining plasma may be equilibrated with various CO_2 tensions. The resulting CO_2 combining curve is, as a rule, flat and shows a smaller CO_2 combining power than does whole blood. Since the only buffer substances present are the plasma proteins, the buffer capacity (see below) of such plasma is very low. Plasma equilibrated with CO_2 without red cells present is called "separated plasma." If we equilibrate whole blood with various CO_2 tensions and spin the red cells under paraffin oil, analyze the plasma for CO_2, and plot the CO_2 concentrations against the p_{CO_2}, we obtain a steep CO_2 combining curve with a CO_2 combining power greater than that of whole blood and with a much greater buffer capacity (Fig. 9).

Plasma which has been equilibrated in the presence of red cells is called "true plasma." It should be noted that the curves of true plasma

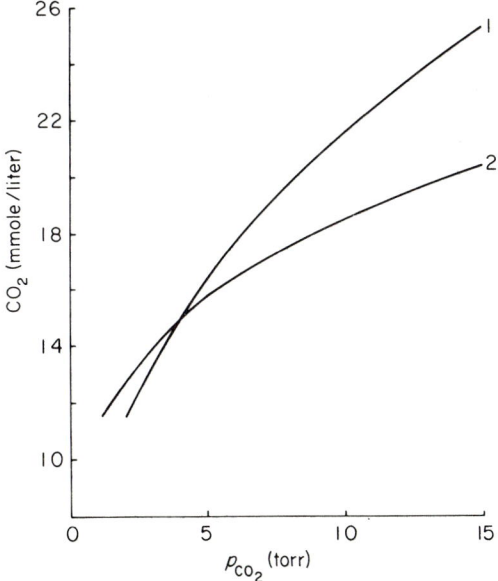

Fig. 10. CO_2 combining curves of true plasma (1) and of separated plasma (2) in the dogfish, Scyliorhinus stellaris. Separation at $p_{CO_2} = 4$ torr (Albers and Pleschka, 1967).

and of separated plasma have one point in common: This is the point of the CO_2 tension where the separation of red cells and plasma has taken place. Examples of the different CO_2 combining curves of true and separated plasma are given in Fig. 10 for the dogfish, *Scyliorhinus stellaris*. Differences between true plasma and separated plasma have been reported for the carp and the trout (Black *et al.*, 1959). No marked differences in the buffer capacity of true and separated plasma are found in the spiny dogfish (Lenfant and Johansen, 1966).

B. The pH of the Blood as Related to CO_2

1. The pH — log p_{CO_2} Diagram

The foregoing sections dealt with the transport of CO_2 by the blood. In this section we shall discuss how CO_2 affects the blood pH. The concentrations of physically dissolved and chemically bound CO_2 are related to the pH by Eq. (12c)

$$\text{pH} = pK_1' + \log\left(\frac{C_T}{Sp_{CO_2}} - 1\right)$$

where pK' is the operational constant which depends on temperature as well as on pH and ionic strength. Total CO_2 (C_T) and p_{CO_2} are linked together by the CO_2 combining curve. It immediately becomes apparent then that for a given p_{CO_2} the pH will be different according to the position of the CO_2 combining curve, i.e., the value of the total CO_2. When calculating the pH from the CO_2 combining curve, the relationship between $[H^+]$ and p_{CO_2} turns out to be a linear one, at least for a limited range of p_{CO_2}. This is valid for mammalian blood (Henderson, 1932) and was demonstrated to be true for fish blood by Root (1931). To avoid the conversion from pH into $[H^+]$ it is common practice to plot pH vs. log p_{CO_2}, which, of course, also results in straight lines. Examples are given in Fig. 11. These lines can be obtained experimentally and are widely used for the indirect determination of p_{CO_2} in blood samples. Two facts are evident from Fig. 11:

(1) As mentioned above, for a given p_{CO_2} the pH may differ from species to species according to the position of the CO_2 combining curve.

(2) There are differences in the slope of the lines. For a given increase in p_{CO_2} some species show a larger decrease in pH indicating a poor buffer capacity, whereas other species show a smaller decrease in pH, indicating a higher buffer capacity.

5. ACID–BASE BALANCE

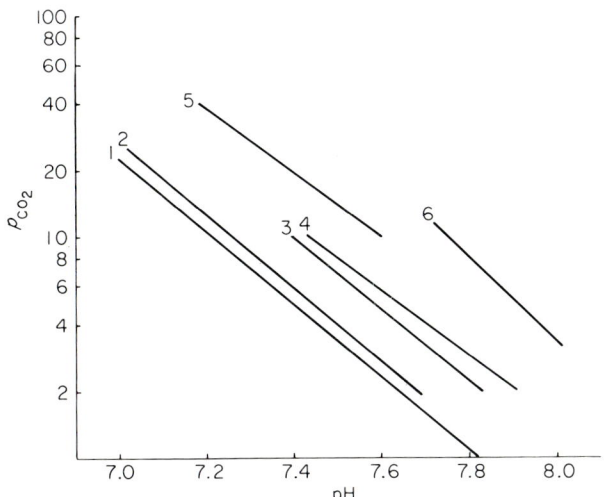

Fig. 11. pH–log p_{CO_2} lines in various fishes and in man. 1, *Opsanus tau* (20°C); 2, *Prionotus carolinus* (20°C); 3, *Scyliorhinus stellaris* (17°C); 4, *Cyprinus carpio* (15°C); 5, *Scomber* (20°C); 6, man (15°C). 1, 2, and 5 from Root (1931); 3 and 6 from Albers and Pleschka (1967); 4 from Ferguson and Black (1941).

2. The Buffer Capacity of Plasma and Blood

The buffer capacity of a buffer solution is quantitatively described by Eq. (14). The chemical reactions described above can be considered as typical examples for buffer reactions if we keep in mind that carbonic acid is the "strong acid" which has to be buffered by the plasma proteins and by hemoglobin. At the physiological pH these proteins act as weak acids which are partly present as dissociated salts thus forming a typical buffer solution. If carbonic acid is added to this system, hydrogen ions recombine with the proteins to form the undissociated "weak acid," leaving the cation together with the bicarbonate ion as fully dissociated salt. Metaphorically speaking, carbonic acid and proteins compete with each other for the available cations to form salts. The resulting compromise depends on the dissociation constant of the buffer substance: the lower the dissociation constant (i.e., the weaker the acid), the more hydrogen ions are bound to the buffer and the more cations are left for the strong acid to form a salt. We shall refer to this relationship when we consider the effect of the oxygenation of hemoglobin on the buffer capacity (see below).

For a quantitative comparison of the buffer action the buffer capacity as described by Eq. (14) has to be modified: Since the concentration of the buffer substances can be different, the buffer capacity often is expressed per unit weight of the buffer substance and denoted β:

$$\beta = \frac{1}{C_P} \frac{d\text{B}}{d\text{pH}} \quad (16\text{a})$$

$$\beta = 2.3 \frac{K[\text{H}^+]}{(K + [\text{H}^+])^2} \quad (16\text{b})$$

Values of β and of $d\text{B}/d\text{pH}$ are listed in Table V. Although β of plasma proteins in some elasmobranchs is higher than that of mammals, $d\text{B}/d\text{pH}$ of separated plasma in these species is about the same as in mammals owing to the lower concentration of plasma proteins (Dill et al., 1932). In most fish, $d\text{B}/d\text{pH}$ of whole blood is much lower than that of mammals. The main reason is the lower concentration of hemoglobin, which seems to have the same β in the dogfish as in man (Albers and Pleschka, 1967). However, there may exist specific differences in the physiochemical properties of the participating buffer substances, as claimed by Lenfant and Johansen (1966). There is an open field for further research which seems to be of great interest since the buffer capacity is considered to be of vital importance and a limiting factor for exercise performance in fish (Hochachka, 1961).

The sum of the cations available for the buffering of CO_2 equals the sum of chemically bound CO_2 and the dissociated moiety of the buffer and is often referred to in the literature as the concentration of "buffer base." The buffer base can be considered as constant as long as no strong acids other than carbonic acid are entering the blood. However, if for instance lactic acid enters the blood, an equivalent amount of cations is converted from the buffer base to the so-called "fixed base." The amount of cations available for the reversible buffer action is thus dimin-

Table V

Buffer Capacity of Whole Blood ($d\text{B}/d\text{pH}$) and Estimated Buffer Capacity per Gram Hemoglobin (β) for Various Fish

Species	t (°C)	Hb (g/liter)	$d\text{B}/d\text{pH}$	β	References
Opsanus tau	20	46	6.7	0.147	Root (1931)
Scomber	20	118	14.8	0.127	Root (1931)
Prionotus	20	57	6.7	0.118	Root (1931)
Cyprinus	15	92	—	0.160	Ferguson and Black (1941)
Salmo gairdneri	15	74	—	0.230	Ferguson and Black (1941)
Mustelus canis	22	30	10–20	—	Ferguson et al. (1938)
Squalus acanthias	11	30	9	0.117	Lenfant and Johansen (1966)
Scyliorhinus	17	31	10	0.190	Albers and Pleschka (1967)

5. ACID–BASE BALANCE

ished. As a consequence, for a given p_{CO_2} less CO_2 can be bound chemically and the CO_2 combining curve will be lowered by an equivalent amount. Likewise the position of the pH $-$ log p_{CO_2} line will be shifted to lower pH values for a given p_{CO_2}. In the medical literature a loss in buffer base is called a "metabolic acidosis" and an increase in buffer base is called a "metabolic alkalosis."

A typical example for a metabolic acidosis in fish is the accumulation of lactic acid in blood and tissues after severe exercise (Black, 1958; Black et al., 1959, 1962). In the rainbow trout, for example, Black et al. (1966a) observed an increase in blood lactate from 5.1 to 69.7 mg % after 15 min of strenuous exercise which would correspond to a decrease in buffer base of 7 mEq/liter.

3. The Effects of Oxygenation of Hemoglobin

As seen in Fig. 7 the CO_2 combining power of oxygenated blood is less than that of reduced blood. The following two factors contribute to this change in the CO_2 combining curve:

(1) Because of the difference in the dissociation constant between oxyhemoglobin and hemoglobin (see below) more bicarbonate is formed at a given p_{CO_2} in a solution containing reduced hemoglobin than in a solution containing oxygenated hemoglobin.

(2) The formation of carbamino hemoglobin is more pronounced with reduced hemoglobin than with oxygenated hemoglobin.

Both factors participate almost equally in the change of the CO_2 combining curve of mammals, although there is still some controversy about the exact quantitative relationship. The effect of the oxygenation on the CO_2 combining curve is one aspect of the so-called Haldane effect. The other and more important aspect is the effect on the buffer capacity and the pH $-$ log p_{CO_2} line. From Eqs. (14) and (16b) it becomes immediately evident that a decrease in K must increase β or dB/dpH. The oxygenation of hemoglobin now increases its dissociation constant appreciably. In horse hemoglobin for instance, K is changed by a factor of about 30 ($pK_{O_2Hb} = 6.68$, $pK_{Hb} = 7.95$). Similar effects are obtained in the blood of other mammals and also in some fish. Reduced hemoglobin, therefore, having a lower K and a higher pK, respectively, must be a stronger buffer substance. Since at the tissues the transfer of CO_2 into the blood takes place when O_2 is released from the oxyhemoglobin, there is a simultaneous increase in the buffer capacity. The importance of this effect is obvious: If the exchange ratio R of CO_2 and O_2 is 0.7, it can be assumed with the figures of pK given above that the CO_2 entering the blood will not cause any change in pH because of the Haldane effect. If $R = 0.8$, about 85% of the hydrogen ions from the carbonic acid are mopped up by the concomitant change in the oxygenation and only 15% have to be buffered according to Eq. (14). As a result the arteriovenous pH difference is low, at least in mammals, where the pH of the mixed venous blood from the right heart has a pH which is about 0.03–0.05 lower than that of the arterial blood. In fish the difference may be higher because of the lower buffer capacity. For instance, in the dogfish the difference in pH between arterial and venous blood has been found to be 0.07 (Baumgarten-Schumann and Piiper, 1969).

4. The Effect of Temperature

As pointed out earlier, temperature affects the dissociation constants of the buffer substances as well as the solubility of CO_2. As a result, the CO_2 combining curves of blood are shifted upward if the temperature is lowered. This is observed in mammalian blood (Harms and Bartels, 1961), in the blood of the skate (Dill et al., 1932), and in the dogfish (Fig. 12a). Since the observed shift is much greater than can be accounted for by the change in solubility, the chemically bound CO_2 must also have been altered by temperature (Fig. 12b).

Wherease a change in the CO_2 combining curve due to changes in the oxygenation of the hemoglobin inevitably affects the relationship

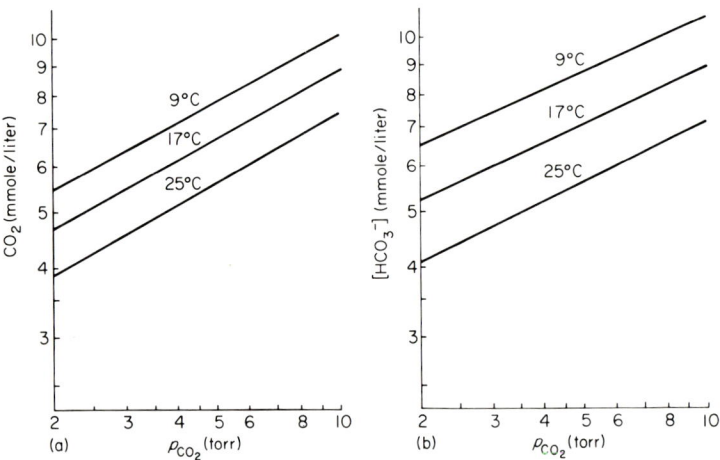

Fig. 12. Effect of temperature on the CO_2 combining curve in the dogfish, Scyliorhinus stellaris. (a) Total CO_2 content of whole blood plotted against p_{CO_2}. (b) Chemically bound CO_2 of true plasma (designated as HCO_3^-) plotted against p_{CO_2}. From Albers and Pleschka (1967). Reprinted by permission of North Holland Publishing Co.

between p_{CO_2} and pH, this does not necessarily happen if the CO_2 combining curve is altered by temperature. If in Eq. (12c)

$$\text{pH} = pK_1' + \log\left(\frac{C_T}{Sp_{CO_2}} - 1\right)$$

total CO_2, (C_T), and S are altered by temperature in the same proportion, no change of the pH − log p_{CO_2} line would occur. This is the case in human blood at physiological CO_2 tensions in the range between 37° and 26°C (Brewin et al., 1955). However, as these authors point out, this constancy of the pH − log p_{CO_2} line is purely fortuitous: Temperature affects the pH − log p_{CO_2} line of both plasma and red cells but in an opposite direction. In man, at the prevailing quantitative relationship between red cell volume and plasma volume, both changes cancel out each other. Therefore, if the hematocrit is changed, the pH − log p_{CO_2} line does not remain unchanged. However, at very low CO_2 tensions and at temperatures between 25° and 9°C the temperature effect on the pH − log p_{CO_2} line of human blood is marked (Albers and Pleschka, 1967). In fish, data are available only for elasmobranchs. In dogfish blood an increase in temperature decreases the pH for a given p_{CO_2}, as shown by Albers and Pleschka (1967). These authors were unable to detect a significant change of $dB/d\text{pH}$ with temperature in dogfish blood. This is in accord with the findings in mammalian blood.

From the changes in the CO_2 combining curve and the accompanying changes in pH it is possible to calculate the heat of dissociation of the functional groups involved in the buffering of CO_2. Albers and Pleschka (1967) arrived at the same value for dogfish blood as for mammalian blood, indicating the participation mainly of imidazole. In contrast the buffering in tissues is almost entirely owing to inorganic and organic phosphate compounds (Netter, 1959), resulting in much lower values for the heat of dissociation. This was confirmed by Mersch (1964) who found for homogenates of rat liver a value of 500 cal/mole as opposed to the value in human blood of 6300 cal/mole.

Because of the profound effects of temperature and the great variability of body temperatures in fish of various habitats, data on the acid–base balance and the CO_2 transport should always include the temperature. Otherwise such data would lose much of its informational value.

A most revealing effect of temperature was observed by Rahn (1967) when he acclimatized carps, turtles, and frogs to various temperatures in the range from 5° to 30°C. With increasing temperature the arterial pH fell in parallel to the pH of neutrality in all species. Thus the difference pH − pN was maintained constant. Since pH − pN measures

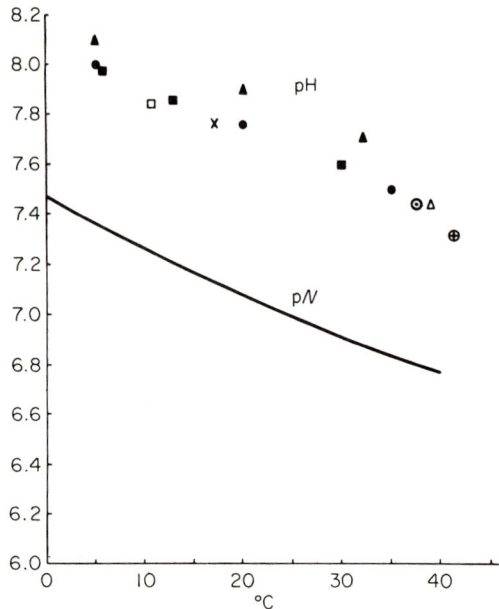

Fig. 13. Arterial pH of various vertebrate species as a function of temperature: (■) carp, (▲) frog, (●) turtle (all data from Rahn, 1967), (□) skate, (×) dogfish, (⊙) man, (△) dog, (⊕) duck (from various references). $pN = pH$ of pure water.

the relative alkalinity (see Section II, E, 2), and moreover, since the same $pH - pN$ is found in many warm-blooded animals (Fig. 13), Rahn (1967) claims the relative alkalinity to be one of the most important quantities for the evolution of respiration.

5. SUMMARY

CO_2 transport may now be summarized briefly as follows:

(1) CO_2 formed by metabolic processes in the tissues diffuses into the blood and reacts with water to form carbonic acid. The reaction is slow in the plasma. Within the red cells it is speeded up by the enzyme carbonic anhydrase.

(2) Carbonic acid dissociates into hydrogen ions, bicarbonate ions, and to a small extent into carbonate ions.

(3) The hydrogen ions recombine with plasma proteins and hemoglobin which represent the buffer substances. Proteins and hemoglobin can be considered as weak acids which are partly present as fully dissociated salts.

(4) Bicarbonate leaves the red cells in exchange for chloride according to Donnan's law.

(5) Carbonic acid also reacts with hemoglobin to form carbamino hemoglobin.

(6) Oxygenation of the hemoglobin decreases the buffer capacity as well as the ability to form carbamino hemoglobin (Haldane effect) in teleosts but not in elasmobranchs.

(7) In the lungs or gills the above-mentioned reactions occur in the opposite direction.

IV. THE INTRACELLULAR pH

The rather complex processes of the CO_2 transport and the buffering of carbonic acid and other acids in the blood are fairly simple when compared with the analogous processes within the tissues. A comparative study of tissue CO_2 in several vertebrates has recently been published by Haning and Thompson (1965). In contrast to the blood, proteins are much less important as buffer substances. Some CO_2 within the tissues seems to be bound in a carbaminolike fashion and cannot be precipitated by barium hydroxide (Conway and Fearon, 1944). There is still some controversy about the chemical meaning of the so-called "barium soluble fraction" of CO_2. Data on this fraction are given for several vertebrate tissues by Haning and Thompson (1965). In the muscle of the channel catfish, *Ictalurus punctatus*, these authors found 70% of the tissue CO_2 to be barium soluble as compared with 77% in rat muscle and 72% in frog muscle. As recently shown by Butler *et al.* (1967) the "barium soluble CO_2" does not result from a carbaminolike compound but from the inhibitory effect of proteins on the precipitation of barium carbonate. As mentioned earlier the buffer capacity of tissues is made up predominantly of inorganic and organic phosphate compounds. The total buffer capacity of tissues can well compete with that of the blood in some organs of mammals (e.g., liver tissue of rats) although in most organs the buffer capacity is lower than in blood. The buffer processes in the interior of a cell are closely linked to so many biochemical reactions that a theoretical approach seems to be impossible.

The intracellular pH has been determined using a variety of methods and has been found to be always lower than that of blood. Chambers *et al.* (quoted by Netter, 1959) reported for the nerve cells of fish a pH of 6.8–6.9, based on a colorimetric indicator method. Electrometric pH measurements have also been employed, yielding results comparable with those of indicator methods or with methods applying the Henderson-Hasselbalch equation for tissue CO_2.

An interesting approach to the intracellular pH was made by Waddell and Butler (1959); it is based on the weak acid DMO which can permeate the cell membrane only as an undissociated acid. Thus, extracellular and intracellular concentrations of the undissociated acid are equal. If the total concentration of DMO is determined in the extracellular fluid (plasma) as well as in the intracellular fluid, the undissociated moiety can be calculated by means of the pK' of DMO and the plasma pH. In the next step of the calculation the dissociated moiety is obtained by simple subtraction of the undissociated fraction from the total concentration. Finally, the intracellular pH is obtained by again applying the Henderson-Hasselbalch equation. Using the glass electrode as well as the DMO method, Robin (1961) determined the pH of the pericardial fluid of the dogfish (*Squalus acanthias*) and found good agreement between both methods. The average pH of the pericardial fluid was about 5.6. Robin *et al.* (1964) determined the intracellular pH with the DMO method in the same species and reported 7.41 as average for the brain and 6.95 for the muscle.

V. CONTROLLING MECHANISMS OF THE ACID–BASE BALANCE

It is not the aim of this section to describe in detail the operation of the controlling mechanisms of the acid–base balance. The reader is referred to the pertinent chapters of these volumes. Instead we shall briefly summarize the basic principles of the regulatory mechanisms involved as can be deduced from the previous sections. All quantities used in the many equations of this chapter can be divided into three groups:

(1) The first group consists of pure physicochemical constants, such as the ionic product of water K_W, the dissociation constants K of the participating acids and buffer substances, the solubility of CO_2, etc. These constants depend on temperature and ionic strength. The organism has to cope with these quantities which are completely independent of biological regulation.

(2) The second group is made up of the concentrations of the substances involved in the acid–base balance except the physically dissolved CO_2 (and the bicarbonate). Among these quantities are the various electrolytes, the plasma proteins, the hemoglobin, and other buffer substances. This group is subjected to regulatory mechanisms of several

organs, especially to the function of excretory organs like the kidney. The "buffer base" can be changed by a redistribution of electrolytes between tissues, extracellular fluid, and plasma. In addition to the regulatory mechanisms in mammals, in fish there is also an exchange possible between the plasma and the surrounding seawater. Robin et al. (1966) and Dejours (1966) point to the possibility of an active excretion of bicarbonate by the gills. The gills also excrete ammonia (Dejours et al., 1968), which in part is exchanged for sodium taken up by the gills (Garcia Romeu and Motais, 1966). Long lasting exposure to asphyxic conditions can produce an increase in hemoglobin, which is not only observed in mammals but also in fish. All changes of variables belonging to the second group are recognized by the evaluation of the CO_2 combining curve or by the pH $-$ log p_{CO_2} line. Generally such changes are long-term.

(3) There is only one quantity in this last group. This is the partial pressure of CO_2 which regulates the physically dissolved CO_2 and, hence, according to the CO_2 combining curve, the pH of the arterial blood. The p_{CO_2} is adjusted by the effective irrigation of the gills since for a given CO_2 production the p_{CO_2} varies inversely with it. Ventilatory changes of p_{CO_2} and pH are almost instantaneous and provide the basis for short-term adaptations required, e.g., during muscular exercise. Changes of the arterial p_{CO_2} can be measured directly.

To put it in another way, we may also say that there are two principally different ways of regulating the acid–base balance. There is the metabolic way, dealing chiefly with the buffer processes of the homogeneous type, and there is the respiratory way, dealing chiefly with the buffer processes of the heterogeneous type. As an open system the fish like all other animals is defended twofold against changes of the acidity of its *milieu intérieur*. This defense, however, appears less powerful in fish than in mammals. Whether this is considered as a lower stage of the phylogenetic development or as an adaptation to the ecological demands of the aquatic life is a question the reader may decide.

REFERENCES

Albers, C. (1962). Die ventilatorische Kontrolle des Säure-Basen-Gleichgewichts in Hypothermie. *Anaesthesist* **11**, 43–51.
Albers, C. (1966). Unpublished data.
Albers, C., and Pleschka, K. (1967). Effect of temperature on CO_2 transport in elasmobranch blood. *Respiration Physiol.* **2**, 261–273.
Albers, C., and Pleschka, K. (1967). Unpublished data.
Albers, C., Pleschka, K., and Spaich, P. (1969). Chloride distribution between red blood cells and plasma in the dogfish (Scyliorhinus canicula). *Respiration Physiol.* **7**, 295–299.

Bartels, H., and Wrbitzky, R. (1960). Bestimmung des CO_2-Absorptionskoeffizienten zwischen 15 und 38°C in Wasser und Plasma. *Arch. Ges. Physiol.* **271**, 162–168.

Baumgarten-Schumann, D., and Piiper, J. (1969). Gas exchange in the gills of resting unanesthetized dogfish (Sciliorhinus stellaris). *Respiration Physiol.* **5**, 317–325.

Becker, E. L., Bird, R., Kelly, J. W., Schilling, J., Solomon, S., and Young, N. (1958). Physiology of marine teleosts. I. Ionic composition of tissue. *Physiol. Zool.* **31**, 224–227.

Black, E. C. (1958). Hyperactivity as a lethal factor in fish. *J. Fisheries Res. Board Can.* **15**, 573–586.

Black, E. C., Chiu, W., Forbes, F. D., and Hanslip, A. (1959). Changes in pH, carbonate and lactate of the blood of yearling Kamloops Trout (Salmo gairdnerii) during and following severe muscular activity. *J. Fisheries Res. Board Can.* **16**, 391–402.

Black, E. C., Robertson Connor, A., Lam, K., and Chiu, W. (1962). Changes in glycogen, pyruvate and lactate in rainbow trout (Salmo gairdnerii) during and following muscular activity. *J. Fisheries Res. Board Can.* **19**, 409–436.

Black, E. C., Manning, G. T., and Hayashi, K. (1966a). Changes in levels of hemoglobin, oxygen, carbon dioxide, pyruvate, and lactate in venous blood of rainbow trout (Salmo gairdnerii) during and following severe muscular activity. *J. Fisheries Res. Board Can.* **23**, 783–795.

Black, E. C., Kirkpatrick, D., and Tucker, H. H. (1966b). Oxygen dissociation curves of the blood of Landlocked Salmon (Salmo salar sebago) acclimated to summer and winter temperatures. *J. Fisheries Res. Board Can.* **23**, 1581–1586.

Black, E. C., Tucker, H. H., and Kirkpatrick, D. (1966c). Oxygen dissociation curves of the blood of Atlantic Salmon (Salmo salar) acclimated to summer and winter temperatures. *J. Fisheries Res. Board Can.* **23**, 1187–1195.

Brewin, E. G., Gould, R. P., Nashat, F. S., and Neil, E. (1955). An investigation of problems of acid-base equilibrium in hypothermia. *Guy's Hosp. Rept.* **104**, 177–214.

Butler, T. C., Poole, D. T., and Waddell, W. J. (1967). Acid-labile carbon dioxide in muscle: Its nature and relationship to intracellular pH. *Proc. Soc. Exper. Biol. Med.* **125**, 972–974.

Conway, E. J., and Fearon, P. J. (1944). The acid-labile CO_2 in mammalian muscle and the pH of the muscle fibre. *J. Physiol. (London)* **103**, 274–289.

Dejours, P. (1966). Respiratory gas exchange of aquatic animals during confinement. *J. Physiol. (London)* **186**, 126–127.

Dejours, P., Armand, J., and Verriest, G. (1968). Carbon dioxide dissociation curves of water and gas exchange of water-breathers. *Respiration Physiol.* **5**, 23–33.

Dill, D. B., Edwards, H. T., and Florkin, M. (1932). Properties of the blood of the skate (Raia oscillata). *Biol. Bull.* **62**, 23–36.

Ferguson, J. K. W., and Black, E. C. (1941). The transport of CO_2 in the blood of certain fresh water fishes. *Biol. Bull.* **80**, 139–152.

Ferguson, J. K. W., Horvath, S. M., and Pappenheimer, J. R. (1938). The transport of carbon dioxide by erythrocytes and plasma in dogfish blood. *Biol. Bull.* **75**, 381–388.

Field, J. B., Elvehjem, C. A., and Juday, C. (1943). A study of the blood constituents of carp and trout. *J. Biol. Chem.* **148**, 261–269.

Fry, F. E. J. (1957). The aquatic respiration of fish. *In* "The Physiology of Fishes" (M. E. Brown, ed.), Vol. 1, pp. 1–63. Academic Press, New York.

Garcia Romeu, F., and Motais, R. (1966). Mise en évidence d'échanges Na^+/NH_4^+ chez l'anguille d'eau douce. *Comp. Biochem. Physiol.* **17**, 1201–1204.
Haning, Q. C., and Thompson A. M. (1965). A comparative study of tissue carbon dioxide in vertebrates. *Comp. Biochem. Physiol.* **15**, 17–26.
Harms, H., and Bartels, H. (1961). CO_2-Dissoziationskurven des menschlichen Blutes bei Temperaturen von 5–37°C und unterschiedlicher O_2-Sättigung. *Arch. Ges. Physiol.* **272**, 384–392.
Hartman, F. A., Lewis, L. A., Brownell, K. A., Shelden, F. F., and Walther, R. W. (1941). Some blood constituents of the normal skate. *Physiol. Zool.* **14**, 476–486.
Harvey, H. W. (1966). "The Chemistry and Fertility of Sea Waters." Cambridge Univ. Press, London and New York.
Henderson, L. J. (1932). "Blut" (German transl. by M. Tennenbaum). Steinkopff, Dresden.
Hochachka, P. W. (1961). The effect of physical training on oxygen debt and glycogen reserves in trout. *Can. J. Zool.* **39**, 767–776.
Hoffert, J. R. (1966). Observations on ocular fluid dynamics and carbonic anhydrase in tissues of lake trout (Salvelinus Namaycush). *Comp. Biochem. Physiol.* **17**, 107–114.
Hoffert, J. R., and Fromm, P. O. (1966). Effect of carbonic anhydrase inhibition on aqueous humor and blood bicarbonate ion in the teleost (Salvelinus Namaycush). *Comp. Biochem. Physiol.* **18**, 333–340.
Larimer, J. L., and Schmidt-Nielsen, K. (1960). A comparison of blood carbonic anhydrase of various mammals. *Comp. Biochem. Physiol.* **1**, 19–23.
Lenfant, C., and Johansen, K. (1966). Respiratory function in the elasmobranch Squalus suckleyi. *Respiration Physiol.* **1**, 13–29.
Maren, T. H. (1962). Ionic composition of cerebrospinal fluid and aqueous humor of the dogfish, Squalus acanthias. II. Carbonic anhydrase activity and inhibition. *Comp. Biochem. Physiol.* **5**, 201–215.
Maren, T. H. (1967). Carbonic anhydrase: Chemistry, physiology and inhibition. *Physiol. Rev.* **47**, 595–781.
Mersch, F. D. (1964). Der Temperatureinfluss auf die CO_2-Dissoziationskurven von Rattenleberhomogenaten. Inaugural Dissertation, Justus Liebig University, Giessen.
Netter, H. (1959). "Theoretische Biochemie." Springer, Berlin.
Pereira, R. S., and Sawaya, P. (1957). Contribution à l'étude de la composition chimique du sang de certains sélaciens du Brésil. *Univ. de Sao Paulo, Fac. Filosof., Cienc. Letras, Zool.* **21**, 85–92.
Prosser, C. L., and Brown, F. A. (1961). "Comparative Animal Physiology." Saunders, Philadelphia, Pennsylvania.
Rahn, H. (1967). Gas transport from the external environment to the cell. *Ciba Found. Symp. Develop. Lung.* pp. 3–29.
Robertson, J. D. (1954). The chemical composition of the blood of some aquatic chordates, including members of the Tunicata, Cyclostomata and Osteichthyes. *J. Exptl. Biol.* **31**, 424–442.
Robin, E. D. (1961). Of men and mitochondria—intracellular and subcellular acid–base relations. *New Engl. J. Med.* **265**, 780–785.
Robin, E. D., Murdaugh, H. V., and Weiss, E. (1964). Acid–base, fluid and electrolyte metabolism in the elasmobranch. I. Ionic composition of erythrocytes, muscle and brain. *J. Cellular Comp. Physiol.* **64**, 409–422.
Robin, E. D., Murdaugh, H. V., and Millen, J. E. (1966). Acid–base, fluid and

electrolyte metabolism in the elasmobranch. III. Oxygen, CO_2, bicarbonate and lactate exchange across the gill. *J. Cellular Physiol.* **67**, 93–100.

Root, R. W. (1931). The respiratory function of the blood of marine fishes. *Biol. Bull.* **61**, 427–466.

Roughton, F. (1964). Transport of oxygen and carbon dioxide. *In* "Handbook of Physiology" (Am. Physiol. Soc., J. Field, ed.), Sect. 3, Vol. I, p. 767. Williams & Wilkins, Baltimore, Maryland.

Severinghaus, J. W., Stupfel, M., and Bradley, A. F. (1956). Variations of serum carbonic acid pK' with pH and temperature. *J. Appl. Physiol.* **9**, 197–200.

Siggaard-Andersen, O. (1965). "The Acid–Base Status of the Blood." Munksgaard, Copenhagen.

Smith, H. W. (1929). Body fluids of elasmobranchs. *J. Biol. Chem.* **81**, 407–419.

Smith, H. W. (1956). "Principles of Renal Physiology." Oxford Univ. Press, New York.

Waddell, W. J., and Butler, T. C. (1959). Calculation of intracellular pH from the distribution of 5,5-dimethyl -2,4-oxazolidinedione (DMO). Application to skeletal muscle of the dog. *J. Clin. Invest.* **38**, 720–729.

Winterstein, H. (1954). Der Einfluss der Körpertemperatur auf das Säure-Basen-Gleichgewicht im Blut. *Arch. Exptl. Pathol. Pharmokol.* **223**, 1–18.

6

PROPERTIES OF FISH HEMOGLOBINS

AUSTEN RIGGS

I. Introduction 209
II. Structural Properties 210
   A. Components and Subunits 210
   B. Primary Structure 217
III. Oxygen Transport 222
   A. Critique of Measurements 223
   B. Cooperativity: Heme–Heme Interaction and Oxygenation-Linked Changes in Aggregation of Subunits 227
   C. Effects of pH and CO_2 229
   D. Temperature Dependence 232
   E. Adaptations of Hemoglobins in Different Fish 235
Acknowledgments 246
References 246

I. INTRODUCTION

Most studies of hemoglobins have been confined to the blood and muscle of mammals. As a result, the knowledge of the functional and structural properties of the hemoglobins of lower vertebrates is quite fragmentary. This preoccupation with mammalian hemoglobins is unfortunate from the standpoint of comparative physiology because the respiratory requirements of lower vertebrates vary over an enormous range; this gives rise to many interesting adaptations. At one extreme, the metabolic requirements for oxygen are so low that an oxygen transport pigment is unnecessary: sufficient oxygen is physically dissolved. Thus the antarctic ice fish (Ruud, 1954) have colorless blood and lack hemoglobin. At the other extreme, metabolically active fish often have such large requirements for oxygen that blood transport hemoglobin is not only necessary but even a slight anemia may be a serious disadvantage. Fish can solve their requirements for oxygen in many ways; the presence of a special blood hemoglobin is only one link in a series of adaptations. This fact should be borne in mind when examining the oxygen transport properties of hemoglobins. Thus oxygen transport depends upon the presence of gills, and/or lungs, their effective surface area, the distribution of capillaries, the rate of fluid pumping by the heart, the number of red blood cells, and the concentration of hemoglobin within them. In addition, recognition needs to be given to the fact that the red blood cells constitute a metabolizing tissue and the state of this metabolism can have important consequences in modifying the oxygen transport function of the hemoglobin. For these reasons it is necessary to consider not only the properties of the hemoglobins in free solution but also their cellular environment and how this environment can modify the function of the hemoglobin.

Extensive surveys of the earlier literature on hemoglobins can be found in the reviews of Prosser and Brown (1961), Manwell (1960), Antonini (1965), Riggs (1965), Braunitzer et al. (1964), Rossi-Fanelli et al. (1964), and Wyman (1964). In addition, data on a variety of hemoglobins are included in two studies on cyclostome and on invertebrate hemoglobins by Manwell (1963b, 1964), which contain much information on fish hemoglobins not suggested by their titles.

Studies of the evolution of hemoglobins are largely based on amino acid sequence studies of mammalian blood and muscle hemoglobins (see Ingram, 1961, 1963). Such data for lower vertebrates exist only for one of the polypeptide chains of carp hemoglobin (Hilse and Braunitzer, 1968) and for the hemoglobin of the lamprey (Braunitzer and Fujiki, 1969). No sequence data have been published on any reptilian or amphibian hemoglobin, and fragmentary structural data exist on only a few fish hemoglobins. Current work in several laboratories indicates that we can look forward to substantial progress in this area in the near future. In the meantime, the reviews by Dixon (1966), Nolan and Margoliash (1968), and Watts (1968) may be consulted for current ideas on the process and mechanism of protein evolution. An annual atlas of protein sequence and structure (Dayhoff, 1969) can be consulted for the current status of structural work.

II. STRUCTURAL PROPERTIES

A. Components and Subunits

1. Nature of Multiple Components

The hemoglobins of vertebrates are remarkably uniform in molecular weight; most have weights near 65,000. The close similarity of the conformation of horse hemoglobin chains (Perutz et al., 1968), sperm whale myoglobin (Kendrew et al., 1960), the hemoglobins of the marine annelid, *Glycera* (Padlan and Love, 1968), and that of the insect *Chironomus* (Huber et al., 1968) lends confidence to the belief that all hemoglobin polypeptide chains, both vertebrate and invertebrate, are arranged similarly and all are derived from a common ancestral form. Therefore, we have every expectation that X-ray diffraction and amino acid sequence studies will show fish hemoglobins to possess many features in common with the better studied mammalian pigments. Nevertheless, this should not obscure the fact that fish hemoglobins possess many unique features, both in their physiology and their primary structure.

Most fish hemolyzates contain multiple components. An example of the considerable variation in number and proportions of hemoglobin components is shown in Fig. 1 from the electrophoretic data of Yamanaka et al. (1965). Other observations of multiple components are summarized in Table I. These data show that hemolyzates with only a single component are quite exceptional. Fish, reptiles, and amphibians generally have much greater multiplicity of hemoglobin components than do mammals or birds (see review by Gratzer and Allison, 1960).

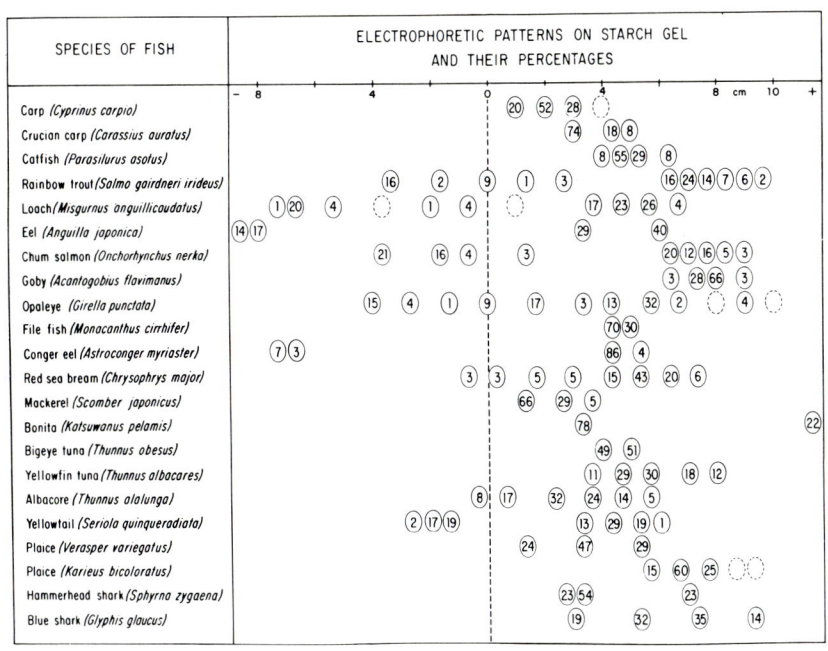

Fig. 1. The electrophoretic patterns obtained from hemolyzates from various species of fish; redrawn from Yamanaka et al. (1965).

Table I
Multiple Components in Various Fish

Fish	No. of components	Reference
Tench, *Tinca tinca*	1–3	Callegarini and Cucchi (1968)
Chinook salmon, *O. tshawytscha*	2	Buhler (1963)
Rainbow trout, *Salmo gairdneri*	3	Buhler (1963)
Herring, *Clupea harengus* (juvenile)	2	Wilkins and Iles (1966)
Herring, *Clupea harengus* (adult)	4	Wilkins and Iles (1966)
Sprat, *Sprattus sprattus*	1–3	Wilkins and Iles (1966)
Pilchard, *C. pilchardus*	2	Wilkins and Iles (1966)
Shad, *Alosa alosa*	4	Wilkins and Iles (1966)

How do so many components arise? Several quite different mechanisms give rise to this multiplicity. A small number of unique polypeptide chains may combine with one another in different ways to produce a larger number of components. In addition, other molecules, large or small, may combine with a hemoglobin molecule and alter its electrophoretic mobility by changing its charge, its conformation, or by inducing a change in the degree of aggregation of subunits. Oxygen itself causes dissociation of several hemoglobins into their subunits (see Section III, E, 1). Oxidation of the heme iron can introduce apparent heterogeneity; partially oxidized tetramers containing four iron atoms may give rise to as many as five electrophoretic components according to whether 0, 1, 2, 3, or 4 of the iron atoms are oxidized (Itano and Robinson, 1956).

Although all normal mammalian blood hemoglobins consist of two types of chain, α and β, of which one form is $\alpha_2\beta_2$, this appears not to be true for all fish hemoglobins. Tetrameric mammalian hemoglobins dissociate into their subunits in very dilute solution and at extremes of pH and ionic strength:

$$\alpha_2\beta_2 \overset{K_1}{\rightleftharpoons} 2\alpha\beta \overset{K_2}{\rightleftharpoons} 2\alpha + 2\beta$$

A high salt concentration appears to increase K_1 but to decrease K_2; K_2 becomes large only below pH 5 (Guidotti *et al.*, 1963). Almost no information on these processes is available for fish hemoglobins. Substantial evidence now exists that solutions of mixtures of two mammalian hemoglobins, $\alpha_2^A\beta_2^A$ and $\alpha_2^B\beta_2^B$, consist of an equilibrium mixture of several forms, including such hybrid species as $\alpha^A\alpha^B\beta_2$, $\alpha^A\alpha^B\beta^A\beta^B$, and $\alpha_2\beta^A\beta^B$. These forms are seldom detected among mammalian hemoglobins by electrophoretic and chromatographic procedures because the separatory procedures usually act on the $\alpha\beta$ units rather than the tetramers (Guidotti *et al.*, 1963). [See Riggs (1965) and Schroeder and Jones (1965) for review of the evidence for these conclusions.] The number of apparent "components" which are observed after starch or acrylamide gel electrophoresis will depend on the relationship between the kinetics and equilibria of dissociation into subunits and the time required for the separatory procedure. Some fish hemoglobins appear to behave quite differently from mammalian hemoglobins in this respect.

Aggregation of tetramers ($\alpha_2\beta_2$, 65,000 MW 4–4.5 S) to larger aggregates (130,000 MW and higher, $s_{20,w} \geq$ 6–7 S) is of widespread occurrence in hemolyzates of reptiles and amphibians (Svedberg and Hedenius, 1934), but such aggregation in fish hemoglobins has only been reported for the dogfish (Chiancone *et al.*, 1966) and constituted only 5–10% of the total hemoglobin. Nevertheless, these polymers, which all appear to involve intermolecular —S—S— bonds, do not appear to modify the oxygen transport function of the hemoglobin (Riggs, 1966; Riggs and Rona, 1969) and do not involve any change in electric charge on the molecule. Therefore, such a mutation may not to be selected severely against. However, the almost universal occurrence in reptiles and amphibians and the almost complete absence from fish and other vertebrates suggest that some selection is at work. One possible function of hemoglobin —SH groups in the hemoglobins of reptiles and amphibians might be as a reservoir of reducing power for maintenance of the iron in the ferrous form.

2. DEVELOPMENTAL CHANGES

Ontogenetic changes in fish hemoglobin components are widespread and have been observed in three species of lamprey (Adinolfi *et al.*, 1959; Manwell, 1963b), the dogfish (Manwell, 1958a, 1963a), the skate (Manwell, 1958b), the teleost, *Scorpaenichthyes* (Manwell, 1957), the herring and spratt (Wilkins and Iles, 1966), and four species of salmon (Vanstone *et al.*, 1964; Hashimoto and Matsuura, 1960b; Koch *et al.*, 1964, 1966; Wilkins, 1968).

The ammocoete larva of the lamprey, *Petromyzon planeri*, has two major components which are replaced after metamorphosis with two different components (Adinolfi *et al.*, 1959). Since the oxy and met forms of lamprey hemoglobin all appear to be monomeric, this finding indicates that the two hemoglobins of the ammocoete are replaced by two new hemoglobins, so a total of four polypeptide chains is probably involved.

Substantial evidence now exists that in a number of fish changes in hemoglobins occur during a large fraction of the life cycle. Thus, Wilkins and Iles (1966) have found that the youngest juveniles of the

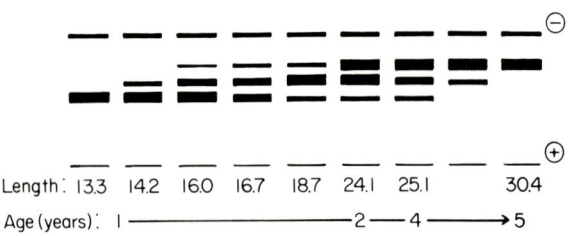

Fig. 2. Changes in the electrophoretic patterns of hemolyzates of the herring, *Clupea harengus*. From Wilkins and Iles (1966).

herring, *Clupea harengus*, have two components. One of these is replaced gradually over a period of 4 years (see Fig. 2). During this time a third component is present which is absent in both the youngest and the fully mature fish. The nature of this component is not known but it is probably a hybrid molecule containing both "juvenile" and "adult" polypeptide chains. If juvenile and adult hemoglobins arise from different cell populations, which is quite probable, this third component would be an artifact of preparation. If, however, the two hemoglobins were present in the same cells the presence of the third component might be physiologically important because just prior to the age of 2 years it appears as the major component. This finding is reminiscent of a hemoglobin component of the bullfrog in metamorphosing tadpoles which was present neither in premetamorphic tadpoles nor in the adult (Herner and Frieden, 1961). Wilkins (1968) has found a similar situation in the growth of the Atlantic salmon, *Salmo salar*, in which a total of 17 hemoglobin components (9 anodic and 8 cathodic) can be identified. Among these components complex ontogenetic changes occur: an increase is followed by a later reduction in the number of fractions. If the hemoglobins are all of the form $\alpha_2\beta_2$ then 7 polypeptide chains would be sufficient, but if forms such as β_4 occur, only 6 polypeptide chains are necessary. If, in addition, some tetramers are composed of four different chains: $\alpha^A\alpha^B\beta^C\beta^D$, then at least 16 bands could be produced by rearranging only 4 chains. These possibilities are not yet resolved. It is interesting that thyroid administration causes changes in the hemoglobin proportions; these changes can be induced during a large portion of the growth of the salmon.

3. Multiplicity of Components by Combination with Other Molecules

   a. *Reaction of NH_2-Terminal Group of the Polypeptide Chain.* Reaction of the NH_2-terminal group with an acetyl or other blocking group will change the net charge on the molecule and thus result in multiplicity of components if the reaction is incomplete. So, for example, the hemoglobin of *Lampetra fluviatilis* consists of two major components which appear to have the same amino acid sequence and differ only at the NH_2 terminus: 20% of the molecules have freely reactive NH_2-terminal proline, while the rest are blocked to reaction with NH_2-terminal reagents (Braunitzer, 1966). Such groups (often acetyl) appear, in hemoglobins, at least, to be added *after* chain synthesis and depend on the presence of an active enzymic acetylation system (Marchis-Mouren and Lipmann, 1965). The presence of this variation, therefore, apparently has nothing directly to do with the genetics of hemoglobin structure itself, although blocking of this group might be important in modifying the role of hemoglobin in CO_2 transport (see Section III, C on the mechanism of the Bohr effect). An acetyl NH_2 terminus would, of course, be genetically determined if the presence or absence of the enzyme system were so determined.

b. Allosteric Effectors. Barcroft showed more than 50 years ago that dialysis of hemolyzates modified the oxygen affinity of the hemoglobin, and he noted that phosphates had particularly large effects on the oxygen equilibrium (see Barcroft, 1928). Had hemoglobin been considered an enzyme, an attempt surely would have been made to determine what was removed by dialysis and to reconstitute by adding back what had been removed. But only recently has it become established that diphosphoglycerate (DPG) is normally present in human hemolyzates in the approximate ratio of one mole of DPG per hemoglobin molecule, and, most important, the DPG combines with the hemoglobin and produces an electrophoretically distinct component, which has a much lower oxygen affinity (Chanutin and Curnish, 1964, 1967; Benesch and Benesch, 1967, 1969). The binding of DPG is primarily to the deoxygenated pigment. Diphosphoglycerate appears, therefore, to act as an "allosteric effector" in controlling the oxygen transport function of hemoglobin. Other organic phosphates, especially ATP, are also present in red cells and have similar effects, but they are quantitatively much less important. Inositol hexaphosphate appears to serve a function in bird and turtle red cells similar to DPG in mammalian cells. Diphosphoglycerate is not removed by dialysis against distilled water and only slowly by dialysis against $0.1\ M$ NaCl. Therefore, these common procedures of "purification" are likely to result in a mixture of two electrophoretically distinct components which will have different oxygenation properties.

The quantity of DPG in the red cell depends on the metabolism; different environmental conditions may result in substantial changes in the quantity present. The point to be made here is that considerable caution is needed where a "new" electrophoretically distinct hemoglobin component is found to be associated with a particular set of ecological conditions. It must be shown whether the component is really unique or whether it resulted from the binding of small molecules such as DPG. This distinction is important because DPG levels do appear to vary with environmental conditions. For example, a man adapted to high altitudes has blood with a higher DPG concentration in the red cells (Lenfant et al., 1968). It is reasonable to expect similar phenomena to occur in fish. Effects of hemolysis and dialysis on the oxygen equilibrium of fish hemoglobins suggest that substances functionally similar to DPG may be present.

Rapoport and Guest (1941) determined the ATP content of the red cells from three species of teleost [unidentified species of catfish, the bullhead, *Ictalurus* (=*Ameiurus*) *melas*, and the black bass, *Micropterus dolomieui*]. They found 55–70% of the total acid-soluble phosphorus to be ATP; no phytic acid was found. The ATP content of dogfish red cells is also substantial (Bricker et al., 1968).

Other complications also occasionally arise to conspire to fool the unwary. So, for example, molecules of the buffer used in electrophoresis may form complexes with the protein, and these complexes will migrate differently from the uncomplexed hemoglobin (Cann and Goad, 1965; Cann, 1966).

Although some of the reports of multiple hemoglobins in fish may well result from the presence of allosteric substances, the widespread occurrence of multiple components with quite different amino acid compositions and functions suggests that extensive gene duplication followed by mutation is a major cause of multiplicity. The almost universal occurrence of multiple components in fish, reptiles, and amphibians but the relative absence from birds and mammals suggests that some important physiological factor may favor multiplicity and that the possession of multiple hemoglobins confers an advantage. Is it because poikilothermic animals are subject to a greater range of environmental oxygen pressures, and hemoglobins are selected which have appropriately different properties? Or is it because the metabolic rate varies considerably, and different rates require different hemoglobins? Or again, do multiple hemoglobins arise because of very little selective pressure? It may be that it is improtant for certain fish to have a hemoglobin, but not particularly critical what kind of a hemoglobin—any hemoglobin might do, within rather wide limits. It may be a gratuitous assumption that every molecular attribute of an animal must be either advantageous or disadvantageous, that the zero on an "advantage" scale is always excluded. Be this as it may, some of the fish with the largest number of components (salmon and trout) are also among the most metabolically active. In such fish the properties of the hemoglobin are quite critical. The oxygen requirements of a salmon on its upstream spawning migration would appear to place a considerable constraint on the variability of the physiological properties of its hemoglobins. The existence of two different hemoglobins in the Chum salmon with entirely different oxygen transport properties suggests that one of these pigments may be of physiological significance under one set of conditions, and the other pigment under a different set. This would mean that only half the pigment is "adapted" at any one time.

B. Primary Structure

The primary structure of a polypeptide chain is the sequence of amino acid residues. This sequence is now believed uniquely to determine the three-dimensional folding: the extent of helical arrangement and determination of the spatial distribution of the amino acid side chains. This distribution determines which residues will be available for the intermolecular interactions responsible for subunit aggregation. Interchain interactions are physiologically important because they form the basis of the interaction between the oxygen-binding sites (hemes) and so are responsible for the shape of the oxygen equilibrium curve. Very little information on these subjects as yet exists for fish hemoglobins. The fact that many of them have oxygen equilibria similar to those of mammalian hemoglobins suggests that the interactions between the subunits may be very similar. The amino acid sequence is known for only two polypeptide chains: the α chain of the carp (Hilse and Braunitzer, 1968) and one component from the lamprey, *Lampetra fluviatilis* (Braunitzer and Fujiki, 1969). Fragmentary data exist on the amino acid composition, NH_2-terminal and COOH-terminal residues for several other fish hemoglobins.

Zuckerkandl *et al.* (1960), in a survey of the hemoglobins from a wide variety of vertebrates, examined the tryptic peptide patterns from four fish: the sheepshead, *Pimelometopon pulcher*, the shark, *Cephaloscyllium uter*, the South American lungfish, *Lepidosiren paradoxa*, and the Pacific hagfish, *Polistotrema stouti*. They found that the patterns for the four fish differed among themselves much more than the mammalian patterns and that the shark and sheepshead differed as much from the lungfish pattern as from the human pattern. They concluded that no large soluble tryptic peptide appeared to exist which was unchanged throughout vertebrate evolution.

Residue No.	42	43	44	45	46	47	48	49
Human	Tyr	Phe	Pro	His	Phe	Asp	Leu	Ser
Carp	Tyr	Phe	Ala	His	Tryp╲	Asp╱	Leu	Ser
					Ala			

Residue No.	54	55	56	57	58	59	60	61	62	63	64	65
Human	Gln	Val	Lys	Gly	*His*	Gly	Lys	Lys	Val	Ala	Asp	Ala
Carp	Pro	Val	Lys	—	*His*	Gly	Lys	Lys	Val╲	Ileu╱	Gly	Ala
									Met			

Residue No.	79	80	81	82	83	84	85	86	87	88	89	90	91	92
Human	Ala	Leu	Ser	Ala	Leu	Ser	Asp	Leu	*His*	Ala	His	Lys	Leu	Arg
Carp	Gly	Leu	Ala	Ser	Leu	Ser	Glu	Leu	*His*	Ala	Ser	Lys	Leu	Arg

Fig. 3. The segments of human and carp α-chain sequences which contain the insertions and deletions and the segments close to the heme-linked histidines. The numbers refer to the human sequence starting from the NH_2-terminal amino acid. Data from Hilse and Braunitzer (1968) and from Dayhoff (1969).

Carp hemoglobin consists of three components (48, 48, and 4%) which are of the form $\alpha_2\beta_2$ (Hilse and Braunitzer, 1968). The α chain appears to be the same for each component: almost no information is as yet available on the β chain.* The NH_2 terminus of the α chain is acetyl-serine; the NH_2 terminus of the β chain is valine. The α chain COOH terminus is identical to that in the human α chain: —Tyr—Arg. Acetylation of the NH_2 terminus may be of some importance in oxygen transport because this group may be partly responsible for the Bohr and CO_2 effects in mammalian hemoglobins (Hill and Davis, 1967; Perutz, et al., 1969; Kilmartin and Rossi-Bernardi, 1969); an acetylated —NH_2 group could not participate in the Bohr effect. Since several hemoglobins with acetylated NH_2 termini do have Bohr effects, other groups must be held responsible for the Bohr effect in these hemoglobins.

The carp α chain contains 142 residues—one more than in the α chain of human hemoglobin. This chain length is the net result of two insertions and one deletion, relative to the human α sequence (see Fig. 3). The first insertion, at position 46–47, occurs in a nonhelical bend and is compensated by a deletion at position 57 in the

* The α, β nomenclature is appropriate here because the amino acid sequence of the α chain of the carp hemoglobin has been shown to be homologous with the α chain of human hemoglobin. However, to avoid confusion, this designation of chains should be used only where sufficient structural information exists to permit a reasonable decision concerning homology. It is quite likely that some chains will be found for which this will prove to be impossible. If so, some other designation will be necessary.

6. PROPERTIES OF FISH HEMOGLOBINS

E helix* so that the distal heme-linked histidine occurs at exactly the same position in each chain. This is important because the bound oxygen molecule occupies the position between the heme iron and residue number 58. The glycine deletion at position 57 puts all the critical residues on the distal side of the heme in the sequence 55-62 in register for both human and carp α chains. The other heme-linked histidine is at position 87. A comparison of the homologous sequences in this region (F helix) shows that very little difference exists (see Fig. 3). The segment of the F helix in the neighborhood of the heme iron is remarkably similar in both human and carp α chains.

The other most critical regions of the hemoglobin molecule are at the points of contact between the α and β subunits. The residues in horse and human hemoglobin which make these contacts have been determined by Perutz et al. (1968) from the results of their X-ray diffraction analysis. An α chain makes two different β-chain contacts which are designated $\alpha_1\beta_2$ and $\alpha_1\beta_1$. Dissociation into dimers is believed to involve cleavage primarily of the $\alpha_1\beta_2$ bonds, but both contacts are probably necessary for the cooperative interactions observed in the oxygenation reaction. The homologous residues of the α chains from

Table II

The Homologous Residues of the $\alpha_1\beta_2$ Contact Region from the
α Chain of Horse, Man, and Carp Compared with the
Corresponding Positions in Lamprey Hemoglobin[a]

Position		Residues			
Residue No.	Helix	Horse	Man	Carp	Lamprey
38	C3	Thr	Thr	Glu	Ala[47]
41	C6	Thr	Thr	Thr	Glu[50]
42	C7	Tyr	Tyr	Tyr	Phe[51]
91	FG3	Leu	Leu	Leu	Phe[107]
92	FG4	Arg	Arg	Arg	Gln[108]
93	FG5	Val	Val	Val	Val[109]
94	G1	Asp	Asp	Asp	Asp[110]
95	G2	Pro	Pro	Pro	Pro[111]
96	G3	Val	Val	Ala	Gln[112]
140	H23	Tyr	Tyr	Tyr	Tyr[146]

[a] These contact positions for the hemoglobins of horse and man are those determined by Perutz et al. (1968). The corresponding positions for carp and lamprey have been determined by homology.

* Hemoglobins so far examined have eight helical segments which are designated A through H, starting at the NH_2 terminus.

man, horse, and carp responsible for these contacts are compared in Table II, together with the corresponding data on lamprey hemoglobin. Evidently the α part of the $\alpha_1\beta_2$ points of contact have changed very little between carp and man: eight of the ten residues are identical; in contrast, only four residues are invariant in the regions in lamprey hemoglobin homologous with the contacts in the human α chain. Of the sixteen α-chain residues of the α part of the $\alpha_1\beta_1$ contact, only seven have remained unchanged between carp and man (Table III).

The dissociation of lamprey hemoglobin from tetramer to monomer upon oxygenation (see Section III, E, 1) may result in part from differences in the nature or position of the residues responsible for the $\alpha_1\beta_2$ and $\alpha_1\beta_1$ contacts in higher organisms. Other residues close to these contacts may exert a large influence as in human hemoglobin Kansas (Bonaventura and Riggs, 1968) which has a neutral substitution in the G helix close to the $\alpha_1\beta_2$ contact. This substitution causes a remarkable oxygenation-induced dissociation from tetramer to dimer. The lamprey sequence appears to have only limited homology in the $\alpha_1\beta_1$ contact

Table III

The Homologous Residues of the $\alpha_1\beta_1$ Contact Region from the α Chains of Horse, Man, and Carp, Compared with the Corresponding Positions in Lamprey Hemoglobin[a]

Position		Residues			
Residue No.	Helix	Horse	Human	Carp	Lamprey
30	B11	Glu	Glu	Gly	Val
31	B12	Arg	Arg	Arg	Lys
34	B15	Leu	Leu	Thr	Thr
35	B16	Gly	Ser	Val	Ser
36	C1	Phe	Phe	Tyr	Thr
103	G10	His	His	Asn	
104	G11	Cys	Cys	His	
106	G13	Leu	Leu	Val	
107	G14	Ser	Val	Val	
111	G18	Val	Ala	Phe	
114	GH2	Pro	Pro	Pro	
117	GH5	Phe	Phe	Phe	
119	H2	Pro	Pro	Pro	
122	H5	His	His	His	
123	H6	Ala	Ala	Met	
126	H9	Asp	Asp	Asp	

[a] These contact positions for the hemoglobins of horse and man are those determined by Perutz et al. (1968). The corresponding positions for carp and lamprey have been determined by homology where possible.

region; I have therefore not attempted to indicate the G and H helical residues in Table III. It is of considerable interest that the only cysteinyl residue in the molecule is complely reactive in oxy or carboxy hemoglobin but is nonreactive in deoxyhemoglobin (Riggs, 1961). This suggests that it may become unavailable because it is in or close to the subunit contact region in the deoxygenated pigment. Alternatively, it might be directed into the central cavity which probably exists in tetramic deoxygenated lamprey hemoglobin.

The hemoglobin of *Lampetra fluviatilis* is polymorphic. According to Braunitzer (1966) this polymorphism results from partial blockage of the NH_2-terminal group: 20% of the NH_2 termini are free; the remainder are blocked. No other difference between the two major components appears to exist. The sequence of *Lampetra fluviatilis* hemoglobin shows that nine additional amino acids are present at the NH_2 terminus (compared with human chains). Only two histidyl residues are present; these are situated on opposite sides of the heme as in other hemoglobins.

Rumen and Love (1963a,b) have isolated six hemoglobins from the lamprey, *Petromyzon marinus*; the components differ substantially in isoelectric point, ability to aggregate upon deoxygenation, amino acid composition (Rumen, 1966), and in their oxygen equilibria (Antonini *et al.*, 1964); some of these data are summarized in Table IV. Recent amino acid analyses on hemoglobin V (Bonaventura, 1968) are in substantial agreement with the earlier determination (Rumen, 1966) except that the number of cysteinyl residues determined as cysteic acid is 1.0 rather than 3. Sequence studies by Li and Riggs (1969) show that hemoglobin V from *Petromyzon marinus* differs from the hemoglobin of *Lampetra fluviatilis* in only 4 positions in peptides which constitute 110 of the 146 residues. Sequence differences in the other five *Petromyzon* hemoglobins have not yet been investigated.

Table IV
Some Properties of the Hemoglobins of the Lamprey, *Petromyzon marinus*[a]

Component	1	2	3	4	5	6
$s_{20,w}$ (deoxy)	2.7	1.9	3.5	4.5	4.5	4.5
Isoelectric point	4.5	4.3	5.4	4.7	5.8	6.0
g/100 g	10.3	7.8	3.5	28.2	40.0	10.3
$\log P_{50}$	0.40	—	0.82	0.98	0.83	0.80
n	1.10	—	1.40	1.55	1.40	1.30

[a] These data were compiled from Rumen and Love (1963a,b), Rumen (1966), and Antonini *et al.* (1964).

The amino acid composition and some NH_2 and COOH terminal amino acid determinations have been made on only a few hemoglobins other than lamprey and carp: those of the eels, *Anguilla japonica* (Yoshioka *et al.*, 1968) and *A. anguilla* (Christomanos and Pavlopulu, 1967), the sea bass, *Serranus gabrilla* (Christomanos, 1964), salmon and trout (Buhler, 1963; Eguchi *et al.*, 1960), lungfish (Oldham and Riggs, 1969), and the coelacanth (Bonaventura and Riggs, 1969). Some of these data are summarized in Table V. Unfortunately, such data are not very illuminating concerning the primary structure, but some features are worthy of mention. All fish examined except the lungfish have about half as much histidine as is found in human hemoglobin. Tyrosine changes hardly at all, and many other amino acids seem to change relatively little. It is curious that one of the components of eel hemoglobin apparently contains six cysteinyl residues while the other component has none. Glutathione is usually present in erythrocytes. Oxidative processes can often result in the formation of a mixed disulfide in certain hemoglobins. Steps to remove this possible contaminant do not appear to have been taken in these studies of eel hemoglobin.

Although Tyr—Arg is the C-terminal sequence of the carp α chain and for the human α chain, arginine has not been reported as the C terminus of any other fish hemoglobin; Tyr—Phe and Tyr—His are the C-terminal sequences for the hemoglobins of both salmon and trout; His—Leu and Leu—His termini have been reported for eel hemoglobin. Valine is the NH_2 terminus of every fish hemoglobin examined except the α chain of carp which is acetyl–serine, but Buhler obtained either 1 or 2 DNP valines (depending on the component) per tetramer of salmon and trout hemoglobins; thus, some of the chains must have NH_2 termini blocked to reaction with FDNB.

III. OXYGEN TRANSPORT

The *in vivo* oxygen–hemoglobin equilibrium is often remarkably sensitive to the metabolic needs of the animal. Most physiological studies of oxygen transport in fish, either by blood, washed erythrocytes, or hemoglobin solutions, fail to distinguish adequately between intrinsic properties of the hemoglobin and the modifications of these properties which result from the intraerythrocyte environment. The detailed study of these relationships is an important task for the future. Since most fish hemolyzates contain multiple hemoglobin components which may interact with one another (see Section III, E, 1), it is important to know whether these components occur in the same cell or whether they are confined to different cells. Some components interact with one another in such a way that the oxygen equilibrium differs from that expected from a noninteracting mixture. Such interaction could only occur *in vivo* if the hemoglobins occurred in the same cells. These questions should now be answerable because techniques have been developed both for the measurement of the electrophoresis of hemoglobin from individual erythrocytes (Matioli and Niewisch, 1965) and the determination of the oxygen equilibrium of single erythrocytes (Huckauf *et al.*, 1969).

From a physiological viewpoint, the most important parameters in the description of the oxygen equilibrium are: shape of the curve (plot of y, the degree of oxygenation, vs. p, the oxygen pressure)—whether sigmoid, hyperbolic, or "undulating"; pH and CO_2 dependence (Bohr effect); the "oxygen affinity," usually expressed in terms of the pressure required for 50% oxygenation; temperature dependence; allosteric effectors, substances (ions and organic molecules which can alter the equilibrium). Ontogenetic variation of such substances may play a crucial role in determining the adaptation of the function of hemoglobin. These five parameters determine the amount of oxygen transported by a given quantity of hemoglobin.

A. Critique of Measurements

Only two procedures are in common use for the determination of the equilibrium between hemoglobin and oxygen: gasometric and spectrophotometric. The gasometric technique involves direct determination of the oxygen content of blood or a hemoglobin solution at a measured oxygen pressure. Thus, it provides unambiguous data of immediate practical use in physiological studies of gas transport. This procedure, as applied to blood, gives no information about the state of the hemoglobin, and is a "black box" technique insofar as the red cell contents are concerned. Furthermore, the gasometric procedure with whole cells does not take into consideration the possible effects of changes in cell shape on oxygen transport (see discussion of shape changes; Riggs, 1965). The spectrophotometric technique is difficult to apply with precision to whole cells and has been most widely used with hemoglobin solutions. Although easier and faster than the gasometric procedure, the spectrophotometric technique has several potential pitfalls. The basic spectrophotometric assumption is that the change in absorbance at any wavelength is linearly related to the fraction of the total number of ferroheme groups oxygenated. This appears to be a valid assumption

Table V
Amino Acid Composition of Fish Hemoglobins

I. Isolated Polypeptide Chains[a]

Component:	Eel (A. japonica) 1 α	1 β	2 α	2 β	Carp α	Sea bass (Serranus gabrilla) α	β
Lysine	12	10	8	10	14	8.8	9.43
Histidine	5	3	3	6	5	4.9	4.66
Arginine	3	4	6	8	3	5.4	6.45
Aspartic acid	16	13	15	6	14	15.6	15.0
Threonine	7	4	6	10	3	7.0	6.3
Serine	8	5	4	8	11	9.8	10.8
Glutamic acid	7	11	13	14	6	11.2	9.57
Proline	5	6	4	6	8	4.4	5.17
Glycine	13	11	11	8	11	10.8	11.3
Alanine	22	18	18	17	17	15.5	15.8
Half cystine					0	0.96	0.75
Valine	14	13	9	9	12	13.8	12.3
Methionine	1	1	1	1	4	1.26	0.79
Isoleucine	7	7	8	8	8	7.65	5.99
Leucine	15	15	14	15	14	14.3	15.66
Tyrosine					4	3.06	3.50
Phenylalanine	7	7	7	8	6	6.81	7.08
Tryptophan					2		

6. PROPERTIES OF FISH HEMOGLOBINS

II. Hemoglobins[b]

Component:	Eel, A. japonica 1	Eel, A. japonica 2	Eel, A. anguilla 1	Chinook salmon 1	Chinook salmon 2	Trout 1	Trout 3	Chum salmon F	Chum salmon S	Sea bass	Lungfish 2	Coelacanth
Lysine	22	18	18.5	23.5	20.5	23.5	20	19.5	22.5	18	18.4	26
Histidine	8	9	9	8	11.5	8.5	11	11	7.5	10	23.6	18
Arginine	7	14	10	8.5	9	7.5	9	8	6	12	9.92	6
Aspartic acid	29	21	31.5	28	26.5	28	29.5	27	26.5	31	21.1	22
Threonine	11	16	16	14	23	14	15.5	13	13.5	13	14.6	20
Serine	13	12	18.5	14	14.5	14	15	16	16	21	17.6	14
Glutamic acid	18	27	21	12	21.5	21.5	21.5	19	12	21	27.4	28
Proline	11	10	11.5	12.5	13.5	23.5	14	13.5	15.5	10	10.7	8
Glycine	14	19	18	25.5	15.5	26	16.5	15.5	26.5	22	20.6	10
Alanine	40	35	32.5	38	32	40.5	34.5	32.5	38	31	28.8	24
Half cystine	0	6	—	1	3	1.5	3	3	1	2	2.15	4
Valine	27	18	25	31.5	27.5	31.5	25	20.5	26.5	26	21.7	24
Methionine	2	2	6.5	8	5	8	6	5	7	2	5.91	2
Isoleucine	14	16	16	16.5	17.5	17.5	16.5	15.5	15	14	11.1	12
Leucine	30	29	26.5	23	30.5	25	30.5	29.5	23	30	30.2	38
Tyrosine	7	6	5.5	6.5	6	7	6	4.5	7.5	7	7.34	6
Phenylalanine	14	15	17	15.5	13	15.5	13	14	15.5	14	13.2	14
Tryptophan	5	4		5	3.5	5	3.5	4.5	6			

[a] Data give estimated number of amino acid residues per chain.
[b] Data give estimated number of residues for an assumed molecular weight of 31,000.

between 264–900 mμ (Enoki and Tyuma, 1964; Anderson and Antonini, 1968). However, the possibility of deviations from linearity should be borne in mind. Many fish hemoglobins are quite sensitive to oxidation. Many investigators of fish hemoglobins have experienced difficulties in maintaining the pigment in the ferrous state after hemolyzing the cells, but it is seldom clear whether this difficulty is owing to an intrinsic tendency toward autoxidation of the hemoglobin or is owing to the presence of oxidizing substances formed upon hemolysis after the disruption of the cellular reducing machinery. If significant methemoglobin forms *during* the oxygenation measurements, spurious data will be produced which cannot properly be corrected because methemoglobin increases the oxygen affinity of the remaining ferrous hemoglobin, even if the methemoglobin content is constant. One explanation for this effect is that oxidation of the heme in one subunit exerts much the same effect on the unoxidized subunits as does *oxygenation*. That is, in a tetramer in which one heme is ferric, the remaining hemes have a higher affinity for oxygen than in completely ferrous unoxygenated hemoglobin. These general problems are discussed in detail elsewhere (Riggs, 1965). It should be pointed out, however, that none of the techniques for removing methemoglobin are without drawbacks, and so the best procedure is to minimize oxidative processes. Enzymic reduction of methemoglobin (Benesch *et al.*, 1964; Rossi-Fanelli *et al.*, 1957) is probably the most satisfactory technique, but it appears not always to be free of problems. One of the lamprey hemoglobin components, for example, remained brown even after enzymic reaction (Antonini *et al.*, 1964).

All studies of whole blood need to consider the intracellular environment of the hemoglobin. Fish erythrocytes are relatively large, nucleated, and carry on a substantial metabolism. Although little in detail is known of the function of this metabolism in fish the presumption is that the overall functions are similar to those of the red cells of other animals. The major functions appear to be: maintenance of the heme iron in the ferrous state, control of a proper ionic environment for optimal function of the hemoglobin, and maintenance of suitable concentrations of substances which serve as allosteric effectors to modify the oxygen affinity of the hemoglobin. But many unsolved problems exist. Why, for example, should it be important for the dogfish, *Squalus acanthias*, to maintain an intracellular Na^+ concentration of only 20 mM compared with the 250 mM extracellular concentration (Bricker *et al.*, 1968)? Dogfish hemoglobin might be particularly sensitive to Na^+ ions, but this possibility has not been studied. In addition, changes in the shape of the red cells are important and often may be linked to the oxygen transport function. It has often been stated that one function of the red cell is to reduce the viscosity of the blood from what it would be if the hemoglobin were free in the blood plasma. Astonishingly, the appropriate experiment was not done until very recently when Schmidt-Nielsen and Taylor (1968) and Cokelet and Meiselman (1968) showed that the viscosity is actually higher after hemolysis. It appears clear that the primary function of the erythrocyte membrane is exactly the same as that of all other cell membranes, i.e., to serve as a barrier to the external environment, and to provide for a special controlled internal environment for the optimal function of the cell.

B. Cooperativity: Heme–Heme Interaction and Oxygenation-Linked Changes in Aggregation of Subunits

The shape of the curve relating degree of oxygenation (y) of a blood hemoglobin to oxygen pressure (p) has long been held to be of crucial importance. Barcroft (1928) early argued that an oxygen equilibrium curve with an S shape was particularly advantageous for animals with an active metabolism because it resulted in a high "unloading" oxygen pressure in the tissues relative to the "loading" pressure in gills or lungs. Hüfner's original description of a hyperbolic curve for human hemoglobin was shown by Bohr to be erroneous (see discussion by Barcroft, 1928). Haldane (1922) suggested that "a man would die on the spot of asphyxia if the oxygen dissociation curve of his blood were suddenly altered so as to assume the form which Hüfner supposed it to have in the living body." The almost hyperbolic oxygen equilibrium of human hemoglobin "Kansas" (Bonaventura and Riggs, 1968) emphasizes that the significance of the S shape of the equilibrium curve has been exaggerated. Confusion arose because a hyperbolic oxygen equilibrium curve was thought always to be associated with a high affinity for oxygen as it is in myoglobin. But this association is by no means always true, especially in fish hemoglobins. Lamprey hemoglobin, for example, has a rather low affinity for oxygen. What is necessary to examine, from the standpoint of oxygen transport adaptation, is not the shape of the curve per se but the influence of this shape on the actual amount of oxygen transported.

Oxygen equilibria are usually plotted in one of three ways: (1), y vs. p; (2), y vs. $\log p$, or (3), $\log y/(1-y)$ vs. $\log p$. Plots (2) and (3) are the most useful from an analytical standpoint. The slope of the last plot defines n in Hills' equation, $y = Kp^n/(1 + Kp^n)$; n is usually evaluated at $y = \frac{1}{2}$. Any value of $n > 1$ is taken to indicate the presence of cooperativity or stabilizing heme–heme interactions between at least some of the O_2-binding sites (hemes); if $n = 1$, the sites are functionally independent of one another. If two or more components are present which differ substantially in oxygen affinity an undulating curve may be produced such that n, at $y = \frac{1}{2}$, may be appreciably less than 1.0. Although values of $n < 1$ have frequently been interpreted as implying "negative heme–heme interactions" (Manwell, 1964), such values can also reflect the presence of multiple components which differ sufficiently in oxygen affinity. This difference may arise experimentally in a number of ways other than the existence of different components with intrinsically different oxygen affinities. So, for example, two hemoglobin components may be present with identical oxygen equilibria but different sensitivities to oxidation of the iron. This situation sometimes results in the formation of significant quantities of methemoglobin in one of the components but not the other during the preparation or during the equilibrium measurements. Since the presence of ferric heme can increase the oxygen affinity of the unoxidized hemes of the same molecule, oxygen equilibrium measurements of a mixture will show a lower apparent value of n than would be found in the absence of oxidation. The oxygen equilibrium curves then appear to have higher than expected oxygenation levels at low pressures. These considerations are mentioned here because most fish hemolyzates have multiple components, and at least some of these differ considerably in sensitivity to oxidation.

The value of n should not be interpreted as reflecting the energy of interaction. Wyman (1964) has shown that high values of n (up to 5) in invertebrate hemoglobins may in fact be associated with a lower overall free energy of interaction than the lower value of n ($\sim$2.8) characteristic of mammalian hemoglobins. He has devised a graphical method for obtaining the overall free energy of interaction (see Fig. 4). Unfortunately, this method requires getting data at both very low and at very high oxygenation levels (<3% and >97%)—values that are often difficult to obtain in practice.

The cooperativity between O_2-binding sites in mammalian hemoglobins is largely independent of pH, at least within the normal physiological range, and is uniform between 10 and 90% saturation ($n \approx 2.8$–3.0). At sufficiently low or high oxygenation levels, as shown in Fig. 4, the value of n approaches unity. For some fish hemoglobins, however, n is a function of pH between 10 and 90% oxygenation and often changes greatly with the degree of oxygenation. This pH dependence differs in sign in different hemoglobins. Thus, for example, in the hemoglobin of the barndoor skate, n increases from 1.2 to 1.8 between pH 6.5 and 7.5 (Manwell, 1958b), whereas in the coelacanth, *Latimeria*, n decreases from 1.6 to 1.1 between pH 7 and 8 (Bonaventura and Riggs, 1969).

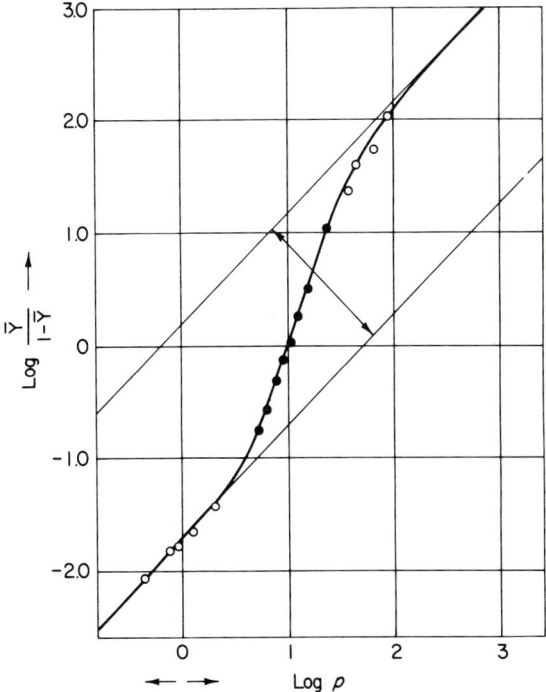

Fig. 4. The Hill plot of the oxygen equilibrium of horse hemoglobin from Wyman (1964). The total free energy of interaction can be obtained by multiplying the distance between the two asymptotes by $RT\sqrt{2}$.

C. Effects of pH and CO_2

An important physiological adaptation of most hemoglobins is the coupling of oxygen and carbon dioxide transport. Thus the CO_2 from tissue metabolism facilitates deoxygenation while oxygenation in the lungs or gills aids in the removal of CO_2. The original "Bohr effect" (Bohr et al., 1904) described the decrease in oxygen affinity brought about by CO_2, but the term is now applied both to the effect of CO_2 and to changes in pH. Part of the CO_2 effect can be explained by changes in pH, but CO_2 also exerts a direct effect which is shown by a CO_2-induced drop in the oxygen affinity even at constant pH. Thus two distinct mechanisms are involved: oxygenation-linked acid groups and CO_2-binding groups. These groups cannot be regarded as mutually exclusive. If certain —NH_2 groups are oxygenation-linked and are responsible for CO_2 binding, their pK values are presumably different in oxy and deoxy hemoglobin. Two kinds of acid group on the protein are functionally linked to oxygenation: one of these becomes more acidic upon oxygenation (alkaline, normal Bohr effect, range pH 6.5–7.5), while the other becomes less acidic (acid or "reverse" Bohr effect, range pH 5.5–6.5). The alkaline Bohr effect is usually the only one which is physiologically important. Although all the groups responsible for these so-called "Bohr" protons have not been unequivocally identified, the acid Bohr effect may involve a carboxyl group and the alkaline Bohr effect either imidazole or the α-NH_2 group. The effective number of protons released during oxygenation varies in different species; the direct effect of CO_2 probably involves the formation of carbamino compounds: —NH_2 + CO_2 → —$NHCOOH$. Recent evidence suggests that CO_2 is bound primarily to the α-NH_2 group of each chain in horse hemoglobin (Kilmartin and Rossi-Bernardi, 1969) and that the C terminal histidyl residue of each β chain is responsible for a major part of the alkaline Bohr effect (Perutz et al., 1969); the α-NH_2 groups of the α chains are believed also to contribute to the Bohr effect.

The number of protons released during oxygenation can be estimated from the dependence of the oxygen equilibrium on pH or directly by differential titration. Wyman (1948, 1964) has shown that r [$\equiv$ (Δ log P_{50})/Δ pH] equals the number of H^+ released per oxygen bound provided that n is independent of pH (i.e., that the "y vs. log p" plots at different values of pH are all parallel to one another). If, however, the shape of the oxygen equilibrium curve changes with pH then the median ligand activity (log p_m) should be used rather than the ligand activity associated with half-saturation (log P_{50}). The minimum change in the pK of acid groups associated with oxygenation is given by the expression (Wyman, 1948):

$$\Delta pK = 2 \log[(1 + r)/(1 - r)]$$

Rossi-Fanelli and Antonini (1960) found that tuna hemoglobin has a remarkably large Bohr effect but that the shape of the curve changed with pH: the measured n drops from 3 to less than 1 when the pH drops from 9 to 6. Indeed, at pH values lower than 6.5, complete saturation is not achieved even at a partial pressure of one atmosphere of oxygen. Brunori (1966) has studied the Bohr effect of this hemoglobin by the differential titration method. He finds 0.86 H^+ released per heme at pH 7.5 compared with 0.56 H^+ for human hemoglobin. His results can be fitted to a model for which he assumes, as in human hemoglobin, only two ligand-linked acid groups, one of which becomes more acidic on oxygenation ($\Delta pK \cong 2.2$) and the other becomes less acidic ($\Delta pK \cong 1.2$). Only the first group is significant in the physiological range of pH, 6.5–7.5. These apparent pK changes are much larger than those found for human hemoglobin.

Lamprey hemoglobin has a very unusual Bohr effect. Wyman (1964) points out that the Bohr effect curve is so steep relative to its amplitude that it cannot be fitted with any choice of independent oxygen-linked acid groups. It must be postulated that stabilizing interactions exist such that ionization or protonation of one group facilitates that of another group. Since this process involves changes in aggregation, it appears quite possible that the groups involved are at or near the points of contact between the subunits. The process might be imagined to be analogous to a zipperlike opening: dissociation into subunits is simultaneously associated with changes in pK values of several acid groups.

The enormous change in oxygen binding below pH 6.5, present in many teleost fish, was first observed by Root (1931) who observed that CO_2 drastically lowered the oxygen saturation of fish bloods at acid pH. This was regarded as an effect on the oxygen capacity as distinct from the oxygen affinity, because oxygen capacity was defined as the amount bound at atmospheric pressure ($\sim$152 mm p_{O_2}). Oxygen pressures as high as 1 atm were used in subsequent work (Green and Root, 1933). This resulted in increased oxygen binding except in the most acid solutions where methemoglobin formed readily. Root and Irving (1941) found that acidified blood from the tautog, *Tautoga onitis*, could be saturated with CO although it was incapable of becoming saturated with oxygen at 150 mm p_{O_2}. Scholander and van Dam (1957) considered whether the Root effect might play a role in the secretion of gases into the swim bladder and whether sufficiently high oxygen pressures would be effective in achieving saturation. They used oxygen pressures as high as 140 atm, but the substantial oxidative effects of such pressures were not investigated. Their results showed that a Root effect was present in some fish bloods at pressures of 140 atm but not in other fish which were sensitive to CO_2 at low oxygen pressures. They concluded that the Root effect was not the basis for oxygen secretion in deep sea fish. However, a correlation does exist: the presence of a functioning swim bladder is often associated with a Root effect in the blood (Fänge, 1966).

The results of Hashimoto *et al.* (1960) indicate that only one of the two major components of the hemoglobin of the chum salmon is characterized by a Root effect; the other component is not affected by pH at all.

Detailed studies of the Root effect in carp hemoglobin have been carried out by Noble *et al.* (1968, 1969). They found (Fig. 5) that carp hemoglobin is only half-saturated with oxygen when equilibrated with air at pH 5.6 and 20°C in 0.05 M phosphate, or citrate. The oxygen

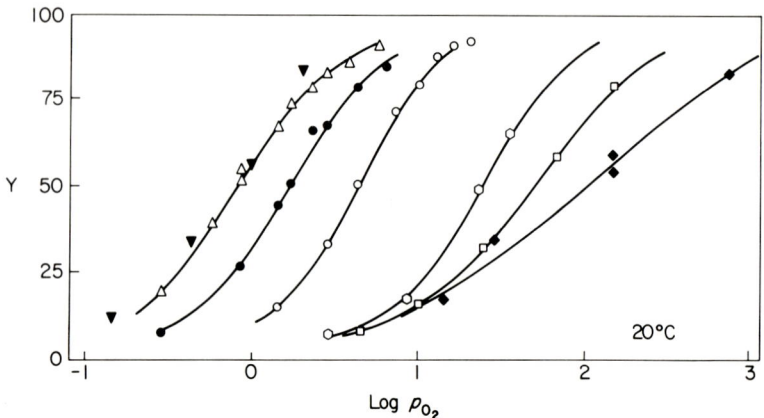

Fig. 5. The oxygen equilibria of carp hemoglobin (Noble et al., 1969): (▼) pH 9, (△) pH 8, (●) pH 7.5, (○) pH 7, (◯) pH 6.5, (□) pH 6, (◆) pH 5.6. The data were obtained at 20°C in 0.05 M phosphate, borate, or citrate or an appropriate mixture depending on the pH. At pH values below pH 6.5 the value of n decreases.

affinity goes down by more than two orders of magnitude as the pH decreases from 9 to 5.6. Kinetic measurements at 2°C show that the rate of dissociation of oxygen also increases by two orders of magnitude at low pH (Fig. 6). Furthermore, the rate of combination with oxygen (k') decreases by a factor of ~3 between pH 8 and 5.5. They have also measured the rate of combination with CO (l') and find that the drop in l' occurs a full pH unit higher than the drop of k'. Thus the Root effect alters the CO and O_2 reactions differently or at least to different extents. This important finding has, as yet, no explanation, but two possibilities should be considered. The ligands could interact differently with an amino acid side group on the distal side of the heme. Alternatively, the bonding of ligand to iron may induce some differences in conformation between the oxy and carboxy forms.

D. Temperature Dependence

The oxygen equilibria of most fish hemoglobins depend on temperature in much the same way that mammalian hemoglobins do: An increase in temperature causes a decrease in oxygen affinity, and the enthalpy values are similar (10–12 kcal/mole). Tuna hemoglobin (Rossi-Fanelli and Antonini, 1960) is an interesting exception: $\Delta H = -1.8$ kcal/mole. The oxygenation data are shown in Fig. 7 which shows that

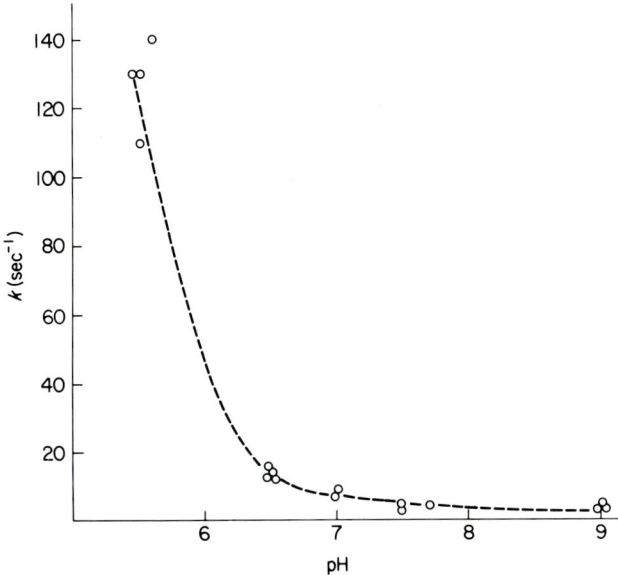

Fig. 6. Measurement of the rate of dissociation of oxygen from carp hemoglobin at 2°C as a function of pH (Noble et al., 1969).

between pH 6.5 and 8.7 the temperature dependence is close to zero. Rossi-Fanelli and Antonini (1960) speculate that this insensitivity to temperature may be a molecular adaptation which would permit the animal to move in waters of greatly changing temperature without changing the oxygen transport properties of the blood. It may be important here that the tuna is not entirely poikilothermic and maintains a body temperature as much as 14°C above the ambient water temperature (Carey and Teal, 1966). They do this by means of a countercurrent system of heat exchange. Large sharks also maintain a temperature above ambient (Carey and Teal, 1969); partial temperature regulation, therefore, may be quite general in large fish.

Wyman (1964) has considered temperature invariance of shape in some detail, and the considerations to follow depend on his analysis. Temperature invariance means that $[\partial \ln p/\partial T]_{\bar{Y},H^+}$ is independent of y, the degree of oxygenation, or that n is independent of T for any y. This means that the heat of introducing one mole of ligand into hemoglobin will be independent of the degree of saturation:

$$\Delta H_y = -RT \left(\frac{\partial \ln y}{\partial \ln T}\right)_{\bar{Y}}$$

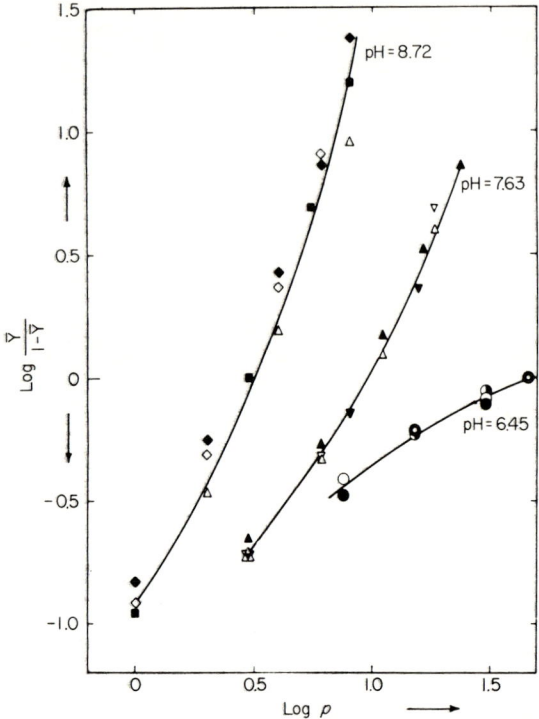

Fig. 7. The oxygen equilibrium of tuna hemoglobin at different temperatures and pH values. Data from Wyman (1964), which were recalculated from Rossi-Fanelli and Antonini (1960). (■), (▽), (◐) at 20°C; (○) at 5°C; (◇) at 6.5°C; (●) at 10°C; (◆) at 10.5°C; (△) at 30°C; (▼) at 35°C.

If all the oxygen-binding sites are the same or have the same ΔH values, then the heat of heme–heme interaction will be zero; i.e., the free energy of interaction would be entirely an entropy effect. Furthermore, the free energy of the $O_2 - H^+$ interaction (Bohr effect) would also be an entropy effect. Wyman points out, however, that the results of Rossi et al. (1963) on human hemoglobin do *not* suggest that the Bohr effect is primarily an entropy effect. Wyman suggests that the explanation may be that the two sets of protons (released in the acid and alkaline ranges) oppose one another: if they have different heats, their degree of overlap must be temperature dependent.

Tuna hemoglobin has a large Bohr effect but apparently no temperature dependence. The heat of ionization of the oxygenation-linked protons should show up but in fact is not measured. Wyman suggests that the heat of oxygenation which is measured by $[\partial \ln p/\partial T]_{\bar{Y}}$ includes only the difference between the heat of ionization of the oxygenation-linked protons and the heats of the buffering groups (tris) in the solution. Tris has a heat of ionization of $\sim 10,000$ cal/mole. If the oxygen-linked groups also have heats of ionization close to this value, the measured heat of ionization might well be independent of pH. This, however, does not explain why the heat of oxygenation is also apparently zero.

E. Adaptations of Hemoglobins in Different Fish

The properties of fish hemoglobins reflect adaptations not only to the metabolic rate but also to the prevailing external oxygen pressure: the hemoglobin must not only be capable of combining with oxygen at the environmental p_{O_2}, but also of unloading the oxygen at pressures appropriate for the tissue metabolism. The overall pattern is that those fish (i.e., trout and mackerel) which are most active generally have hemoglobins with lower affinities for oxygen (high P_{50}) than do less active fish. Furthermore, these hemoglobins usually have the largest Bohr effects. Thus those fish with the highest oxygen requirements and the largest CO_2 production have hemoglobins adapted both to deliver oxygen at relatively high pressures and to facilitate removal of CO_2. Although a large Bohr effect helps release oxygen in the tissues, it can also prevent adequate binding of oxygen in the gills. Under the stress of emergency and violent muscular activity the lowered oxygen binding which results from the increase in lactic acid can be catastrophic, and it may be one of the major causes of death in hyperactive fish (Black, 1958). The Bohr effect, deemed an advantage under normal circumstances, may thus prevent adequate oxygen from reaching the tissues.

Fish which live in a low oxygen environment, and which lack special adaptations (lung, mouth, or gut) for air breathing, usually have hemoglobins with high oxygen affinities (low P_{50}). Pelagic fish live in a relatively constant environment with a high p_{O_2} (~ 100–160 mm) where the p_{CO_2} is usually less than 1 mm. These fish usually have hemoglobins with relatively low oxygen affinities. Variation of the oxygen transport properties of the bloods of these fish depends primarily on metabolic requirements. Under environmental conditions which prevent equilibration with the atmosphere, however, the p_{CO_2} may become relatively high: this situation occurs under the arctic pack ice (Kelley, 1968). Oxygen transport properties of the hemoglobins of fish living under these circumstances have not been studied, but the total quantity of hemoglobin in the blood of arctic fishes is frequently considerably reduced; much more O_2 is physically dissolved (Scholander and van Dam, 1957).

Although the oxygen transport properties of bloods are frequently described in terms either of metabolic requirements or of environmental oxygen pressures, these factors cannot be separated from one another. The most active fish must inhabit a high oxygen environment; they could not otherwise obtain sufficient oxygen for their activity. Air-breathing fish, of course, can bypass their low oxygen aqueous environment.

These conclusions concerning adaptation are largely based on studies of blood by many individuals over the last five decades; much of the data has been reviewed by Manwell (1960) and by Prosser and Brown (1961). For the most part, the intrinsic properties of hemoglobin are not adequately distinguished from the modified properties in the red blood cell. Krogh and Leitch (1919), in the first comparative study of fish bloods, clearly anticipated the possibility that fish red cells might contain allosteric substances which could control the oxygen transport properties:

> We believe that the adaptation of fish blood must be brought about by some substance or substances present along with the haemoglobin within the corpuscles, and we wish to point out the general significance of the haemoglobin being enclosed in corpuscles surrounded by semipermeable membranes. By this arrangement just that chemical environment can be secured which is most suitable for the respiratory function of the haemoglobin in that particular organism, while at the same time the chemical composition of the blood plasma can be adapted, as it must needs be, to the general requirements of the body cells . . . the possession of semipermeable red corpuscles is therefore a necessary condition for utilizing to the full wonderful respiratory properties of the haemoglobin.

Although this was written fifty years ago, we still know little about these relationships. In the following discussion an attempt will be made to distinguish between what are presumed to be properties of hemoglobin and those of blood or red cell suspensions. This distinction has often not been made in the physiological literature which is rife with descriptions of "hemoglobin–oxygen dissociation curves" when whole blood is really being described.

1. CYCLOSTOMES

From the standpoint of cooperativity of O_2 binding, lamprey hemoglobin is one of the most instructive because subunit aggregation is functionally linked to deoxygenation. The first measurements (Wald and Riggs, 1951) on hemoglobin from *Petromyzon marinus* showed that n was usually close to unity, but it was sometimes significantly higher. Because the molecule was then believed always to be monomeric, with only a single heme, values higher than 1.0 were attributed to experimental errors. Briehl (1963), Love and Rumen (1963), and Rumen and

Love (1963b) showed, however, that deoxygenated lamprey hemoglobin is at least partly aggregated and that the oxy form is monomeric. Thus, the anomalously high n value was found to be associated with an oxygenation-induced dissociation of subunits. This phenomenon is widespread, and it has been found in an abnormal human hemoglobin (Bonaventura and Riggs, 1968) and in many invertebrate hemoglobins (Kitto and Bonaventura, 1969).

Since dilution favors dissociation into subunits, it might be expected that n would be concentration dependent. Clearly, sufficient dilution should result in complete dissociation into monomeric units for which n should be unity. Briehl's data for lamprey hemoglobin (Briehl, 1963) indicate that n increases with degree of oxygenation. Insufficient data exist to permit an unambiguous conclusion concerning the possible changes of n with concentration, but no doubt exists about changes of n with degree of oxygenation. This change is, however, only apparent at oxygenation levels above 50%, and it becomes pronounced only above 80%. Antonini et al. (1964) have also measured this equilibrium, but the experiments were apparently confined to the region below 80%; thus, they did not observe the dependence of n on y. Wyman (1964) has analyzed the effects of changes in aggregation on the oxygen equilibrium in some detail. He showed that if the oxygenated molecules are wholly monomeric and the deoxygenated molecules are all tetramers, the resulting oxygen equilibria will be quite asymmetric with $n = 1.6$ at the midpoint. The essential feature of Wyman's model is that a large range of oxygen pressure exists in which the monomers are 100% oxygenated, and the tetramers are not oxygenated at all. In this model, the oxygen equilibria for the tetramers and monomers are each assumed to have $n = 1$. The model predicts a tenfold change in concentration to result in a change in log P_{50} of about 0.75, which is not far from that found by Briehl (1963).

Rumen and Love (1963a) have found that the six components of *Petromyzon marinus* hemoglobin differ from one another in their aggregation characteristics. Components 4, 5, and 6 appear to aggregate to tetramers upon deoxygenation but component 2 does not aggregate at all and components 1 and 3 display intermediate behavior (Table IV). Antonini et al. (1964) determined the oxygen equilibrium of each component (except 2) from *Petromyzon marinus* (Table IV). Components 1 and 4 appear to have significantly different oxygen equilibria from components 3, 5, and 6. However, these differences may have resulted from variations in the concentration dependence of the equilibrium which, in turn, would reflect differences in subunit dissociation equilibria. In addition, component 1 was reported to be brown even after enzymic reduction so the possibility of denatured artifacts cannot be excluded. The data of Antonini *et al.* (1964) on the unfractionated *Petromyzon marinus* hemolyzate were obtained at a much lower protein concentration than the data of Briehl (1963) and Wald and Riggs (1951). When the latter data are adjusted for this difference according to the concentration dependence found by Briehl, all three sets of data are in substantial agreement. Briehl showed that an increase in concentration from about 0.1 to 3.6% was accompanied by an increase in log P_{50} from about 0.8 to 2.7—an eightyfold increase in O_2 pressure required for half-saturation. This enormous effect of concentration should serve warning to those inclined to impart physiological significance to a value obtained from an experiment at a single (dilute) protein concentration. Unfortunately, virtually all published data on hemoglobin solutions of fish hemoglobins do not consider the possible importance of concentration dependence. Such "single-value" studies may be of considerable significance in other ways, but they cannot be used to imply anything about the physiological adaptation of the hemoglobin in its *in vivo* function.

Rumen and Love (1963b) demonstrated that the different components of lamprey hemoglobin, when mixed and deoxygenated, will form hybrid molecules. Component 1 can form a dimer with components 2, 4, and 5, and component 3 will combine with 4 and with 5 to form tetramers of the form $(3-4)_2$ and $(3-5)_2$. Appraisal of the physiological importance of such complexes must await determination of the oxygen equilibria of the appropriate mixtures. This would only be significant if different components occurred in the same cell.

Manwell (1958c, 1963b) has measured the oxygen equilibria of the hemoglobins (both in solution and in red cells) of the adult and ammocoete larva of *Petromyzon marinus*, and of the Atlantic and Pacific hagfishes, *Myxine glutinosa*, and *Eptatretus* (*=Polistotrema*) *stouti*. He states that no significant difference occurs in cyclostomes between oxygen equilibria obtained in solution and in red cells. As discussed in Section III, E, 2, it seems that considerations of the Donnan equilibrium would require that a difference exists between cell and solution in any hemoglobin with a large Bohr effect.

In contrast to the lamprey hemoglobins, those of the hagfishes are devoid of significant Bohr effect (except below pH 6.5), so equality of measurements of cell and solution is possible with these fish. The hagfish hemoglobins also differed from those of the lampreys in oxygen affinity: *Eptatretus* hemoglobin in solution had a P_{50} value of only 1.8 mm at 18°C, and 3–4 mm in cells, in contrast to values 3–5 times as great for the lampreys.

2. ELASMOBRANCHS

The early studies by McCutcheon (1947) indicated that the hemoglobins of three species of shark possessed higher oxygen affinities ($P_{50} \approx 7$ mm) than found for six species of ray ($P_{50} \approx 13$–15 mm). The oxygen equilibria were not quite hyperbolic: $n \cong 1.5$–1.6 for the ray hemoglobins and 1.2–1.3 for the sharks. These measurements at 25.5°C, pH 7.4, and $M/30$ phosphate were made on very dilute solutions of hemoglobin; thus, their relationship to *in vivo* or intraerythrocyte function cannot be evaluated.

More recently, Manwell (1958a, 1963a) has compared the oxygen equilibria of erythrocyte suspensions and hemoglobin solutions from the dogfish, *Squalus suckleyi*. He found that hemoglobin solutions (concentration, 1.5–2.0%) mirrored the properties of the red cell suspensions. He found that no significant difference exists between the oxygen equilibria of fresh, buffered hemolyzates, of crystallized, dialyzed preparations, or of red cell suspensions. Suspension of the red cells in either 0.5 or 1.0 M urea was without effect on the oxygen equilibrium. The urea present in dogfish blood is therefore unlikely to have any direct effect on oxygen transport. These facts do not suggest the presence of any intracellular modifier of oxygen transport. For the adult pigment, $n = 1$ and the Bohr effect is low ($r = 0.34$). The fetal pigment possessed a higher oxygen affinity both in cells and in solution than did the adult hemoglobin. This was shown to depend on the presence of a unique fetal hemoglobin. Tryptic peptide patterns showed the presence of two peptides absent in patterns from the adult. The fetal hemoglobin occurs during most of the 22–23-month gestation period.

The absence of significant differences between cell and solution is surprising. If the Bohr effect were really zero then the apparent equality of the oxygen affinity inside and outside the cell would be understandable, but Manwell's data indicate that the Bohr effect is not zero. Therefore it is necessary to suppose that some difference between the oxygen affinity of red cells and hemoglobin solutions must exist at every pH except at the isoelectric point because of the Donnan equilibrium. However, recent experiments by Lenfant and Johansen (1966) failed to show any effect of CO_2 or pH on the oxygen equilibrium of intact blood from the same species.

Albers and Pleschka (1967) failed to find any significant Haldane effect (effect of oxygen saturation on CO_2 content of blood) in studies of blood from three elasmobranchs (*Scyliorhinus stellaris, Torpedo ocellata*, and *Mustelus mustelus*). Since the Haldane effect is but the necessary thermodynamic consequence of the Bohr effect (Wyman, 1948), these bloods presumably have hemoglobins with oxygen equilibria essentially independent of pH.

Lenfant and Johansen (1966) make the important point that blood samples should be obtained from "free swimming undisturbed animals with the least possible trauma." This suggestion is given added weight by the finding that the P_{50} of rabbit blood rises 2.7 mm when 100 μg aldosterone per kilogram body weight is injected (Bauer and Rathschlag-Schaefer, 1968). Of course, one cannot extrapolate from rabbit to fish, but the observation indicates that changes in hormone balance can influence the oxygen affinity.

In sharp contrast to the dogfish, the hemoglobin of the barn-door skate, *Raja binoculata*, has a large Bohr effect—in both red cells and in solution which extends from pH 6.5 to 10.5 (Manwell, 1958b). No other hemoglobin is known which has a Bohr effect which extends over such a range. The red cell suspension has a significantly lower oxygen affinity than the hemoglobin solution. The largest difference between cell and solution occurs at pH 6.5 and is smaller at higher pH. Therefore, the measured Bohr effect is larger in the red cell suspension than in the solution when measured between pH 7.0 and 7.5. Although these differences may be explained by the presence of allosteric substances in the red cell, other explanations are not excluded: the two sets of experiments were not carried out under identical conditions. The large Bohr effect requires that at any pH above the isoelectric point the red cell will have a lower oxygen affinity than possessed by hemoglobin in solution at the extracellular pH.

Manwell found that the cooperativity is pH dependent: $n = 1.2$ at pH 6.5 and increases to $\sim$1.6–1.7 at pH 7.5. Although this held true for both red cell and solution from the adult, hemolysis appears to abolish cooperativity in the embryonic hemoglobin. These curious features strongly suggest a unique embryonic hemoglobin, but the possibility cannot be excluded that the measured difference in oxygen affinity may result in whole or in part from some allosteric substance.

3. TELEOSTS

The properties of the hemoglobins of teleost fishes cover an enormous range in oxygen affinity, size of Bohr effect, and temperature dependence. One might suppose that these variations reflect the large differences in metabolic activity and in the available environmental oxygen, and this appears to be generally true. However, many seeming paradoxes exist. For example, most teleost hemoglobins have oxygen equilibria which depend strongly on temperature: the P_{50} rises about 1 mm for each degree centrigrade the temperature rises, but the oxygen equilibrium of tuna hemoglobin is almost completely independent of temperature. This peculiarity is discussed in an earlier section. Generally, marine teleosts live in a more uniform environment than their freshwater relatives: the p_{O_2} is usually high and the p_{CO_2} low (often less than 1 mm). Nevertheless, their hemoglobins show just as much variation in oxygen affinity as those of freshwater fish, which suggests the importance of metabolic variations. Rather than catalog these variations, it will, perhaps, be more instructive to examine the hemoglobins of certain fishes in detail.

Salmon hemoglobins have been studied more extensively than almost any other fish hemoglobin. The oxygen equilibrium of the whole hemolyzate near pH 7.0 often appears "unduatory," and claculations of n at 50% oxygenation show evidence for reduced cooperativity or apparent "negative heme–heme interaction." Hashimoto et al. (1960) show that this is only apparent and that the departure is wholly owing to the presence of two different components. These, designated F (55%) and S (45%), from the hemoglobin of the chum salmon, *Oncorhynchus keta*, have been isolated. The components have completely different oxygen equilibria and substantially different physical properties (see Table VI). Combination of the sedimentation and diffusion data yields a substantial apparent difference in molecular weight: 72,000 for component S, about 11,000 higher than that for component F. These measurements, carried out at 0.2% in hemoglobin concentration, need to be confirmed by sedimentation equilibrium measurements. If subunit dissociation were ap-

Table VI

Summary of the Properties of the Two Hemoglobins of the Chum Salmon[a]

Component	F	S
Proportions	55%	45%
Heat sensitivity	Unstable	Stable
Isoelectric point	6.4	6.7
$s_{20,w}$	4.11 S	3.88 S
D	6.6×10^{-7}	4.9×10^{-7}
Oxygen equilibrium:		
Bohr effect	Large	None
Effect of phosphate	Oxygen affinity down	No effect on oxygen affinity
Temperature dependence	Large	Small

[a] Data assembled from Hashimoto et al. (1960), Hashimoto and Matsuura (1959a,b, 1960a), and Hashimoto and Matsuura (1962).

preciable, the molecular weight could not be calculated by a simple determination of the S and D values.

Component F has a lower oxygen affinity ($P_{50} \cong 30$ mm) at "physiological pH" than any other teleost fish reported (most others are apparently under 20 mm) and a very large Bohr effect (Fig. 8), in contrast to component S which has a much higher affinity and no Bohr effect. This remarkable difference accounts completely for the observation that the oxygen equilibria are biphasic at pH 7.0–7.2, and that at pH 6.8 the oxygenation level at 100 mm p_{O_2} is only 45% which is the proportion that is component S. They found no change in the oxygen affinity for either component between 0.2 and 0.8% in protein concentration. The shapes of the oxygen equilibria for both components are sigmoid with an n value of ~2.4. Perhaps the most interesting observation is that the two components are affected very differently by phosphate. The oxygen affinity of component F, but not S, decreases when the phosphate is increased from 0.03 M to 1.0 M. This curious finding suggests that intracellular phosphates may be particularly important in controlling the

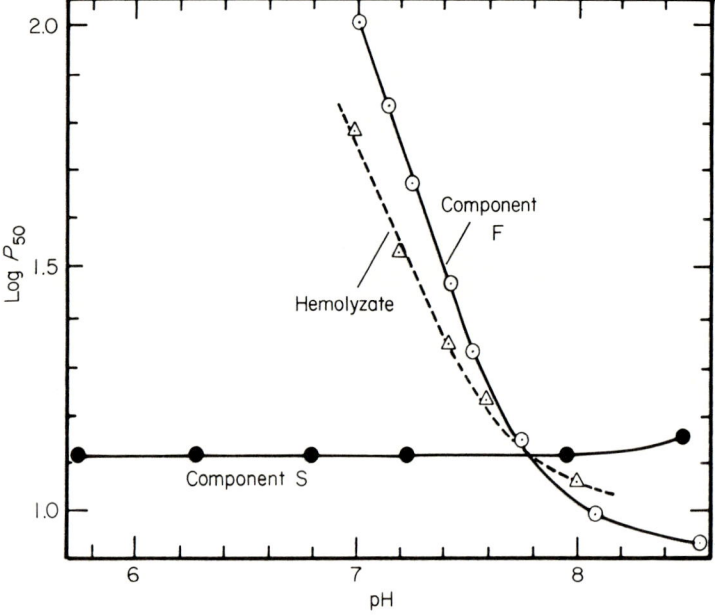

Fig. 8. Comparison of the Bohr effects of components F and S of the hemoglobin of the chum salmon, obtained at 15°C in 0.1 M phosphate (Hashimoto et al., 1960). The value of ($\Delta \log P_{50}$)/(Δ pH) for component F reaches a value of 1.25, as high as known for any hemoglobin.

oxygen affinity. Although phosphate concentrations as high as 1 M were used, substantial effects were observed at much lower concentrations. Below 0.1 M in phosphate the value of $(\Delta \log P_{50})/(\Delta \log [\text{Phos}])$ is about 0.32 which indicates that about 1.3 phosphate ions are released for each tetramer oxygenated. At much higher concentrations, more is bound. The oxygen equilibrium of component F is much more sensitive to temperature than is that of component S.

Two hemoglobins with properties as different as those of F and S might reflect an adaptation to changes in physiology or to environment. However, Hashimoto and Matsuura (1960b) found no correlation of the proportions with place of capture, temperature of water, body length, or sex. Although they found no clear evidence for an ontogenetic change in proportions, such changes have been found in other species of salmon (Vanstone et al., 1964; Wilkins, 1968; see Section II, A, 2).

Yoshioka et al. (1968) have isolated the two major hemoglobins from the eel, *Anguilla japonica*, and have determined the oxygen equilibrium. Most of their results are quite similar to the earlier measurements of Yamaguchi et al. (1962) on the eel and on the loach (Yamaguchi et al., 1963). Indeed it is remarkable that in salmon, eel, and loach two hemoglobins are present with such similar properties. Components I and II of the eel hemoglobin are each composed of two kinds of polypeptide chain and possess no chain in common. Just as with the salmon hemoglobin, one component (II) has a large Bohr effect, while the other (I) has none. They report that mixtures of I and II give oxygen equilibria which give no evidence of interaction between the components. They have also compared the oxygen equilibrium of erythrocytes suspended in an isotonic buffer of pH 7.0 both with a hemolyzate and with a 3:7 mixture of components I and II; all three sets of data were essentially indistinguishable. This result is very hard to understand because the pH inside the erythrocytes cannot be the same as that outside. The isoelectric points are 8.08 and 5.96 for components I and II, respectively (Yoshioka et al., 1968). Component I has no Bohr effect and so can be ignored for the present purposes, but component II has a large Bohr effect. A difference in pH, between the inside and outside of the red cell, would not cause any change in the oxygen equilibrium of component I, but component II would be negatively charged and so the pH would be substantially lower inside the red cell than outside. Their data show that the minimum oxygen affinity occurs close to pH 7; any shift in pH from 7 would therefore cause an increase in affinity, although this would not be detected if the $\log P_{50}$ vs. pH plot has a sufficiently broad maximum. Steen and Turitzin (1968) found that the degree of oxygenation of erythrocytes and of hemolyzed blood from the eel, *Anguilla vulgaris*, differ substantially at a p_{O_2} of 150 mm. The red cells were only 50% saturated with O_2 at an extracellular pH of 7.0, whereas the hemolyzate was at least 95% saturated at this pH. These results appear to be inconsistent with those of Yoshioka et al. (1968). Although different species of eel were used in the two studies, this fact still does not explain why Yoshioka et al. did not observe any difference between the hemolyzate and the intact cells.

Manwell et al. (1963) have studied the oxygen equilibria of hemolyzates obtained from several hybrid sunfish. They report that these experiments form a possible molecular basis of "hybrid vigor." The primary basis for this conclusion was the oxygen equilibrium of the hemolyzate from F_1 hybrids obtained by crossing "warmouth" and "green" sunfish (*Chaenobryttus gulosus* X *Lepomis cyanellus*). Their starch gel patterns showed that two new components were present in addition to those of the parents. They did not determine the proportions of these components nor did they determine their nature. The data show that the individual points are slightly displaced from those of the parental hemolyzates. They plotted their data as the log ratio of oxy- to deoxyhemoglobin vs. log p_{O_2}. The slope of such a plot gives n, a measure of cooperativity. The hybrid curve appears slightly steeper, but this depends primarily on only two points that appear to differ from the average of points for the parental strains by no more than about 0.5% oxygenation. The double log plot magnifies this difference. The other points are very close to one of the parental curves, and any differences appear to be within experimental error. The conclusion that it has been "uncontestably established" that "larger heme–heme interactions" are present therefore is unwarranted.

Extensive studies of air-breathing fishes have been carried out by Johansen, Lenfant, and colleagues. Their study of the obligate air breather, the electric eel, *Electrophorus electricus*, is particularly instructive (Johansen et al., 1968). Their studies of lungfishes will be discussed in the next section. The adult eel utilizes extensive gas exchange in the mouth; the gills are degenerate. The blood of the electric eel has an enormous Bohr effect ($r = 0.77$)—one of the largest found in any fish. The oxygen affinity of the blood is not unusually high ($P_{50} = 14$ mm at pH 7.4), but the oxygen capacity of the blood is quite high. No studies have yet been carried out on hemoglobin solutions.

The high Bohr effect appears to be an adaptation to deal with the high CO_2 tension of *Electrophorus* blood. Johansen et al. point out that a high tension of CO_2 is characteristic of air-breathing fish which utilize a bimodal gas exchange. The CO_2 is also elevated because blood passes directly from the mouth to the venous circulation.

Johansen (1966) has also studied the air-breathing teleost, *Symbranchus marmoratus*. The oxygen equilibrium shows no sigmoid shape, no Root effect, and a moderate Bohr effect. This remarkable fish utilizes its gills for air breathing above the surface of marsh water which is low in oxygen but high in CO_2.

4. DIPNOI AND LATIMERIA

Extensive studies of the oxygen transport properties of the bloods of the lungfishes, *Protopterus* sp., *Neoceratodus* sp., and *Lepidosiren* have been carried out by Lenfant et al. (1966–1967) and Lenfant and Johansen (1968). Oldham and Riggs (1969) have compared the properties of hemoglobin solutions from *Protopterus aethiopicus* and from *P. annectans*. Bonaventura and Riggs (1969) have studied the oxygen

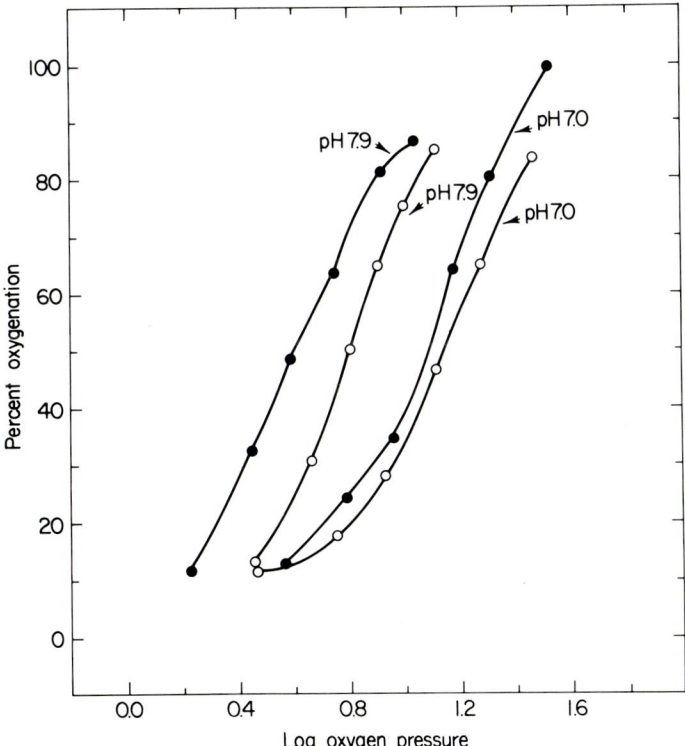

Fig. 9. The oxygen equilibria of the hemoglobins from the two closely related lungfish, *Protopterus annectans* (open circles) and *Protopterus aethiopicus* (closed circles) (Oldham and Riggs, 1969). Measurements made in 0.1 M phosphate at 25.5°C.

equilibrium of *Latimeria* hemoglobin. All of these fish possess hemoglobins with large Bohr effects, but the magnitudes differ: that of the blood of *Neoceratodus* is highest, followed by *Protopterus* and *Lepidosiren*. Although *Neoceratodus* has the largest Bohr effect and low blood CO_2, and *Protopterus* has high blood CO_2 and a smaller Bohr effect, the difference in Bohr effect is not large. These bloods have sigmoid oxygen equilibria, but the work of Oldham and Riggs (1969) on *Protopterus* hemoglobin shows that n is pH dependent, and rises from 1.2–1.4 below pH 6.5 to 2.0–2.4 above pH 7.5. Of particular interest is the fact that the two species of *Protopterus*, *P. aethiopicus* and *P. annectans*, have hemoglobins which differ not only in primary structure but also in oxygen affinity, which is considerably higher in *P. aethiopicus* than in *P. annectans* (see Fig. 9). This may be associated with the fact that the *P. aethiopicus* specimens were obtained from the deep water lakes of Uganda, whereas the *P. annectans* fish were obtained from marshland in Ghana. Both hemoglobins are composed of at least five components which are formed from four chains.

In contrast to the hemoglobins of *Protopterus*, Bonaventura and Riggs (1969) have found that *Latimeria* hemoglobin consists largely of a single component whose oxygen equilibrium shows a substantial Bohr effect ($r = 0.55$ between pH 7 and 8). In addition, the sigmoid coefficient n decreases from 1.6 to 1.1 between pH 7 and 8, in contrast to *Protopterus* hemoglobin for which n increases with pH.

ACKNOWLEDGMENTS

I thank Dr. Robert Noble for valuable discussions and for permission to use certain data prior to publication. I also thank Dr. Shanti Aggarwal and Dr. Joseph Bonaventura for reading the review and providing valuable suggestions.

Original work described here was supported in part by NIH Grant GM-05818 and by the Robert A. Welch Foundation.

REFERENCES

Adinolfi, M., Chieffi, G., and Siniscalco, M. (1959). Haemoglobin pattern of the cyclostome *Petromyzon planeri* during the course of development. *Nature* **184**, 1325.

Albers, C., and Pleschka, K. (1967). Effect of temperature on CO_2 transport in elasmobranch blood. *Resp. Physiol.* **2**, 261.

Anderson, S., and Antonini, E. (1968). The binding of carbon monoxide by human hemoglobin. *J. Biol. Chem.* **243**, 2918.

Antonini, E. (1965). Interrelationship between structure and function in hemoglobin and myoglobin. *Physiol. Rev.* **45**, 123.

Antonini, E., Wyman, J., Bellelle, L., Rumen, N., and Siniscalco, M. (1964). The oxygen equilibrium of some lamprey hemoglobins. *Arch Biochem. Biophys.* **105**, 404.

6. PROPERTIES OF FISH HEMOGLOBINS

Barcroft, J. (1928). "The Respiratory Function of the Blood, Part II, Haemoglobin," pp. 48 and 96. Cambridge Univ. Press, London and New York.

Bauer, C., and Rathschlag-Schaefer, A.-M. (1968). The influence of aldosterone and cortisol on oxygen affinity and cation concentration of the blood. *Resp. Physiol.* **5**, 360.

Benesch, R., and Benesch, R. E. (1967). The effect of organic phosphates from the human erythrocyte on the allosteric properties of hemoglobin. *Biochem. Biophys. Res. Commun.* **26**, 162.

Benesch, R., and Benesch, R. E. (1969). Intracellular organic phosphates as regulators of oxygen release by haemoglobin. *Nature* **221**, 618.

Benesch, R., Benesch, R. E., and MacDuff, G. (1964). Spectra of deoxygenated hemoglobin in the Soret region. *Science* **144**, 68.

Black, E. C. (1958). Hyperactivity as a lethal factor in fish. *J. Fisheries Res. Board Can.* **15**, 573.

Bohr, C., Hasselbalch, K., and Krogh, A. (1904). Ueber einen in biologischer beziehung wichtigen Einfluss, den die Kohlensäurespannung des Blutes auf dessen Säuerstoffbindung übt. *Skand. Arch. Physiol.* **16**, 402.

Bonaventura, J. (1968). Unpublished results.

Bonaventura, J., and Riggs, A. (1968). Hemoglobin Kansas, a human hemoglobin with a neutral amino acid substitution and an abnormal oxygen equilibrium. *J. Biol. Chem.* **243**, 980.

Bonaventura, J., and Riggs, A. (1969). Unpublished results.

Braunitzer, G. (1966). Phylogenetic variation in the primary structure of hemoglobins. *J. Cellular Physiol.* **67**, Suppl. 1, 1.

Braunitzer, G., Hilse, K., Rudloff, V., and Hilschmann, N. (1964). The hemoglobins. *Advan. Protein Chem.* **19**, 1.

Braunitzer, G., and Fujiki, H. (1969). Zur Evolution der Vertebraten, Die Konstitution und Tertiärstruktur des Hämoglobins des Flussneunauges. *Naturwissensch.* **56**, 322.

Bricker, N., Guerra, L., Klahr, S., Beauman, W., and Marchena, C. (1968). Sodium transport and metabolism by erythrocytes of the dogfish shark. *Am. J. Physiol.* **215**, 383.

Briehl, R. (1963). The relation between the oxygen equilibrium and aggregation of subunits in lamprey hemoglobin. *J. Biol. Chem.* **238**, 2361.

Brunori, M. (1966). The carbon monoxide Bohr effect in hemoglobin from *Thunnus thynnus*. *Arch. Biochem. Biophys.* **114**, 195.

Buhler, D. (1963). Studies on fish hemoglobins, Chinook salmon and Rainbow trout. *J. Biol. Chem.* **238**, 1665.

Callegarini, C., and Cucchi, C. (1968). Intraspecific polymorphism of hemoglobin in *Tinca tinca*. *Biochim. Biophys. Acta* **160**, 264.

Cann, J. (1966). Multiple electrophoretic zones arising from protein-buffer interaction. *Biochemistry* **5**, 1108.

Cann, J., and Goad, W. (1965). Theory of zone electrophoresis of reversibly interacting systems. Two zones from a single macromolecule. *J. Biol. Chem.* **240**, 1162.

Carey, F., and Teal, J. (1966). Heat conservation in tuna fish muscle. *Proc. Natl. Acad. Sci. U.S.* **56**, 1464.

Carey, F., and Teal, J. (1969). Mako and Porbeagle: Warmbodied sharks. *Comp. Biochem. and Physiol.* **28**, 199.

Chanutin, A., and Curnish, R. (1964). Factors influencing the electrophoretic patterns of red cell hemolysates analyzed in cacodylate buffers. *Arch. Biochem. Biophys.* **106**, 433.

Chanutin, A., and Curnish, R. (1967). Effect of organic and inorganic phosphates on the oxygen equilibrium of human erythrocytes. *Arch. Biochem. Biophys.* **121**, 96.

Chiancone, E., Vecchini, P., Forlani, L., Antonini, E., and Wyman, J. (1966). Dissociation of hemoglobin from different animal species into subunits. *Biochim. Biophys. Acta* **127**, 549.

Christomanos, A. (1964). Zur Kenntnis des Hämoglobins des Sägebarsches (*Serranus gabrilla*). II. *Enzymologia* **27**, 199.

Christomanos, A., and Pavlopulu, C. (1967). Zur Konstitution der Hämoglobine der Süsswasserschildkrote *clemys caspica rivulata* und Die des Aals *Anguilla anguilla*. *Folia Biochim. Biol. Graeca* **4**, 24.

Cokelet, G., and Meiselman, H. J. (1968). Rheological comparison of hemoglobin solutions and erythrocyte suspensions. *Science* **162**, 275.

Dayhoff, M. (1969). "Atlas of Protein Sequence and Structure." Natl. Biomed. Res. Found., Silver Spring, Maryland.

Dixon, G. (1966). Mechanisms of protein evolution. *Essays Biochem.* **2**, 148.

Eguchi, H., Hashimoto, K., and Matsuura, F. (1960). Comparative studies on two hemoglobins of salmon. III. Amino acid composition. *Bull. Japan. Soc. Sci. Fisheries* **26**, 810.

Enoki, Y., and Tyuma, I. (1964). Further studies on hemoglobin-oxygen equilibrium. *Japan. J. Physiol.* **14**, 280.

Fänge, R. (1966). Physiology of the swimbladder. *Physiol. Rev.* **46**, 299.

Gratzer, W., and Allison, A. (1960). Multiple haemoglobins. *Biol. Rev.* **35**, 459.

Green, A., and Root, R. (1933). The equilibrium between hemoglobin and oxygen in the blood of certain fishes. *Biol. Bull.* **64**, 383.

Guidotti, G., Konigsberg, W., and Craig, L. (1963). On the dissociation of normal adult human hemoglobin. *Proc. Natl. Acad. Sci. U.S.* **50**, 774.

Haldane, J. (1922). "Respiration," p. 72. Yale Univ. Press, New Haven, Connecticut.

Hashimoto, K., and Matsuura, F. (1959a). Comparative studies on two hemoglobins of salmon. *Bull. Japan. Soc. Sci. Fisheries* **24**, 724.

Hashimoto, K., and Matsuura, F. (1959b). Comparative studies on two hemoglobins of salmon. II. Crystallization and some physical properties. *Bull. Japan. Soc. Sci. Fisheries* **25**, 465.

Hashimoto, K., and Matsuura, F. (1960a). Multiple hemoglobins in fish. II. *Bull. Japan. Soc. Sci. Fisheries* **26**, 354.

Hashimoto, K., and Matsuura, F. (1960b). Comparative studies on two hemoglobins of salmon. V. Change in proportion of two hemoglobins with growth. *Bull. Japan. Soc. Sci. Fisheries* **26**, 931.

Hashimoto, K., and Matsuura, F. (1962). Comparative studies on two hemoglobins of salmon. VI. N-terminal amino acid. *Bull. Japan. Soc. Sci. Fisheries* **28**, 914.

Hashimoto, K., Yamaguchi, Y., and Matsuura, F. (1960). Comparative studies on two hemoglobins of salmon. IV. Oxygen dissociation curve. *Bull. Japan. Soc. Sci. Fisheries* **26**, 827.

Herner, A., and Frieden, E. (1961). Biochemical changes during anuran metamorphosis. VIII. Changes in the nature of red cell proteins. *Arch. Biochem. Biophys.* **95**, 25.

Hill, R., and Davis, R. (1967). The pK of specific groups of proteins. I. The α-amino group of the α chain of human CO-hemoglobin. *J. Biol. Chem.* **242**, 2005.

Hilse, K., and Braunitzer, G. (1968). Die Aminosäuresequenz der α-Ketten der beiden Hauptkomponenten des Karpfenhämoglobins. *Z. Physiol. Chem.* **349**, 433.

Huber, R., Formanek, H., and Epp, O. (1968). Kristallstrukturanalyse des Met-Erythrocruorins bei 5, 5 Å Auflösung. *Naturwissenschaften* **55**, 75.

Huckauf, H., Hutten, H., and Waldeck, F. (1969). Beitrag zur mikrophotometrischen O_2-Sättigungsbestimmung an einzelnen erythrocyten. *Arch. Ges. Physiol.* **305**, 190.

Ingram, V. (1961). "Hemoglobin and its Abnormalities." Thomas, Springfield, Illinois.

Ingram, V. (1963). "The Hemoglobins in Genetics and Evolution." Columbia Univ. Press, New York.

Itano, H., and Robinson, E. (1956). Demonstration of intermediate forms of carbonmonoxy- and ferri-hemoglobin by moving boundary electrophoresis. *J. Am. Chem. Soc.* **78**, 6415.

Johansen, K. (1966). Air breathing in the teleost *Symbranchus marmoratus*. *Comp. Biochem. Physiol.* **18**, 383.

Johansen, K., Lenfant, C., and Schmidt-Nielsen, K. (1968). Gas exchange and control of breathing in the electric eel *Electrophorus electricus*. *Z. Vergleich. Physiol.* **61**, 137.

Kelley, J. (1968). Carbon dioxide in the seawater under the arctic ice. *Nature* **218**, 862.

Kendrew, J., Dickerson, R., Standberg, B., Hart, R., Davies, D., Phillips, D., and Shore, V. Structure of myoglobin. A three-dimensional Fourier synthesis at 2 Å. resolution. (1960). *Nature* **185**, 422.

Kilmartin, J. V., and Rossi-Bernardi, L. (1969). Inhibition of CO_2 combination and reduction of the Bohr effect in haemoglobin chemically modified at its α-amino groups. *Nature* **222**, 1243.

Kitto, G., and Bonaventura, J. (1969). Deoxygenation induced aggregation of some invertebrate hemoglobins. *Federation Proc.* **28**, 867.

Koch, H., Bergstrom, E., and Evans, J. (1964). The microelectrophoretic separation on starch gel of the haemoglobins of *Salmo salar L*. *Mededel. Koinkl. Vlaam. Acad. Wetenschap., Kl. Wetenschap.* **26**, No. 3, 9.

Koch, H., Bergstrom, E., and Evans, J. (1966). A size correlated shift in the proportion of the haemoglobin components of the Atlantic salmon (*Salmo salar* L.) and of the sea trout (*Salmo trutta* L.) *Mededel Koinkl. Vlaam. Acad. Wetenschap. Kl. Wetenschap.* **28**, 1.

Krogh, A., and Leitch, I. (1919). The respiratory function of the blood in fishes. *J. Physiol. (London)* **52**, 288.

Lenfant, C., and Johansen, K. (1966). Respiratory function in the elasmobranch *Squalus suckleyi* G. *Resp. Physiol.* **1**, 13.

Lenfant, C., and Johansen, K. (1968). Respiration in the African lungfish *Protopterus aethiopicus*. I. Respiratory properties of blood and normal patterns of breathing and gas exchange. *J. Exp. Biol.* **49**, 437.

Lenfant, C., Johansen, K., and Grigg, G. (1966–1967). Respiratory properties of blood and pattern of gas exchange in the lungfish *Neoceratodus forsteri*. *Resp. Physiol.* **2**, 1.

Lenfant, C., Torrance, J., English, E., Finch, C., Reynafarje, C., Ramos, J., and Faura, J. (1968). Effect of altitude on oxygen binding by hemoglobin and on organic phosphate levels. *J. Clin. Invest.* **47**, 2652.

Li, S. L., and Riggs, A. (1969). Unpublished results.

Love, W., and Rumen, N. (1963). Heme–heme interaction in lamprey hemoglobin—an explanation. *Biol. Bull.* **125**, 353.
McCutcheon, F. (1947). Specific oxygen affinity of hemoglobin in elasmobranchs and turtles. *J. Cellular Comp. Physiol.* **29**, 333.
Manwell, C. (1957). Alkaline denaturation of hemoglobin of postlarval and adult *Scorpaenichthys marmoratus*. *Science* **126**, 1175.
Manwell, C. (1958a). A "fetal-maternal shift" in the ovoviviparous spiny dogfish *Squalus suckleyi* (Girard). *Physiol. Zool.* **31**, 93.
Manwell, C. (1958b). Ontogeny of hemoglobin in the skate *Raja binoculata*. *Science* **128**, 419.
Manwell, C. (1958c). On the evolution of hemoglobin. Respiratory properties of the hemoglobin of the California hagfish, *Polistotrema stouti*. *Biol. Bull.* **115**, 227.
Manwell, C. (1960). Comparative physiology: Blood pigments. *Ann. Rev. Physiol.* **22**, 191.
Manwell, C. (1963a). Fetal and adult hemoglobins of the spiny dogfish, *Squalus suckleyi*. *Arch. Biochem. Biophys.* **101**, 504.
Manwell, C. (1963b). The blood proteins of cyclostomes. A study in phylogenetic and ontogenetic biochemistry. *In* "Biology of Myxine" (A. Brodal and R. Fänge, eds.), p. 372. Oslo Univ. Press, Oslo.
Manwell, C. (1964). Chemistry, genetics, and function of invertebrate respiratory pigments—configurational changes and allosteric effects. *In* "Oxygen in the Animal Organism" (F. Dickens and E. Neil, eds.) pp. 91–93. Pergamon Press, London.
Manwell, C., Baker, C. A., and Childers, W. (1963). The genetics of hemoglobin in hybrids. I. A molecular basis for hybrid vigor. *Comp. Biochem. Physiol.* **10**, 103.
Marchis-Mouren, G., and Lipmann, F. (1965). On the mechanism of acetylation of fetal and chicken hemoglobins. *Proc. Natl. Acad. Sci. U.S.* **53**, 1147.
Matioli, G., and Niewisch, H. (1965). Electrophoresis of hemoglobin in single erythrocytes. *Science* **150**, 1824.
Noble, R., Parkhurst, L., and Gibson, Q. (1968). The kinetic basis for the Root effect in hemoglobin from *Cyprinus carpio*. *Biophys. J.* **8**, A-102.
Noble, R., Parkhurst, L., and Gibson, Q. (1969). Unpublished results.
Nolan, C., and Margoliash, E. (1968). Comparative aspects of primary structures of proteins. *Ann. Rev. Biochem.* **37**, 727.
Oldham, J., and Riggs, A. (1969). Unpublished results.
Padlan, E., and Love, W. (1968). Structure of the haemoglobin of the marine annelid worm, *Glycera dibranchiata* at 5.5 Å resolution. *Nature* **220**, 376.
Perutz, M., Muirhead, H., Cox, J., and Goaman, L. (1968). Three-dimensional Fourier synthesis of horse oxyhaemoglobin at 2.8 Å resolution: The atomic model. *Nature* **219**, 131.
Perutz, M. F., Muirhead, M., Mazzarella, L., Crowther, R. A., Greer, J., and Kilmartin, J. V. (1969). Identification of residues responsible for the alkaline Bohr effect in Haemoglobin. *Nature* **222**, 1240.
Prosser, C., and Brown, F. A. (1961). "Comparative Animal Physiology," 2nd ed. Saunders, Philadelphia, Pennsylvania.
Rapoport, S., and Guest, G. (1941). Distribution of acid-soluble phosphorus in the blood cells of various vertebrates. *J. Biol. Chem.* **138**, 269.
Riggs, A. (1961). Unpublished results.
Riggs, A. (1965). Functional properties of hemoglobins. *Physiol. Rev.* **45**, 619.

Riggs, A. (1966). Hemoglobin polymerization. Polymerization of vertebrate hemoglobins: Mechanism and effect on properties. In "International Symposium on Comparative Hemoglobin Structure" (A. Christomanos and D. J. Polychronakos, eds.), p. 126. Triantafylo, Thessaloniki, Greece.
Riggs, A., and Rona, M. (1969). The oxygen equilibria and aggregation behavior of polymerizing mouse hemoglobins. *Biochim. Biophys. Acta* **175**, 248.
Root, R. (1931). The respiratory function of the blood of marine fishes. *Biol. Bull.* **61**, 427.
Root, R., and Irving, L. (1941). The equilibrium between hemoglobin and oxygen in whole hemolyzed blood of the tautog with a theory of the Haldane effect. *Biol. Bull.* **81**, 307.
Rossi, L., Chipperfield, J., and Roughton, F. (1963). The effect of temperature on the titration curves of human oxygenated and reduced haemoglobin. *Biochem. J.* **87**, 33P.
Rossi-Fanelli, A., and Antonini, E. (1960). Oxygen equilibrium of haemoglobin from *Thunnus thynnus*. *Nature* **186**, 895.
Rossi-Fanelli, A., Antonini, E., and Mondovi, B. (1957). Enzymic reduction of ferrimyoglobin. *Arch. Biochem. Biophys.* **68**, 341.
Rossi-Fanelli, A., Antonini, E., and Caputo, A. (1964). Hemoglobin and myoglobin. *Advan. Protein. Chem.* **19**, 73.
Rumen, N. (1966). A comparison of sea lamprey (*Petromyzon marinus*) and mammalian hemoglobins. In "International Symposium on Comparative Hemoglobin Structure" (A. Christomanos and D. J. Polychronakos, eds.), p. 134. Triantafylo, Thessaloniki, Greece.
Rumen, N., and Love, W. (1963a). The six hemoglobins of the sea lamprey (*Petromyzon marinus*. *Arch. Biochem. Biophys.* **103**, 24.
Rumen, N., and Love, W. (1963b). Some hybrids of deoxygenated sea lamprey hemoglobins (*Petromyzon marinus*). *Acta Chem. Scand.* **17**, 222.
Ruud, J. (1954). Vertebrates without erythrocytes and blood pigment. *Nature* **173**, 848.
Schmidt-Nielsen, K., and Taylor, C. (1968). Red blood cells: Why or why not? *Science* **162**, 274.
Scholander, P., and van Dam, L. (1957). The concentration of hemoglobin in some cold water arctic fishes. *J. Cellular Comp. Physiol.* **49**, 1.
Schroeder, W., and Jones, R. (1965). Some aspects of the chemistry and function of human and animal hemoglobins. *Fortschr. Chem. Org. Naturstoffe* **23**, 113.
Steen, J., and Turitzin, S. (1968). The nature and biological significance of the pH difference across red cell membranes. *Resp. Physiol.* **5**, 234.
Svedberg, T., and Hedenius, A. (1934). The sedimentation constants of the respiratory proteins. *Biol. Bull.* **66**, 191.
Thompson, K. (1969). The biology of the lobe-finned fishes. *Biol. Rev.* **44**, 91.
Vanstone, W., Roberts, E., and Tsuyuki, H. (1964). Changes in the multiple hemoglobin patterns of some Pacific salmon, genus *Oncorhynchus*, during the Parr-Smolt transformation. *Can. J. Physiol. Pharmacol.* **42**, 697.
Wald, G., and Riggs, A. (1951). The hemoglobin of the sea lamprey, *Petromyzon marinus*. *J. Gen. Physiol.* **35**, 45.
Watts, D. (1968). Variation in enzyme structure and function: the guidelines of evolution. *Advan. Comp. Physiol. Biochem.* **3**, 1.
Wilkins, N. (1968). Multiple haemoglobins of the Atlantic salmon (*Salmo solar*). *J. Fisheries Res. Board Can.* **25**, 2651.

Wilkins, N., and Iles, T. (1966). Haemoglobin polymorphism and its ontogeny in herring (*Clupea harengus*) and sprat (*Sprattus sprattus*). *Comp. Biochem. Physiol.* **17**, 1141.

Wyman, J. (1948). Heme proteins. *Advan. Protein Chem.* **4**, 407.

Wyman, J., Jr. (1964). Linked functions and reciprocal effects in hemoglobin: A second look. *Advan. Protein Chem.* **19**, 223.

Yamaguchi, K., Kochiyama, Y., Hashimoto, K., and Matsuura, F. (1962). Studies on multiple hemoglobins of eel. II. Oxygen dissociation curve and relative amounts of components F and S. *Bull. Japan. Soc. Sci. Fisheries* **28**, 192.

Yamaguchi, K., Kochiyama, Y., Hashimoto, K., and Matsuura, F. (1963). Studies on two hemoglobins of loach. II. Oxygen dissociation curve. *Bull. Japan. Soc. Sci. Fisheries* **29**, 180.

Yamanaka, H., Yamaguchi, K., and Matsuura, F. (1965). Starch gel electrophoresis of fish hemoglobins. I. Usefulness of cyanmethemoglobin for the electrophoresis. *Bull. Japan. Soc. Sci. Fisheries* **31**, 827.

Yoshioka, M., Hamada, K., Okazaki, T., Kajita, A., and Shukuya, R. (1968). Hemoglobins from erythrocytes of the eel, *Anguilla japonica*. II. Further studies on the properties of the two hemoglobins. *J. Biochem. (Tokyo)* **63**, 70.

Zuckerkandl, E., Jones, R., and Pauling, L. (1960). A comparison of animal hemoglobins by tryptic peptide pattern analysis. *Proc. Natl. Acad. Sci. U.S.* **46**, 1349.

7

GAS EXCHANGE IN FISH

D. J. RANDALL

I. Introduction	253
II. Gas Exchange between Blood and Water	254
A. Gas Exchange across the Gills	254
B. Gas Exchange across the Skin	276
III. Gas Exchange between Blood and Tissues	278
IV. Methods of Analysis of the Transfer of Gases	279
A. Introduction	279
B. Transfer Factor	280
C. The Effectiveness of Gas Transfer	281
D. Capacity Rate Ratio	282
E. Ventilation–Perfusion Ratio	282
V. The Effects of Various Parameters on Gas Exchange	283
A. Temperature	283
B. Hypoxia	284
C. Exercise	284
References	286

I. INTRODUCTION

Studies of gas exchange in fish have concentrated on the movement of oxygen and carbon dioxide across the gills. In comparison the transfer of gases between blood and tissues in fish has received very little attention except for a few special exchange sites that have been studied in detail. These are the rete mirabile and gas gland of the teleost swim bladder (see chapter by Steen, this volume) and, to a lesser extent, the rete in the choroid layer of the eye of teleosts (Wittenberg and Wittenberg, 1962).

Fish breathe either seawater, freshwater, or air or some combination of these media. There are a number of surfaces in fish for the exchange of gases between blood and the medium. The primary site, the gills, is

designed for gas exchange between blood and water. There are many
accessory respiratory surfaces, which in general are associated with the
movement of fish from an hypoxic aquatic to an aerial environment and
are designed for gas exchange between blood and air. The problems
an animal has to face when breathing seawater or freshwater are similar,
although the differences in ionic content, osmolarity, and gas solubility require small changes in the functioning of the gas exchange system. The
problems associated with aquatic gas exchange, however, are different
from those associated with aerial gas exchange. This chapter is confined
to a discussion of gas transfer between water and blood and between
blood and the tissues. Air breathing in fishes is discussed in the chapter
by Johansen, this volume.

Oxygen is brought into close contact with the gills by the bulk flow
of water. Oxygen diffuses across the gills into the blood down a gradient
of between 40 and 100 mm Hg P_{O_2}. Oxygen is transported in the blood
from the gills to the capillaries; it then diffuses across the capillary walls
into the tissues. Gas gradients between blood and tissues in fish have not
been measured, but tissue gas tensions are probably in the range of
1–15 mm Hg P_{O_2} and 3–15 mm Hg P_{CO_2}. The partial pressure of oxygen
in the environment is dissipated as oxygen molecules move from the
water into the tissues via the blood. The largest pressure drop, and
hence the major resistance to the movement of oxygen, occurs at the
gills between water and blood.

II. GAS EXCHANGE BETWEEN BLOOD AND WATER

Gas exchange between blood and water occurs across the gills and
the general body surface. The surface area of the gills is between 10 and
60 times that of the rest of the body (Parry, 1966) and is the more important site for gas exchange. Much more is known about the movement
of oxygen and carbon dioxide across the gills, and therefore gas exchange
across the gills is discussed separately from and in greater detail than
gas exchange across the skin.

A. Gas Exchange across the Gills

Gas exchange across the gills of fishes has been reviewed by Krogh
(1941), Black (1951), and Fry (1957). More recently, Hughes and
Shelton (1962), Hughes (1964, 1967), Lenfant and Johansen (1966),
Rahn (1966a,b, 1967), Garey (1967), Kylstra et al. (1967), Hanson

(1967), Randall et al. (1967), Robin and Murdaugh (1967), Dejours et al. (1968), and Piiper and Baumgarten-Schumann (1968a,b) have discussed aquatic gas exchange in detail; they have shown how the design of the exchanger can be related to the properties of the media on either side of the respiratory epithelium, in this case blood and water. The diffusion of gases in both tissues and water is extremely slow (Table I), and the design of the gas exchange system is such that diffusion is kept to a minimum and largely restricted to the movement of oxygen and carbon dioxide across the gill epithelium. Gas molecules are delivered to or removed from the gill epithelium by the bulk flow of water and blood.

Except for some larval forms, gills are ubiquitous in fish. The fine structure of the gills of teleosts has been described by Hughes and Grimstone (1965), Rhodin (1964), Newstead (1967), and Hughes and Datta Munshi (1968). Other groups of fish have received less attention, references and a general description of the anatomy of the gills of elasmobranchs, cyclostomes, and Dipnoi can be found in Daniel (1922), Fry (1957), Chapman et al. (1963), and Johansen and Strahan (1963).

The gills form a sievelike structure placed in the path of the respiratory water flow. The secondary lamellae form the side walls of this sieve (Fig. 1) and probably represent the major respiratory portion of the gill structure (Hughes, 1966a; Muir and Hughes, 1969). The total surface area of the secondary lamellae is about 5 cm^2/g body weight (Gray, 1954; Hughes, 1966a). There is a countercurrent (van Dam, 1938; Hughes and Shelton, 1962) or multicapillary (Piiper and Schumann, 1967) arrangement of the flows of blood and water on either side of the gill epithelium; the epithelium is usually between 1 and 5 μ in thickness. The ratio of the flows of blood and water is somewhere between 1:10 (Piiper and Schumann, 1967; Garey, 1967) and 1:80 (Stevens and Randall, 1967b). The ratios of the content per mm Hg partial pressure of both oxygen and carbon dioxide in water and blood are between 1:10 and 1:20 (Black et al., 1966; Beaumont and Randall, 1968).

1. DIMENSIONS OF THE GILLS

A number of workers (see Muir, 1969, for reference) have measured the dimensions of the gills of teleost fish and estimated the total surface area of the secondary lamellae, which is generally considered to represent the anatomical respiratory surface area. The surface area of the gills discussed below refers to the total surface area of the secondary lamellae.

The average surface area of the gills of teleosts, compiled from the

Table I
Various Gas Coefficients and Some Other Constants[a]

Temp (°C)	Gas solubility coefficients (ml/liter/mm Hg)						Water vapor pressure (mm Hg)	Viscosity of water (poise, dynes/cm²/sec)
	Freshwater			Seawater (salinity 34–35‰)				
	O_2	CO_2	CO_2/O_2	O_2	CO_2[b]	CO_2/O_2		
0	0.0647	2.254	35	0.0497	1.90	38	4.58	0.0175
5	0.0567	1.874	33	0.0433	1.57	36	6.54	0.0149
10	0.0505	1.571	31	0.0390	1.34	35	9.21	0.0128
15	0.0455	1.341	30	0.0353	1.16	33	12.79	0.0112
20	0.0414	1.155	28	0.0324	1.00	31	17.54	0.0098
25	0.0381	0.999	26	0.0300	0.88	29	23.76	0.0087
30	0.0351	0.875	25	0.0281	0.77	28	31.82	0.0078
37	0.0322	0.683	21	0.0256	—	—	47.07	

[a] Data from West (1965), Hodgman (1943), Gameson and Robertson (1955), Truesdale and Gameson (1957), Strickland and Parsons (1965), Altman and Dittmer (1964), Kohn (1965), and Kaye and Laby (1958).

[b] Extremely variable, only approximate values given.

"True" diffusion coefficient D (Washburn, 1929)
Gases in water D(cm²sec⁻¹)
- (1) O_2 at 18°C 1.9×10^{-5}
- (2) CO_2 at 18°C 1.7×10^{-5}

NaCl in 0.1 M aqueous solution
- (3) NaCl at 18°C 1.24×10^{-5}

Permeation coefficients D' (Krogh, 1919)
Gases in water D'(ml/min/cm²/cm/760 mm Hg)
- (4) O_2 at 20°C 3.4×10^{-5}
- (5) CO_2 at 20°C 78.3×10^{-5}
- (6) Ratio $D'_{O_2}:D'_{CO_2}$ 1:23

Calculated permeation coefficients

Permeation coefficients can be calculated from (1) and (2) assuming the temperature coefficient of D is 2% per °C and that $D' = D\alpha$, where α is the solubility coefficient in milliliters gas (STPD) dissolved per milliliter of water at a partial pressure of 1 atm (Dittmer and Grebe, 1958).

O_2 at 20°C = 3.75×10^{-5} ml/min/cm²/cm/760 mm Hg
CO_2 at 20°C = 93.2×10^{-5} ml/min/cm²/cm/760 mm Hg
Ratio $D'_{O_2}:D'_{CO_2}$ = 1:25

Other permeation coefficients (ml/min/cm²/cm/760 mm Hg $\times$ 10⁵)

Substance	Temp. (°C)	O_2	CO_2	CO_2/O_2
Frog connective tissue	20	1.15	41.0	36
Frog muscle	20	1.4	52.0[c]	37

[c] From Wright (1934) assuming a temperature coefficient of 1% per °C for D'. Other values from Krogh (1941).

7. GAS EXCHANGE IN FISH   257

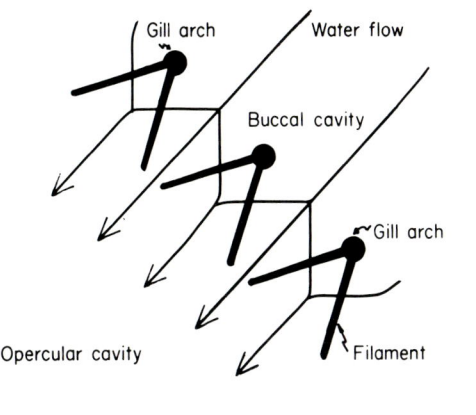

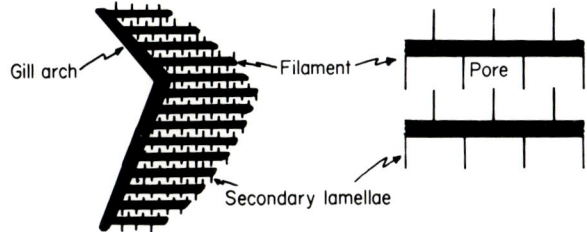

Side view of gill arch

Fig. 1. Diagram illustrating the general structure of the teleost gill arch, possible arrangement of the secondary lamellae in forming pores through which water flows, and the general pattern of water flow over the gills. The arrows in the top diagram illustrate the pattern of water flow over the gills. Water passes either over the filaments and through the pores formed by the secondary lamellae or between the tips of filaments. Water passing between the tips of the filaments will not be involved in gas transfer (see text) and will constitute a portion of the water shunt, the anatomical dead space volume. Adapted from Hughes (1966a).

data of Hughes (1966a) and Gray (1954) on 31 species of teleost fish, is 4.9 cm^2/g body weight. There is considerable variability around this average value. Gray (1954) recorded a gill area of 11.58 cm^2/g for the mackerel, whereas that of the goosefish is only 1.43 cm^2/g body weight (Hughes, 1966a). Smaller fish have a larger gill area per unit weight than heavier fish of the same species (Muir, 1969), the area increasing with weight to a power of about 0.85 (Fig. 2). Most active fish have a larger gill area than sluggish forms (Gray, 1954; Hughes, 1966a; Steen and Berg, 1966). The surface area of the gills in tuna (Muir and Hughes, 1969) is much larger than normally encountered in fish and approaches

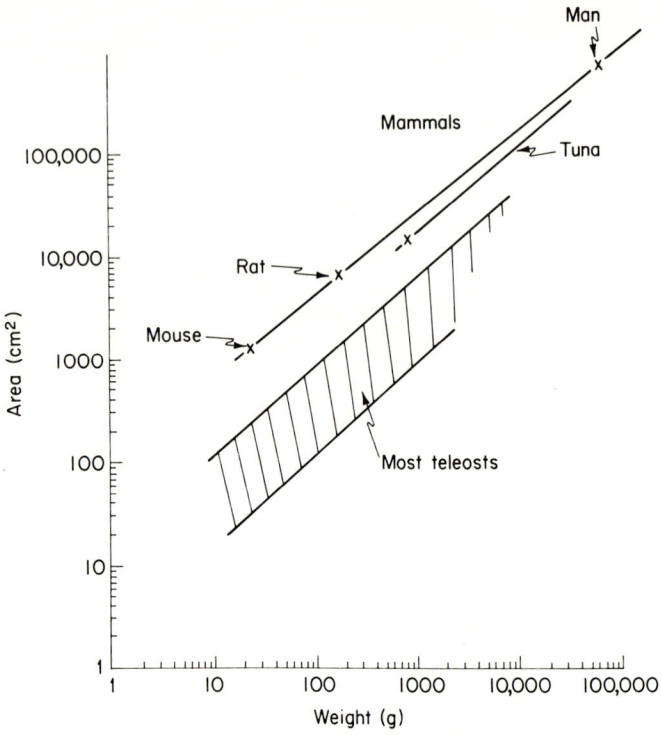

Fig. 2. The relationship between respiratory surface area and body weight for teleosts and a few mammals. The shaded area covers most of the recorded values for gill area in teleost. Redrawn and adapted from Muir (1969).

the lung surface area of a terrestrial mammal (Fig. 2). This is undoubtedly related to the high levels of sustained activity that can be maintained by tuna. In general, however, the area of the respiratory surface is smaller in fish than in terrestrial mammals. This may be related to either the lower rates of oxygen consumption in fish or the dual function of the gills. The gills are an important site for ion and water exchange as well as gas exchange (see chapter by Conte, Volume I). An increase in the size of the gills increases ion and water exchange as well as gas exchange. Fish in freshwater may have to restrict the surface area of their gills in order to reduce ion and water exchange, and it is possible that the large surface area of the gills of tuna could only have evolved under the reduced osmotic load of seawater.

The surface area of the gills tends to be reduced in air-breathing fish (Dubale, 1951). In these fish the secondary lamellae are often far apart, and unlike non-air-breathing fish, the lamellae do not collapse or stick together when the fish is in air (Saxena, 1958, 1959). Many air-breathing fish cannot live by aquatic respiration even in well-oxygenated water (Carter, 1957). The climbing perch, Anabas, an air-breathing fish, has a gill area of about 1.44 cm^2/g (Dubale, 1951) and the respiratory epithelium is about 20 μ thick (Hughes and Datta Munshi, 1968). Gas exchange across the gills of this fish cannot be very significant, and the gills probably function primarily in ion exchange. The area is presumably reduced to prevent loss of oxygen to the water and to restrict ion and water movement.

The values presented and discussed above are anatomical areas of the gills. The functional area need not, and probably does not, equal the anatomical area. The anatomical area is a measure of the maximum possible functional area. Hughes (1966a) estimated that the respiratory surface was between 60 and 70% of the total lamellar surface. This value was based on the area of blood channels in the secondary lamellae. However not all the blood pathways may be utilized at any instant in time, and the functional area may be smaller than 60–70% of the lamellar surface area. There are alternate nonrespiratory pathways that shunt blood past the secondary lamellae which can be used to decrease lamellar blood flow (Steen and Kruysse, 1964). The following equation from Hughes (1966a) relates the functional area A' to the oxygen uptake $\dot{V}_{O_2}$ across the gills, d' the diffusion distance between blood and water, D' the oxygen permeation coefficient (Table I), and ΔP_{O_2} the oxygen gradient between blood and water across the gill epithelium:

$$A' = \frac{\dot{V}_{O_2} \cdot d' \cdot 760}{D' \cdot \Delta P_{O_2}}$$

If one assumes that D' is the same as that for frog connective tissue (Table I) and d' is 2 μ, then the oxygen uptake and gas gradient data of Stevens and Randall (1967a,b), when applied to the above equation, indicate that the functional surface area of a resting trout is only 20% of the measured lamellar area reported for this fish by Hughes (1966a).

The thickness of the gill epithelium is usually between 1 and 5 μ (Newstead, 1967), but under certain circumstances the liberation of mucus from cells in the gills may increase the diffusion distance between blood and water. The production of mucus cells is probably regulated by the hormone prolactin (Ball, 1969), and mucous is released under a variety of conditions including the movement of fish from seawater to freshwater (Lam, 1968). The release of mucus may increase the diffusion distance for both ions and water as well as gases. The discussion between Hughes, Strang, Reid, and Pattle is germane to this topic. Pattle points out that "One of the functions of mucus at respiratory surfaces may be to prevent excessive diffusion of water whilst allowing adequate diffusion of gases" (page 108, "The Development of the Lung" de Reuck and Porter, eds., 1967). Nothing is known about the rates of diffusion of water, ions, or gases through the mucus produced by the skin and gills of fishes.

2. WATER FLOW OVER THE GILLS

A unidirectional flow of water over the gills is produced by the action of muscular pumps (see chapter by Shelton, this volume). The pioneering work of van Dam (1938) indicated that between 60 and 80% (percent utilization) of the oxygen in the water passing over the gills is utilized by the fish. The more recent studies of Saunders (1961, 1962), Garey (1967), Holeton and Randall (1967b), Piiper and Schumann (1967), and Stevens and Randall (1967b) have shown that percent utilization is variable between fish and in the same fish under different conditions.

The cost of ventilating the gills with such a dense medium is generally considered to be high (Hughes and Shelton, 1962) and may be as much as 30% of the total oxygen uptake (Schumann and Piiper, 1966). Garey (1967), however, has pointed out that the high density of water is offset by a low resistance to flow through the gills and has calculated that the cost of breathing in carp is only 2% of the total oxygen uptake, a value similar to that for man (see also chapter by Shelton, this volume).

There are probably differences in the cost of breathing among fish. Fish, like carp and tench, in still water, must pump water over their gills. These fish probably have the highest cost of ventilation, and so they tend to have low ventilation volumes (ventilation volume is the volume of water passing over the gills per unit of time) and high percent utilization of oxygen in the water (Garey, 1967; Baumgarten and Randall, 1967). Fish swimming rapidly or fish that can maintain position in fast flowing water need only open their mouths to ventilate their gills. Gill ventilation in this case is related to swimming speed, water flow rate, resistance to water flow through the gills, and the gape of the mouth and operculum (Muir and Kendall, 1968). At high swimming speeds the problem may be to reduce water flow over the gills rather than maintain an adequate rate of ventilation. In these fish the cost of ventilating the gills is probably small and related to the increased resistance to forward motion produced by passing water over the gills. Salmonids can maintain position in fast flowing water using only frictional forces between the fish and the bottom. As water flow is increased the amplitude of breathing decreases; residual breathing movements are always present, however, even in high flows of water supersaturated with oxygen (Fairey, 1966). In this instance, ventilation of the gills is largely a by-product of maintaining position in a stream; hence, the cost of breathing is probably not an important factor in the total energy budget of the fish, ventilation volumes are high, and the percent utilization of oxygen is low (Baumgarten and Randall, 1967). The remora stops breathing in water velocities greater than about 60 cm/sec (Muir and Buckley, 1967), ventilation volume is probably regulated at high velocities by altering the gape of the mouth. At low velocities the fish actively ventilates its gills. This animal rides on the body of sharks and uses the swimming efforts of its host to ventilate its gills. There are no breathing movements in mackeral and tuna (Hall, 1930), and forward motion of the fish through the water acts to ventilate the gills. Muir and Kendall (1968) refer to this as "ram" ventilation. Thus some fish can probably maintain a high ventilation volume at low cost while others must continually pump water over their gills. Those animals that can maintain a high ventilation volume will have a low percent utilization of oxygen from the water. The advantage of a low percent utilization is that the water P_{O_2} at the respiratory surface remains high along the whole length of the secondary lamellae.

If the respiratory surface area is infinitely large, the diffusion distance between blood and water infinitely small, and gas exchange across the gills passive, then, as there is a countercurrent arrangement of the flows of blood and water in teleosts it is theoretically possible for the P_{O_2} in water leaving the gills to be in equilibrium with that of venous blood. In practice, however, the P_{O_2} in water leaving the gills is not in equilibrium with that of venous blood in either teleosts (Garey, 1967; Holeton and Randall, 1967b; Stevens, 1968b) or elasmobranchs (Piiper and Baumgarten-Schumann, 1968b). Thus not all the oxygen that can be removed from the water is utilized by the animal. There are two possible explanations for this: first, there could be a large diffusion resistance across the gills, and, second, some of the water may not come into close contact with the gills and be shunted past the gills. This volume of water can be considered as a water shunt (V_D shunt) and expressed as a percentage of the total ventilation volume.

If the diffusion resistance across the gills is negligible, then the magnitude of the water shunt can be calculated from the following equation:

$$V_D \text{ shunt} = \dot{V}_G \frac{(P_{E_{O_2}} - P\text{veq}_{O_2})}{P_{I_{O_2}}}$$

where $\dot{V}_G$ is the total gill ventilation, $P_{I_{O_2}}$ the partial pressure of oxygen in inspired water, $P_{E_{O_2}}$ that in expired water, and $Pveq_{O_2}$ that in water having the same P_{O_2} as venous blood entering the gills. If the expired water has the same P_{O_2} as venous blood then both the diffusion resistance and the water shunt are zero. In practice, however, $P_{E_{O_2}}$ is always greater than $Pveq_{O_2}$ and there is probably both a diffusion resistance and a water shunt. The magnitude of the actual water shunt can be calculated from the following equation:

$$V_D \text{ shunt} = \dot{V}_G \frac{(P_{E_{O_2}} - Pveq_{O_2} - \Delta P_{O_2})}{P_{I_{O_2}}}$$

where ΔP_{O_2} is the oxygen gradient between blood and water across the gill epithelium. Using measurements of the dimensions of the respiratory surface, ΔP_{O_2} can be calculated by rearranging an equation from Hughes (1966a):

$$\Delta P_{O_2} = \dot{V}_{O_2} \cdot \frac{d' \cdot 760}{D'A}$$

where $\dot{V}_{O_2}$ is the oxygen uptake, d' is the thickness of the gill epithelium, A is the area of the secondary lamellae, and D' is the Krogh permeation coefficient for oxygen in the gill epithelium [assumed to be the same as that for frog connective tissue (Krogh, 1941)]. The calculated oxygen gradient across the gill epithelium (2 μ thick) is between 2 and 8 mm Hg at standard rates (Fry, 1957) of oxygen consumption. This gradient will obviously be larger if the functional area of the gills is decreased or the diffusion distance is increased (see Section II, A, 1).

Direct measurements of ΔP_{O_2} have not been made, and the actual gradient that exists across the gill epithelium will depend on the functional rather than the anatomical dimensions of the respiratory surface. The functional area of the gills will depend on the extent of gill vascularization and on the number of capillaries open to blood flow. There are no adequate estimates of the functional area of the gills but it will be less than the anatomical area. In the absence of any measurements, if one assumes that ΔP_{O_2} is 20 mm Hg, then the water shunt in the trout based on the data of Stevens and Randall (1967b) is 60% of the total ventilation volume.

Water flow over the gills can therefore be divided into a series of separate volumes or flows. First, there is that portion of the water flow which contains oxygen that passes into the blood, which is the respiratory water flow or volume. Second, there is some water that is not brought into close contact with the respiratory epithelium, which is the water shunt. Finally, there is the remainder of the water flow, which can be considered as a residual volume, the magnitude of which will be determined by Pv_{O_2}.

The water shunt can be further divided into component volumes. These are a diffusion dead space volume $V_{D\ diffO_2}$, an anatomical dead space volume $V_{D\ anatO_2}$, and finally a distribution dead space volume $V_{D\ distO_2}$. They are described below.

a. Diffusion Dead Space. Water is probably in contact with the respiratory surface for about 1 or 2 sec (see below), and since the rate of diffusion of gases in water is slow (Krogh, 1941) water must be brought into close contact with the gills if it is to exchange gases with the blood. Distances between lamellae may be so large and/or flow through the lamellae so rapid that there is not sufficient time for all water to reach equilibration with the blood. In this case there will be persistent gradients in the water, and these may be considered as representative of a diffusion dead space volume.

b. Distribution Dead Space. If ventilation of the pores formed by the secondary lamellae is high, or if the pores are ventilated unequally, more oxygen may be delivered to all or part of the respiratory surface than is required to saturate the blood; hence, there may be a distribution dead space associated with unequal ventilation and perfusion of the gills.

c. Anatomical Dead Space. Water flowing over the gills of teleosts may pass through the pores formed by the secondary lamellae (Fig. 1) or spill between the edges of the filaments. Only water passing through the pores will be involved in gas exchanged, and water passing between the ends of the filaments can be considered as part of the water shunt. Hughes (1966a) referred to this portion of the water flow as the anatomical dead space volume.

Thus water flow over the gills can be divided into a series of component volumes as follows:

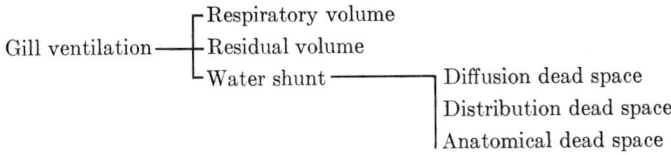

The percent utilization of oxygen is a measure of the size of the respiratory volume and is an inverse measure of the size of the combined residual volume and water shunt.

The size of the water shunt varies in fish, as mentioned above, and may be as high as 60% of the total ventilation volume in the trout. The relative contribution of the anatomical, diffusion, and distribution dead space volume to the water shunt is difficult to assess. The size of the anatomical dead space will be related to the arrangement of the filaments to each other and to the ventilation volume. The anatomical dead space is probably larger at high water flow rates. The position of the filaments is controlled by muscles at their base (Bitjel, 1949) which could play a role in regulating the size of the anatomical dead space (Pasztor and Kleerekoper, 1962). The author has observed movements of trout gills during normal breathing movements, and these may be caused by changing water velocities or by the action of the muscles at the base of the filaments. Whatever the cause, the effect of altering the relative position of filaments to each other must be to change the size of the anatomical dead space, which is probably very variable with time in a single fish, as well as between different species of fish. There appears to be a tendency to enlarge supporting structures and fuse parts of the gill in fish exposed to high ventilation volumes. In tuna, the secondary lamellae of adjacent filaments are used to form a compact sievelike gill structure (Muir and Kendall, 1968). This presumably helps to maintain flow between the secondary lamellae and reduces the size of the anatomical dead space, but it must also increase the resistance to flow through the gills. Fusion of gill parts, however, need not always be associated with the maintenance of flow between secondary lamellae. Hughes (1966b) has suggested that fusion in *Amia* serves to prevent collapse of the gill sieve when the animal is in air.

Kylstra *et al.* (1967) have analyzed gas transfer in water-breathing dogs and in the gills of fish. They related the size of the diffusion dead space ($V_{D\ diffO_2}$) in the gills of fishes to the rate of diffusion (D), the time for diffusion (t), and the distance over which diffusion takes place (a) and have derived the following equation which permits the evaluation of the diffusion dead space for oxygen at the gills (Kylstra *et al.*, 1967):

$$V_{D'\ diffO_2} = \frac{1}{1 + 3D \cdot t/a^2} \times \dot{V}_G$$

The volume of water contained within the pores of the gills can be calculated by multiplying the surface area of the lamellae (A) by half the distance (d) between successive secondary lamellae. The time (t) that a particular volume of water is in contact with the respiratory surface can be determined from the equation

$$t = 2\dot{V}_{G\ pore}/Ad$$

where $\dot{V}_{G\ pore}$ is the water flow between the secondary lamellae. The time (t) can be determined for a variety of values of $\dot{V}_G$ using the anatomical data of Hughes (1966a). Values for t in the resting trout are of the order of 0.4–2 sec depending on ventilation volume. The diffusion dead space can be calculated (Fig. 3) from the equation of Kylstra et al. (1967); D is 1.43×10^{-5} cm^2/sec at 5°C (from Table I, assuming a temperature coefficient at 2%/°C). The diffusion distance (a) is taken to be half the distance between successive secondary lamellae on a filament. The calculated values of diffusion dead space are small in trout and carp, largely because of the very small distances between secondary lamellae. Large water shunts past the gills cannot therefore be explained in terms of a large diffusion dead space. In the trout the diffusion dead space is only 7% $\dot{V}_G$ (Fig. 3) even when $\dot{V}_G$ is 10.3 ml/sec ($t = 0.4$ sec) and all the water passes between the secondary lamellae. A large water shunt can only be explained in terms of a large anatomical or distribution dead space.

A low percent utilization of oxygen from the water may therefore be

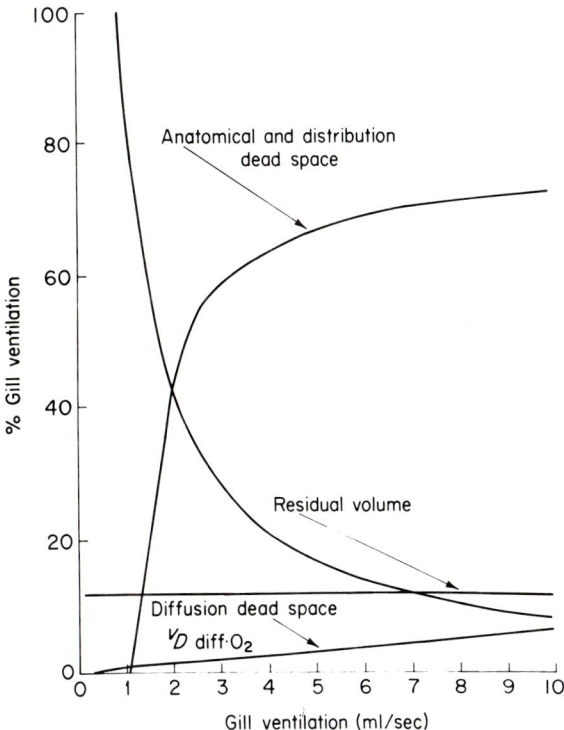

Fig. 3. The effect of ventilation volume on the size of the anatomical ($V_{D\ \text{anat}O_2}$) and distribution ($V_{D\text{dist}O_2}$) dead space, the residual dead space, and the diffusion dead space ($V_{D\text{diff}O_2}$) assuming an oxygen uptake ($\dot{V}_{O_2}$) of 0.47 ml/min and a venous oxygen tension (P_{VO_2}) of 20 mm Hg.

associated with either a high venous oxygen tension (Pv_{O_2}), a large anatomical dead space, or a large distribution dead space. Large differences between Pv_{O_2} and $P_{E_{O_2}}$ (expired water oxygen tension) may be associated with either a large anatomical or distribution dead space. The size of the anatomical dead space can be reduced by maintaining contact between the tips of adjacent filaments and lamellae, thus forcing water between the secondary lamallae. The distribution dead space could be reduced by ensuring maximum blood flow to those portions of the gill receiving maximum water flow. There is no evidence at present that this occurs in fish. Measured values for $P_{E_{O_2}} - Pv_{O_2}$ are 100 mm Hg in the trout (Stevens and Randall, 1967b) 40 mm Hg in the carp (Garey, 1967), and 46 mm Hg in the dogfish (Baumgarten-Schumann and Piiper, 1968), indicating a large water shunt in these fish.

It is important to note that, although there may be large differences between the mean P_{O_2} values for blood and water across the gills and arterial P_{O_2} values may be very different from those in inspired water, this does not indicate that the gills are not effective in transferring oxygen between water and blood. In fact, blood leaving the gills is usually between 85 and 95% saturated with oxygen.

3. Blood Flow through the Gills

In fish, unlike mammals, the respiratory and systemic circulations are in series rather than in parallel. Blood ejected from the heart is conveyed by the ventral aorta to the afferent branchial arteries supplying each gill arch. Blood flows from the afferent to the efferent branchial arteries through the capillary bed of the gills. These efferent branchial vessels join to form the dorsal aorta through which blood passes to the general body circulation. Figure 4 illustrates the vascularization of the gills of a salmonid. There are a number of alternate pathways of varying distance from the water interface, for the passage of blood through the gills (Steen and Kryusse, 1964; Hughes and Grimstone, 1965; Newstead, 1967; Datta Munshi and Singh, 1968). Blood may flow through either a few or all of the secondary lamaellae on each filament. Some blood may bypass the lamellae, flowing through capillaries joining afferent and efferent vessels within the filaments. Alterations in the distribution and volume of blood flow through these channels will change the functional surface area of the gills and the diffusion distance between blood and water, thus affecting the capacity of the gills to transfer gases. At low rates of oxygen uptake in the eel some blood is probably shunted past the secondary lamellae, and blood leaving the gills is not fully saturated.

7. GAS EXCHANGE IN FISH

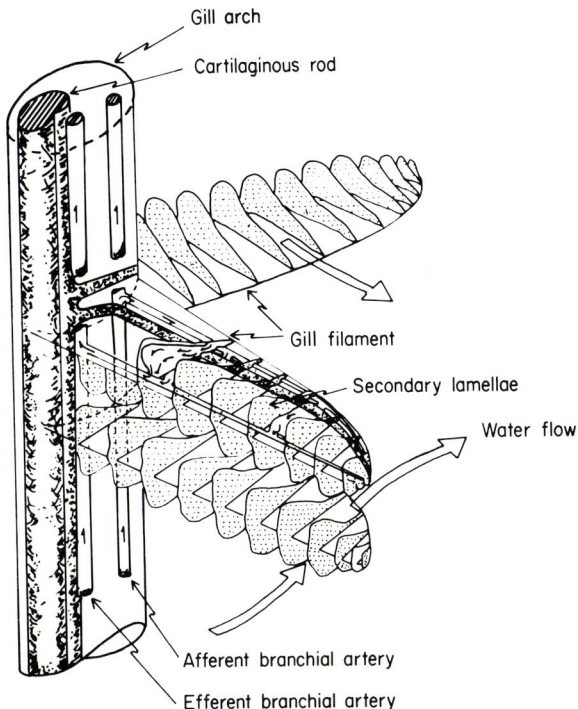

Fig. 4. A diagram to illustrate the pattern of blood flow through and water flow over the gills of a teleost fish. Unpublished diagram by F. Conte.

Thus the eel can increase the arterial–venous oxygen difference by raising arterial saturation as well as by lowering venous saturation (Steen and Kruysse, 1964). There are probably gill blood shunts in elasmobranchs (Piiper and Baumgarten-Schumann, 1968a) and Dipnoi (Johansen and Hanson, 1968) as well as in teleosts.

Capillaries in the secondary lamellae are close to the water interface and are undoubtedly more involved in gas transfer than those in the gill filament. Steen and Kryusse (1964) have shown that adrenaline increases lamellar blood flow and the percent saturation of the blood leaving the gills of the eel. Adrenergic receptors are present in fish gills, and catecholamines are known to cause a marked vasodilation of the gills (Keys and Bateman, 1932; Östlund and Fänge, 1962). Exercise in salmonids is associated with an increase in the level of circulating catecholamines (Nakano and Tomlinson, 1967) and a rise in dorsal aortic blood pressure (Randall and Stevens, 1967). This rise in dorsal aortic blood pressure is not seen after α-adrenergic receptor blockade with phenoxybenzamine. There is a decrease in the resistance to blood flow in both the respiratory and systemic circulations during exercise and an increase in the transfer factor of the gills for oxygen (see Section IV, B). The circumstantial evidence cited above indicates that increased levels of catecholamines (adrenaline and noradrenaline) alter blood flow through the gills in a way that either increases the functional surface area of the gills or decreases the diffusion distance in order to increase the rate of gas exchange across the gills.

Acetylcholine increases the resistance to blood flow through the gills and decreases lamellar blood flow (Steen and Kruysse, 1964). Hypoxia in the water flowing over the gills of trout increases the resistance to blood flow through the gills (Holeton and Randall, 1967a) but does not impair the capacity of the gills to transfer gases.

The pattern of capillaries in the secondary lamellae of the tuna appear to be different from the rest of teleosts. The capillaries are not divided into respiratory and nonrespiratory vessels, but all blood passes through the secondary lamellae (Muir, 1970). Tuna have typical afferent and efferent vessels, but each filament afferent gives off about 20 lamellar afferents to each secondary lamellae. Each of the lamellar afferents subdivides forming a large number of blood channels in the secondary lamellae, all of which are respiratory. The respiratory surface area of the gills of the tuna may be regulated by altering the number of lamellar afferents open to blood flow. In some tuna species there are valvelike flaps in the filament afferent that may play a role in regulating blood flow to the lamellae.

The capillaries in teleost gills are wide enough to allow the passage of nucleated red blood cells. The erythrocytes are about 11 μ long and 6.5 μ wide in tuna (Muir, 1967), similar in size to those of *Scomberomorus* (Bastos, 1966). The diameter of the capillaries appears to be less than that of the erythrocyte, which becomes sausage-shaped as it is forced through the gill capillaries. Red blood cells in fish swell markedly if blood CO_2 levels increase (Ferguson and Black, 1941; Holeton and Randall, 1967b).

The importance of countercurrent exchange in fishes has been stressed by a number of investigators (van Dam, 1938; Hazelhoff and Evenhuis, 1952; Hughes and Shelton, 1962). Higher oxygen levels in arterial blood than efferent water have been recorded in both teleosts (Holeton and Randall, 1967b) and elasmobranchs (Piiper and Baumgarten-Schumann, 1968a) indicating the presence of a functional countercurrent arrangement of the flows of blood and water. However, in a number of instances (Saunders, 1962; Lenfant and Johansen, 1966), the recorded oxygen level in arterial blood was below that of the efferent water. These observations led Lenfant and Johansen (1966) to question the presence of a countercurrent system in elasmobranchs. Robin and Murdaugh (1967) discussed the effects of various arrangements of the flows of blood and water on gas exchange and concluded that a countercurrent arrangement of the flow was not present in elasmobranchs. The morphology of the elasmobranch gill is in close agreement with the multicapillary arrangement of the circulation (Fig. 5) and Piiper and Schumann (1967) used a serial multicapillary arrangement of the flows of blood and water to explain the phenomenon of P_{O_2} levels in arterial blood

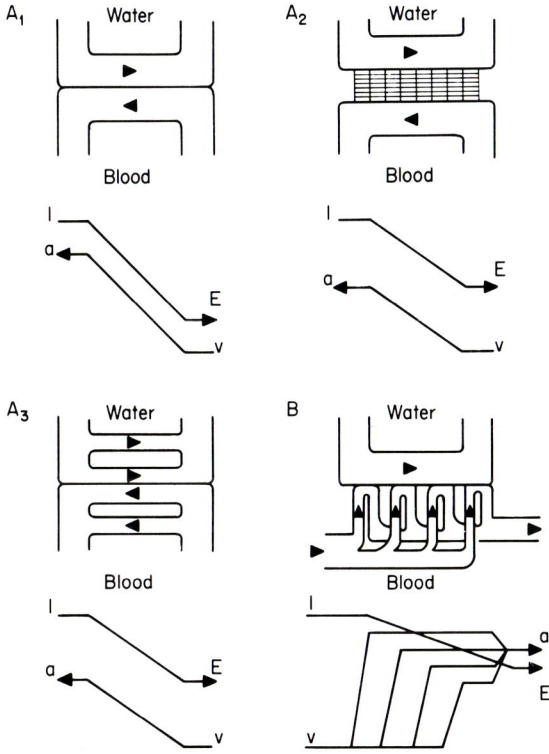

Fig. 5. Diagrams to illustrate possible arrangements of water and blood flows in the gills. A_1 simple countercurrent arrangement of the flows; A_2 as in A_1 but with a large diffusion resistance between blood and water; A_3 countercurrent system with blood and water shunts which bypass the respiratory surface; B multicapillary arrangement. A combination of A_{1-3} probably occurs in teleosts, whereas a multicapillary system is most likely to occur in elasmobranchs. The slopes of the lines beneath each diagram represent the changes in oxygen tension in blood and water. I, inspired water P_{O_2}; E, expired water P_{O_2}; a, arterial P_{O_2}; v, venous P_{O_2}. Adapted from Piiper and Schumann (1967) and Baumgarten-Schumann and Piiper (1968).

of the dogfish that were higher than in expired water. Thus, both teleosts and elasmobranchs can have arterial oxygen tensions higher than that in expired water; however, in teleosts, these are the result of a countercurrent arrangement of flows, whereas in elasmobranchs a serial multicapillary system may predominate. These high arterial oxygen tensions are also explicable in terms of the active transport or exchange diffusion of oxygen across the gill epithelium (see the discussion on pages 104–105 and 388–391 in the "Development of the Lung," de Reuck and Porter, eds., 1967). Although active transport or exchange diffusion of oxygen cannot be excluded, there is no evidence for the involvement of these processes in oxygen transport across the gills.

4. Gas Content in Blood and Water

The concentration of CO_2 and O_2 in both seawater and freshwater is extremely variable (Table I). The amount of oxygen dissolved in water per mm Hg varies with the temperature and ionic content of the water. All the oxygen in water is in physical solution; CO_2 in blood and water and O_2 in blood is not simply in physical solution, and the gas content per mm Hg varies with the partial pressure of the gas. The relationship between gas content and the partial pressure of gas in solution is described by a series of CO_2 and O_2 dissociation curves of water and blood (Figs. 6 and 7). Carbon dioxide is between 20 and 30 times more soluble in water than oxygen. Fully saturated blood contains between 10 and 20 times more oxygen, than water at the same partial pressure. Arterial blood contains twice as much CO_2 as O_2 even though the partial pressure of O_2 may be 40 times that of CO_2.

The rate of gas exchange across a respiratory epithelium depends on the dimensions of the epithelium, the concentration gradient, and the diffusion coefficient of the gas. The diffusion coefficient of carbon dioxide is only slightly less than that for oxygen, and the concentration gradient for free carbon dioxide is similar to that for physically dissolved oxygen in the reverse direction. Consequently, the exchange of oxygen and carbon dioxide across the gills occurs at more or less the same rate. Although the concentration gradients for carbon dioxide and oxygen across the gills are similar, the tension gradient for carbon dioxide across the gills is much less than that for oxygen in the reverse direction, because the solubilities of oxygen and carbon dioxide in blood and water are different.

The oxygen capacity of fish blood is variable between species. The antarctic icefish have no erythrocytes or hemoglobin in their blood (Ruud, 1954) and an instance of a carp without hemoglobin has

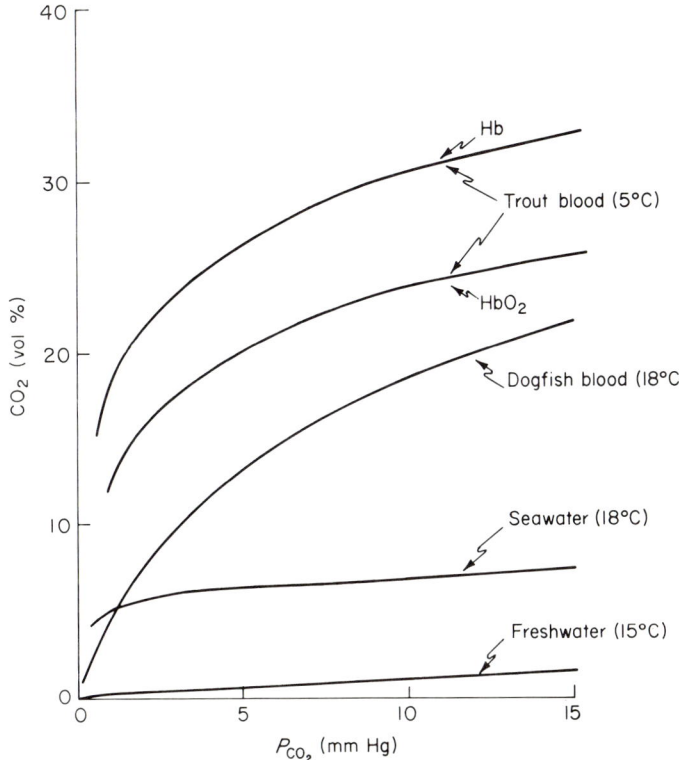

Fig. 6. A variety of CO_2 dissociation curves. Seawater and dogfish data from Piiper and Baumgarten-Schumann (1968b); freshwater = tap water, Vancouver, B.C., Canada; trout data from Stevens (1968b).

been reported by Schlicher (1927). Using carbon monoxide poisoning of the hemoglobin, it has been shown that trout, *Salmo gairdneri*, can exist without functional hemoglobin as long as the temperature is below 5°C (Holeton, 1968; see also Anthony, 1961; Nicloux, 1923). At these temperatures the oxygen demands of the tissues can be met by that in physical solution in the blood; at higher temperatures hemoglobin is required to increase the oxygen carrying capacity of the blood to meet the increased oxygen requirements of the fish. The oxygen capacity of most fish blood is between 4 and 10 vol %. Tuna have blood oxygen capacities that are much higher than most fish and hemoglobin levels of up to 20 g/100 ml of blood (Klawe *et al.*, 1963). Hemoglobin content of the blood and hematocrit increase in response to hypoxia in fish as in mammals (Phillips, 1947; Chiba, 1965). Hemoglobin, hematocrit, and the number of red blood cells increases, but the volume of each

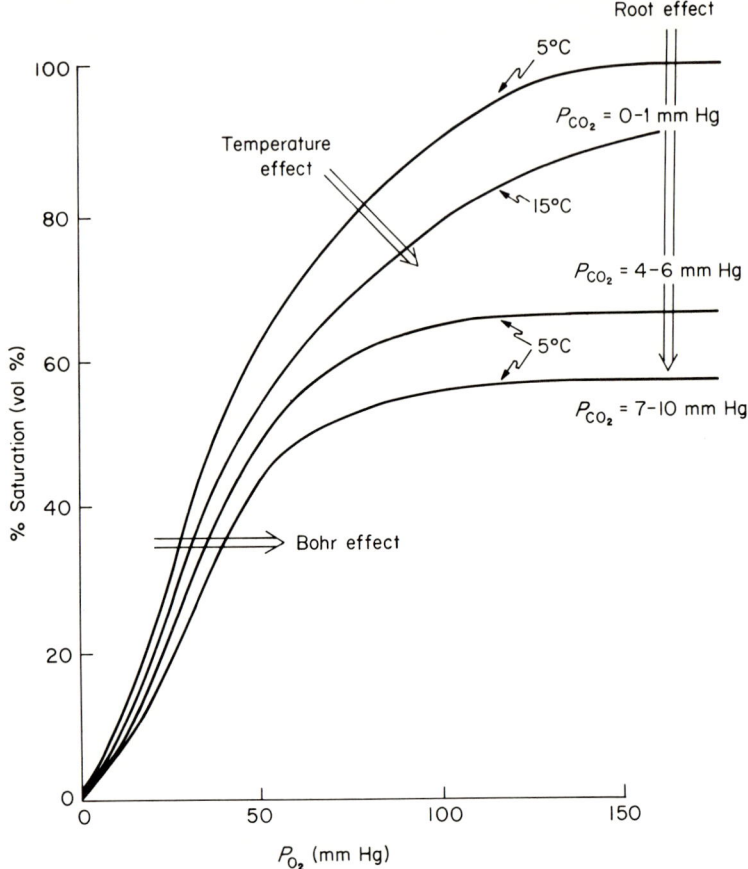

Fig. 7. Oxygen dissociation curves for the blood of trout (*Salmo gairdneri*) at different CO_2 levels and different temperatures (Beaumont, 1968).

erythrocyte decreases with an increase in temperatures in salmonids (Dewilde and Houston, 1967; Miles and Smith, 1968). Thus hemoglobin levels and the red blood cell surface area available for gas exchange increase in the summer, correlated with a rise in temperature and activity and hence the oxygen requirements of the fish.

If the solubility of a gas in solution is high, or if many of the gas molecules in solution are in a bound form, large volumes of the gas can enter or leave the solution without causing large changes in partial pressure. Therefore, the form of the oxygen or carbon dioxide dissociation curve has an effect on the rate of change of the gradient across the gills when gases are diffusing from one solution to another. The steeper the slope of the dissociation curve, the smaller the change in partial pressure per unit volume of gas exchanged.

Fish blood has a high affinity for oxygen and is 95% saturated at relatively low partial pressures of oxygen compared with pressures required to saturate mammalian blood. Decreases in temperature increase the oxygen affinity of the blood (Grigg, 1967), except in tuna, where the oxygen affinity of hemoglobin is largely independent of temperature (Rossi-Fanelli and Antonini, 1960; see Riggs, this volume, for further details). The P_{50} (partial pressure for 50% O_2 saturation) is 17 mm Hg (10°C) in the dogfish (Lenfant and Johansen, 1966), 38 mm Hg (5°C) in the trout (Beaumont, 1968), and 4 mm Hg (10°C) in the carp (Garey, 1967). At low P_{CO_2} levels (2 mm Hg) carp blood is 95% saturated at P_{O_2} tensions as low as 25 mm Hg. This means that the P_{O_2} of carp blood will not rise above 25–30 mm Hg until the hemoglobin is fully loaded, provided the rate of reaction of oxygen with hemoglobin is equal to or more rapid than the rate of entry of oxygen into the blood. The amount of oxygen that enters the blood will be increased if the respiratory area is increased, the diffusion distance decreased, or the oxygen activity gradient across the gills increased. The oxygen gradient can be increased either by lowering the blood P_{O_2} or by increasing the water P_{O_2}. A hemoglobin with a high affinity for oxygen will maintain low P_{O_2} levels in the blood until the hemoglobin is fully saturated. High ventilation volumes with a low percent utilization of oxygen will maintain a P_{O_2} in water which is close to ambient along the whole length of the secondary lamellae. A countercurrent arrangement of the flows of blood and water in teleosts also enables the fish to maintain a large O_2 gradient between blood and water along the whole length of the secondary lamella. The mean oxygen pressure gradient across the gills is similar in carp and trout. The carp, however, has a high utilization of oxygen from the water and a hemoglobin with a very high affinity for oxygen, whereas the trout has a low utilization of oxygen from water passing over the gills and a hemoglobin with a lower affinity for oxygen. The maintenance of large oxygen gradients enables the fish to utilize a smaller respiratory area for a given oxygen uptake and so reduce ion and water exchange across the gills.

Blood oxygen dissociation curves have been described for a number of fish (Krogh and Leitch, 1919; Root, 1931; Dill *et al.*, 1932; Willmer, 1934; Black and Irving, 1938; Hall and McCutcheon, 1938; Root *et al.*, 1939; Black, 1940; Irving *et al.*, 1941; Root and Irving, 1941; Ferguson and Black, 1941; Black and Black, 1946; Fish, 1956; Fry, 1957; Burke, 1965; Black *et al.*, 1966; Swan and Hall, 1966; Lenfant and Johansen,

1966; Lenfant et al., 1966–1967; Garey, 1967; Griggs, 1967; Hunn, 1967; Johansen et al., 1966; Johansen and Lenfant, 1967; Piiper and Baumgarten-Schumann, 1968b).

Recently, Forster and Steen (1969) have measured the half-time for some oxygen hemoglobin reactions in the blood of eels, but apart from these data very little is known about oxygen hemoglobin reaction times and these may play a role in limiting the rate of gas exchange across the gills. In teleost fish an increase in CO_2 content or a decrease in pH of the blood causes not only a reduction in the affinity of hemoglobin for oxygen (Bohr effect) but also a reduction in the oxygen carrying capacity of the blood (Root effect). Unlike the blood of terrestrial mammals, teleost hemoglobin never becomes fully saturated once the P_{CO_2} is more than 1–2 mm Hg (Fig. 7). The magnitude of this "Root effect" is largest at CO_2 tensions between 1 and 5 mm Hg and decreases with an increase in P_{CO_2} above this range. If the CO_2 tension in blood fully saturated with oxygen is increased, oxygen is released from the hemoglobin (Root-off shift), and if the CO_2 tension is lowered the reverse occurs (Root-on shift). The half-time for the eel Root-off shift at $23°–25°C$, produced by an increase in P_{CO_2}, is 87 msec, whereas the half-time for the reverse Root-on shift is 9 sec (Forster and Steen, 1969). The asymmetrical reaction velocities of the Root shift are important in unloading gases in the swim bladder (Berg and Steen, 1968), but a slow Root-on shift will impair oxygen transfer across the gills. The time course of the Root-on shift is related to the reaction velocity of the uncatalyzed CO_2 hydration–dehydration reaction in the plasma. The presence of a CO_2 sink, for example, water passing over the gills, will cause a more rapid reduction of plasma and intracellular P_{CO_2} than recorded under the conditions of Forster and Steen's experiments, and hence the rate of the Root-on shift will be more rapid as blood passes through the gills. Blood leaving the gills is usually 95% saturated with oxygen in trout (Stevens and Randall, 1967b). Incomplete saturation of arterial blood has been recorded in the eel (Steen and Kruysse, 1964) but this was probably related to the utilization of venous shunts in the gills rather than the result of a slow Root-on shift.

Acclimation to high CO_2 levels in the water has been reported to depress the magnitude of Root effect (Eddy and Morgan, 1969). This acclimation process is associated with an increase in hemoglobin levels with little change in blood oxygen capacity. These observations require further investigation.

Teleosts generally have a marked Bohr and Root effect, both of which are absent or small in elasmobranchs (Lenfant and Johansen, 1966). There is no Root effect in Dipnoi, but there is a pronounced Bohr effect (Lenfant et al., 1966–1967). The reader is referred to the chapter by Riggs, this volume, for further information on this subject.

The much higher CO_2 than O_2 solubility in water means that if the respiratory quotient is around unity, the changes in CO_2 tension in the water as it passes over the gills are small and of the order of a few mm Hg whereas the changes in P_{O_2} are large. Under these circumstances the ratio of the P_{O_2} to P_{CO_2} changes will be equal to the ratio of the solubilities of O_2 and CO_2 in water, that is, between 20 and 30:1. This effect of solubility on the changes in partial pressure is described by the $O_2:CO_2$ diagram of Rahn (1966a).

The relationship between the amount of CO_2 in solution and the partial pressure of the gas is not linear in either blood or water. The CO_2 dissociation curves are curvilinear in water below 2 mm Hg P_{CO_2}, particularly in seawater or highly carbonated freshwater (Fig. 6). Carbon dioxide entering water is buffered by a carbonate-bicarbonate system reducing the magnitude of the P_{CO_2} increase (Dejours et al., 1968). As water passes over the gills of teleosts the excretion of ammonium ions increase the CO_2 buffering capacity of water, further reducing the magnitude of the rise in P_{CO_2} in efferent water. This latter effect is probably not seen in elasmobranchs because they are ureotelic and not ammonotelic (Piiper and Baumgarten-Schumann, 1968b).

Carbonic anhydrase has been located in fish red blood cells and the gill epithelium (Maren, 1967) and bicarbonate is exchanged for chloride by an exchange diffusion mechanism across the gill epithelium of freshwater teleosts (Maetz and Garcia Romeu, 1964). Thus because of the presence of carbonic anhydrase and an exchange diffusion mechanism, bicarbonate as well as CO_2 enters the water passing over the gills (Fig. 8). If this is so then the slow formation of CO_2 from bicarbonate in water will delay the rise in P_{CO_2} until the water has left the respiratory surface. A body of water is in contact with the gill epithelium for about a second, the time required for the formation of CO_2 from bicarbonate in water is of the order of several seconds; hence, the major portion of the rise in P_{CO_2} will occur after the water has left the respiratory surface. The relative importance of the buffering system, the excretion of ammonium ions, and the slow formation of CO_2 from bicarbonate, in water, in maintaining the CO_2 gradient across the gills will depend on the water flow rate and the ambient temperature, but all tend to reduce or retard the rise in water P_{CO_2} and maintain the P_{CO_2} gradient between blood and water.

The rate of formation of CO_2 in the red blood cell, and hence the rate of excretion of CO_2, will depend upon the presence of carbonic anhydrase (Fig. 8). The activity of carbonic anhydrase in fish blood

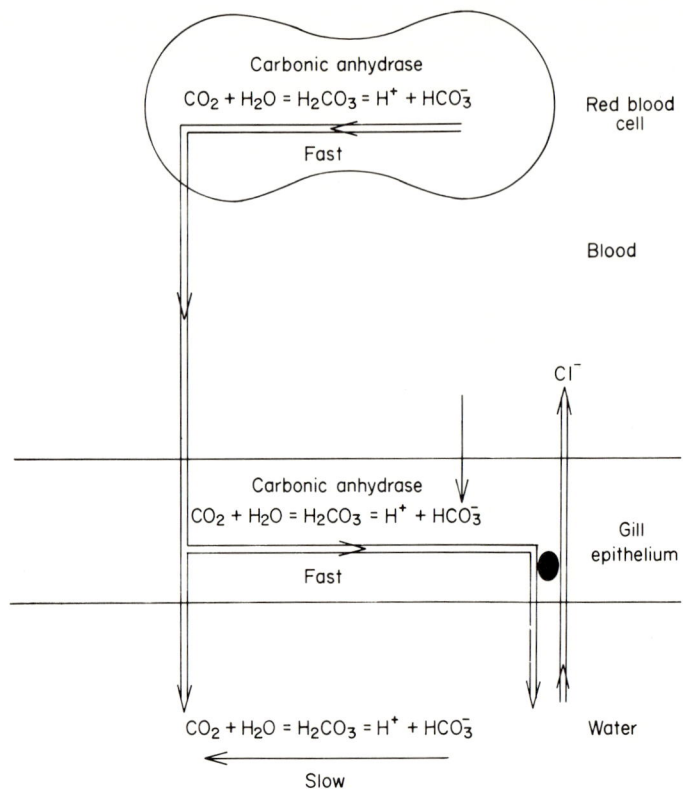

Fig. 8. Possible pathway for the elimination of CO_2 across the gills of freshwater teleost fish (see text).

is only about 2% of that in mammals, and as tentatively suggested by Maren (1967) the CO_2 gradient that does exist across fish gills may result from a limitation in the rate of formation of CO_2 in the blood. Inhibition of carbonic anhydrase with acetozolamide causes an increase in P_{CO_2} in the blood (Hoffert and Fromm, 1966). Seasonal changes in the level of carbonic anhydrase have been observed in the plaice, *Hippoglossoides platessoides* (Pocklington, 1968), and these changes may be important in regulating CO_2 excretion across the gills on a long-term basis.

B. Gas Exchange across the Skin

The amount of gas transfer between blood and water occurring across surfaces other than the gills varies considerably from fish to fish.

Krogh (1904) showed that eels have a cutaneous oxygen uptake in air that is up to 60% of the total oxygen uptake in water at the same temperature. His eels were prevented from ventilating their gills. Eels normally pump air over their gills when out of water. Berg and Steen (1965) have shown that the oxygen uptake in air is reduced to between 40 and 63% of that in water and from 31 to 47% of the oxygen enters the blood via the gills when the fish is in air. At higher temperatures there is a larger reduction in oxygen uptake when the animal is in air compared with the oxygen uptake in water and a smaller percentage of the total oxygen uptake enters the body via the gills. Berg and Steen observed that the longitudinal fin becomes more red when the eel is in air indicating a vasodilation of skin capillaries. In water, 85–90% of the oxygen uptake occurs across the gills. The eel has a well-developed cutaneous circulation and often comes out on land, to steal peas, according to Day (1880) who also reports a case of an eel "which lived upwards of 31 years in a well and was then choked by a frog that was larger than it could swallow" (Day, 1880, p. 243).

Jakubowski (1963) has carried out an extensive study of the structure and vascularization of the skin of a variety of teleosts. He reports values of between 0.5 and 1.5 cm^2/g body weight for the surface area of blood vessels in the skin of a variety of teleosts, as well as ratios for the surface area of gill to skin capillaries, ranging from 3:1 in the pond loach to 10:1 in the carp and flounder. Although the area of capillaries in the skin is large, it is important to note that the capillaries are covered by an epidermis, ranging in thickness from 31–38 μ in the flounder to 263 and 338 μ in the eel and pond loach, respectively. The differences in thickness largely result from differences in size of mucous and goblet cells within the epidermis. The number of mucous cells and the thickness of the skin are probably controlled by prolactin (Ball, 1969). By comparison the gill epithelium is usually between 1 and 5 μ in thickness.

Jakubowski could not find any relationship between the extent of the skin vascularization, thickness of the epidermis, and the amount of oxygen uptake occurring through the skin. Gas exchange across the skin will not only depend on the extent of skin vascularization but also on the gas gradients, diffusion distances, and the blood flow and number of capillaries open in the skin. Nothing is known about many of the parameters which affect gas exchange across the skin.

There appears to be considerable variability in the extent to which fish respire through their skin. In many fish with thick, poorly vascularized skins, cutaneous respiration is probably not important. Obvious exceptions to this are larval fishes in which the gills have not yet developed. Here cutaneous respiration must be very significant (Fry, 1957).

III. GAS EXCHANGE BETWEEN BLOOD AND TISSUES

There are very few direct measurements of tissue gas tensions in fish. Haning and Thompson (1965) have recorded tissue P_{CO_2} levels of 9.2 mm Hg in the catfish. Gas tensions in venous blood are an indication of tissue gas partial pressures; however, there is a gas gradient between the tissues and blood, the magnitude of which is unknown. Oxygen and carbon dioxide levels in mixed venous blood appear to be variable both between species and within a single species (Garey, 1967; Holeton and Randall, 1967b; Piiper and Baumgarten-Schumann, 1968a). Carbon dioxide tensions in trout venous blood are about 5 mm Hg, doubling during activity. Values are generally lower for carp and dogfish. The P_{O_2} in mixed venous blood is 3.2 mm Hg in the carp (Garey, 1967), 10 mm Hg in the dogfish (Piiper and Baumgarten-Schumann, 1968a), and 19 mm Hg in the trout (Stevens and Randall, 1967b). In the absence of direct measurements one might assume that tissue P_{CO_2} is between 3 and 15 mm Hg and P_{O_2} is between 1 and 15 mm Hg.

Almost nothing is known of the relative size of, and blood flow to, the various capillary beds in the systemic circulation of fish. Red myotomal muscle contains about 3 times as much blood and has 3 times the number of capillaries per unit weight as white myotomal muscle (Stevens, 1968a). White myotomal muscle, used during violent burst responses, operates anaerobically. P_{O_2} levels are presumably much lower, and P_{CO_2} levels much higher, than those in red myotomal muscle, which operates aerobically. Stevens (1968a) has estimated the volume distribution of blood to parts of the body of rainbow trout. He demonstrated that, except for a diminished spleen blood volume, the distribution of blood was unaffected by violent swimming. The retention of lactic acid in white myotomal muscle for a considerable period after violent exercise or hypoxia (Leivestad et al., 1957) is enhanced by the small number of capillaries and may be aided by an ischemia during and after exercise or hypoxia.

Gas exchange across the rete mirabile and gas gland of the swim bladder has been studied in detail. These studies are reviewed in the chapter by Steen, this volume. Studies of other systemic capillary beds are sparse and poorly documented.

Some teleosts have a rete mirabile in the choroid layer of the eye.

High oxygen levels in the vitreous humor are associated with the presence of a rete, and in some cases the oxygen tensions are in excess of one atmosphere (Wittenberg and Wittenberg, 1962). Carbonic anhydrase and bicarbonate levels in fish eyes are in excess of those in the plasma (Hoffert, 1966; Hoffert and Fromm, 1966; Maren, 1967). The concentration of bicarbonate is possibly associated with the formation of the aqueous humor and the maintainence of intraocular pressure in fish as well as mammals (Maren, 1967; Hoffert, 1966). Bicarbonate accumulation in the aqueous humor of the rabbit is owing to formation from CO_2 and not the transport of the ion (Maren, 1967). In fish the rete could maintain CO_2 levels in the eye that are higher than those in the plasma, which are only of the order of 2–5 mm Hg, and therefore assist in bicarbonate concentration and the maintenance of intraocular pressure. Therefore, high oxygen levels in the eye may not be functionally significant but only a byproduct of the presence of a rete.

Some fish are viviparous, and in a few instances a complex system exists for exchanging material between adult and young. The anatomy of circulatory structures associated with development of viviparity in fish have been described (see chapter by Hoar, Volume III), but nothing is known of the exchange processes in these maternal fetal connections.

IV. METHODS OF ANALYSIS OF THE TRANSFER OF GASES

A. Introduction

The $O_2:CO_2$ diagram has been used extensively to analyze gas exchange in mammals (Rahn and Fenn, 1956; West, 1965). In aquatic respiration the relationship between P_{CO_2} and P_{O_2} in expired water is not linear, as in air, for a given respiratory quotient and a series of ventilation volumes. This nonlinearity results from the CO_2 buffering capacity of the water (Dejours et al., 1968) and the slow rate of formation of CO_2 from bicarbonate, both of which tend to reduce the P_{CO_2} in water in contact with the respiratory surface. In fish, oxygen tensions on either side of the gills do not approach an equilibrium point, and some of the factors that determine the rate of CO_2 removal appear to be different from those that govern O_2 uptake. This, plus the general paucity of relevant data, makes an extensive analysis of gas exchange in fish using the $O_2:CO_2$ diagram difficult. No system for the graphical analysis of gas transfer in fish, similar to the $O_2:CO_2$ diagrams, has been developed. Although there is no adequate method for the extensive analysis of the interaction of several parameters, a few investigators have attempted to analyze certain aspects of gas exchange in fish.

B. Transfer Factor

The transfer factor of the gills for a particular gas is a measure of the ability of the gills to transfer that gas per unit gradient. The transfer factor of the gills for oxygen T_{O_2} is defined as

$$T_{O_2} = \frac{\dot{V}_{O_2} \text{ (oxygen uptake)}}{\Delta F_{O_2} \text{ (mean oxygen gradient across gill epithelium)}}$$

Randall et al. (1967) calculated the mean gradient using the following equation

$$P_{O_2} = \tfrac{1}{2}(P_{I_{O_2}} + P_{E_{O_2}}) - \tfrac{1}{2}(Pa_{O_2} + Pv_{O_2})$$

where $P_{I_{O_2}}$ is the inspired oxygen level, $P_{E_{O_2}}$ the expired oxygen level, Pa_{O_2} the arterial oxygen tension, and Pv_{O_2} the venous oxygen level. Piiper and Baumgarten-Schumann (1968a), criticizing this method on the grounds that the mean gradient was affected by the slope of the blood oxygen dissociation curve, used a modified Bohr integration technique to determine T_{O_2}. Uptake was divided into a series of steps, and then the T_{O_2} for each step calculated. The mean partial pressures for each step were obtained using an oxygen dissociation curve. The total T_{O_2} value was obtained by summation of the individual T_{O_2} values for each of the steps of $\dot{V}_{O_2}$. They found that the differences were small in the calculated values of T_{O_2} and T_{CO_2}, obtained for dogfish using the method of Randall et al. (1967) and the Bohr integration technique.

The transfer factor of the gills for oxygen (T_{O_2}) in the resting trout is 0.0056 ml/min/mm Hg/kg, increasing by a factor of five during moderate swimming (Randall et al., 1967). The ratio of T_{CO_2} (the transfer factor of the gills for carbon dioxide) to T_{O_2} is about 20:1 decreasing to 14:1 during activity (Stevens, 1968b). That is, the transfer factor of the gills for oxygen increases more than the transfer factor for CO_2 during swimming in the trout. The reported T_{O_2} for the dogfish is 0.0080 ml/min/mm Hg/kg (Piiper and Baumgarten-Schumann, 1968a), a value very similar to that for the resting trout. The ratio of T_{CO_2} to T_{O_2} is 56:1 in the dogfish (Piiper and Baumgarten-Schumann, 1968a).

The transfer factor may be altered by changing the functional area of the gills or diffusion distance between blood and water by changing the pattern of blood or water flow over the respiratory surface. The thickness of the boundary layer of water at the gill surface may also be altered and affect the transfer factor, but this effect is probably not important (see Section II, A, 2, c). The velocity of the various hemoglobin reactions and the rate of bicarbonate formation will also affect the respective transfer factor of the gills for oxygen and carbon dioxide. The relative importance of these parameters in determining the size of the transfer factor has yet to be investigated.

Hughes and Shelton (1962) and Hughes (1964) introduced a term analogous to the transfer factor. This term "the number of transfer units" is determined by the ratio between the capacity of the gills for transfer of gases and the load imposed on the system by the blood flow (Hughes, 1964). In practical terms it is difficult to assign a numerical value to this term, and the transfer factor appears to be a more suitable estimate of the transfer capacity of the gills.

C. The Effectiveness of Gas Transfer

This term introduced by Hughes and Shelton (1962) is a measure of the relative ability of the system to transfer a particular gas. The effectiveness of gas transfer is the ratio of the actual rate of gas exchange to the maximum rate of gas exchange possible. The maximum rate of oxygen removal from water occurs (assuming that no active processes are involved) when water leaving the gills is in equilibrium with venous blood entering the gills.

The effectiveness of gas exchange can be applied to the removal of oxygen from water, the uptake of oxygen by the blood, the removal of CO_2 from the blood and the uptake of CO_2 by water, this has been done for trout (Randall et al., 1967) and dogfish (Piiper and Baumgarten-Schumann, 1968a). In both trout and dogfish the effectiveness of oxygenating the blood is high, ranging from 79% in the unanesthetized dogfish to 95–100% in the trout. The effectiveness of CO_2 removal from the blood is not so high (35–60%) in trout or dogfish, but in both cases it is considerably more effective than CO_2 removal from human blood. The effectiveness of CO_2 uptake by the water is between 38 and 43% in dogfish and 4–6% in trout. The low effectiveness of oxygen removal from water (11–30%) passing over the trout gills is related to the very high ventilation volumes and the low utilization of oxygen recorded in this animal (Randall et al., 1967). The difference between the effectiveness of oxygen removal from the water and the percent utilization of oxygen is that effectiveness takes into consideration the venous P_{O_2}. Effectiveness, therefore, is a more accurate estimate of the efficiency of oxygen removal from the water. Comparison of the effectiveness of gas transfer between fish and man (Randall et al., 1967; Piiper and Baumgarten-Schumann, 1968a) indicate that the CO_2 uptake by and O_2 removal from the medium and O_2 uptake by the blood is similar in man and fish.

D. Capacity Rate Ratio

Hughes and Shelton (1962) compared the capacity rates of water ($\dot{V}_G \cdot \alpha w_{O_2}$) and blood ($\dot{Q}_G \cdot \alpha b_{O_2}$) and discussed the relevance of a changing capacity rate ratio of water and blood on gas exchange across the gills. ($\dot{Q}_G$ = gill blood flow; $\dot{V}_G$ = gill water flow; αb_{O_2} = oxygen content of blood per mm Hg; αw_{O_2} = oxygen solubility coefficient in water.) The oxygen content of blood is not linearly related to P_{O_2} but changes in proportion to the slope of the oxygen dissociation curve. The shape of the dissociation curve is affected by P_{CO_2}, pH, and temperature of the blood. The solubility of oxygen in water is affected by temperature and ionic content of the water. The value of using capacity rate ratio in an analysis of gas exchange in fish is limited by the large number of variables that must be measured before capacity rate ratio can be determined. The capacity rate ratio of water to blood in the resting and active trout is about 5 (Randall et al., 1967).

E. Ventilation–Perfusion Ratio

Although the effects of changes in the ventilation–perfusion ratio on gas exchange have been extensively studied in mammals, very little attention has been paid to the effects of variations in this ratio in fish. The $\dot{V}_G/\dot{Q}$ ratio is extremely variable in fish and ranges from 10:1 in the carp and dogfish (Piiper and Schumann, 1967; Garey, 1967) to 80:1 in the trout (Stevens and Randall, 1967b). The ratio increases in the trout during hypoxia (Holeton and Randall, 1967b). There has been no extensive analysis to date of the effect of changing $\dot{V}_G/\dot{Q}$ ratios on gas exchange in fish.

Although there is little concrete data on changes in the ratio of blood to water flow at the gills, there are a number of observations on the relationship between heartbeat and breathing movements in fish (Randall, 1968). The heart beats during a particular phase of the breathing cycle in many fish (Hanson, 1967; Satchell, 1960), and there may be a 1:1, 1:2, or 1:3 ratio between heart and breathing rates. In the trout, synchrony between a phase of the breathing cycle and the heart beat occurs during hypoxia (Randall, 1966; Randall and Smith, 1967). The eel often stops breathing for several seconds or breathes through only one side of the gills; during these periods of reduced or no breathing the heart rate slows (Randall, 1962). A type of Cheyne-Stokes breathing often occurs in quiet, inactive teleosts. As the breathing rate oscillates heart rate also changes; the rate slows as breathing is reduced or absent (Labat et al., 1962; Peyraud and Serfaty, 1964). Thus there appears to be a correlation between heart and breathing rate in many fish under a variety of conditions, indicating some overall regulation of the ventilation–perfusion ratio.

V. THE EFFECTS OF VARIOUS PARAMETERS ON GAS EXCHANGE

A. Temperature

The major effect of temperature on gas exchange is to change the oxygen requirements of the fish. A change from 10° to 20°C increases the standard oxygen uptake in the goldfish by 254% (Fry and Hart, 1948); the amount of oxygen in the water decreases by only 18% and is offset by a decrease in viscosity of water and a more rapid rate of diffusion of gases. Krogh's permeation coefficients increase by 1%/°C (Table I). The fish must therefore adapt its respiratory system to meet the increased oxygen demand as temperature rises.

There are a large number of studies demonstrating the effect of temperature on the oxygen uptake of both resting and active fish (Brett, 1964; Fry, 1957). The increase in oxygen uptake with temperature is associated with an increase in cardiac output (Randall, 1968) and ventilation volume (Davis, 1968). The peripheral resistance to blood flow decreases with increasing temperature (Davis, 1968), reducing the amount of work required of the heart to maintain a given cardiac output as the temperature is raised.

The rate of oxygen transfer across the gills is limited either by the cost of moving blood and water past the respiratory surface or by the magnitude of the transfer factor of the gills for oxygen. Brett (1964) showed that maximum levels of sustained activity of salmon at temperatures above 15°C could be raised by supersaturating the water with oxygen. These observations indicate that, under the conditions of his experiments at least, either the cost of delivering oxygen to the gills or the transfer factor of the gills for oxygen is limiting oxygen uptake.

B. Hypoxia

The effects of hypoxia on O_2 uptake and CO_2 removal have been studied in the trout (Holeton and Randall, 1967a,b) and the carp (Garey, 1967). A reduction of the oxygen level in the water causes an increase in breathing rate and amplitude, a decrease in heart rate, but an increase in stroke volume of the heart in the trout. Cardiac output varies little but the ventilation–perfusion ratio increases because of a large increase in water flow over the gills. The transfer factor of the gills for oxygen is increased, resulting in a decrease in the mean pressure gradient for oxygen between blood and water across the gills. The partial pressure of oxygen in venous blood falls as the animal utilizes the venous oxygen reserve. The end result of the increase in transfer factor and ventilation–perfusion ratio and the decrease in venous oxygen content is that oxygen uptake is maintained in the face of a decreased amount of oxygen in the water (Holeton and Randall, 1967b). Oxygen uptake falls at low oxygen levels in the water, and there is an increased production of lactate by the fish. The exact level at which oxygen uptake falls varies between fish. Teleosts appear to be able to regulate oxygen uptake over a wide range of oxygen levels in the water; elasmobranchs appear to be less able in this respect. Both groups become more active in hypoxic conditions and attempt to leave the oxygen-depleted environment.

An increase in ventilation volume associated with a decreased heart rate appears to be a general response of fish to hypoxia. The increase in ventilation volume maintains the delivery of oxygen to the respiratory surface as the oxygen content falls. The decreased heart rate is offset by an increase in stroke volume in the trout such that there are only small changes in cardiac output. During hypoxia, therefore, the pattern of blood flow, rather than the cardiac output, changes. The changing pattern of blood flow through the gills may, as suggested by Satchell (1960), augment gas exchange.

C. Exercise

There have been a large number of studies on the changes in oxygen uptake with exercise in fish (Brett, 1964; Fry, 1957). Stevens and Randall (1967a,b) and Randall et al. (1967) have investigated gas exchange across the gills of trout during swimming. Moderate exercise in the trout is associated with an increase in cardiac output, ventilation volume, and the gill transfer factor for oxygen. The increase in cardiac output is largely the result of an increase in stroke volume.

Although both water and blood flow increases, the ventilation–perfusion ratio remains constant. The transfer factor of the gills for oxygen increases in phase with the changes in oxygen uptake in such a way that the gas gradient across the gills remains constant. Hence, the oxygen partial pressures in afferent and efferent blood and water vary little throughout the period of exercise (Fig. 9).

The maximum oxygen uptake is 4–8 times the resting consumption. Maximum oxygen uptake increases in salmon as the temperature rises to 15°C and then varies little for further increases of temperature (Brett, 1964). Supersaturation of the water increases the maximum oxygen uptake, indicating that at temperature above 15°C either the cost of ventilating the gills or the transfer factor of the gills for oxygen is limiting oxygen uptake.

An increase in the oxygen transfer factor in the trout is associated with an increased rate of sodium exchange and also possibly a net influx of water into the body across the gills (Stevens, 1968a; Baumgarten-

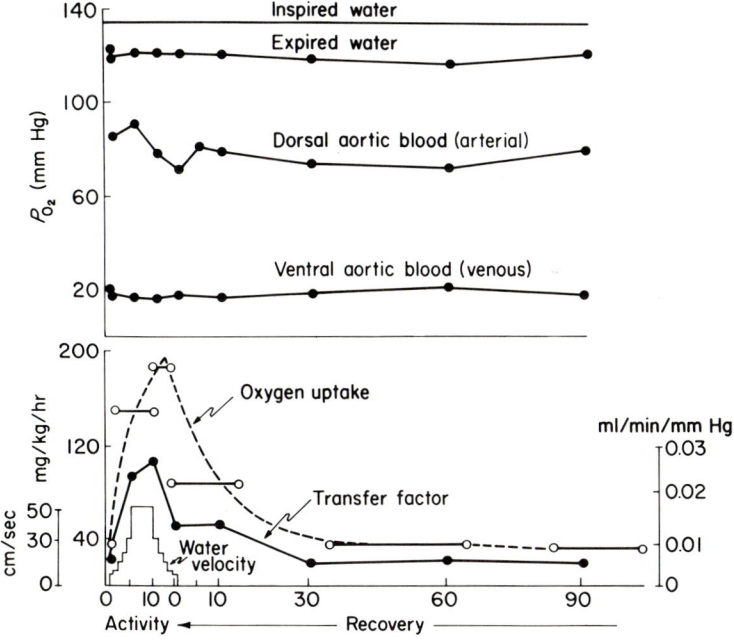

Fig. 9. Change in oxygen uptake, transfer factor for oxygen, and afferent and efferent blood and water oxygen tensions in the trout. At the lowest water velocities the fish is stationary; when water velocity is increased the fish swims to maintain position in the stream of water. Adapted from Stevens and Randall (1967a,b) and Randall et al. (1967).

Schumann et al., 1969). The changes in transfer factor are probably in part related to changes in the pattern of blood flow in the gills caused by increased levels of circulating catecholamines (see Section II, A, 3). The increased flux of sodium across the gills of trout during swimming in freshwater is probably alleviated by a rise in blood cortisol levels (Donaldson and McBridge, 1967) which causes a reduction in sodium excretion in the kidney (Mayer et al., 1967).

ACKNOWLEDGMENTS

I would like to thank Dr. David Jones for reading and criticizing this manuscript, which was prepared while the author was a John Simon Memorial Guggenheim Fellow in the Zoology Department, University of Bristol. I would also like to thank Professor G. M. Hughes for permission to use the facilities of his department during that period.

REFERENCES

Altman, P. L., and Dittmer, D. S., eds. (1964). "Biology Data Book." Fed. Am. Soc. Exptl. Biol., Washington, D.C.
Anthony, E. H. (1961). Survival of goldfish in presence of carbon monoxide. *J. Exptl. Biol.* **38**, 105–125.
Ball, J. N. (1969). The pituitary: Prolactin and growth hormone. In "Fish Physiology" (W. S. Hoar and D. J. Randall, eds.), Vol. II, pp. 207–240. Academic Press, New York.
Bastos, J. R. (1966). Sobre a Serie vermelha do sangue de *Scomberomorus maculatus* (Mitchell). *Arg. Est. Biol. Mar. Univ. Fed. Ceara* **6**, 39–45.
Baumgarten, D., and Randall, D. J. (1967). Unpublished observations.
Baumgarten-Schumann, D., and Piiper, J. (1968). Gas exchange in the gills of resting unanesthetized dogfish. (*Scyliorhinus stellaris*). *Resp. Physiol.* **5**, 317–325.
Baumgarten-Schumann, D., Randall, D. J., and Malyusz, M. (1969). Gas exchange versus ion exchange across the gills of fishes. In preparation.
Beaumont, C. (1968). Unpublished data.
Beaumont, C., and Randall, D. J. (1968). Unpublished data.
Berg, J., and Steen, J. B. (1965). Physiological mechanism for aerial respiration in the eel. *Comp. Biochem. Physiol.* **15**, 469–484.
Berg, T., and Steen, J. B. (1968). The mechanism of oxygen concentration in the swimbladder of the eel. *J. Physiol. (London)* **195**, 631–638.
Bitjel, J. H. (1949). The structure and the mechanism of movement of gill filaments in teleostii. *Arch. Neerl. Zool.* **8**, 1–22 and iii.
Black, E. C. (1940). The transport of oxygen by the blood of freshwater fish. *Biol. Bull.* **80**, 215–229.
Black, E. C. (1951). Respiration in fish. *Publ. Ontario Fisheries Res. Lab.* **71**, 91–111.
Black, E. C., and Black, V. S. (1946). Oxygen and CO_2 dissociation curves of the blood of Atlantic salmon, etc. *Federation Proc.* **5**, 8.
Black, E. C., and Irving, L. (1938). The effect of hemolysis upon the affinity of fish blood for oxygen. *J. Cellular Comp. Physiol.* **12**, 255–262.
Black, E. C., Kirkpatrick, D., and Tucker, H. H. (1966). Oxygen dissociation curves

of the blood of atlantic salmon (*Salmo salar*) acclimated to summer and winter temperatures. *J. Fisheries Res. Board Can.* **23**, 1187–1195.

Brett, J. R. (1964). The respiratory metabolism and swimming performance of young sockeye salmon. *J. Fisheries Res. Board Can.* **21**, 1183–1226.

Burke, J. D. (1965). Oxygen affinities and electrophoretic patterns of hemoglobins in trout and basses from Virginia. *Medical College Virginia Quart.* Spring, 1965 pp. 16–21.

Carter, G. S. (1957). Air breathing. In "The Physiology of Fishes" (M. E. Brown, ed.), Vol. 1, pp. 65–79. Academic Press, New York.

Chapman, C. B., Jensen, D., and Wildenthal, K. (1963). On circulatory control mechanisms in the Pacific hagfish. *Circulation Res.* **12**, 427–440.

Chiba, K. (1965). A study of the influence of oxygen concentration on the growth of juvenile common carp. *Bull. Freshwater Fisheries Res. Lab.* **15**, 35–47.

Daniel, J. F. (1922). "The Elasmobranch Fishes." Univ. of California Press, Berkeley, California.

Datta Munshi, J. S., and Singh, B. N. (1968). On the micro-circulatory system of the gills of certain freshwater teleostean fishes. *J. Zool.* **154**, 365–376.

Davis, J. (1968). The influence of temperature and activity on certain cardiovascular and respiratory parameters in adult sockeye salmon. M.Sc. Thesis, University of British Columbia.

Day, F. (1880). "The Fishes of Great Britain and Ireland," Vol. II. Williams & Norgate, London.

Dejours, P., Armand, J., and Verriest, G. (1968). Carbon dioxide dissociation curves of water and gas exchange of water breathers. *Resp. Physiol.* **5**, 23–33.

Dewilde, M. A., and Houston, A. H. (1967). Heamatological aspects of the thermoacclimatory process in the rainbow trout (*Salmo gairdneri*). *J. Fisheries Res. Board Can.* **24**, 2267–2281.

Dill, D. B., Edwards, H. T., and Florkin, M. (1932). Properties of the blood of the skate. *Biol. Bull.* **62**, 23–36.

Dittmer, D. S., and Grebe, R. M., eds. (1958). "Handbook of Respiration." Saunders, Philadelphia, Pennsylvania.

Donaldson, E. D., and McBridge, J. R. (1967). The effects of hypophysectomy in the rainbow trout *Salmo gairdneri* (Rich) with special references to the pituitary-interrenal axis. *Gen. Comp. Endocrinol.* **9**, 93–101.

Dubale, M. S. (1951). A comparative study of the extent of gill surface in some representative Indian Fishes and its bearing on the origin of air-breathing habit. *J. Univ. Bombay.* [N.S.] **19B**, 6–13.

Eddy, F. B., and Morgan, R. I. G. (1969). Some effects of carbon dioxide on the blood of rainbow trout (*Salmo gairdneri*, Richardson). *J. Fish. Biol.* **1**, 361–372.

Fairey, R. (1966). Unpublished observations.

Ferguson, J. K. W., and Black, E. C. (1941). The transport of CO_2 in the blood of certain fresh-water fishes. *Biol. Bull.* **80**, 139–152.

Fish, G. R. (1956). Some aspects of the respiration of six species of fish from Uganda. *J. Exptl. Biol.* **33**, 186–195.

Forster, R. E., and Steen, J. B. (1969). The rate of the 'Root Shift' in eel red cells and eel haemoglobin solution. *J. Physiol.* (*London*) **204**, 259–282.

Fry, F. E. J. (1957). The aquatic respiration of fish. In "The Physiology of Fishes" (M. E. Brown, ed.), Vol. 1, pp. 1–63. Academic Press, New York.

Fry, F. E. J., and Hart, J. S. (1948). The relation of temperature to oxygen consumption in the goldfish. *Biol Bull.* **94**, 66–77.

Gameson, A. L. H., and Robertson, K. G. (1955). The solubility of oxygen in pure water and seawater. *J. Appl. Chem.* **5**, 502.

Garey, W. F. (1967). Gas exchange, cardiac output and blood pressure in free swimming carp (*Cyprinus carpio*). Ph.D. Dissertation, University of New York, Buffalo, New York.

Gray, I. E. (1954). Comparative study of the gill area of marine fishes. *Biol. Bull.* **107**, 219–225.

Grigg, G. C. (1967). Some respiratory properties of the blood of four species of antarctic fishes. *Comp. Biochem. Physiol.* **23**, 139–148.

Hall, F. G. (1930). The ability of the common mackeral and certain other marine fishes to remove dissolved oxygen from sea water. *Am. J. Physiol.* **93**, 417–421.

Hall, F. G., and McCutcheon, F. H. (1938). The affinity of hemoglobin for oxygen in marine fishes. *J. Cellular Comp. Physiol.* **11**, 205–212.

Haning, Q. C., and Thompson, A. M. (1965). A comparative study of tissue carbon dioxide in vertebrates. *Comp. Biochem. Physisol.* **15**, 17–26.

Hanson, D. (1967). Cardiovascular dynamics and aspects of gas exchange in chondrichthyes. Ph.D. Thesis, University of Washington, Seattle, Washington.

Hazelhoff, E. H., and Evenhuis, H. H. (1952). Importance of the countercurrent principle for oxygen uptake in fishes. *Nature* **169**, 77.

Hodgman, C. D., ed. (1943). "Handbook of Chemistry and Physics," 27th ed. Chem. Rubber Publ. Co., Cleveland, Ohio.

Hoffert, J. R. (1966). Observations on ocular fluid dynamics and carbonic anhydrase in tissues of lake trout (*Salvelinus namaycush*). *Comp. Biochem. Physiol.* **17**, 107–114.

Hoffert, J. R., and Fromm, P. O. (1966). Effect of carbonic anhydrase inhibition on aqueous humor and blood bicarbonate ion in the teleost (*Salvelinus namaycush*). *Comp. Biochem. Physiol.* **18**, 333–340.

Holeton, G. F. (1968). Personal communication.

Holeton, G. F., and Randall, D. J. (1967a). Changes in blood pressure in the rainbow trout during hypoxia. *J. Exptl. Biol.* **46**, 297–305.

Holeton, G. F., and Randall, D. J. (1967b). The effect of hypoxia upon the partial pressure of gases in the blood and water afferent and efferent to the gills of rainbow trout. *J. Exptl. Biol.* **64**, 317–327.

Hughes, G. M. (1964). Fish respiratory homeostatis. *Symp. Soc. Exptl. Biol.* **18**, 81–107.

Hughes, G. M. (1966a). The dimensions of fish gills in relation to their function. *J. Exptl. Biol.* **45**, 177–195.

Hughes, G. M. (1966b). Species variation in gas exchange. *Proc. Roy. Soc. Med.* **59**, 494–500.

Hughes, G. M. (1967). Evolution between air and water. *Ciba Found. Sym., Develop. Lung* pp. 64–80.

Hughes, G. M., and Datta Munshi, J. S. (1968). Fine structure of the respiratory surfaces of an air breathing fish, the climbing perch *Anabas testudinens* (Bloch). *Nature* **219**, 1382–1384.

Hughes, G. M., and Grimstone, A. V. (1965). The fine structure of the secondary lamellae of the gills of *Gadus pollachius*. *Quart. J. Microscop. Sci.* **106**, 343–553.

Hughes, G. M., and Shelton, G. (1962). Respiratory mechanisms and their nervous control in fish. *Advan. Comp. Physiol. Biochem.* **1**, 275–364.

Hunn, J. B. (1967). "Bibliography on the Blood Chemistry of Fishes," pp. 1–32. Bur. Sport Fisheries Wild Life, U.S. Govt. Printing Office, Washington, D.C.

Irving, L., Black, E. C., and Safford, V. (1941). The influence of temperature upon the combination of oxygen with the blood of trout. *Biol. Bull.* **80**, 1–17.
Jakubowski, M. (1963). Budowa i unaczynienie skory glowacza (*Cottus gobio* L.) i czarnomorskiego skarpa [*Rhombus maeoticus* (Pall)]. The structure and vascularization of the skin of the river bullhead (*Cottus gobio* L.) and Black Sea turbot [*Rhombus maeoticus* (Pall)]. *Acta Biol. Crakov., Ser. Zool.* **6**, 159–175.
Johansen, K., and Hanson, D. (1968). Functional anatomy of the hearts of lungfishes and amphibians. *Am. Zoologist* **8**, 191–210.
Johansen, K., and Lenfant, C. (1967). Respiratory function in the South American Lungfish *Lepidosiren paradoxa* (Fitz). *J. Exptl. Biol.* **46**, 205–218.
Johansen, K., and Strahan, R. (1963). The respiratory system of *Myxine glutinosa*. In "Biology of Myxine" (A. Brodal and R. Fänge, eds.), pp. 352–371. Scandinavian Univ. Books, Oslo.
Johansen, K., Grigg, G. C., and Lenfant, C. (1966). Respiratory properties of the blood and responses to diving of the platypus *Onithorhynchus anatinus* (Shaw). *Comp. Biochem. Physiol.* **18**, 597–608.
Kaye, G. W. C., and Laby, T. H. (1958). "Tables of Physical and Chemical Constants," 12th ed. Longmans, Green, New York.
Keys, A., and Bateman, J. B. (1932). Branchial responses to adrenaline and pitressin in the eel. *Biol. Bull.* **63**, 327–336.
Klawe, W. L., Barrett, I., and Klawe, B. M. H. (1963). Haemoglobin content of the blood of six species of Scombroid fishes. *Nature* **198**, 96.
Kohn, P. G. (1965). Tables of some physical and chemical properties of water. *Symp. Soc. Exptl. Biol.* **19**, 3–16.
Krogh, A. (1904). Some experiments on the cutaneous respiration of vertebrate animals. *Archs. Skand. Physiol.* **16**, 348–357.
Krogh, A. (1919). The rate of diffusion of gases through animal tissues, with some remarks on the coefficient of invasion. *J. Physiol. (London)* **52**, 391–408.
Krogh, A. (1941). "The Comparative Physiology of Respiratory Mechanisms." Univ. of Pennsylvania Press, Philadelphia, Pennsylvania.
Krogh, A., and Leitch, I. (1919). The respiratory function of the blood in fishes. *J. Physiol. (London)* **52**, 288–300.
Kylstra, J. A., Paganelli, C. V., and Rahn, H. (1967). Some implications of the dynamics of gas transfer in water breathing dogs. *Ciba Found. Symp., Develop. Lung* pp. 34–58.
Labat, R., Peyraud, C., and Serfaty, A. (1962). Réactions électrocardiographiques et operculaires provoquées par stimulation lumineuse au par effet d'approche chez les téléostéens marius. *J. Physiol. (Paris)* **54**, 591–598.
Lam, T. (1968). Personal communication.
Leivestad, H., Anderson, H., and Scholander, P. F. (1957). Physiological response to air exposure in codfish. *Science* **126**, 505.
Lenfant, C., and Johansen, K. (1966). Respiratory function in the elasmobranch *Squalus suckleyi* G. *Resp. Physiol.* **1**, 13–29.
Lenfant, C., Grigg, G. C., and Johansen, K. (1966–1967). Respiratory properties of blood and pattern of gas exchange in lungfish. *Resp. Physiol.* **2**, 1–21.
Maetz, J., and Garcia Romeu, F. (1964). The mechanism of sodium and chloride uptake by the gills of a fresh-water fish Carassius auratus. II. Evidence for NH_4^+/NA^+ and HCO_3^-/Cl^- exchange. *J. Gen. Physiol.* **47**, 1209–1226.
Maren, T. H. (1967). Carbonic anhydrase: Chemistry, physiology and inhibition. *Physiol. Rev.* **47**, 595–781.

Mayer, N., Maetz, J., Chan, D. K. O., Forster, M., and Chester Jones, I. (1967). Cortisol, a sodium excreting factor in the eel (*Anguilla anguilla* L.) adapted to sea water. *Nature* **214**, 1118–1120.
Miles, H. M., and Smith, L. S. (1968). Ionic regulation in migrating juvenile coho salmon, *Oncorhynchus Kisutch*. *Comp. Biochem. Physiol.* **26**, 381–398.
Muir, B. S. (1967). Personal communication.
Muir, B. S. (1969). Gill dimensions as a function of fish size. *J. Fisheries Res. Board Can.* **26**, 165–170.
Muir, B. S. (1970). Contribution to the study of blood pathways in teleost gills. *Copeia* 19–28.
Muir, B. S., and Buckley, R. M. (1967). Gill ventilation in *Remora remora*. *Copeia* 581–586.
Muir, B. S., and Hughes, G. M. (1969). Gill dimensions for three species of tuna. *J. Exptl. Biol.* **51**, 271–285.
Muir, B. S., and Kendall, J. I. (1968). Structural modifications in the gills of tunas and some other oceanic fishes. *Copeia* 388–398.
Nakano, T., and Tomlinson, N. (1967). Catecholamine and carbohydrate concentrations in Rainbow trout (*Salmo gairdneri*) in relation to physial disturbance. *J. Fisheries. Res. Board Can.* **24**, 1701–1715.
Newstead, J. D. (1967). Fine structure of the respiratory lamellae of teleostean gills. *Z. Zellforsch. Mikroskop. Anat.* **79**, 396–428.
Nicloux, M. (1923). Action de l'oxyde de carbone sur les poissons et capacité respiratoire du sang de ces animaux. *Compt. Rend. Soc. Biol.* **89**, 1328–13331.
Östlund, E., and Fänge, R. (1962). Vasodilation by adrenaline and noradernaline and the effects of some other substances on perfused gills. *Comp. Biochem. Physiol.* **5**, 307–309.
Parry, G. (1966). Osmotic adaptation in fishes. *Biol. Rev.* **41**, 392–444.
Pasztor, V. M., and Kleerekoper, H. (1962). The role of the gill musculature in teleosts. *Can. J. Zool.* **40**, 785–802.
Peyraud, C., and Serfaty, A. (1964). Le rythme respiratoire de la carp (*Cyprinus Carpio* L.) et ses relations avec le taux de d'oxygène dissous dans le biotope. *Hydrobiologia* **23**, 165–178.
Phillips, A. M., Jr. (1947). The effect of asphyxia upon the red cell content of fish blood. *Copeia* 183–186.
Piiper, J., and Baumgarten-Schumann, D. (1968a). The effectiveness of O_2 and CO_2 exchange in the gills of the dogfish (*Scyliorhinus stellaris*). *Resp. Physiol.* **5**, 338–349.
Piiper, J., and Baumgarten-Schumann, D. (1968b). Carriage of O_2 and CO_2 by water and blood in gas exchange of the dogfish (*Scyliorhinus stellaris*). *Resp. Physiol.* **5**, 326–337.
Piiper, J., and Schumann, D. (1967). Efficiency of oxygen exchange in the gills of the dogfish. (*Scyliorhinus stellaris*). *Resp. Physiol.* **2**, 135–148.
Pocklington, P. (1968). B.Sc. Honours Thesis, Dalhousie University.
Rahn, H. (1966a). Aquatic gas exchange: Theory. *Resp. Physiol.* **1**, 1–12.
Rahn, H. (1966b). Evolution of gas exchange in vertebrates. *In* "Studies in Pulmonary Physiology, Mechanics, Chemistry and Circulation of the Lung," Vol. 2, pp. 42–54. Aerospace Med. Res. Lab.
Rahn, H. (1967). Gas transport from the external environment to the cell. *Ciba Found. Symp., Development of the Lung* pp. 3–23.

Rahn, H., and Fenn, W. O. (1956). "A Graphical Analysis of the Respiratory Gas Exchange." Am. Physiol. Soc., Washington, D.C.
Randall, D. J. (1962). Unpublished observations.
Randall, D. J. (1966). The nervous control of cardiac activity in the tench (*Tinca tinca*) and the goldfish (*Carassius auratus*). *Physiol. Zool.* **39**, 185–192.
Randall, D. J. (1968). Functional morphology of the heart in fishes. *Am. Zoologist* **8**, 179–189.
Randall, D. J., and Smith, J. C. (1967). The regulation of cardiac activity in fish in a hypoxic environment. *Physiol. Zool.* **40**, 104–113.
Randall, D. J., and Stevens, E. Don. (1967). The role of adrenergic receptors in cardiovascular changes associated with exercise in salmon. *Comp. Biochem. Physiol.* **21**, 415–424.
Randall, D. J., Holeton, G. F., and Stevens, E. Don. (1967). The exchange of oxygen and carbon dioxide across the gills of rainbow trout. *J. Exptl. Biol.* **46**, 339–348.
de Reuck, A. V. S., and Porter, R., eds. (1967). *Ciba Found. Symp., Development of the Lung.* Churchill, London.
Rhodin, J. A. G. (1964). Structure of the gills of the marine fish pollack (*Pollachius virens*). *Anat. Record* **148**, 420.
Robin, E. D., and Murdaugh, H. V. (1967). Quantitative aspects of vertebrate gas exchange. *Ciba Found. Symp., Develop. Lung* pp. 85–98.
Root, R. W. (1931). The respiratory function of the blood of marine fishes. *Biol. Bull.* **61**, 427–456.
Root, R. W., and Irving, L. (1941). The equilibrium between haemoglobin and oxygen in whole and haemolyzed blood of the tautog, with a theory of the Haldane effect. *Biol. Bull.* **81**, 307–323.
Root, R. W., Black, E. C., and Irving, L. (1939). The effect of hemolysis upon the combination of oxygen with the blood of some marine fishes. *J. Cellular Comp. Physiol.* **13**, 303–313.
Rossi-Fanelli, A., and Antonini, E. (1960). Oxygen equilibrium of haemoglobin from *Thunnus thynnus*. *Nature* **186**, 895.
Ruud, J. T. (1954). Vertebrates without erythrocytes and blood pigment. *Nature* **173**, 848–850.
Satchell, G. H. (1960). The reflex co-ordination of the heart beat with respiration in the dogfish. *J. Exptl. Biol.* **37**, 719–731.
Saunders, R. L. (1961). The irrigation of the gills of fishes. I. Studies of the mechanism of branchial irrigation. *Can. J. Zool.* **39**, 637–653.
Saunders, R. L. (1962). The irrigation of the gills in fishes. II. Efficiency of oxygen uptake in relation to respiratory flow, activity and concentration of oxygen and carbon dioxide. *Can. J. Zool.* **40**, 817–862.
Saxena, D. B. (1958). Extent of the gill surface in the teleosts *Heteropneustes fossilis* Block and *Clarias Batrachus* Linn. *Proc. Natl. Acad. Sci., India* **B28**, 258–263.
Saxena, D. B. (1959). Extent of gill surface in the teleost *Rita Rita* (Hamilton). *Zool. Soc. India, Proc. 1st All India Congr. Zool.* Part 2, pp. 165–168.
Schlicher, J. (1927). Vergleichendphysiologische Untersuchungen der Blutkörperchenzahlen bei Knochenfischen. *Zool. Jahrb.* **43**, 121–200.
Schumann, D., and Piiper, J. (1966). Der Sauerstoffbedarf der Atmung bei Fischen nach Messungen an der narkotisierten Schleie (*Tinca tinca*). *Arch. Ges Physiol.* **288**, 14–26.

Steen, J. B., and Berg, T. (1966). The gills of two species of heamoglobin free fishes compared to those of other teleosts—with a note on severe anemia in an eel. *Comp. Biochem. Physiol.* **18**, 517–526.

Steen, J. B., and Kruysse, A. (1964). The respiratory function of teleostean gills. *Comp. Biochem. Physiol.* **12**, 127–142.

Stevens, E. Don. (1968a). The effect of exercise on the distribution of blood to various organs in rainbow trout. *Comp. Biochem. Physiol.* **25**, 615–625.

Stevens, E. Don. (1968b). Cardiovascular dynamics during swimming in fish, particularly rainbow trout (*Salmo gairdneri*). Ph.D. Thesis, University of British Columbia.

Stevens, E. Don, and Randall, D. J. (1967a). Change in blood pressure, heart rate and breathing rate during moderate swimming activity in rainbow trout. *J. Exptl. Biol.* **46**, 307–315.

Stevens, E. Don, and Randall, D. J. (1967b). Changes in gas concentrations in blood and water during moderate swimming activity in rainbow trout. *J. Exptl. Biol.* **46**, 329–337.

Strickland, J. D. H., and Parsons, J. R. (1965). A manual of seawater analysis. *Fisheries Res. Board Can.* Bull. 125.

Swan, H., and Hall, F. G. (1966). Oxygen-hemoglobin dissociation in *Protopterus aethiopicus*. *Am. J. Physiol.* **210**, 487–489.

Truesdale, G. A., and Gameson, A. L. H. (1957). The solubility of oxygen in saline water. *J. Conseil, Conseil Perm. Intern. Exploration Mer* **22**, 163–166.

van Dam, L. (1938). On the utilization of oxygen and regulation of breathing in some aquatic animals. Dissertation, Gröningen.

Washburn, E. W., ed. (1929). "International Critical Tables," Vol. V. McGraw-Hill, New York.

Weast, R. C., ed. (1964). "Handbook of Chemistry and Physics," 45th ed. Chem. Rubber Publ. Co., Cleveland, Ohio.

West, J. B. (1965). "Ventilation/Blood Flow and Gas Exchange." Blackwell, Oxford.

Willmer, E. N. (1934). Hydrogen ion concentration of blood of South American fish. *J. Exptl. Biol.* **11**, 283–306.

Wittenberg, J. B., and Wittenberg, B. A. (1962). Active secretion of oxygen into the eye of fish. *Nature* **194**, 106–107.

Wright, C. I. (1934). The diffusion of carbon dioxide in tissues. *J. Gen. Physiol.* **17**, 657–676.

8

THE REGULATION OF BREATHING

G. SHELTON

I. Introduction	293
II. The Respiratory Pump	294
A. Teleosts	295
B. Elasmobranchs	310
C. Cyclostomes	314
III. Central Factors in the Regulation of Breathing Patterns	316
A. Experimental Techniques for Investigating Central Respiratory Mechanisms	317
B. The Site and Extension of the Respiratory Center	319
C. The Functional Organization of the Respiratory Center	322
IV. The Role of Mechanoreceptors in Respiratory Regulation	326
V. The Chemical Regulation of Respiration	328
A. Oxygen	328
B. Carbon Dioxide	342
C. Respiratory Homeostasis in Resting and Active Fish	346
VI. The Relationship between Ventilation and Perfusion	347
References	352

I. INTRODUCTION

In this chapter performance of the systems involved in ventilation of gills with water will be considered, especially in relation to the way in which the basic patterns of activity are produced and coordinated. The adaptive modification of these patterns in response to changes in internal and external environment will also be discussed. To a lesser extent, consideration will be given to perfusion of the gills with blood in relation to ventilation and gas exchange. Events at the level of the cell are excluded (see chapters by Hochachka, Volume I, and Hochachka and Somero, Volume VI) even though they set the level of oxygen consumption and carbon dioxide production and thus create the demands to be met by the ventilatory system in its widest sense. Nor is this chapter concerned with details of the gas exchange process since this, too, is discussed elsewhere (see chapter by Randall, "Gas Exchange in Fish," this volume). The broader problem to be discussed here is essentially that of the regulation of ventilation and perfusion so that the gas exchanger is able to function and, as far as possible, meet the needs of the cells. Previous articles in which these aspects of fish respiration have been reviewed are those of Black (1951), Fry (1957), Hughes and Shelton (1962), and Hughes (1964).

If a multistage system such as that involved in the transfer of respiratory gases between environment and cell enzymes is to operate satisfactorily, then control becomes a matter of some complexity. Prosser (1955) has pointed out that an organism can survive in an altered environment in different ways. At one extreme it may change its internal state with the environment (physiological adjustment), or alternatively it may regulate the internal state at a constant level over wide environmental fluctuations (physiological regulation). The control systems responsible for physiological regulation seldom result in complete homeostasis so that adjustment and regulation invariably occur together, particularly at the extremes of the regulated range. This is true of the fish respiratory system, and although environmental changes result in adaptive regulation of an immediate or longer term type (see Hughes, 1964) these are frequently accompanied by adjustments in metabolism.

II. THE RESPIRATORY PUMP

In the gill breathing vertebrates there is a surprising diversity of mechanisms whose function is one of maintaining a flow of water over the respiratory surface. Water is usually taken in through the mouth and, in those fish possessing it, the spiracle, is passed over the gills, and leaves through separate branchial or common opercular openings on either side of the body. This unidirectional flow pattern is thought to be the most effective way of bringing the environment into intimate contact with the gill epithelium since anatomical dead space is not necessarily a characteristic of such a system, although it may in fact exist (Saunders, 1961; Hughes, 1966; chapter by Randall, "Gas Exchange in Fish," this volume). It has also been claimed that unidirectional flow reduces the work of ventilation as compared to a tidal flow system (Hughes and Shelton, 1962). This generalization may be open to question since an elastic recovery system can prevent undue energy loss in reciprocating or tidal systems. In fact, the adult lamprey possesses a tidal ventilation system, the water being taken in and ejected through the external openings of the gills (Roberts, 1950).

Many of the differences in pumping mechanism such as the presence of velum, spiracle, and operculum depend in the first place on the phylogenetic position of the fish. Other differences of a less obvious type are related to habitats in which the fish live. Details of many of these differences will be given in the following account, although the teleosts will be treated in greatest detail because they are the most fully documented group of fish.

A. Teleosts

1. THE PATTERN OF WATER FLOW

a. General Features of the Breathing Movements and Pressures. It is probably true to say that the recent developments in the study of fish breathing movements began with the work of Woskoboinikoff and Balabai (1936, 1937) and van Dam (1938). These workers introduced the concept of a continuous gill curtain separating buccal and opercular cavities. They also suggested that water flow over the gills was essentially a continuous process even though the water entered and left the overall pumping system in a discontinuous manner. Woskoboinikoff and Balabai concluded that two pumps, one in front and one behind the gill curtain, were responsible for maintaining the flow. Attempts to substantiate these claims were made by measuring pressures in the two parts of the system by means of simple water manometers. The views of these three workers were summarized and extended by Henschel (1939) in a generalized scheme of teleost breathing mechanisms.

Modern manometric methods were applied to the problem by Hughes and Shelton (1958) working on trout, *Salmo*, roach, *Rutilus*, and tench, *Tinca*. Pressure measurements in buccal and opercular cavities showed that the gills offered appreciable resistance to water flow so that a differential pressure was always found, usually with the gradient from buccal to opercular cavity. Together with records of the breathing movements obtained from cine films, the pressure records offered good evidence of a dual mechanism responsible for maintaining water flow (Fig. 1). The breathing cycle was accordingly divided into two major phases as the opercular suction pump (phase 1) or the buccal pressure pump (phase 3) predominated, with periods of transition when the

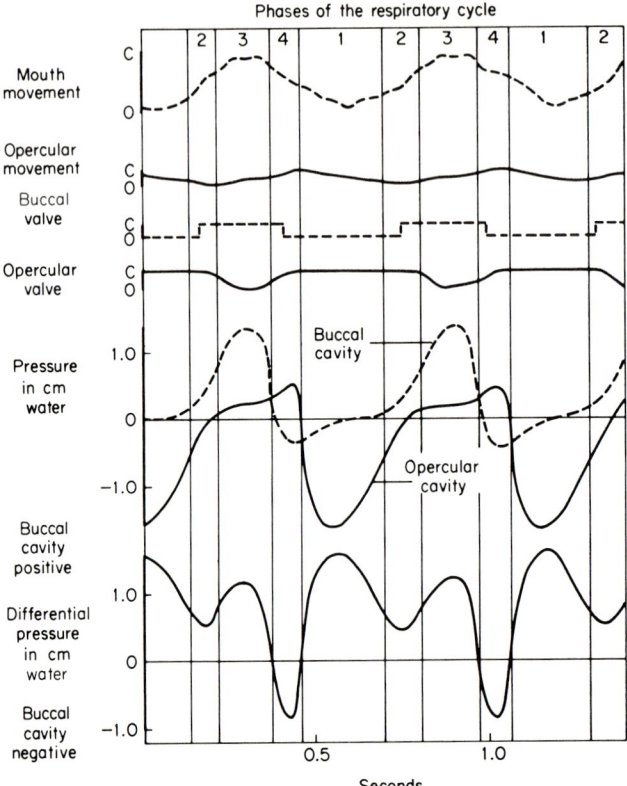

Fig. 1. Trout (70 g). The breathing movements of the mouth and operculum, together with associated pressure changes in the buccal and opercular cavities. The differential pressure between these cavities is shown in the bottom trace. O and C indicate the opened and closed positions of the mouth, operculum, and their associated valves. From Hughes and Shelton (1958).

differential pressure was reduced (phase 2) or reversed (phase 4). During phase 1, the mouth was open, water was entering the buccal cavity and passing over the gills to the opercular cavities whose valves were closed. In phase 3, the mouth and buccal valves were closed, water was passing through the gills from the contracting buccal cavity and leaving the system through the opercular openings.

The concept of a dual pump, although a useful one, arises from pressure measurements made on either side of a gill resistance, relates primarily to water flow through the gills, and has little anatomical basis since there is mechanical interaction throughout the system. This point will be returned to later. However, the concept is a useful one provided that the full extent of interaction between pumps in producing flow into the system, through the gills, and out to the environment is appreciated.

The experiments of Hughes and Shelton (1958) were done on anesthetized animals held in a clamp which, although not interfering with the breathing, was somewhat restricting. The pressures were measured via fine metal tubes passed through the mouth and opercular openings. These unnatural circumstances may have affected the overall respiratory behavior of the fish since this is notoriously sensitive to interference of all types (Fry, 1957). Saunders (1961) avoided these difficulties by implanting flexible polyethylene cannulae in the buccal cavity through the ethmoid region of the skull and in the opercular cavity through the cleithrum. The cannulae were then connected to the manometer system but were sufficiently flexible to permit free movement of the unanesthetized fish. This cannulation technique of Saunders, or modifications of it, has since been used extensively (e.g., Holeton and Randall, 1967a; Smith *et al.*, 1967). Saunders did not measure the characteristics of his manometers, and it seems possible that their frequency response was low. However, in the case of *Catostomus*, the sucker, and *Ictalurus*, the bullhead, the results differ only in details from those of Hughes and Shelton (1958). In the case of *Cyprinus*, the carp, the differences were marked since the resting animal was found to breathe intermittently with a double excursion of pressure in both cavities for each cycle of activity (Fig. 2a). There is some resemblance to the cough recorded in the roach by Hughes and Shelton. The heavy breathing pattern was also different and showed no gradient reversal in phase 4 (Fig. 2b).

b. Ecological Variation. Baglioni (1907) suggested that marine teleosts fell into four ecological categories between which there were fairly clear differences in respiratory mechanism. The variation was mainly in the degree of development of the branchiostegal apparatus below the opercular flap. Hughes (1960b) recorded breathing movements and pressure in a range of marine teleosts and, in general, confirmed the validity of Baglioni's classification.

In his first group are pelagic fish which never rest on the bottom and in which the operculum is well developed with a small branchiostegal apparatus. Hughes (1960b) found a great deal of variation in the pressure characteristics of fish in this category. In wrasse, *Crenilabrus*, and herring, *Clupea*, the two pumps were balanced (Fig. 3a), but the buccal pump predominated in the horse mackerel, *Trachurus*, and the opercular pump in the whiting, *Gadus*. There has been very little work

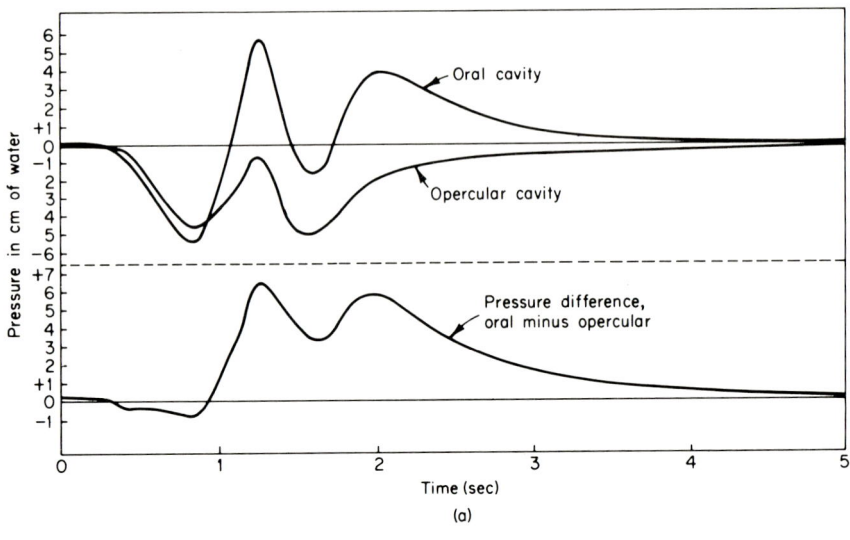

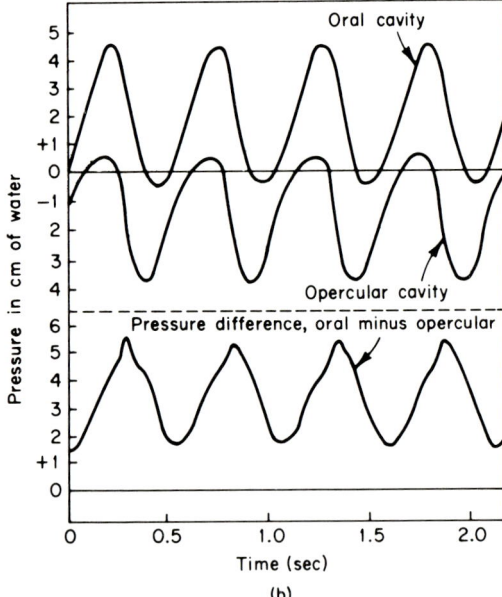

Fig. 2. Carp. Pressure changes in the buccal and opercular cavities, together with differential pressure curves. The records were taken from (a) quietly breathing fish and (b) fish stimulated to breathe heavily by increasing the carbon dioxide content of the water. From Saunders (1961). Reproduced by permission of the National Research Council of Canada from the Can. J. Zool. 39, 637–653 (1961).

done on the mechanics of ventilation in pelagic fish which are moving, although clearly this is of the greatest significance. Devices in which water currents can be maintained, such as the Brett (1964) respirometer or the Bainbridge (1963) fish wheel, make this type of investigation

8. THE REGULATION OF BREATHING 299

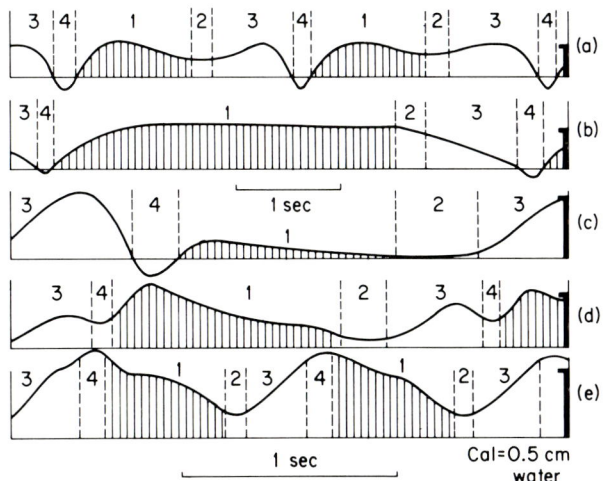

Fig. 3. Time course of the differential pressure between buccal and opercular cavities during the respiratory movements of five marine teleosts: (a) *Crenilabrus melops*, (b) *Callionymus lyra*, (c) *Conger conger*, (d) *Pleuronectes platessa*, (e) *Microstomus kitt*. A positive differential indicates that the pressure in the buccal cavity exceeds that in the opercular cavity. The respiratory cycle is divided into the four phases described in the text. Same time scales throughout except for record from *Callionymus*. After Hughes (1960b).

possible. It has been known for some time that adequate oxygenation of the blood in mackerel depends on forward swimming (Hall, 1930), and it has been assumed that the gills could not be ventilated in stationary fish. Many other fish have been observed to reduce their breathing movements as they move forward. For example, Brett (quoted by Stevens and Randall, 1967b) found that breathing stopped in sockeye salmon during violent exercise. It seems reasonable that animals living in a dense fluid should exploit this method of ventilation which may have little or no effect on the total drag of the body. The tuna ventilates its gills entirely by means of its movement through the water and regulates the stream by the degree of mouth opening (Muir, 1969). Brown and Muir (1969) suggest that the drag of the gill system is some 5–10% of the total drag in tuna swimming at 66 cm/sec. An ingenious examination of the problem was carried out by Muir and Buckley (1967) in *Remora*, a fish which is a poor swimmer but which moves rapidly through the water when riding on sharks. In still water this fish ventilated its gills by balanced buccal and opercular pumps but, when placed in a water current, breathing slowed down and eventually stopped at a water velocity between 60 and 80 cm/sec. *Remora* then adjusted ventilation by variation of the cross sectional area of mouth, buccal cavity, and opercular openings.

Fish in the second group of Baglioni's classification are largely bottom-living animals in which the branchiostegal apparatus is of greater importance in breathing. In general there is a relative increase in the size of the opercular cavity and in the predominance of the opercular pump in fish of this group (Fig. 3b). There is also a tendency to reduce the size of the exhalent openings, to increase the duration of the inspiratory phase, and therefore to eject water over a short period of time and at high velocity. These features can be seen in several fish (Henschel, 1941; Willem, 1947; Hughes, 1960b); *Callionymus* is a good example (Fig. 3b).

Baglioni's group III animals are those in which the branchiostegal apparatus is best developed and are true benthic forms. Hughes (1960b) examined two pleuronectid fish and again found a dominant opercular pump, although it was not obviously, different from that of group II fish. The main point of interest was the absence of a reversed differential pressure in phase 4 (Fig. 3d and 3e), and Hughes attributed this to an active branchiostegal valve mechanism. However, a number of types of fish have been found to possess patterns of this sort (Hughes, 1960b; Saunders, 1961; Muir and Buckley, 1967), and it may be that timing of the movements is of more significance than active valve mechanisms.

The relative importance of upper and lower opercular cavities in pleuronectids is a matter of some interest. Hughes (1960b) cited the earlier work and concluded from his own experiments that water left through both opercular openings. He also demonstrated that the gill area was identical on the two sides. A channel connecting the two opercular cavities was described by Yazdani and Alexander (1967) who found that, although both gills were ventilated, water left only from the upper operculum. This view was corroborated by Arnold (1967) who observed the respiratory currents when the fish were in still and in moving water. He found that water left only from the upper operculum even when the ventilation movements were large, provided the fish was in contact with the substrate. Actively swimming plaice, observed in a flume, breathed through both opercular openings.

The fish in group IV are a heterogenous collection of families between which there is no clear ecological relationship. They are characterized by the loss of the branchiostegal apparatus, but this is not sufficient to impose any uniformity in breathing mechanism. The buccal pump (Fig. 3c) or opercular pump may be dominant in different species.

c. *The Gill Resistance.* It was suggested that the concept of the dual pump rests on a gill curtain offering appreciable resistance to water flow. This resistance must be known before differential pressure records can be used to assess flow patterns and rates through the gills. Although analysis would be much easier if gill resistance remained constant, it seemed from the outset that this would be unlikely. The geometry of the gills changes constantly during a single breathing cycle since they are bridging a gap of changing dimensions and are attached ventrally to the basihyal which also moves rhythmically. Parts of the gill sieve could thus be alternately exposed and protected or, more simply, there could be variations of the dimensions in the pores through which the water flows.

Interdigitation of secondary lamellae could, as Hughes (1966) points out, have an enormous effect on resistance. In addition, it is uncertain that contact can be maintained under all conditions between the tips of gill filaments from adjacent gill arches. A number of workers have suggested that the gill filaments do separate during some stage in the opercular cycle. Saunders (1961) reported separation as the operculum was maximally abducted, Hughes (1961a) during opercular adduction, and Pasztor and Kleerekoper (1962) during all phases but principally during adduction. The latter workers were satisfied that the movements were not a result of the force of the ventilating currents but were produced by rhythmic contraction of gill adductor muscles. They also suggested that the filament tips parted to protect the gills against excess water flow, although it is by no means clear that an animal could produce a gill damaging current. In tuna and some other oceanic fishes there is considerable fusion of gill filaments and secondary lamellae to give a more solidly constructed hemibranch (Muir and Kendal, 1968). Even in these forms, which usually rely on "ram" ventilation associated with fast swimming, water can still bypass the gills if the filament tips part.

Such determination of gill resistance as have been made support the conclusion that it cannot remain constant. Hughes and Shelton (1958) reported experiments on tench in which the ventilation volume was changed by adding carbon dioxide to the water. As the volume increased so the ratio (mean pressure)/(minute volume) decreased, and this was taken as strong evidence that the gill resistance was decreasing. It also seemed reasonable to conclude that if resistance changed with overall flow rate it could quite easily vary within a single cycle since the pressure gradient was not constant. Later experiments in which the gills of deeply anesthetized fish were artificially perfused (Hughes and Shelton, 1962) gave results in which the resistance varied in an unpredictable way, sometimes rising, sometimes falling, and sometimes remaining constant with increased flow (Fig. 4a). It was quite common for the resistance to remain constant over certain parts of the pressure-

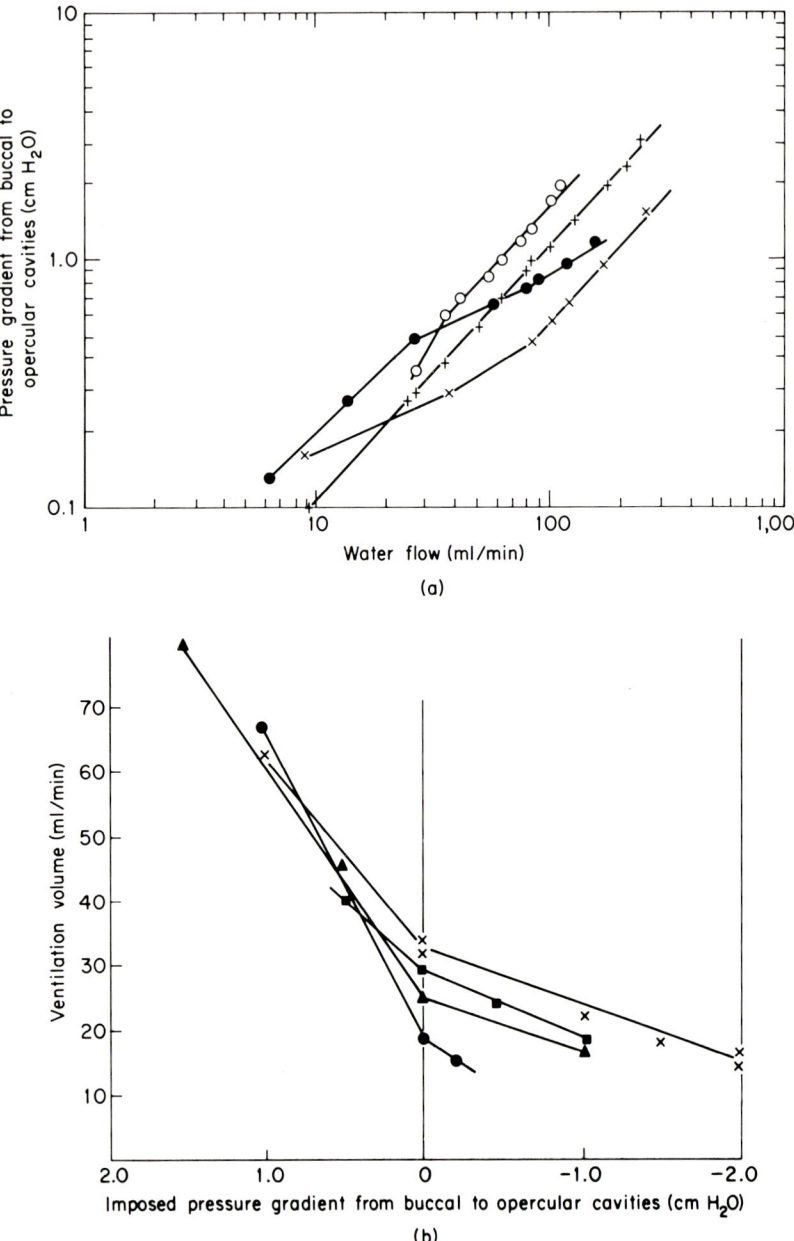

Fig. 4. The relationship between water flow over the gills and experimentally applied pressure gradients. (a) In tench (4 animals each of approximately 100 g) which were anesthetized to the point where breathing movements had ceased. The

flow range, but invariably discontinuities were found in the relationship. The most usual pattern was one in which the gill resistance remained constant down to a certain flow rate, below which the resistance increased gradually as the flow decreased. The values for resistance obtained in these experiments agreed reasonably well with those from the actively ventilating tench. Increase in resistance with decreasing ventilation could be explained in terms of change in gill geometry, for example, in the position of the filament tips. The discontinuity in the relationship could result from some new factor such as size of mouth and opercular openings, assuming greater importance at high flow rates.

Experiments on the relationship between ventilation and differential pressure across the gills have also been done on *Callionymus* (Hughes and Umezawa, 1968b). Because of the convenient anatomy of this animal it is possible to glue rubber connections around the opercular openings and impose external changes of pressure across the gills of otherwise normally breathing fish. Hughes and Umezawa found a change in slope of the pressure–volume relationship at the point where the imposed pressure gradient was zero (Fig. 4b). Larger changes in ventilation were produced by increasing the gradient from buccal to opercular cavities than were seen when similar gradients were imposed in the reverse direction. The change in slope is not necessarily to be interpreted in terms of a change in gill resistance because the fish is actively pumping water and may respond to the imposed gradient. A change in activity may be more important than a change in gill resistance in this case.

d. The Work of Breathing. Breathing muscles do work against three main categories of force, the magnitude of which may vary greatly with the environment in which an animal lives. The flow resistive forces offered by the gill curtain of fish or by the airways in a mammalian lung will depend upon the viscosity of water and air, respectively (these are related in a ratio of approximately 55:1); similarly inertial forces will depend on the mass of water or air (densities related approximately 840:1) and of the tissues being accelerated. Only the third category, the elastic forces developed in the tissues of the ventilating system, may

---

mouth was held open by means of a glass tube. Slopes different from 1 on the log–log plot indicate changing gill resistance. From Hughes and Shelton (1962). (b) In *Callionymus* (4 animals each of approximately 100 g) which were breathing normally. The applied pressure changes were therefore not the only pressures produced in the system, and a reversed gradient could be applied without reversing the flow. The change in slope at zero applied pressure need not reflect a change in gill resistance since the animal may change its breathing pattern. From Hughes and Umezawa (1968b).

be independent of the medium and related more directly to the nature of the ventilating system. These considerations have led many workers to conclude that the cost of breathing in aquatic organisms is very high (Hughes and Shelton, 1962), although precise figures are not usually available.

In man the total energy needed for breathing can be measured by determining the overall oxygen consumption at different forced ventilation levels (Otis, 1964; Glauser et al., 1967). The cost in resting man is about 2% of the overall oxygen consumption, although there is some variation in different workers' results. Pressure–volume measurements have been used to determine the mechanical work of breathing (Milic-Emili and Petit, 1960), and comparison of the two types of determination gives efficiencies in respiratory musculature of 19–25%.

There are difficulties in persuading fish to ventilate their gills at different levels by means which will not otherwise affect oxygen consumption. Conditions of carbon dioxide excess or oxygen lack will produce the necessary changes but it is known, for example, that oxygen depletion ultimately limits uptake. Figures given by van Dam (1938), who measured ventilation in eels and trout by enclosing their opercular region in a chamber separated from the mouth, enable one to calculate that the eel uses 10–12% and the trout 20% of its oxygen uptake for the work of breathing. There are too few determinations to give any degree of confidence in these calculations, added to which the fish, although fastened, may have been variably active. However, calculations based on the results of other workers (Saunders, 1962; Holeton and Randall, 1967b) also give figures for metabolism of breathing muscles between 10 and 15% of the resting oxygen uptake. Increased ventilation certainly has a marked effect on oxygen uptake as Beamish (1964c) demonstrated. He determined standard oxygen consumption in fish by extrapolating the relationship between oxygen uptake and activity back to zero activity. These extrapolations showed that the standard level of oxygen uptake increased in goldfish, carp, and trout, as they breathed more vigorously in lowered oxygen tensions (Fig. 16). Surprisingly, carbon dioxide had no effect on standard oxygen consumption (Beamish, 1964d). Since ventilation volume was not measured, however, a more quantitative relationship cannot be determined from these results. Schumann and Piiper (1966) enclosed tench in the same type of opercular chamber as used by van Dam in order to determine both ventilation volume and oxygen uptake. Spontaneous changes in ventilation were used to determine the cost of breathing together with measurements of oxygen uptake during artificial ventilation of animals paralyzed by succinylcholine. The results showed that 18–43% of the total oxygen uptake at resting ventilation volumes and 44–69% at ventilation volumes three times resting were used by the respiratory musculature.

Unfortunately, calculations on the mechanical work of breathing give somewhat different answers. In fact, simultaneous pressure and volume measurements have not been made. The separate determination of volume changes in buccal and opercular cavities is a difficult proposition, and, although volume changes of the whole system could be more easily measured (from cine films, for example), no data are available. However, making some simplifications, it is possible to use the ventilation volume as a measure of the latter as Alexander (1967) has pointed out. If the system is represented by a simplified model (Fig. 5) then it is clear that on inspiration the work done is

$$W_i = -P_i(V + v) + (P_i - p_i)v$$

and on expiration

$$W_e = (P_e - p_e)V + p_e(V + v)$$

where V and v are the volume changes of buccal and opercular cavities, respectively, P_i and p_i the pressures in these cavities on inspiration, and P_e and p_e the pressures on expiration. If it is assumed that the differential pressure across the gills remains constant throughout the cycle so that

$$P_i - p_i = P_e - p_e = D$$

the total work done per cycle is

$$W = (-P_i + D + p_e)(V + v)$$

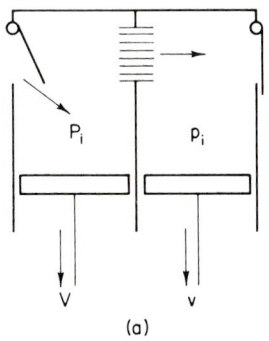 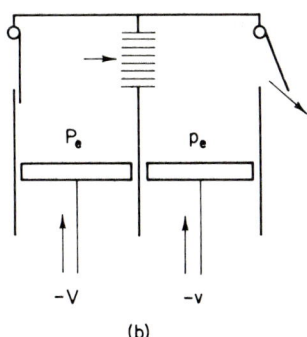

Fig. 5. Model of ventilating system showing buccal and opercular pumps operating during (a) inspiration and (b) expiration. The volumes of the buccal and opercular cavities are changed through the ranges V and v, respectively. The pressures associated with these volume changes are P_i and p_i (negative with respect to the outside) on inspiration, and P_e and p_e (positive with respect to the outside) on expiration. From Alexander (1967).

The work done in breathing a unit volume of water would be $-P_i + D + P_e$. If the data available in work cited previously are used, this total would usually be between 1×10^6 and 2×10^6 ergs/liter of water ventilated. Assuming 20% efficiency in the respiratory muscles, 0.025–0.05 ml of oxygen would be needed to pump 1 liter of water. Under normal conditions this would represent only 0.5–1% of the oxygen taken from that amount of water by the fish. Considerable increases in these figures are found if the calculations are repeated for active fish or for animals in a poorly aerated environment (Alexander, 1967).

The fact that the pressures in all parts of the system vary during both inspiration and expiration will make the real system less efficient than the model. In addition, there is often considerable reflux through the mouth during expiration and through the opercular openings during inspiration. This means that the ventilation volume is not the same as the overall volume changes. It seems very doubtful that these factors could account for the discrepancy between oxygen consumption measurements and the calculations.

2. The Pump Musculature and Skeleton

Several accounts have been published of the head skeleton and its musculature in teleosts, some of which refer specifically to ventilation (Ballintijn and Hughes, 1965; Henschel, 1941; Hughes and Ballintijn, 1968; Hughes and Shelton, 1962; Kirchhoff, 1958). Interactions occur between the skeleton, muscles, and tendons so that movements of one component tend to cause complementary movements in many others. Because of these interactions it is very difficult to determine the precise role of any muscle in the overall pattern. Anatomical studies, extirpation of muscles, and movement recordings all provide valuable information, but the recent studies of Ballintijn and Hughes (1965) and Hughes and Ballintijn (1968) using electromyography give the most convincing picture of the sequences of muscle contraction. The level of electrical activity in a muscle can also be used as a guide to the relative contribution of that muscle in the total mechanism, and in *Callionymus* Hughes and Ballintijn (1968) have demonstrated a clear relationship between stroke volume and electrical activity as shown by the height of the integrated electromyogram.

The following brief account is based on a generalized scheme of the teleost respiratory musculature (Fig. 6). Details of the musculature obviously vary from species to species as will their mode of action (Hughes and Ballintijn, 1968). One of the interesting conclusions of Ballintijn and Hughes (1965) is that the same breathing pattern is maintained

8. THE REGULATION OF BREATHING

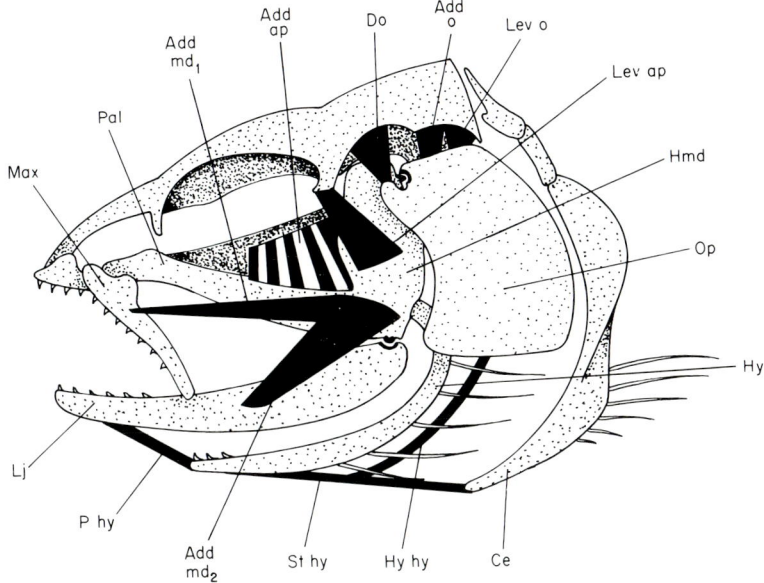

Fig. 6. Lateral view of the skull and important respiratory muscles in a typical teleost: Max, maxilla; Pal, palatine; Add md_1 and md_2, adductor mandibulae (two divisions); Add ap, adductor arcus palatini; Do, dilator operculi; Add o, adductor operculi; Lev o, levator operculi; Lev ap, levator arcus palatini; Hmd, hyomandibula; Op, operculum; Hy, hyoid; Ce, cleithrum; Hy hy, hyohyoideus; St hy, sternohyoideus; P hy, protractor hyoidei; and Lj, lower jaw.

even when different sets of muscles are involved. Thus the change from shallow to deep breathing with no other differences in the overall pattern is achieved by recruiting new, and quite important, muscles and not merely by increasing activity in the ones already working. In a system as labile as this it would be unreasonable to expect uniformity.

a. Expiration. During expiration the mouth and opercular flaps are closing, and as the whole system decreases in volume, water leaves through the opercular openings. The muscles involved are as follows (see Fig. 6):

(1) Adductor mandibulae, closing the mouth
(2) Adductor arcus palatini, adducting the palatal complex and hyomandibula, thus laterally decreasing the volume of the buccal cavity
(3) Adductor operculi, adducting the operculum and so decreasing the volume of the opercular cavity
(4) Protractor hyoideus, protracting the hyoid and thus adducting the ventral end of the hyomandibula

(5) Hyohyoideus, adducting the operculum and folding up the branchiostegal apparatus
(6) Levator operculi, adducting and lifting the dorsal border of the operculum. This has the effect of rotating the operculum on its articulation with the hyomandibula so that the operculum base is retracted. A ligament running from this region to the angular causes a lowering of the jaw at the end of expiration

b. Inspiration. During inspiration the whole system is increasing in volume and, since the opercular openings are closed by their valves, water is taken in through the mouth. The muscles involved are as follows (see Fig. 6):
(1) Levator arcus palatini, abducting the palatal complex and hyomandibula by rotating the latter outward on its articulation with the skull
(2) Sternohyoideus, retracting the hyoid arch thus expanding the branchial arches and abducting the hyomandibula
(3) Dilator operculi, abducting the operculum by pivoting it on its articulation with the hyomandibula

c. Interaction of Elements. Because the hyomandibula is articulated with the neurocranium, the quadrate, the operculum, and the more ventral hyoid elements, it forms a vital coupling element in the interactions between the skeletal and muscular components. The lower end of the hyomandibula moves freely and the mechanical linkages are such that its abduction is accompanied by retraction of the more basal elements. This expands both the buccal and opercular cavities, opens the mouth and, because the quadrates are also moved out laterally, widens the mouth. The branchial arches also expand. In the opposite sense, closure of the mouth will protract the hyoid which in turn adducts the hyomandibula and quadrate and swings in the operculum. A fuller account of this type of coupling is given by Ballintijn and Hughes (1965).

The precise timing of activity in seven of the main respiratory muscles of the trout as determined by the electromyographic studies of Ballintijn and Hughes (1965) is shown in Fig. 7. It might seem reasonable to suppose that in a smoothly oscillating system there would be temporal dispersion of overall activity throughout the cycle, and overlap between different units, certainly as inspiration took over from expiration and vice versa. This appears to be the case over most of the cycle as Fig. 7 shows, but there is a period during the final stages of opercular abduction when little activity is found in any muscles. It may be that elastic elements continue the movements smoothly during this period. It is interesting that all the muscles in Fig. 7 are active over at least two

8. THE REGULATION OF BREATHING

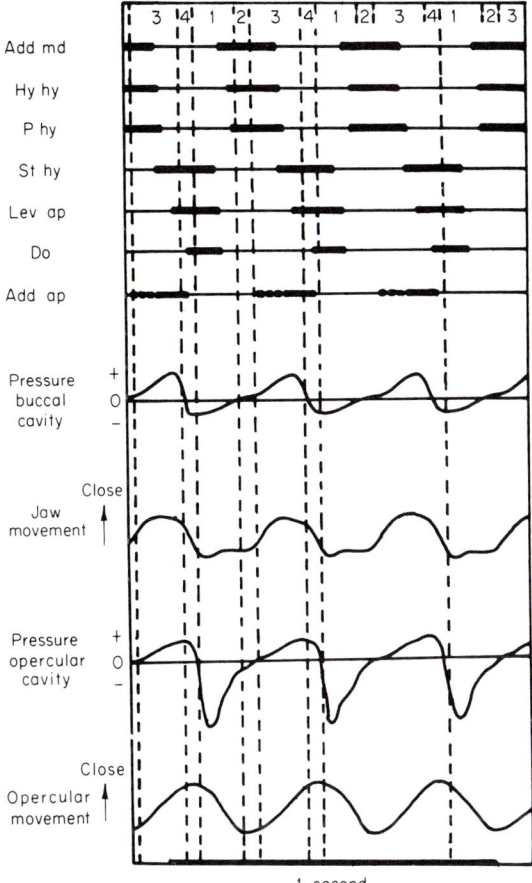

Fig. 7. Breathing movements and pressure changes in the buccal and opercular cavities of the trout, together with a diagrammatic representation of the time relations of activity in seven respiratory muscles (see Fig. 6 legend for definitions). The respiratory cycle is divided into the four phases described in the text. After Ballintijn and Hughes (1965).

and sometimes three of the four phases proposed by Hughes and Shelton (1958). This is desirable to produce the continuity of movement as already mentioned, but in fact the overlapping activity does not usually occur during the transitional phases. It is clear from this that there are very few muscles, perhaps none, that can be conveniently and specifically ascribed to buccal pressure pump or opercular suction pump as defined by manometric studies.

In general the most important muscles in the trout are the levator arcus palatini, the adductor arcus palatini (which also incorporates the adductor operculi in this animal), and the adductor mandibulae. During shallow breathing these muscles only are active, the sternohyoidens, protractor hyoideus, and hyohyoideus being brought into action as the breathing deepens. The dilator operculi only operates during hyperventilation.

B. Elasmobranchs

1. THE PATTERN OF WATER FLOW

The nature of the relationship between water flow and the gill filaments in elasmobranch fish is less well defined than it is in teleosts. There are no direct observations of the gill configuration during breathing although it has been assumed that filaments from adjacent gill arches are, to some extent, in contact. The fact that a measurable pressure gradient exists from orobranchial to parabranchial cavities (Hughes, 1960a) suggests that this is the case. Since the visceral arches themselves are participating in the pumping process to a much greater extent than in teleosts, there can be little doubt that changes in gill position and hence gill resistance occur during the breathing cycle. Flow direction over filaments and lamellae, particularly in relation to blood flow through them, is also poorly established. Robin and Murdaugh (1967), using indirect methods of calculating oxygen uptake and the amount of oxygen removed from the ventilation stream, came to the conclusion that countercurrent gas exchange did not occur in dogfish gills. However, other investigations (Hanson and Martin, 1967; Piiper and Schumann, 1967; Baumgarten-Schumann and Piiper, 1968) have shown that in many cases the arterial oxygen tension was higher than the mean expired oxygen tension. This can be explained by some sort of countercurrent system, although the path of water across the secondary lamellae is ultimately obstructed by the median septum. The possibility that water must flow along the length of the filaments in elasmobranchs has led Piiper and Schumann (1967) to propose a multicapillary model which is also effective in giving a negative value for the difference between expired and arterial oxygen tensions.

Anatomical differences between teleost and selachian are marked, since in the latter the respiratory current passes in through the spiracle as well as the mouth and out through a variable number of gill slits not covered by an operculum. The flow pattern from mouth and spiracle is curious. Water entering the spiracle of a dogfish will leave through the three anterior gill slits of the same side, whereas water entering through

8. THE REGULATION OF BREATHING

the mouth leaves through the last three slits. There is clearly not much turbulence in the orobranchial cavity. The pattern is not so obvious in the skate.

These problems and differences notwithstanding, the general picture of gill ventilation in selachians as it emerges from studies of movement and pressure does not appear to be radically different from that seen in the teleosts (Hughes, 1960a). A pressure gradient from orobranchial to parabranchial cavities exists for most of the respiratory cycle in dogfish and for the whole cycle in skates (Figs. 8 and 10). Again four phases of activity are recognizable. A parabranchial suction pump predominates in phase one and an orobranchial pressure pump in phase three, and there are transitional phases in between. These distinctions are most obvious on the differential pressure curve. There are no direct measurements of gill resistance in selachians, and the problem of determining relative volume changes in different parts of the system is made even more difficult than in teleosts because of the large number of parabranchial chambers. Since the latter are small in volume compared with the orobranchial chamber, it might seem likely that the changes in volume would also be small. The implication of this would be that the suction pump mechanism would be relatively unimportant in selachians. However, the pressure measurements do not bear this out if it can be assumed that the gill resistance does not change too drastically during the breathing cycle. Even if the resistance does change it seems

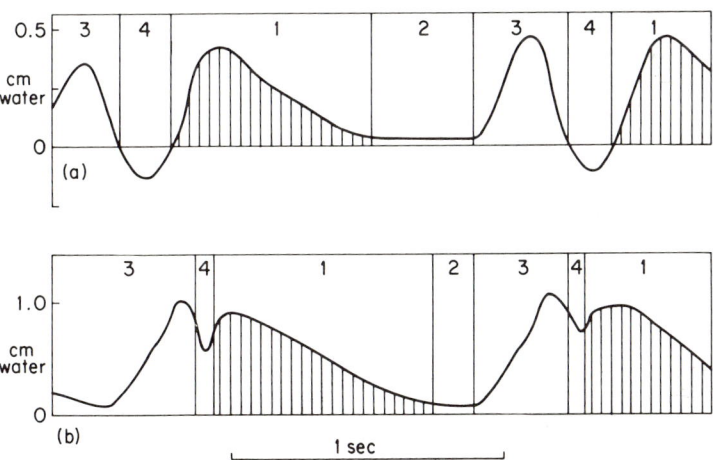

Fig. 8. Differential pressures between orobranchial and third parabranchial cavities of *Scyliorhinus* (a) and *Raia* (b). A positive differential pressure indicates that the orobranchial pressure is greater than the parabranchial pressure. The respiratory cycle is divided into the four phases described in the text. From Hughes (1960a).

likely that it will fall during the suction pump phase since this is the time when the gill chamber is expanding in volume. Under these conditions the flow of water during the suction phase would be considerable. The differential pressure curves obtained by Hughes (1960a) suggest that the parabranchial suction pump is at least as important as the orobranchial pressure pump in dogfish and of greater importance in the skate. Teichmann (1959), who measured the pressures achieved before and after functional disturbance of different parts of the system, claimed that the pressure pump dominated in *Scyliorhinus*, there was a balance in *Mustelus*, and the suction pump was of greater significance in *Torpedo*. In this work the time course of the ventilation pressures was not recorded. The whole relationship between pressure and flow is an extremely difficult problem which is still virtually unresolved in these animals.

The dominance of the parabranchial suction pump in skate and rays was correlated with the bottom-living habit by Hughes (1960a). The absence of gradient reversal in the differential pressure curves of these animals was also associated with their ecology and was thought to be the result of active closure of the external openings to the gill slits. This could ensure that no sand entered the gills by reflux through the external gill openings. Water intake took place through the spiracle when the animals were partially buried in sand, but both mouth and spiracles were used if the animals were swimming or resting with the snout raised. The more pelagic selachians also showed respiratory adaptations. Some sharks stopped their breathing movements when swimming actively and ventilated their gills simply by opening their mouths in the water stream. It may be that, as in teleosts of similar habits, ventilation is controlled by the extent to which the mouth is opened, but so far this type of regulation has not been confirmed by observation.

2. The Pump Musculature and Skeleton

The skeleton of the dogfish head is represented diagrammatically in Fig. 9. The head muscles of selachians are, with one or two exceptions, poorly differentiated but those that are defined are shown in Fig. 9, where the arrows on each muscle give some indication of the effect produced by contraction. With the exception of the hypobranchials (coracohyoideus, coracobranchialis), the muscles shown are adductor or adductorlike in function. Their contraction reduces the volume of orobranchial and parabranchial cavities. In addition to the muscles

8. THE REGULATION OF BREATHING 313

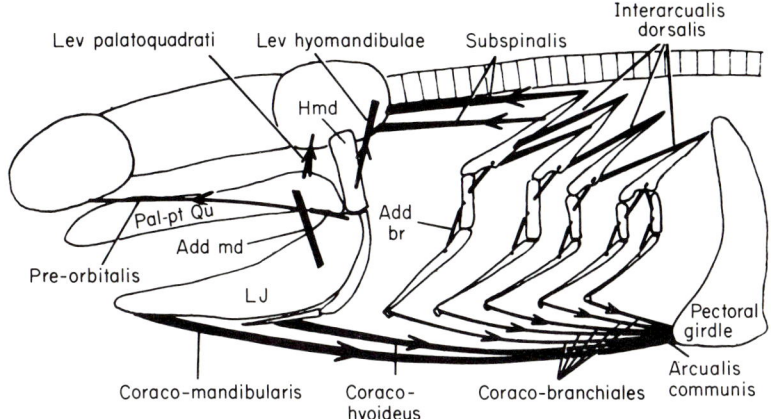

Fig. 9. Lateral view of the skull, visceral skeleton, and main respiratory muscles of the dogfish. The superficial constrictor muscles are not shown. Add br, adductor branchialis; Add md, adductor mandibulae; Pal-pt Qu, palato-pterygoid; LJ, lower jaw. From Hughes and Ballintijn (1965).

appearing in the diagram, there is a sheet of superficial constrictor muscle covering the whole head region and being particularly well developed in the branchial region. The constrictor system is formed of several overlapping sheets, each sheet associated primarily with a branchial arch (Lighttoller, 1939).

The timing of activity in the respiratory muscles is shown in Fig. 10, based on the electromyographic results obtained by Hughes and Ballintijn (1965). It is clear from this figure that most of the activity occurred as first the orobranchial and then the parabranchial cavities decreased in volume. Following the period of major activity, the orobranchial cavity began to expand slowly and passively. When the dogfish was breathing quietly the whole of this part of the cycle was a result of elastic recovery. If the dogfish was made to hyperventilate, only a short period of purely elastic recovery occurred, followed by a more rapid expansion caused by contraction of the hypobranchial musculature. Even when these muscles were active there was usually a pause after their contraction before the next cycle began. The only muscle to be excited during the whole of this period was the adductor mandibulae in which a certain amount of tonic activity probably served to check the opening of the mouth (Fig. 10). Hughes and Ballintijn (1965) conclude that contraction of the constrictor muscles, reducing the volume of both orobranchial and parabranchial cavities, is the prime mover in normal ventilation and that recovery is largely the result of elastic recoil of the visceral skeleton.

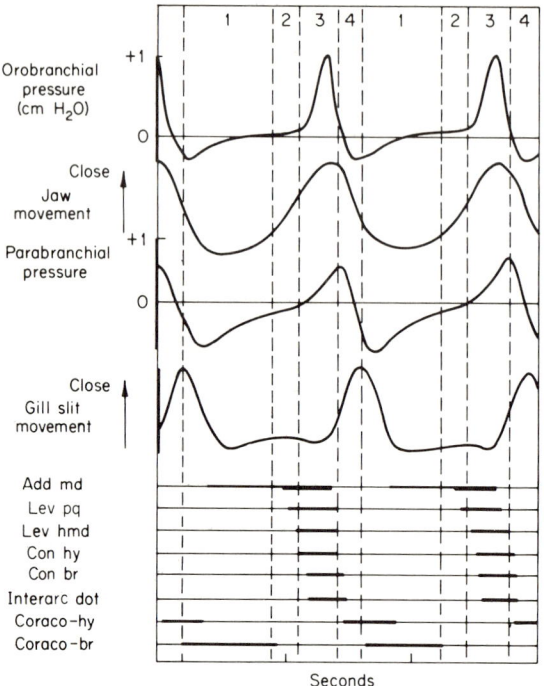

Fig. 10. Dogfish. Pressures in the orobranchial and parabranchial cavities in relation to movements of lower jaw and gill region. The main phases of activity in eight respiratory muscles are also shown. No indication of activity level is given except in the case of the adductor mandibulae which is more active when the line is thicker. The respiratory cycle is divided into the four phases described in the text. After Hughes and Ballintijn (1965). Add md, adductor mandibulae; Lev pq, levator palatoquadrati; Lev hmd, levator hyomandibulae; Con hy, constrictor hyoideus; Con br, constrictor branchiales; Interarc dot, interarcualis dorsalis; Coraco-hy, coracohyoideus; Coraco-br, coracobranchiales.

C. Cyclostomes

An account of the skeleton and musculature of the branchial region in the lamprey, *Lampetra*, has been given by Roberts (1950). The gills are found in pouchlike chambers supported laterally by the adjacent branchial arch skeleton. The pouches are also supported top and bottom by extensions from the branchial arches forming two lateral commissures. The whole complex forms an unjointed network, the enclosed volume of which can be changed by bending reentrant curves in the cartilage of which it is made up. The important muscles are the constrictors running from the dorsal to ventral part of the branchial skeleton. In addition, a pair of diagonal muscles span the dorsal and ventral reentrant curves of each branchial arch; there are some smaller muscles in the interbranchial septa and the gill sacs. The constrictor and diagonal muscles were shown, by observation and experiments in which electromyograms were recorded, to function together in expiration. No muscles were active during the inspiration of water, which in the adult lamprey occurs through the external gill openings. Just as in the dogfish it appears that expansion of the gill chambers is caused by elastic recoil, although Roberts (1950) noted that this was not a very forceful movement in the cartilaginous branchial skeleton of *Lampetra*.

Gill ventilation in myxinoids is very different from that described in the lamprey. The respiratory tract is quite complex as Fig. 11 shows. Water is taken in through the single nostril, propelled down the pharynx by the rolling and unrolling action of a velum working in a velar chamber, and eventually leaves through paired branchial apertures formed by the union of efferent ducts from the gills of either side. A pharyngocutaneous duct which connects the pharynx directly to the outside, thus short circuiting the gills, also empties into the common opening. Normally this duct is closed off by a sphincter (Johansen and Hol, 1960), but it occasionally opens to clear the pharynx of large particulate matter.

The pumping action of the velum has been established by studies on preserved material, by examination of live specimens (Strahan, 1958), and by cineradiography (Johansen and Hol, 1960). The velar skeleton and musculature is well described by Strahan (1958), and a full review is given by Johansen and Strahan (1963). The velum pulsates rhythmically, and Fig. 12 shows four successive stages in the velar cycle. In Fig. 12A the velum is at rest, and Figs. 12B and 12C show successive stages in the unrolling process. At this time the velar chamber would begin to decrease in volume and the velum would roll up. This activity,

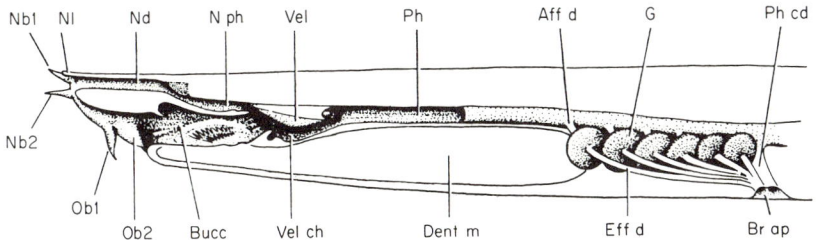

Fig. 11. Respiratory tract of *Myxine glutinosa*. Diagrammatic half-section of anterior part of body: Aff d, afferent gill duct; Br ap, common branchial and pharyngo-cutaneous aperture of left side; Bucc, buccal cavity; Dent m, dental muscles; Eff d, efferent gill duct; G, gill; Nb1 and Nb2, first and second nasal barbels; Nd, nasal duct; Nl, nasal lip; N ph, nasopharyngeal duct; Ob1 and Ob2, first and second oral barbels; Ph, pharynx; Ph cd, pharyngo-cutaneous duct; Vel, velum; Vel ch, velar chamber. From Johansen and Strahan (1963).

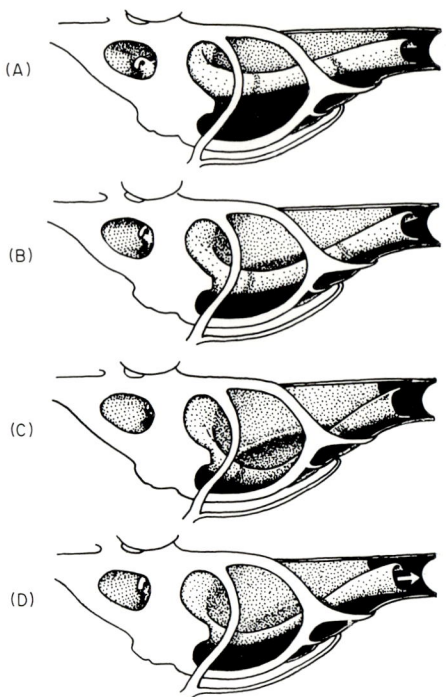

Fig. 12. Four successive stages in the velar cycle. The velar chamber is shown from the left side with the left pharyngeal wall removed: (A) resting; (B) velar scroll beginning to unroll; (C) velar scroll unrolled to full extent; and (D) velar scroll beginning to move dorsally, arrow showing direction of water flow. From Strahan (1958).

which is shown in Fig. 12D, propels water backward as indicated by the arrow and draws water in anteriorally through the nostril. It has been demonstrated by radiography that constrictor muscles of the pharynx and branchial regions also take part in pumping water. The gill pouches themselves contract rhythmically, and synchronized activity can be seen in the sphincters of afferent and efferent gill ducts operating in such a way as to ensure unidirectional flow.

III. CENTRAL FACTORS IN THE REGULATION OF BREATHING PATTERNS

Electrical discharges recorded, as described above, in the respiratory musculature of the head may be considered to be closely representative of activity passing down the motor nerves from its origins within motor neurons of brain stem nuclei. The fact that motor activity is not synchronous in all these final pathways but occurs in different phases of the breathing cycle in a very regular and specific way underlines the complexity of the respiratory oscillator. This mechanism must be capable of maintaining an appropriate basic rhythm and in addition be able to initiate activity of different duration and pattern to individual muscles at different times in a breathing cycle. The processes which lead up to the complex and highly integrated discharges in the respiratory musculature can be considered in two convenient but not totally independent categories. In the first place it is necessary to understand the interaction of large numbers of neural elements, most of them situated centrally, whose function is to initiate the basic respiratory process and produce the final coordinated motor pattern. The aggregation of neurons within the medulla oblongata involved in this function is usually referred to as the respiratory center, although implications in such a term of a fixed group of neurons with precise and restricted anatomical location are unfortunate (Hughes and Shelton, 1962; Wang and Ngai, 1964). In addition, there are many processes involved in respiratory homeostasis. That is to say, the basic process must be modified in an appropriate way to meet the demands of changes in internal state such as the transition from exercise to rest or of changes in external environment of great range and variety. The second category is thus one of feedback processes which ensure that the motor pattern produced by the center is appropriate to the conditions. The two categories will be considered in turn.

A. Experimental Techniques for Investigating Central Respiratory Mechanisms

Investigations of neural coordinating mechanisms rely in general upon the three classical methods of destruction, stimulation, and detection of activity. Their use in experiments on the central coordination of respiration in a variety of animals has led to the present position in which a great deal of information (and controversy) exists about the large-scale arrangement and interaction of neuron groups, but very little is known about the role of individual neurons; for example, expiratory and inspiratory groups which are functionally, if not morphologically, distinct form the conceptual basis of many hypotheses about the medullary respiratory center in mammals. In some cases they are considered adequate to coordinate rhythmic breathing (Salmoiraghi, 1963);

in others they are thought to interact with further neuron groups such as apneustic and pneumotaxic centers before normal activity is produced (Wang and Ngai, 1963; Kahn and Wang, 1967). The differences and problems result from studying a system of dispersed yet interacting neurons with methods which to some extent are all inadequate.

The extent of damage caused by transection or electrical coagulation is seldom easy to determine. In addition, it is not clear whether the changes in breathing always result from the destruction of regions or neurons exclusively associated with this particular function. Thus Wang and Ngai (1964), following a number of earlier workers, found that pontine transection provided evidence for both a pneumotaxic center and apneustic center in the pontine region of cat brains. However, Hoff and Breckenridge (1949), using similar techniques on the dog, were led to conclude that apneusis was the result of a generalized activity in the facilitator system of the reticular formation. They thought that pontine transection produced the necessary facilitator–suppressor imbalance to give inspiratory cramps. Nor are stimulation experiments free from these limitations. Stimulus spread, using either chemical or electrical methods, can be controlled with care but not, so far, to the point where drugs can be administered into the immediate environment of a neuron or a single unit can be stimulated electrically (Salmoiraghi, 1962; Liljestrand, 1953). Furthermore, the precise status of the region stimulated, whether exclusively respiratory or more general in function, whether neuron or interconnecting nerve tract, is often in doubt. In the medulla of mammals, for example, electrical stimulation has always led to a wider and, to some extent, a different area being implicated in respiratory control than determinations made by recording electrodes. The recording technique, on the other hand, can only be used to identify neurons as respiratory when they are rhythmically active and synchronized precisely with the breathing movements. Cells, discharging in ways different from this may be involved in respiratory coordination but cannot be recognized. Moreover, the technique has provided little information about neuron interaction because single neurons or, with less selective electrodes, single groups of neurons only have been studied. Some workers have tried to combine the stimulation and recording techniques. For example, Gill and Kuno (1963) studied the effect of medulla stimulation on activity in phrenic motor neurons, Keder-Stepanova and Ponomarev (1965) stimulated the medial region of the medulla and examined the effect this had on respiratory activity in more laterally situated neurons, and Hori (1966) recorded the changes produced in respiratory center cells by mesencephalic stimulation. Multielectrode preparations recording from several respiratory neurons, or stimulating some and recording from others, should provide more information on neuron interaction. The practical difficulties are obviously considerable and interpretation is not easy as studies on the integrative properties of decapod heart ganglia have shown. Even though the anatomy and physiology of the interconnections in this potentially simple, nine-neuron group are well established (Hartline, 1967), a complete analysis of the ganglion's rhythmic activity is not yet possible.

B. The Site and Extension of the Respiratory Center

The teleost medulla is capable of coordinating rhythmic breathing movements independently of the rest of the brain and spinal cord. In this respect the respiratory centers of fish and mammals are similar, since it is usually agreed that in the latter basic control of respiratory rhythmicity is a property of the medulla. (Wang and Ngai, 1964; Salmoiraghi, 1963). However, in fish, no extramedullary centers have been identified, although there is, of course, no differentiated pontine region. Transections immediately in front of the motor nuclei of the Vth and VIIth cranial nerves did not substantially change the breathing movements of tench (Shelton, 1959). Similarly, transections through the posterior parts of the Xth motor nucleus, just behind the facial lobe,

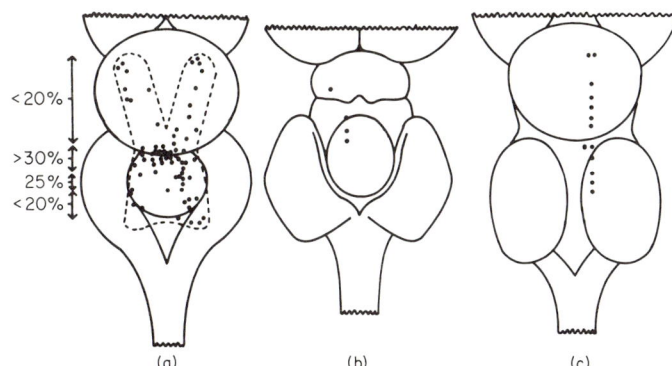

Fig. 13. Respiratory regions of the teleost brain: (a) dorsal view of the tench medulla showing the regions where successful electrode insertions were made. Broken line delimits the proposed respiratory area. The percentages of total electrode insertions which were successful at various levels within this area are indicated, and give some idea of the relative density of respiratory neurons. From Shelton (1961). The regions from which activity has been recorded in (b) the carp medulla (Woldring and Dirken, 1951) and (c) the goldfish medulla (von Baumgarten and Salmoiraghi, 1962) are given for comparison.

suggested that the more caudal regions of the central nervous system were inessential. The fish center thus appears to extend over a wide area of the medulla below the cerebellum and facial lobes (Fig. 13). No recognizable subunits, either anatomical or functional, were located in the area.

These general conclusions are supported by a number of different investigations in which electrical activity was recorded in the brain of teleosts (von Baumgarten and Salmoiraghi, 1962; Hukuhara and Okada, 1956; Shelton, 1961; Woldring and Dirken, 1951). All the activity was recorded from the area outlined above, although the search was often extended over much wider areas of the brain. The results of the recording experiments are also in quite good agreement as to the more detailed location of active site (Fig. 14). The motor neurons within the nuclei of the Vth, VIIth, IXth, and Xth cranial nerves were frequently found to be active. This is not surprising since these are the neurons whose axons innervate the musculature of the head and gill arches. It is perhaps worth making the point that motor neurons are not usually considered to be part of the respiratory center in mammals, to some extent because they are anatomically distinct from the medullary center. The work of Eccles et al. (1962), Sumi (1963), and Sears (1964) has shown that integrative processes go on in the spinal cord, down to the level of

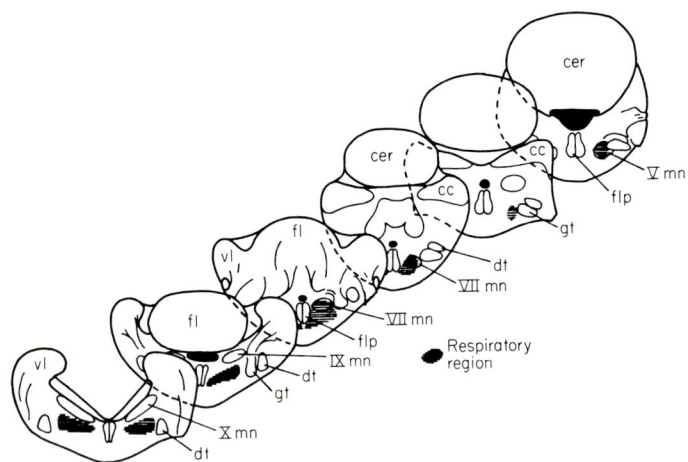

Fig. 14. Distribution of active respiratory sites within the tench medulla. Series of cross sections, taken from just behind the facial lobe and running forward to the midcerebellar region, with the respiratory regions indicated. cc, crista cerebellis; cer, cerebellum; dt, descending root of Vth cranial nerve; fl, facial lobe; flp, fasciculus longitudinalis posterior; gt, secondary gustatory tract; mn, motor nuclei of cranial nerves; vl, vagal lobe. From Hughes and Shelton (1962).

the motor neurons, so that there is only slight justification for making a distinction on functional grounds. The central respiratory drive potentials (Eccles et al., 1962) recorded intracellularly in spinal motor neurons are almost certainly derived directly or via interneurons from activity within the medullary center. To the extent that the latter is independently rhythmic it is therefore possible to treat it separately.

In the teleosts examined, rhythmic discharges of a respiratory nature also occurred in the smaller cells of the motor nuclei (von Baumgarten and Salmoiraghi, 1962) and in the reticular cells around the nuclei (Shelton, 1961). There was evidence that the active cells were not uniformally distributed throughout the respiratory area. In general, the region which was most likely to result in a successful insertion of the searching electrode was immediately behind the cerebellum slightly to either side of the midline (Shelton, 1961). If the search was carried on anterior or posterior to this region, the chances of making a successful electrode insertion were decreased. This could account for some of the discrepancies in different investigators' results. The full extent of the area defined would depend very greatly on the number of electrode insertions made and, if the distribution varies between animals, on the number of fish studied. There was, in fact, reason to suggest that the respiratory area was not constant from fish to fish or in the same fish at different times (von Baumgarten and Salmoiraghi, 1962; Shelton, 1961). For example, it was often the case that activity once located could not be relocated by positioning the electrode in the same region at a later time. Von Baumgarten and Salmoiraghi (1962) suggested that the extent of the area depended upon the depth of respiration. Certainly there is good evidence that the behavior of neurons can alter with this type of variation. Electrodes chronically implanted in the brains of tench or African perch detected changes both in activity, and in the number of units active, when breathing was stimulated by lack of oxygen or high levels of carbon dioxide (Young, 1969). It seems very likely that the number of neurons, if not the general area they occupy in the medulla, reflects changes in ventilation, both tending to increase or decrease together.

There is very little information available on the central coordination of respiration in elasmobranchs. The early work of Hyde (1904) and Springer (1928), based on ablation and transection experiments, has not been extended. Hyde believed that independent segmental units associated with the VIIth, IXth, and Xth nuclei existed in the skate medulla but this was questioned by Springer. Both workers showed that the medulla was capable independently of coordinating rhythmic breathing, although Springer claimed that some supramedullary influences were important. Satchell's more recent transection experiments (1959) demonstrated that the mid- and forebrains could be removed without causing marked change in the breathing pattern of the dogfish. After transection the medulla was more susceptible to reflex inhibition, but in the absence of reflex effects the medulla maintained rhythmic activity.

C. The Functional Organization of the Respiratory Center

It has been assumed in the foregoing account that the population of respiratory neurons in the fish medulla is a large one. The fact that a number of widely separate, active sites with several neurons at each site can be found in a single fish suggests that this is so. Since the population is a changing one, however, it is very difficult to reach an accurate estimate of its size. Indeed, on this evidence it could be quite small if one is prepared to accept rapid shifts from one locus to another. Fortunately, it is possible to show that there is a little more stability than this and some units can be held for hours, or even days, without showing substantial change (Shelton and Young, 1965). Furthermore, it is possible to destroy respiratory neurons around the tip of a recording electrode by electrocoagulation and for rhythmic breathing to go on normally (Shelton, 1961). Localized coagulation of this type was never effective in bringing breathing to a halt, other than momentarily, even though as many as five such sites were destroyed in a single fish. This argues in the favor of a large rather than a small population.

1. INTRINSIC RHYTHMIC ACTIVITY IN THE RESPIRATORY CENTER

The early work of Adrian and Buytendijk (1931) is often cited as demonstrating the autonomous nature of the fish center. They recorded slow potential changes from the medulla of a completely isolated goldfish brain. The similarity in frequency between these potentials and breathing movements of an intact fish was slender evidence for the former's respiratory function. More recently, Schadé and Weiler (1959) also detected slow potentials on the surface of the goldfish medulla. In the intact animals they occurred in rhythm with the breathing movements and persisted after the fish had been paralyzed by Intocostrin injection. Although the precise nature of these slowly changing surface phenomena is still in doubt, their connection with some aspect of respiratory activity seems now to be well established. Using needle electrodes, Hukuhara and Okada (1956) demonstrated autonomous activity more directly within the cells of catfish and carp respiratory centers. The rhythmic bursts of action potentials, which identified the cells as respiratory in the intact fish, continued after medulla isolation by nerve and brain transection.

As further evidence of the center's independence of mechanoreceptor feedback, a number of workers have described the effects of paralyzing doses of neuromuscular blocking agents such as curare, succinylcholine, and gallamine. In such preparations not only can rhythmic proprioceptive feedback be eliminated but also the ventilation can be controlled entirely by the experimenter. There seems to be little doubt that the respiratory neurons of mammals continue to fire rhythmically after paralysis of the musculature (Cohen and Wang, 1959; Salmoiraghi and von Baumgarten, 1961), although some workers have shown a change in the pattern of activity (von Baumgarten and Kanzow, 1958). In fish there seems to be more variation. Von Baumgarten and Salmoiraghi (1962) sometimes recorded an unaltered discharge and sometimes a slight increase in the frequency of rhythmic bursts from respiratory cells of goldfish after succinylcholine paralysis. Satchell (1959) found a similar increase in the frequency of bursts in efferent fibers from the dogfish center after administration of curare. Precisely opposite results were described by Serbenyuk (1965) working on carp. Paralysis of the respiratory muscles produced either by curare injection or division of the motor nerves at first caused no change in respiratory neuron activity. After a few minutes the normal rhythmic pattern disappeared and was replaced by activity of a different type at a much lower frequency. Serbenyuk (1965) suggested that sensory activity from a variety of sources could affect cells of the center by raising both their general excitability and their frequency of oscillation. Removal of any one of his "reflexogenic zones," such as muscle proprioceptors, had a depressant effect on the center more or less equivalent to the removal of any other of the zones, such as gill or swim bladder receptors.

These results on immobilized fish are not easy to interpret for several reasons. They depend on a pre- and postparalysis comparison of activity from a single recording site within the medulla. It has therefore been difficult to decide whether the changes are representative of the center as a whole. Ballintijn (1969) has recently described three categories of response in respiratory neurons after gallamine induced paralysis. He found that some cells continued to fire rhythmically, some stopped firing completely, and a third category stopped but could be started again by appropriate, but not necessarily rhythmic, proprioceptor stimulation. Clearly the relationships that different cells have to the central oscillator and to the proprioceptive feedback are quite variable (Fig. 15). Moreover, immobilization of an animal affects many systems in addition to the sensory feedback from muscles. For example, the pattern of water

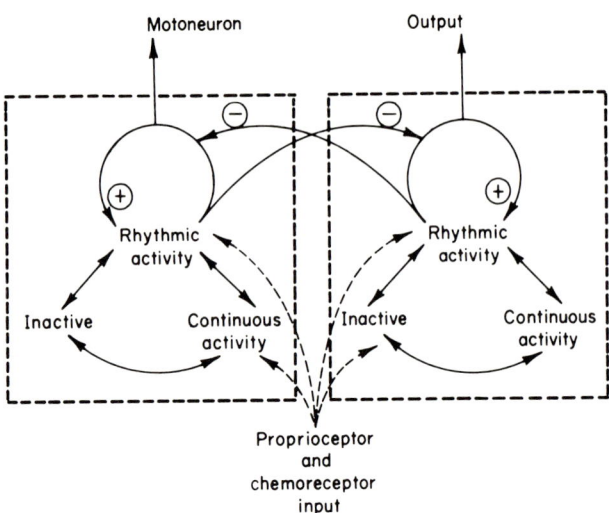

Fig. 15. Diagram showing a hypothetical functional arrangement of the respiratory center. Two populations of rhythmically active neurons are shown, each population being self reexciting and having an inhibitory action on the opposing population. The rhythmically active cells are influenced by sensory input and directly control motoneuron output. In addition, the populations contain cells whose activity may be or may become nonrhythmic, depending on the nature of the proprioceptor and chemoreceptor input.

flow through the gills is obviously changed, as is the level of oxygen and carbon dioxide in both gills and blood. Shelton (1966) has done some preliminary work to examine the effect of such changes in tench paralyzed by curare or succinylcholine. No respiratory units were found in the medulla that were not susceptible to alteration in the flow of water through the gills from a cannula placed in the mouth. Some discharged rhythmically when the flow was stopped, others only when it was restarted. Usually neurons were silent when the rhythmic activity was not produced, although occasionally a continuous discharge was seen. A smooth transition from rhythmic to nonrhythmic state could usually be produced by appropriate adjustment of flow. The influence of different gas tensions in the water has not been fully examined. Although there is little doubt that rhythmic discharges can be detected in the paralyzed fish, the evidence that the functional integrity of the respiratory center is preserved is not convincing. In fact, the reverse seems to be the case in that different parts behave in quite different ways. It would be interesting to know how this variation affects motor activity to the respiratory musculature.

2. The Production of Rhythmic Activity

A satisfactory theory cannot yet be produced to account for the generation of the normal breathing rhythm. A single pacemaker neuron or neuron group cannot be responsible since breathing cannot be stopped by destruction in any particular region. There must therefore be interaction between a number of dispersed respiratory neurons. Since two fairly distinct neuron populations have been described in mammals (Burns and Salmoiraghi, 1960; Salmoiraghi and Burns, 1960a,b) and in fish (Shelton, 1961), it has been suggested that the interaction takes the form of alternating and reciprocally inhibiting discharges between the two groups (Fig. 15). A parallel can be drawn with the two triodes in a multivibrator circuit: when one becomes maximally conducting the other is cut off. Each group is thought to be self reexciting so that activity increases in individual neurons and spreads throughout the group. In order to explain how the balance alternates it must be supposed that activity is in some way self-limiting; in the multivibrator analogy there must be some phenomenon which corresponds to the extreme positions of the triodes and to the discharge of the coupling condensers setting off the opposing cycle. In some way the active group must reach saturation and its activity thereafter decline. It is possible to suggest a number of ways in which this could happen. Salmoiraghi and von Baumgarten (1961) described a small but progressive accommodation of the postsynaptic membrane in both inspiratory and expiratory neurons of the cat. As a burst of activity proceeded, slightly greater depolarization of a neuron membrane was required to initiate an action potential. If all the cells of an inspiratory or expiratory group behaved in this way there would be a substantial fall in excitability throughout an activity cycle and eventually it would be brought to an end. These observations, which were made on very few cells because of the difficulty of impaling respiratory neurons with intracellular electrodes, do not exclude other ways of terminating an inspiratory or expiratory discharge.

These suggestions are extremely speculative and raise a number of difficulties. In the mammal there is evidence conflicting with that given above to show that rhythmicity arises, not by an equivalent interaction of two opposing neuron groups, but by periodic inhibition of a primary, tonic activity occurring in the inspiratory neurons. Centers other than the expiratory and inspiratory groups of the medulla are thought to be involved in these processes (Wang *et al.*, 1957; Kahn and Wang, 1967; Wyss, 1954). In fish there seem to be no neuron groups other than the two medullary ones, but one of these groups (that which fires as the mouth and opercula are adducted) is numerically very much the dominant one. It may be that two opposing groups are unnecessary for oscillation; certainly there is no *a priori* reason why a single neuron group should not show alternating periods of activity and quiescence. At some level of integration bursts of activity, which do alternate in a general way, must be produced since the discharges from motor neurons are known to be of this nature. None of the suggestions made so far accounts satisfactorily for the extensive variation in behavior occurring in different neurons in both the normal and the immobilized animal. This variety is particularly difficult to explain in a scheme which assumes two fairly uniform and opposing populations. Some of the components in the oscillator must be different to the extent that they can become silent or continuously active under some conditions without causing complete disintegration of the rhythmic system (Fig. 15).

IV. THE ROLE OF MECHANORECEPTORS IN RESPIRATORY REGULATION

Although the evidence reviewed above suggests that the respiratory center can function in a rhythmic manner in the absence of feedback from mechanoreceptors, there is no doubt that they have a significant part to play in the normal fish. In elasmobranchs it has been known for some time that the breathing rate is in some way related to the volume of water which is allowed to flow over the gills from cannulae in the mouth or spiracles (Lutz, 1930; Ogden, 1945). Satchell (1961) confirmed these results and showed that the response was produced as a result of mechanoreceptor stimulation by the movements of ventilation. It was totally independent of the oxygen content of the inspired water.

Sharks decrease their breathing rate by prolonging the period of oro- and parabranchial expansion and extending the pause after this inspiratory period; the opposite occurs when the breathing rate goes up. Satchell (1959) showed that stimulation of the branchial branches of the IXth and Xth cranial nerves, or the prespiracular branch of the VIIth, stopped or slowed breathing by increasing the duration of the pause. Inflation of a balloon inside the pharynx produced a similar result. It was suggested that these experiments demonstrated the existence of inhibitory pathways to the respiratory center in the branchial nerves. It also seemed clear that the pathways originated in pharyngeal mechanoreceptors, and support for this came from recordings of sensory discharges in the branchial nerves during contraction of oro- and parabranchial cavities. Satchell and Way (1962) examined activity from the pharyngeal mechanoreceptors in more detail. They were found on the branchial processes lining the inner opening of the gill slits and were slowly adapting sense organs with both static and dynamic characteristics. The effective stimulus was thought to be displacement of the pharyngeal wall rather than water flow itself. The branchial processes were thus deflected by being pressed against the gill bar in front. Satchell and Way suggested that the receptors were involved in a breath by breath regulation of breathing rate, the duration of the inspiratory period and the following pause being determined by the level of activity during the preceding expiratory period. Thus in an almost completely curarized animal the breathing rate became largely independent of imposed ventilation volume, suggesting that the mechanoreceptors were not excited by abnormally small breathing movements. In the normal animal, however, the decrease in rate as ventilation was progressively restricted resulted from greater excitation of the mechanoreceptors because the amplitude of breathing increased. The original observations of the relationship between water flow and breathing rate must now be explained in terms of a change in amplitude of breathing. The receptors and pathways involved in this response are not known. The inhibitory reflex is unusual in that the receptors are excited during orobranchial contraction and decrease breathing rate by delaying the onset of the next expiration phase after a longer inspiration and pause.

Section of the branchial branches of the IXth and Xth cranial nerves in the shark caused an increase in breathing rate; inspiration was quicker and the pause disappeared (Satchell, 1959). This correlates well with the proposed inhibitory role of the mechanoreceptors. In teleost fish the effects of IXth and Xth nerve section are not so well established. Shelton (1959) showed that the immediate change caused by sectioning the nerves in tench was an increase in amplitude of the breathing movements. The nature of the receptors and the function of the afferent pathways involved in this response were not defined. Powers and Clark (1942) had earlier done similar experiments and had found that breathing failed after section of nerves IX and X. The gill blood vessels were also sectioned in their experiments. Effects other than the direct ones of nerve section may therefore have been complicating the results, although the respiratory center seems to be quite resistant to circulatory failure (Serbenyuk, 1965). Serbenyuk et al. (1959) and Serbenyuk (1965) found that rhythmic activity persisted in the center after section of the branchial nerves or novocaine anesthesia of the gills in carp, although there was a considerable change in breathing pattern. There can be little doubt that important afferents, probably from mechanoreceptors as well as other sense organs, are carried in the branchial nerves of teleosts although their role cannot at present be precisely defined.

V. THE CHEMICAL REGULATION OF RESPIRATION

The metabolic rate in fish, as measured by oxygen consumption, can be influenced by a large number of internal or external factors (Fry, 1957). Temperature and size (Job, 1955; Beamish and Mookherjii, 1964; Beamish, 1964a), pollution of various types (Erichsen Jones, 1964), season (Beamish, 1964b), and activity (Brett, 1964; Stevens and Randall, 1967b) have been examined in some detail. Although they all change the animal's demand for oxygen there is no evidence, except perhaps in the case of changes in activity (Stevens and Randall, 1967a), that they are directly involved in the modifications of ventilation and perfusion which result in the demand being met. It seems likely that factors such as these operate primarily, although not exclusively, at the level of the cell. Other changes, most notably of oxygen concentration, carbon dioxide concentration, and pH level in the ventilation stream and/or the blood, also influence oxygen consumption but in a more complex way. All three factors may be involved to varying degrees in mechanisms important in the regulation of gas exchange. The degree of internal change in the oxygen, carbon dioxide, or pH levels as a result of environmental fluctuations in these parameters would therefore depend on the characteristics of the total regulatory system. Variations in metabolic rate in this case would be a function of the metabolic cost of regulation, the overall efficiency of regulation, and ultimately the direct effect of any changes at the tissue level, although these would be small if the regulation were efficient. It is probable that regulatory systems of this general type, sensitive to respiratory gas concentrations within the body, and perhaps in the environment as well, would be directly involved in the responses which cope with the respiratory requirements of the cells under a wide variety of different conditions, including those of temperature, size, season, and activity mentioned above.

A. Oxygen

1. EFFECT OF OXYGEN CONCENTRATIONS ON OXYGEN CONSUMPTION AND GILL VENTILATION

a. Respiratory Dependence. Progressive depletion of oxygen in the water breathed by a fish at first produces no change (respiratory independence) but eventually leads to a greater and greater restriction of

8. THE REGULATION OF BREATHING

the animal's oxygen consumption (respiratory dependence). Activity is usually included in the concept of respiratory dependence thus increasing its ecological significance by giving some insight into the equilibrium between environment and active fish. Determinations can be made over the whole range of activities shown by the animal, but for comparative purposes the maximally active and the standard (resting) level are most useful (Fig. 16). Fry (1947, 1957) pointed out that a decline in oxygen consumption of a maximally active fish would indicate a similar decline in activity if the animal were to remain in oxygen balance. This would be a progressive effect from the "incipient limiting level" at which reduced oxygen concentration just produced a change to the "level of no excess activity" where oxygen consumption equaled the standard rate. In the case of a resting animal, independence should be maintained at a steady level over a wider range of oxygen concentrations, and at lower levels the consumption might be expected to go up as the work of ventilation increased. Fry (1947) suggested that the cost of ventilation could continue to rise until the standard oxygen consumption equaled the reduced active rate.

It has been widely accepted that limitations in the gas exchange system are largely responsible for the appearance of respiratory dependence. There are, however, a number of other factors which will complicate a possible relationship of this type. Rates of oxygen consumption, both at the active and standard levels, are different for different species of fish. Suggestions have been made that the species having the highest rates are also those which show dependence in the highest concentrations of oxygen (Fry, 1957; Basu, 1959) so that the onset of dependence is related to the ability of the fish to be active. Determinations of active oxygen consumption made for any particular species by different authors (see, for example, Job, 1955; Basu, 1959; Saunders, 1962; Beamish, 1964c) show considerable variation, and it is difficult to substantiate this suggestion, reasonable though it seems. The standardization of activity levels in the experimental animals constitutes a major difficulty in these determinations. Another difficulty is that fish such as the dragonet, *Callionymus*, and the toadfish, *Opsaunus*, which are inactive, bottom-living forms, show dependence at oxygen values only just below air saturation (Hall, 1929; Hughes and Umezawa, 1968b).

Anaerobiosis and acclimation to different oxygen tensions may also affect respiratory dependence curves. Beamish (1964c) demonstrated that at very low levels of oxygen a fish's standard rate of oxygen consumption increased as predicted, but never to the active level (Fig. 16). Furthermore, the increase was usually less in animals acclimated to the low tensions concerned than in those acclimated to air saturation

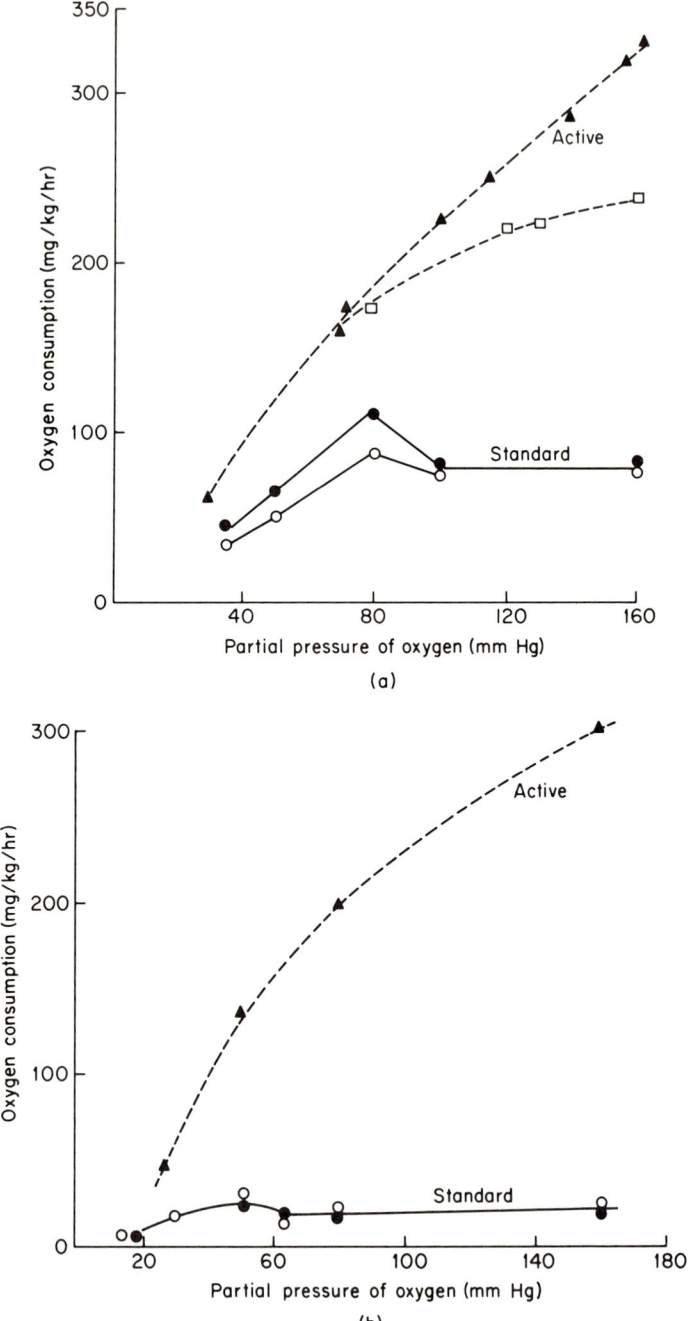

values. Beamish concluded that anaerobic respiration occurred in reduced oxygen concentrations. He was able to show that fish which were active to the same extent consumed less oxygen at lowered tensions than they did in air-saturated water. Fish acclimated to low oxygen tensions not only had lower standard rates of metabolism at those tensions but also showed reduced consumption for any level of activity when compared with their fellows acclimated at air saturation values. The acclimation appears to have some influence on the extent to which a fish can indulge in anaerobic respiration. The accumulation of lactic acid in the muscles and bloodstream of trout and sunfish during exercise (Black et al., 1962) and hypoxia (Heath and Pritchard, 1965) and its subsequent return to normal during a prolonged period of elevated oxygen consumption are well established. Blazka (1958) produced evidence for anaerobiosis during hypoxia in the crucian carp with no oxygen debt or accumulation of lactate.

b. Ventilation Volume and Utilization. Although the factors outlined above complicate the relationship between oxygen concentration, activity, and respiratory exchange, it is very likely that the gills and ventilating mechanism do tend to become limiting during respiratory dependence. The failure may result from deficiences in the measurement and control systems or from physical limitations on the part of the respiratory pump and exchanging surfaces. The augmented volume of water must be sufficient not only to cope with the reduced oxygen content but also to compensate for the increased work involved in ventilation and the reduced utilization of oxygen in the ventilation stream. Indeed, it has been argued (Hughes and Shelton, 1962) that the increased metabolic cost of breathing and the fall in utilization may together determine the effective upper limit for gill ventilation rather than the capacity of the pump itself. This suggestion was based on an unsubstantiated estimate of ventilation work and an extrapolation (shown by the dotted line in Fig. 17) of van Dam's data (1938) from a single trout recovering from exercise in air-saturated water. The fall in utilization in this case

Fig. 16. The effect of changes in partial pressure of environmental oxygen on standard and active oxygen consumption of fish: (a) trout, *Salvelinus fontinalis.* 10°C. Active rates taken from (▲) Basu (1959) and (□) Job (1955). Standard rates taken from Beamish (1964c); (●) values for fish acclimated to air saturation and (○) fish acclimated to each of the partial pressures of oxygen used. The standard rates significantly higher at low oxygen tensions for those fish acclimated to air saturation. (b) Carp, *Cyprinus carpio.* 10°C. Active and standard rates taken from Beamish (1964c). Symbols for standard rates as before; no significant difference exists in this case between the two differently acclimated groups. Reproduced by permission of the National Research Council of Canada from the *Can. J. Zool.* **42**, 355–366.

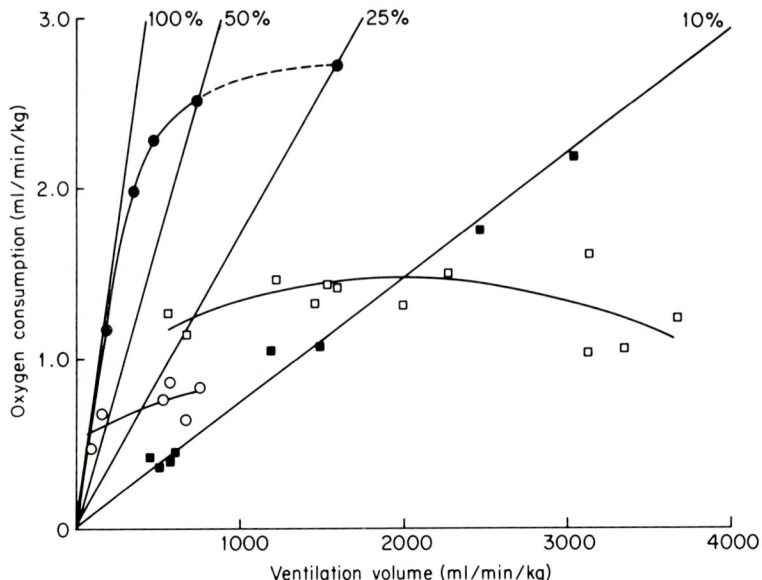

Fig. 17. Relationship between oxygen consumption and ventilation volume in fish under different conditions. (●) Trout (van Dam, 1938) recovering from exercise in aerated water at 12°C. The lines indicate relationships between ventilation and oxygen consumption which would exist in fully aerated water at this temperature for values of utilization of 100, 50, and 25%. van Dam's results are extrapolated to the 25% utilization line as explained in the text. (○) Eel (van Dam, 1938) during hypoxia at 18°C. (■) Trout (Stevens and Randall, 1967b) before, during, and after a 10-min period of exercise in aerated water at 5°C. The 10% utilization line is drawn for fully aerated water at this temperature. (□) Trout (Holeton and Randall, 1967b) during reduction of environmental oxygen to a partial pressure of 30 mm Hg at 15°C.

could very well lead to a maximum oxygen intake being reached at ventilation volumes 8–10 times the resting level, but the conclusion rests precariously on results which van Dam himself regarded as a "first orientation."

At this point it may be useful to digress briefly and discuss the measurement of ventilation and utilization, their usefulness in assessing respiratory function, and their relationships to one another and to other parameters which have been used in respiratory studies on aquatic animals. Ventilation has been measured by three methods, all of which have inherent faults. The first method depends on separating inspired from expired water, usually by using a rubber membrane stretched across part of the head (van Dam, 1938; Hughes and Shelton, 1958) or glued over the external gill opening (Piiper and Schumann, 1967;

Hughes and Umezawa, 1968a,b). Using suitable siphons the expired water can be collected for measurement. The method has the disadvantage of restricting the fish and placing a load on the muscles of the branchial pump. A considerable effect of this type of system on the oxygen consumption of *Callionymus* has been demonstrated by Hughes and Knights (1968). Ventilation can be calculated from measurements of the oxygen consumption of a fish if the oxygen content of the inspired and expired water is also known. The method is suitable for work on freely moving fish from which samples can be taken through buccal and opercular catheters (Saunders, 1962; Holeton and Randall, 1967a). Inaccuracies will occur if the samples are not representative, and relatively small changes in oxygen concentration in considerable volumes of water are not easily measured. A dye dilution technique has been described by Millen *et al.* (1966). A bolus of dye is injected from a syringe into the buccal or orobranchial cavity, and continuous sampling from opercular or parabranchial chambers into a densitometer gives a dye dilution curve. From the downslope of this curve the ventilation volume can be calculated. For this technique to be accurate, the injection and sampling sites must be suitably located and the mixing of the dye, once injected should be complete and preferably instantaneous. Millen *et al.* (1966) achieved good mixing by implanting a stirrer in the orobranchial cavity.

Utilization (U) is the ratio between oxygen extracted from the ventilation stream and oxygen contained in the inspired water, often expressed as a percentage:

$$U = \frac{P_{I_{O_2}} - P_{E_{O_2}}}{P_{I_{O_2}}} \times 100$$

where P is the partial pressure of gas, in I the inspired water, and E the expired water. Its greatest use is probably in examining ventilation and ventilation control because at known oxygen concentrations it relates oxygen consumption and ventilation in a direct way without involving the determination of any parameters in the circulatory system. The relationship between ventilation volume and utilization is not a simple one. This point is illustrated in Figs. 17 and 18 in which data obtained by Holeton and Randall (1967b) and Stevens and Randall (1967b) from the trout, *Salmo gairdneri*, are plotted. Thus the exercising trout maintained a steady utilization of 10% and showed a progressively increasing oxygen consumption as ventilation increased in air-saturated water. The hypoxic trout also increased its ventilation over a similar range, but it had a higher overall utilization falling to about 20% and

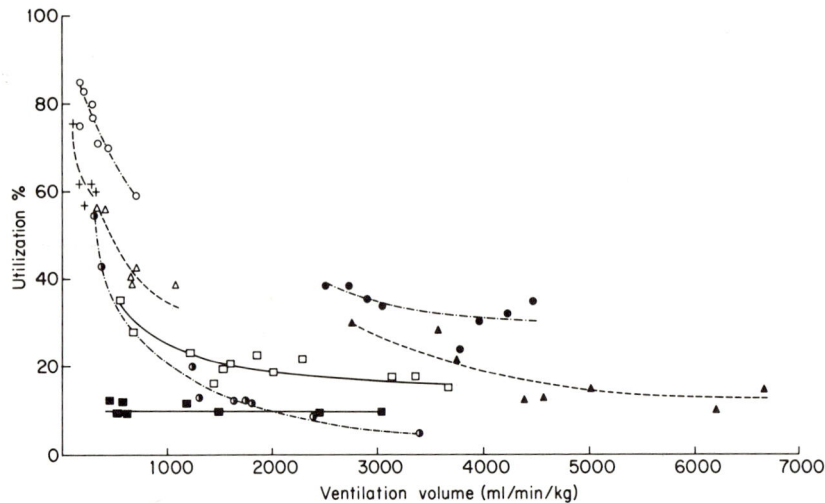

Fig. 18. Relationship between utilization and ventilation volume in fish under different environmental conditions. The conditions which led to increased ventilation are specified below. Carp (Saunders, 1962) 20°C, (○) hypoxia; (●) exercise; and (◐) carbon dioxide excess and hypoxia. Sucker (Saunders, 1962) 20°C; (△) hypoxia and (▲) exercise. Trout (Holeton and Randall, 1967b) 15°C, (□) hypoxia; and (Stevens and Randall, 1967b) 5°C, (■) exercise. Dogfish (Baumgarten-Schumann and Piiper, 1968) 16°C, (+) recovery from activity.

eventually may have shown a declining uptake of oxygen as the environmental oxygen fell. Although utilization usually falls as ventilation increases, the extent of the fall depends on the factors responsible for the ventilation change and the way they affect the oxygen requirements of the animal and the oxygen content of the medium. Saunders (1962) measured utilization in fish which were exercised or exposed to low levels of oxygen in the presence or absence of carbon dioxide all of which affect oxygen requirement and/or oxygen content differently. The utilization and ventilation changes were also different as his results given in Fig. 18 and Table I show. The variation between species shown in Fig. 18 and Table I are also very striking.

It is clear that utilization will be affected by any change which alters the characteristics of the total exchange system. Thus a falling oxygen concentration, in the absence of compensatory changes, has an adverse effect on utilization because the oxygen diffusion gradient goes down at the same time as increased water velocity reduces diffusion time and probably alters gill geometry (Lloyd, 1961; Hughes and Shelton, 1962; Hughes, 1966). In fact, it is extremely likely that some compensation does occur as ventilation increases, thus making the relationship between ventilation and utilization even more complicated. A measure of the gas exchange capacity of the respiratory surface is given by the transfer factor which is the constant of proportionality relating total gas exchange and the mean partial pressure gradient between water and blood. Randall et al. (1967) have shown that the transfer factor in trout increases gradually during hypoxia, and to a lesser extent during exercise, so that the gas exchange capacity of the gills must also be increased.

Although utilization is determined by a combination of ventilation volume and the gas exchange properties of the gills (see chapter by Randall, "Gas Exchange in Fish," this volume, for a detailed discussion), it is so all-embracing as to say little about the details of the exchange. For example, it takes no account of the role of the bloodstream in gas exchange. In this respect effectiveness (E), which is the ratio between actual and maximum possible gas transfer in a countercurrent exchange system (Hughes and Shelton, 1962), can give a more useful measure. Effectiveness can be calculated for oxygen with reference to the transfer either from water or into blood, and similar figures can be obtained for carbon dioxide moving in the opposite direction (Randall et al., 1967; Piiper and Baumgarten-Schumann, 1968b). The relationship between utilization and the effectiveness of oxygen removal from the ventilation stream would be

$$E_{W_{O_2}} = \frac{\dot{V}_{O_2} \text{ (actual)}}{\dot{V}_{O_2} \text{ (maximum)}}$$

$$= \frac{\dot{V}_G}{\dot{V}_G} \cdot \frac{\beta_{O_2}}{\beta_{O_2}} \cdot \frac{(P_{I_{O_2}} - P_{E_{O_2}})}{(P_{I_{O_2}} - P_{\bar{v}_{O_2}})} \times 100 = U \frac{P_{I_{O_2}}}{(P_{I_{O_2}} - P_{\bar{v}_{O_2}})}$$

where P is the partial pressure of gas in I (inspired water), E (expired water), and $\bar{v}$ (mixed venous blood; $\dot{V}_G$ is the volume flow of water over the gills in unit time, $\dot{V}_{O_2}$ the oxygen uptake in unit time, and β_{O_2} the solubility coefficient of oxygen in water. In this case transfer which is 100% effective can be achieved by a system having no water shunt, a very high transfer factor (low diffusion resistance), and any countercurrent arrangement which brings venous blood and outgoing water into contact. If the effectiveness of oxygen uptake by the blood is also to be 100%, the requirements are more stringent still, since a perfect countercurrent arrangement then becomes necessary. In addition, the flow of water and blood must be appropriately balanced so that outgoing blood is either in equilibrium with incoming water or fully saturated. For complete utilization the same conditions will obtain as for effective withdrawal of oxygen from water, but in addition the venous blood must be devoid of oxygen.

Table I

Comparative Data on Ventilation Volume, Utilization, and Oxygen Consumption, Including the Changes Produced in Response to Low Oxygen Concentrations and Exercise

Experimental animal		Oxygen content of inspired water (ml/liter)	Ventilation volume (ml/kg/min)	Stroke volume (ml/kg)	Breathing rate (No./min)	Utilization (%)	Oxygen consumption (ml/kg/min)	Reference
Elasmobranchs								
Scyliorhinus stellaris (16°C, 2180 g)	Mean	6.7	195	4.8	41	62	0.64	Baumgarten-Schumann and Piiper (1968)
	Range		124–322			73–60	0.51–1.02	
Squalus suckleyi (11°C, 3500 g)		6.5	500			25	0.71	Lenfant and Johansen (1966)
Teleosts (Oxygen-induced changes)								
Eel, *Anguilla* (17°C, 400 g)		6.6	89	5.6	16	82	0.48	van Dam (1938)
		2.1	762	23.8	32	53	0.83	
Trout, *Salmo* (10°C, 900 g)		7.6	150	1.8	82	80	0.90	van Dam (1938)
		2.1	519	5.2	100	63	0.67	
Trout, *Salmo* (15°C, 400 g)		6.8	556	6.9	80	35	1.27	Holeton and Randall (1967b)
		1.8	3350	31.3	107	18	1.05	

8. THE REGULATION OF BREATHING

Trout, *Salmo* (12°C, 115 g)	7.3	—	95	—	1.20	Marvin and Heath (1968)
Tench, *Tinca* (15°C, 70 g)	2.2	—	43	—	0.43	Randall and Shelton (1963)
	6.2	—	30	—	—	
Sucker, *Catostomus* (20°C, 250 g)	0.4	—	43	—	—	
	5.2	344	Increase →	—	1.00	Saunders (1962)
		12,950	1.6 × increase	—	—	
Bullhead, *Ictalurus* (20°C, 186 g)	5.1	291	—	56	1.00	Saunders (1962)
		7420	—	7	—	
Carp, *Cyprinus* (20°C, 160 g)	6.0	187	—	68	0.96	Saunders (1962)
		710	—	13	—	
				85		
				59		

Teleosts (Exercise-induced changes)

Trout, *Salmo* Pre-exercise (5°C, 400 g) During exercise	7.2	594	7.71	77	10	0.43	Stevens and Randall (1967b)
	7.2	3042	30.7	99	10	2.18	
Sucker, *Catostomus* (20°C, 209 g)	5.3	6080	—	—	18	5.75	Saunders (1962)
Carp, *Cyprinus* (20°C, 237 g)	6.4	3520	—	—	33	7.38	Saunders (1962)

The data at present available do not allow firm conclusions to be reached about the effective upper limit of ventilation which would restrict the availability of oxygen to the body. If the needs of the whole body, except for the respiratory musculature, are considered then falling utilization, increased work of breathing, and limitations in the stroke volume and rate of operation of the branchial pump all operate to determine that upper limit. Little is known of the work factor which in any case will differ in moving and nonmoving fish. Even if a simpler view is taken and the changes in the work of breathing are ignored the situation is still far from simple. Since ventilation–utilization relationships are variable, the balance between factors will change with conditions. In none of the experiments (van Dam, 1938; Saunders, 1962; Holeton and Randall, 1967b) in which hyperventilation was the result of low environmental oxygen did utilization decrease more rapidly than ventilation increased (Fig. 18), as would have been the case if utilization alone had been limiting. In order to maintain oxygen consumption at a steady level it is necessary for the ventilation volume to go up at the same rate as the product of oxygen concentration and utilization declines. This just failed to be the case in the trout examined by Holeton and Randall (1967b) which showed signs of dependence below oxygen tensions of 80 mm Hg. It is significant that ventilation was not maximal but increased almost by a factor of two as the oxygen concentration was decreased further. The inference from this is that the control systems may also participate in the failure to maintain oxygen availability before the pump has reached its maximum capacity.

2. Coordination of the Responses to Changes in Oxygen Concentration

The evidence cited above shows that in the majority of teleosts examined depletion of oxygen in the medium results in an increased ventilation volume over a wide range. Other reactions, for example, avoidance behavior, have also been described in a number of teleosts (Erichsen Jones, 1952; Whitmore et al., 1960); but it seems likely that the respiratory response is the primary one. The situation is not so clearly resolved in the case of elasmobranchs. According to Ogden (1945) a decrease of environmental oxygen produced no change in the ventilation volume of the dogfish, Mustelus, although the animal struggled violently under these conditions. Satchell (1961) showed that perfusion of dogfish, Squalus acanthias, gills with completely deoxygenated water caused only a very slight increase in amplitude, together with a decrease in the rate of breathing. Because of the extreme nature of the stimulus this work did not rule out the possibility of more appropriate ventilation responses to moderate anoxia. However, the more recent work of Hughes and Umezawa (1968a) showed only a small increase in breathing rate in *Scyliorhinus canicula* during a gradual reduction of oxygen tension in the inspired water. Similarly, Baumgarten and Piiper (1969) found only small changes of ventilation volume in *Scyliorhinus stellaris* under these conditions. Because simple experimental procedures have so far been ineffective in causing substantial and reproducible modifications of ventilation in elasmobranchs very little is known of its range and control. There is obviously not a fixed ventilation volume; Baumgarten-Schumann and Piiper (1968) reported marked changes in resting dogfish (see also Table I) and commented that the animals were not nearly at their maximum ventilation levels. They attributed the ventilation changes they observed to repayment of an oxygen debt after excitement and activity. Whether some product of metabolism constitutes the effective stimulus to ventilation change in elasmobranchs cannot be decided on present evidence. It is possible that the level of activity in the animal is in some way directly coupled to breathing, but this cannot be the whole explanation since an inactive fish shows variations in ventilation. The whole field of respiratory regulation in elasmobranch fish is poorly understood and clearly requires closer study, particularly on freely breathing preparations.

In those fish in which a fall in oxygen concentration does result in increased ventilation, the environmental changes must influence the respiratory center in such a way as to cause changes in the motor output and increase the rate and/or the amplitude of breathing. The relationship between rate and amplitude has been shown to vary considerably from species to species (Table I). Very little is known of the underlying neurophysiology directly responsible for these alterations in breathing. Modifications in the oscillatory activity of cells making up the respiratory center are obviously produced by incoming sensory information related to environmental change. The receptors for this information could be in one or more of three general regions (Fig. 19):

(1) On the outside of the animal, on the walls of the pharynx and the surface of the gills. They would detect oxygen concentrations in the water of the environment or ventilation stream.

(2) On arterial blood vessels containing blood after oxygenation in the gills or in tissues closely associated with arterial blood. They would detect oxygen concentration in the bloodstream.

(3) On blood vessels carrying venous blood or in tissues closely associated with venous blood. They would detect oxygen concentration in the blood after it has passed through the tissues of the body. This

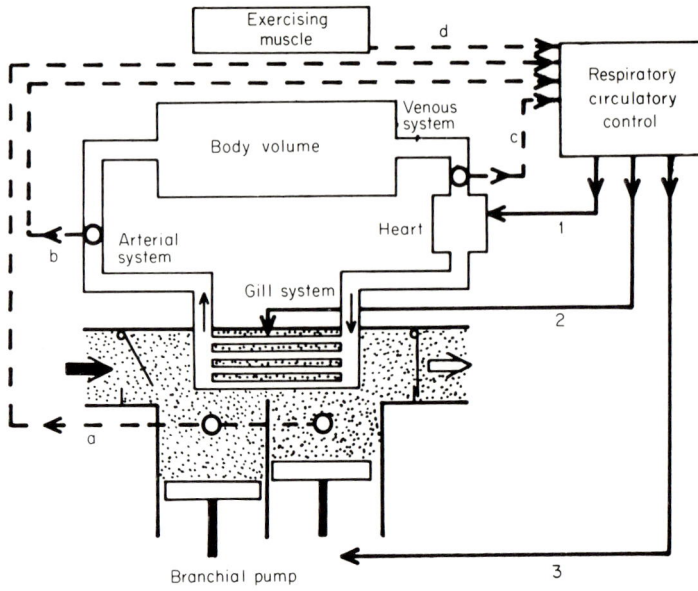

Fig. 19. Diagram of the peripheral relationships of control systems regulating ventilation, perfusion, and gas exchange in fish. Three possible receptor sites for gas tension changes are shown, located in (a) ventilation stream, (b) arterial, and (c) venous blood as suggested in the text. Receptors detecting exercise levels (d) may also be important. Information provided by these sensory channels may lead to the modification of (1) cardiac output, (2) functional gill area, and (3) ventilation volume, as shown.

region would include the ventral aorta and afferent branchial arteries, taking in sites which are homologous with those where carotid and aortic glomi are found in the mammal.

It is not easy to decide either experimentally or theoretically which of these possible sites is in use in the fish. Satchell (1961), working on dogfish, and Randall and Smith (1967), working on rainbow trout, claimed that oxygen receptors were to be found in the buccal cavity or on the gills. The evidence in both cases was that ventilation with oxygen-poor water caused a reflex slowing of the heart which was still seen after the ventral aorta had been clamped. The receptors detecting the low oxygen concentration must have been directly available to the ventilation stream since no blood could flow to more remote sites. Hughes and Umezawa (1968b) also proposed that the receptors for bradycardia induced by oxygen lack were located on the gills. In addition, Hughes and Ballintijn (1968) thought that the increased breathing amplitude during hypoxia resulted from activation of receptors which detected oxygen concentrations in the inspired water. De Kock (1963) has described "end buds" in the buccal cavity and on the gills of trout and has suggested that they have a chemoreceptor function. However, control based solely on detection of environmental oxygen would entail a precise relationship between different concentrations of oxygen and the corresponding ventilation volumes with no feedback adjustment. Even supposing that this relationship could be modified appropriately at different levels of exercise, it would be unlikely that such a system could give stable control of oxygen supply under all conditions, especially with varying gill geometry and perfusion.

It is probable that such stability can only be conferred by a control system with receptors in the bloodstream, and the situation in mammals argues strongly for some such system. If the receptors were located in the vessels of the gills, control would be susceptible to local variations in geometry and perfusion rates. Integration of activity from receptors throughout the gills would overcome this difficulty. More satisfactorily, the receptors could be situated downstream from the gills so as to measure oxygen concentration in mixed arterial blood. The pseudobranch has frequently been considered as a region where receptors might occur. Indeed, Laurent (1967) demonstrated a sensitivity to changes in oxygen and carbon dioxide concentrations by recording from the nerve of an isolated and perfused tench pseudobranch. This does not mean that all the receptors need be located in a single site. Shelton (in Hughes and Shelton, 1962) showed that bilateral section of the IXth cranial nerves did not abolish the ventilatory response to oxygen lack although it was considerably modified. This experiment was repeated with section of the Xth cranial nerves and IXth and Xth nerves together, and in all cases the fish was able to respond to low oxygen levels. There is thus clear evidence that oxygen receptors exists outside the areas innervated by the IXth and Xth cranial nerves.

A suggestion that the receptor system could be located in the venous blood system was made by Taylor *et al.* (1968), who based their work on an analysis of the equilibria between environment and blood under a variety of conditions. These workers described a digital computer simulation of the gas exchange system in trout, using the results of Randall and his collaborators. Equations were derived to describe oxygen transport and storage in all parts of the system, together with three feedback equations which controlled ventilation volume, heart output, and gill area. All the equations could be simultaneously solved for blood and environmental gas tensions, ventilation volume, and heart output, by using different input driving values for metabolic rate or inspired oxygen tension. Realistic simulations could not be achieved if arterial or venous carbon dioxide tensions were used as reference factors in the feedback equations. Equally, the use of arterial oxygen tensions produced responses in the model which did not correspond with those of the fish. However, if venous oxygen tension was used as a reference to control ventilation, blood flow, and gill area, the model responded in a way which resembled the resting and exercising trout and had a qualitative similarity to the hypoxic animal. The model has some shortcomings, for example, in relating cardiac output directly to ventilation, but in spite of these it is reasonably consistent with experimental evidence. It is not yet clear that the venous location for the receptor system is justified. The dissociation curve would normally be steepest at venous oxygen tensions, and there may be disadvantages in locating the error detector where the error is smallest.

There are therefore reasons to believe that oxygen receptors may be found in the ventilation stream and in arterial and venous blood. Much of the evidence needs more detailed experimental examination but, if confirmed, points to a control system of greater complexity than hitherto supposed.

B. Carbon Dioxide

1. Effects of Carbon Dioxide Concentrations on Oxygen Consumption, Gill Ventilation, and Utilization

The precise role played by changes in carbon dioxide concentration, either in the environment or within the body of a fish, in regulating breathing is not yet understood. Various effects of increased environmental carbon dioxide have been described. At very high levels the gas is toxic and completely inhibits oxygen consumption. Below these levels fish survive for varying periods of time, although the ability of an animal to extract oxygen from the environment is reduced to an extent which depends on the amount of carbon dioxide present (Black et al., 1954) or on hydrogen ion concentration (Anthony, 1961). Basu (1959) found that the rate of oxygen consumption by active suckers, bullhead, and carp fell as carbon dioxide levels increased, the logarithm of the consumption rate being linearly related to carbon dioxide tensions up to 50 mm Hg. The same linear relationship was found for trout, using carbon dioxide tensions up to 25 mm Hg (Fig. 20). The results on carp were confirmed by Beamish (1964d), who also examined the effects of similar ranges of carbon dioxide on the standard rates of oxygen consumption. He found that carbon dioxide usually produced no significant change in the standard metabolic rate (Fig. 20).

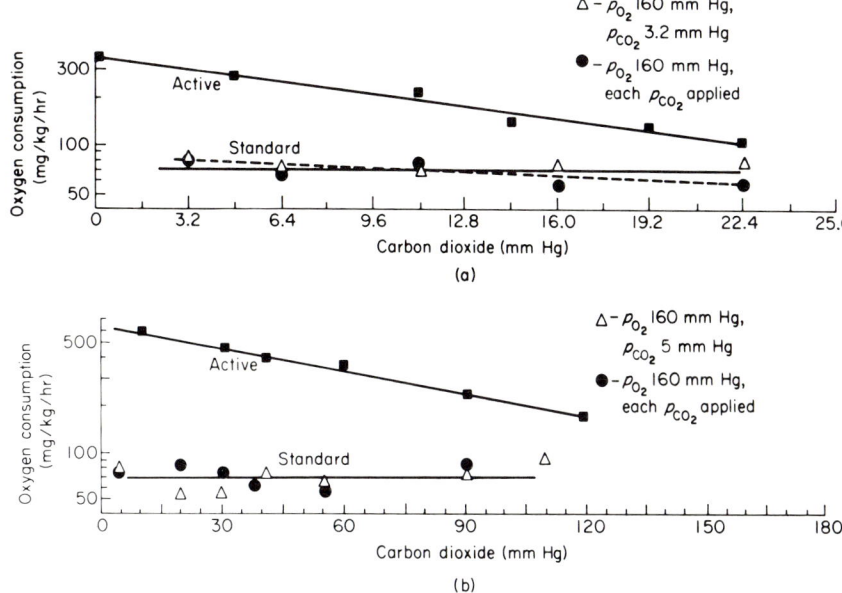

Fig. 20. Standard and active rates of oxygen consumption in (a) brook trout at 10°C and (b) carp at 25°C in relation to elevated carbon dioxide tensions in the environment. Oxygen tension maintained at 160 mm Hg. In plotting the standard rates, closed circles represent results from fish acclimated to each tension of carbon dioxide applied, and open triangles results from fish acclimated to low levels of carbon dioxide. In the trout acclimated to each carbon dioxide tension applied, there is a small but significant fall in oxygen consumption as carbon dioxide increases (interrupted line). From Beamish (1964d). Reproduced by permission of the National Research Council of Canada from the *Can. J. Zool.* **42**, 847–856.

It is not easy to reconcile these results with those of other workers who examined the direct effect of carbon dioxide on gill ventilation. The early experiments of Olthof (1934) and Meyer (1935) showed that carbon dioxide stimulated an increase in breathing frequency in minnow, perch, and stargazer. In the tench (Randall and Shelton, 1963) and trout (van Dam, 1938) the amplitude of breathing movements went up, causing significant changes in ventilation volume in both cases. Saunders (1962) also described increased ventilation in the sucker, bullhead, and carp (Fig. 18) but only with high tensions (25–56 mm Hg) of carbon dioxide. At more moderate tensions (15–35 mm Hg) there was an initial increase in ventilation with a later return to normal. The utilization of oxygen was found to be quite substantially reduced in the presence of carbon dioxide (van Dam, 1938; Hughes and Shelton, 1962; Saunders, 1962) as Fig. 18 shows. The rate at which utilization fell with increased ventilation was also increased in the presence of carbon dioxide, compared with a solely hypoxic stimulus (Saunders, 1962). The effect of these ventilation and utilization changes on the work of breathing might reasonably be expected to increase the standard rate of oxygen consumption. That they did not do so in Beamish's experiments (1964d) could be explained by his suggestion that carbon dioxide does not in fact stimulate the truly resting fish to hyperventilate. Certainly many fish become excitable and active if carbon dioxide is introduced into the environment and measurements on resting animals become difficult.

The significance of these responses can best be examined in the light of some of the physical properties of carbon dioxide. In an important contribution to the theory of aquatic gas exchange, Rahn (1966) demonstrated that because of the differences in solubilities of oxygen and carbon dioxide in water the change in carbon dioxide tension between inspired and expired water must necessarily be small (Fig. 21). Dejours et al. (1968) went on to show that carbon dioxide differences were even smaller than Rahn had predicted if fish breathed in well-aerated, high carbonate water, because of the buffering action of the carbonate–bicarbonate system. For example, the carbon dioxide tension in water expired by a goldfish at 25°C was increased by less than 1 mm Hg. The relationships vary in different types of water, but in all cases the tensions are low and are not substantially increased by gas exchange at the gills. Carbon dioxide tensions in the tissues and bloodstream of fish would also be expected to be low, and this has been repeatedly confirmed by experiment (Lenfant and Johansen, 1966; Holeton and Randall, 1967b; Stevens and Randall, 1967b; Baumgarten-Schumann and Piiper, 1968).

Low tensions in the blood are also attributable to the high permeation coefficient of carbon dioxide (see chapter by Randall, "Gas Exchange in Fish," this volume, Table I). The diffusion gradients between blood and water are very much smaller for carbon dioxide than they are for oxygen. Indeed, Kylstra et al. (1967) have derived diffusion equations for a gill model from which they suggest that diffusion dead space for carbon dioxide may be negligible; that is to say that persisting gradients of tension within exchange units would be very small, assuming realistic values for diffusion time and distance. This would not be the case for oxygen, however, and Kylstra et al. argue that a major difference of this type would relate the rates of oxygen uptake and carbon dioxide elimination volume in quantitatively different ways. The regulation of ventilation and oxygen uptake by reference to arterial tensions of carbon dioxide would therefore be extremely unsatisfactory.

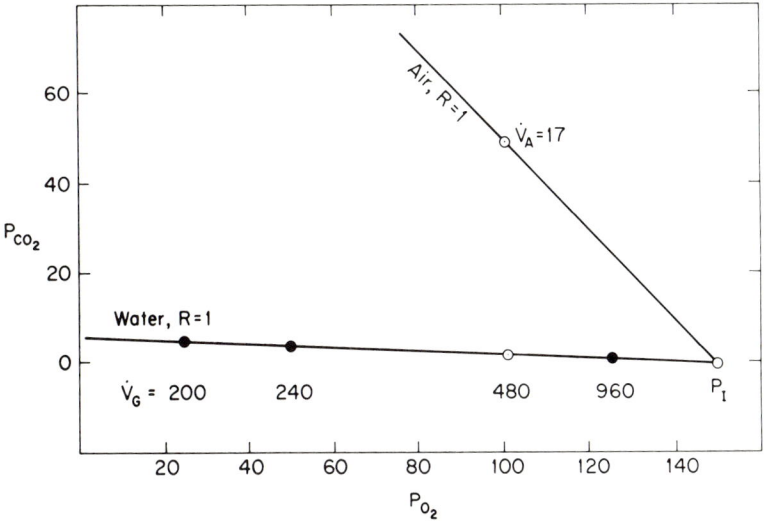

Fig. 21. Oxygen–carbon dioxide tension diagram showing the simultaneous oxygen and carbon dioxide tensions which occur when air is breathed (air, $R=1$ line) and when water is breathed at a temperature of 20°C (water, $R=1$ line); R is the exchange ratio, $\dot{V}_{CO_2}/\dot{V}_{O_2}$, and the $R \cdot Q$ is assumed to be one. The inspired oxygen and carbon dioxide tensions are the same for both water and air. The numbers, $\dot{V}_G$, indicate the gill ventilation required, in ml/min, for every 1 ml of oxygen taken up by the aquatic animal when the expired tensions are at the four levels plotted. Alveolar ventilation, $\dot{V}_A$, in ml/min, need be at a very much lower level for similar quantities of oxygen to be taken up, as the diagram shows. From Rahn (1966).

It can be argued that carbon dioxide concentrations as high as 50 mg/liter do occur in some polluted waters (Hynes, 1960) and that increased ventilation would assist in reducing the gradient from blood to environment under such conditions (Hughes, 1964). But the gradient is very small in any case so that hyperventilation would not affect the blood concentration significantly. This being so, Randall and Shelton (1963) suggested that the response to environmental carbon dioxide was inappropriate. The ventilation volume increased beyond the point that was necessary to satisfy the animal's oxygen demands unless the oxygen concentration in inspired water was simultaneously reduced to very low levels. This superfluous increase in ventilation would account for the large fall in utilization that was usually attributed to the effect of carbon dioxide on affinity of the blood. It seems more likely that the carbon dioxide response is coupled primarily to increased levels in the tissues or in the venous blood system, changes which would normally signify increased metabolism. The tension changes would be relatively small, and the evidence on responses to low level effects is conflicting. Moreover, a control system of this type could not be universally effective and would, for example, be unsatisfactory during hypoxia. There may be some role for it, however, in conjunction with an oxygen control system such as is known to exist.

2. COORDINATION OF THE RESPONSES TO CHANGES IN
CARBON DIOXIDE CONCENTRATION

Very little is known about the sensory mechanisms involved in the response to carbon dioxide. The relative importance of carbon dioxide tension or pH change as the effective stimulus has not been established. Receptors could be in the three general areas suggested earlier for oxygen receptors, but the arguments above suggest that environmental and arterial blood receptors would have very limited use in respiratory control. Carbon dioxide sensitivity has been shown in the pseudobranch (Laurent, 1967). However, Shelton (in Hughes and Shelton, 1962) demonstrated that substantial ventilatory changes to carbon dioxide persisted after total denervation of the gill area by section of the IXth and Xth cranial nerves. Injection of small quantities of saline containing carbon dioxide into some regions of the teleost medulla caused changes in the breathing movements, but the precise nature and location of the stimulated area was not decided.

C. Respiratory Homeostasis in Resting and Active Fish

The validity of experiments on the effects of changes in environmental gas concentrations on some species of fish is occasionally questioned if that species is only found in well-aerated environments. One justification for such work is that the changes may elicit responses from the same overall system as would be activated by minor internal fluctuations in the resting animal, or major alterations in demand as the animal became active. The results at present available would suggest that homeostasis is achieved in the resting animal by monitoring oxygen tensions in the blood with carbon dioxide not being involved in any significant role. The precise location of the oxygen receptors is not settled.

Differences in the level of activity are the main cause of short-term variation in oxygen consumption; the relationship has been well documented by Brett (1964). Adjustments in ventilation which meet increased oxygen demand during exercise are also accompanied by increasing transfer factor as Randall et al. (1967) have shown in the trout. They

8. THE REGULATION OF BREATHING 347

may or may not involve a decrease in utilization (Fig. 18). By virtue of the increased transfer factor, active trout maintain a steady utilization (Stevens and Randall, 1967b), whereas less active fish such as carp and suckers show a marked fall in utilization as ventilation goes up (Saunders, 1962).

The regulation of ventilation during exercise presents some problems to the physiologist if not to the fish. The changes in both arterial and venous oxygen tensions are very small, whereas ventilation changes are marked (Stevens and Randall, 1967b). This impressive stability of tension might be expected in a well-designed feedback control system. However, in hypoxic conditions there are, necessarily, larger fluctuations in blood oxygen tensions; but not until these tensions become very low are they accompanied by ventilation changes as large as in the active fish (Holeton and Randall, 1967b). The same feedback system could not operate in both cases since the error in oxygen tension is differently related to ventilation volume. During hypoxia, the tension changes are smallest in venous blood, which is why the computer simulation of Taylor *et al.* (1968) for both exercising and hypoxic trout works best with the error detector in the venous circuit. Stevens and Randall (1967a) suggest that ventilation in swimming fish could be regulated by the changes in muscular activity in some way. It is well known that there is a direct effect of exercise on respiratory activity in mammals (Dejours, 1964). Another possibility might be that the increased levels of carbon dioxide in the active tissues and venous blood could act in conjunction with the oxygen receptor system, for example, to reset the level of feedback to work with a smaller error than during hypoxia. In the absence of further experimental evidence, oxygen, carbon dioxide, and muscular activity must be assumed to have some part to play in the regulation of ventilation during exercise (Fig. 19).

VI. THE RELATIONSHIP BETWEEN VENTILATION AND PERFUSION

In order for gas exchange to occur, water and blood must be brought into intimate contact. The provision for this at the level of the gill lamella is discussed elsewhere (see chapter by Randall, "Gas Exchange in Fish," this volume). In addition, there must be an adequate flow of both blood and water to transport the respiratory gases to and from the exchanging surface. The problems involved in the provision and control of blood flow to the gills, especially in relation to the ventilation volume and gas exchange have been discussed by Hughes and Shelton (1962), Hughes

(1964), Rahn (1966), Randall et al. (1967), and Piiper and Baumgarten-Schumann (1968a,b). Hughes and Shelton (1962) introduced the concept of capacity–rate ratio in an attempt to quantify some aspects of the relationship. This ratio can be determined as the quotient of the changes in oxygen tension in water and blood as they pass through the gill, viz.:

$$\text{Capacity–rate ratio for oxygen} = \frac{\dot{V}_G}{\dot{Q}} \cdot \frac{\beta_{O_2}}{\gamma_{O_2}} = \frac{P_{a_{O_2}} - P_{\bar{v}_{O_2}}}{P_{I_{O_2}} - P_{E_{O_2}}}$$

where P is the partial pressure of gas in a (arterial blood), $\bar{v}$ (mixed venous blood), I (inspired water), and E (expired water); $\dot{V}_G$ is the volume flow of water over the gills in unit time, $\dot{Q}$ the volume flow of blood through the gills in unit time, β_{O_2} the solubility coefficient of oxygen in water, and γ_{O_2} the slope of the oxygen dissociation curve for blood between the arterial and venous points. Hughes and Shelton suggested that capacity–rate ratios of approximately unity were likely because oxygen transfer (carbon dioxide does not constitute a problem because of its high solubility in both blood and water) could then be achieved with minimum total flow of water and blood. This would be desirable if large amounts of energy were expended in pumping the two media. Since the solubility of oxygen in blood is very much higher than it is in water, a capacity–rate ratio of one would argue for high ventilation:perfusion ($\dot{V}_G/\dot{Q}$) ratios in fish. Rahn (1966) also reached this conclusion.

The precise value for both capacity–rate and ventilation:perfusion ratios in different fish will depend on many factors such as gill area, affinity of blood and oxygen, and activity characteristics of the animal. A considerable range of capacity–rate ratios have been found in practice, with values from 5.0 in resting and exercising trout (Randall et al., 1967) to 0.4 in dogfish (Baumgarten-Schumann and Piiper, 1968) and carp (Garey, 1967). The ratio is high in trout largely because of the relatively enormous ventilation volume ($\dot{V}_G/\dot{Q} = 70$) and this is accompanied by a correspondingly low oxygen utilization of 10%. Conversely, the low values in carp and dogfish are the result of much lower ventilation volumes ($\dot{V}_G/\dot{Q} = 12$ in carp, 9 in dogfish) with high utilizations of 60% and 50–70% respectively. Further details of these relationships are discussed by Randall ("Gas Exchange in Fish," this volume).

In the resting elasmobranch, evidence of a centrally coordinated relationship between breathing and heartbeat has existed for a long time. The early work was confirmed by Satchell (1960) when he showed that the heart rate was always some simple fraction, e.g., one-half, one-third, or one-fourth of the breathing rate. He suggested that the coordination was achieved by a reflex mechanism, proprioceptors in the pharynx

8. THE REGULATION OF BREATHING

being connected to inhibitory fibers running in the cardiac branch of the vagus. The reason for such a precise relationship, as opposed to a more general one between bulk flows as outlined above, was thought to be one of providing synchronization of high flow rates in the two pulsatile systems. The outcome would be a moment-to-moment balance of ventilation and perfusion and an increase in efficiency of gas exchange. This proposal was reexamined by Hughes and Shelton (1962) who combined Hughes' results (1960a) on ventilation in *Scyliorhinus* with those of Satchell on the circulatory system. The temporal relationships of blood pressure and water pressure differentials across the gills are shown in Fig. 22. If it is assumed that flow is proportional to differential pressure in both systems, it can be seen that maximum flow of blood coincides roughly with the important events of a single respiratory cycle. Because the heart rate is always less than the breathing rate it is difficult to draw any firm conclusions about the total value of synchronization to the fish.

Teleosts in well-aerated environments do not show the same exact coordination between heartbeat and breathing movements (Serfaty and

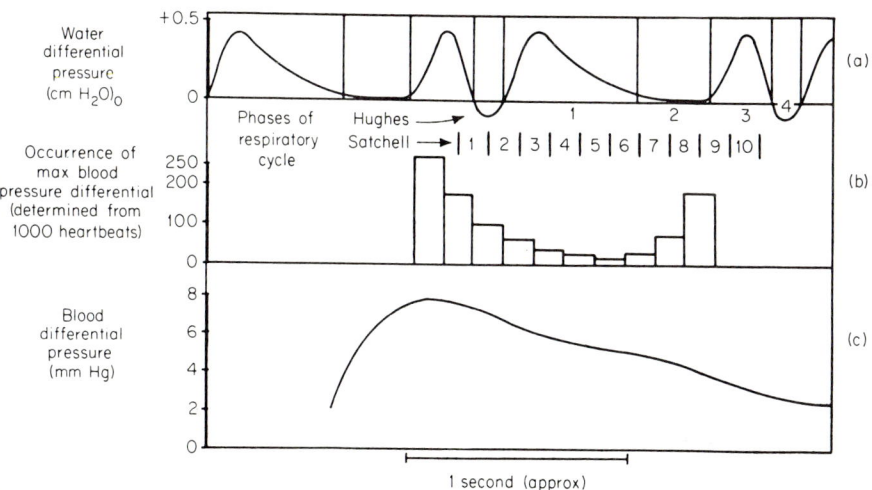

Fig. 22. Time relationships between the respiratory cycle and heartbeat in a dogfish. (a) Differential water pressure between orobranchial and third parabranchial cavities. From Hughes (1960a). (b) The occurrence, in ten phases of a respiratory cycle, of the maximum blood pressure differential across the gills. From Satchell (1960). (c) Differential blood pressure across the gill circulation produced by a single heartbeat. The curve is plotted so that the maximum differential coincides with the period of maximum occurrence as shown in the histogram. From Satchell (1960).

Raynaud, 1957; Hughes, 1961b; Hughes and Umezawa, 1968b), although there is often a greater tendency for the heart to beat during mouth closing than in other phases of breathing in the tench (Shelton and Randall, 1962). Although the relationship is not so clear cut, there is still a great deal of evidence for control in teleosts. Shelton and Randall (1962), working on eels and tench, and Peyraud and Serfaty (1964), working on carp, showed that periodic bursts of breathing movements were accompanied by an increase in heart rate, whereas the rate fell in periods of apnoea between bursts. A very clear synchrony developed in trout if the oxygen content of the environment was reduced, and, as in the dogfish, it was shown that the receptors were in the branchial region and the efferent fibers were cardio-inhibitory ones carried in the vagus (Randall, 1966; Randall and Smith, 1967). As in the case of elasmobranchs, it was suggested that phasic cardiac inhibition allowed the heart to beat during the period of maximum water flow over the gills.

Different environmental conditions and variations in activity place different demands on heart output just as they do on the gill ventilating mechanism, although the responses of the two systems need not necessarily be similar in nature. Bradycardia occurs as a response to hypoxia in both elasmobranchs (Satchell, 1961) and teleosts (Randall and Shelton, 1963; Serfaty *et al.*, 1965; Labat, 1966) and is attributable to an increase in vagal tone in both cases. Randall and Shelton (1963) suggested that the bradycardia represented a fall in cardiac output, but Holeton and Randall (1967b) have recently shown that this is not so. In the trout, hypoxia produced no substantial change in cardiac output although there was a shift from high rate–low stroke volume to low rate–high stroke volume (Fig. 23). While the gills can maintain complete saturation of the hemoglobin during hypoxia, no change in cardiac output is necessary to maintain a uniform oxygen consumption. The significance of the bradycardia is not clear. If the arterial oxygen tensions fall below the level for complete saturation, then the heart output must go up, or there must be greater extraction in the tissues giving lowered venous oxygen tensions. The latter seems to be the case in trout.

Exercise causes substantial increase in oxygen consumption and will therefore require greater perfusion rates to transport the oxygen from gills to tissues. Increased cardiac output and ventilation volume should proceed together in the absence of any very great tension changes in the different parts of the system. This was in fact the case in the trout (Stevens and Randall, 1967b), heart rate and stroke volume both going up to produce the increased cardiac output. The control of this response

8. THE REGULATION OF BREATHING

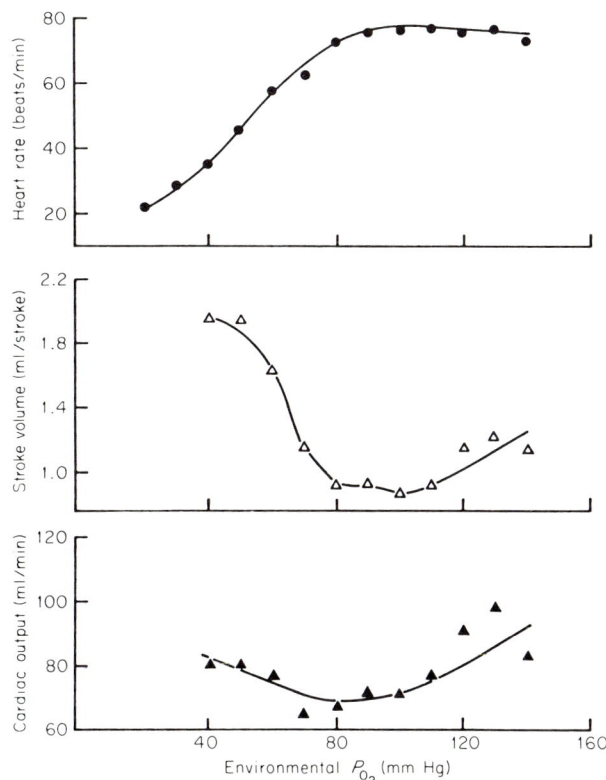

Fig. 23. The effect of hypoxia on cardiac output, stroke volume, and heart rate in the rainbow trout. Although the hypoxia results in a bradycardia, the cardiac output is maintained by a compensating increase in stroke volume. From Holeton and Randall (1967b).

is as difficult to understand as that of the ventilation change already discussed. Ventilation and cardiac output are not directly coupled together as the hypoxic animal shows. There must therefore be some specific signal in the exercising animal, and it seems most likely that it is a direct monitoring of the exercise itself which results in the increased heart output (Fig. 19). Exercise will also increase venous return to the heart, but this can hardly be responsible for the total effect. Oxygen tensions cannot constitute the effective sensory signals because they change most in the hypoxic animal where the cardiac output changes least. The only evidence which implicates oxygen receptors is that which shows the deoxygenation of the gill region produces a bradycardia, not an increase in rate. The efferent side of the cardiac response during exercise also presents problems. It has been widely accepted that there are no sympathetic fibers innervating the fish heart (Couteaux and Laurent, 1957). The whole nervous supply was thought to be cholinergic and inhibitory, although recent electron microscope studies by Yamauchi and Burnstock (1968) suggest that nerves liberating catecholamines do innervate the fish heart. Stevens and Randall (1967a) found that injections of atropine had no effect either on the heart rate of resting trout or on the responses during exercise. They concluded that increased output as a result of exercise was aneurally mediated. Injection of adrenalin produced cardiovascular changes which were similar to those observed during increased activity (Randall and Stevens, 1967). Some fish may regulate heart rate by changes in vagal tone, but it now seems likely that a great deal of cardiovascular adjustment in swimming, and perhaps in resting, fish is accomplished by changes in the level of circulating catecholamines. The source of these catecholamines and the mechanism regulating their release have not been identified. In this, as in all fields of respiratory and circulatory physiology, there have been tremendous advances in recent years only to reveal further problems for experimental analysis.

REFERENCES

Adrian, E. D., and Buytendijk, F. J. J. (1931). Potential changes in the isolated brainstem of the goldfish. *J. Physiol. (London)* 71, 121–135.
Alexander, R. McN. (1967). "Functional Design in Fishes." Hutchinson, London.
Anthony, E. H. (1961). Survival of goldfish in presence of carbon monoxide. *J. Exptl. Biol.* 38, 109–25.
Arnold, G. P. (1967). The behaviour of the plaice in water currents. Ph.D. Dissertation, University of East Anglia, Norwich.
Baglioni, S. (1907). Der Atmungsmechanismus der Fische. Ein Beitrag zur vergleichenden Physiologie des Atemrhythmus. *Z. Allgem. Physiol.* 7, 177–282.
Bainbridge, R. (1963). Caudal fin and body movement in the propulsion of some fish. *J. Exptl. Biol.* 40, 23–56.
Ballintijn, C. M. (1969). The influence of proprioception upon respiratory neurones in the medulla oblongata of fishes. *Acta Physiol Pharmacol. Neerl.* 15, 25–26.
Ballintijn, C. M., and Hughes, G. M. (1965). The muscular basis of the respiratory pumps in the trout. *J. Exptl. Biol.* 43, 349–362.
Basu, S. P. (1959). Active respiration of fish in relation to ambient concentrations of oxygen and carbon dioxide. *J. Fisheries Res. Board Can.* 16, 175–212.
Baumgarten, D., and Piiper, J. (1969). In press.
Baumgarten-Schumann, D., and Piiper, J. (1968). Gas exchange in the gills of resting unanaesthetised dogfish (*Scyliorhinus stellaris*). *Resp. Physiol.* 5, 317–325.
Beamish, F. W. H. (1964a). Respiration of fishes with special emphasis on standard oxygen consumption. II. Influence of weight and temperature on respiration of several species. *Can. J. Zool.* 42, 177–188.
Beamish, F. W. H. (1964b). Seasonal changes in the standard rate of oxygen consumption of fishes. *Can. J. Zool.* 42, 189–194.

Beamish, F. W. H. (1964c). Respiration of fishes with special emphasis on standard oxygen consumption. III. Influence of oxygen. *Can. J. Zool.* **42**, 355–366.
Beamish, F. W. H. (1964d). Respiration of fishes with special emphasis on standard oxygen consumption. IV. Influence of carbon dioxide and oxygen. *Can. J. Zool.* **42**, 847–856.
Beamish, F. W. H., and Mookherjii, P. S. (1964). Respiration of fishes with special emphasis on standard oxygen consumption. I. Influence of weight and temperature on respiration of goldfish. *Can. J. Zool.* **42**, 161–175.
Black, E. C. (1951). Respiration in fishes. *Univ. Toronto Biol. Ser.* **59**; *Publ. Ontario Fisheries Res. Lab.* **71**, 91–111.
Black, E. C., Fry, F. E. J., and Black, V. S. (1954). The influence of carbon dioxide on the utilisation of oxygen by some freshwater fish. *Can. J. Zool.* **32**, 408–420.
Black, E. C., Connor, A. R., Lam, K. C., and Chiu, W. G. (1962). Changes in glycogen, pyruvate, and lactate in rainbow trout during and following muscular activity. *J. Fisheries Res. Board Can.* **19**, 409–436.
Blazka, P. (1958). The anaerobic metabolism of fish. *Physiol. Zool.* **31**, 117–128.
Brett, J. R. (1964). The respiratory metabolism and swimming performance of young sockeye salmon. *J. Fisheries Res. Board Can.* **21**, 1183–1226.
Brown, C. E., and Muir, B. S. (1969). Analysis of ram ventilation in fish with application to skipjack tuna. In preparation.
Burns, B. D., and Salmoiraghi, G. C. (1960). Repetitive firing of respiratory neurones during their burst activity. *J. Neurophysiol.* **23**, 27–46.
Cohen, M. J., and Wang, S. C. (1959). Respiratory neuronal activity in the pons of the cat. *J. Neurophysiol.* **22**, 33–50.
Couteaux, R., and Laurent, P. (1957). Observations au microscope électronique sur l'innervation cardiaque des Téléostéens. *Bull. Assoc. Anatomistes* (*Paris*) **98**, 230–234.
Dejours, P. (1964). Control of respiration in muscular exercise. In "Handbook of Physiology" (Am. Physiol. Soc., J. Field, ed.), Sec. 3, Vol. I, pp. 631–648. Williams & Wilkins, Baltimore, Maryland.
Dejours, P., Armand, J., and Verriest, G. (1968). Carbon dioxide dissociation curves of water and gas exchange of water breathers. *Resp. Physiol.* **5**, 23–33.
de Kock, L. L. (1963). A histological study of the head region of two salmonids with special reference to pressor and chemoreceptors. *Acta Anat.* **55**, 39–50.
Eccles, R. M., Sears, T. A., and Shealy, C. N. (1962). Intracellular recording from respiratory motoneurons of the thoracic spinal cord of the cat. *Nature* **193**, 844–846.
Erichsen Jones, J. R. (1952). The reactions of fish to water of low oxygen concentration. *J. Exptl. Biol.* **29**, 403–415.
Erichsen Jones, J. R. (1964). "Fish and River Pollution." Butterworth, London and Washington, D.C.
Fry, F. E. J. (1947). Effects of the environment on animal activity. *Univ. Toronto Studies, Biol. Ser.* **55**; *Publ. Ontario Fisheries Res. Lab.* **68**, 1–62.
Fry, F. E. J. (1957). The aquatic respiration of fish. In "The Physiology of Fishes" (M. E. Brown, ed.), Vol. 1, pp. 1–63. Academic Press, New York.
Garey, W. F. (1967). Gas exchange, cardiac output, and blood pressure in free swimming carp (*Cyprinus carpio*). Ph.D. Dissertation, University of New York, Buffalo.
Gill, K. P., and Kuno, M. (1963). Excitatory and inhibitory actions on phrenic motoneurones. *J. Physiol.* (*London*) **168**, 274–289.

Glauser, S. C., Glauser, E. M., and Rusy, B. F. (1967). Gas density and the work of breathing. *Resp. Physiol.* **2**, 344–350.

Hall, F. G. (1929). The influence of varying oxygen tensions upon the rate of oxygen consumption in marine fishes. *Am. J. Physiol.* **88**, 212–218.

Hall, F. G. (1930). The ability of the common mackerel and certain other marine fishes to remove dissolved oxygen from sea water. *Am. J. Physiol.* **93**, 417–421.

Hanson, D., and Martin, A. W. (1967). Countercurrent gas exchange in gills of dogfish. *Federation Proc.* **26**, 442.

Hartline, D. K. (1967). Impulse identification and axon mapping of the nine neurones in the cardiac ganglion of the lobster. *J. Exptl. Biol.* **47**, 327–341.

Heath, A. G., and Pritchard, A. W. (1965). Effects of severe hypoxia on carbohydrate energy stores and metabolism in two species of freshwater fish. *Physiol. Zool.* **38**, 325–334.

Henschel, J. (1939). Der Atmungsmechanismus der Teleosteer. *J. Conseil Perm. Intern. Exploration Mer.* **14**, 249–260.

Henschel, J. (1941). Neue Untersuchungen über den Atemmechanismus mariner Teleosteer. *Helgoländer Wiss. Meeresuntersuch.* **2**, 244–278.

Hoff, H. E., and Breckenridge, C. G. (1949). The medullary origin of respiratory periodicity in the dog. *Am. J. Physiol.* **158**, 157–172.

Holeton, G. F., and Randall, D. J. (1967a). Changes in blood pressure in the rainbow trout during hypoxia. *J. Exptl. Biol.* **46**, 297–305.

Holeton, G. F., and Randall, D. J. (1967b). The effect of hypoxia upon the partial pressure of gases in the blood and water afferent and efferent to the gills of rainbow trout. *J. Exptl. Biol.* **46**, 317–327.

Hori, T. (1966). Facilitation and inhibition of the medullary respiratory neurones. *Japan. J. Physiol.* **16**, 436–449.

Hughes, G. M. (1960a). The mechanism of gill ventilation in the dogfish and skate. *J. Exptl. Biol.* **37**, 11–27.

Hughes, G. M. (1960b). A comparative study of gill ventilation in marine teleosts. *J. Exptl. Biol.* **37**, 28–45.

Hughes, G. M. (1961a). How a fish extracts oxygen from water. *New Scientist* **11**, 346–348.

Hughes, G. M. (1961b). Gill ventilation in fishes. *Rept. Challenger Soc.* **3**, No. 13.

Hughes, G. M. (1964). Fish respiratory homeostasis. *Symp. Soc. Exptl. Biol.* **18**, 81–107.

Hughes, G. M. (1966). The dimensions of fish gills in relation to their function. *J. Exptl. Biol.* **45**, 177–195.

Hughes, G. M., and Ballintijn, C. M. (1965). The muscular basis of the respiratory pumps in the dogfish (*Scyliorhinus canicula*). *J. Exptl. Biol.* **43**, 363–383.

Hughes, G. M., and Ballintijn, C. M. (1968). Electromyography of the respiratory muscles and gill water flow in the dragonet. *J. Exptl. Biol.* **49**, 583–602.

Hughes, G. M., and Knights, B. (1968). The effect of loading the respiratory pumps on the oxygen consumption of *Callionymus lyra*. *J. Exptl. Biol.* **49**, 603–615.

Hughes, G. M., and Shelton, G. (1958). The mechanism of gill ventilation in three freshwater teleosts. *J. Exptl. Biol.* **35**, 807–823.

Hughes, G. M., and Shelton, G. (1962). Respiratory mechanisms and their nervous control in fish. *Advan. Comp. Physiol. Biochem.* **1**, 275–364.

Hughes, G. M., and Umezawa, S. I. (1968a). Oxygen consumption and gill water flow in the dogfish *Scyliorhinus canicula*. *J. Exptl. Biol.* **49**, 557–564.

Hughes, G. M., and Umezawa, S. I. (1968b). On respiration in the dragonet *Callionymus lyra. J. Exptl. Biol.* **49,** 565–582.
Hukuhara, T., and Okada, H. (1956). On the automaticity of the respiratory centers of the catfish and the crucian carp. *Japan. J. Physiol.* **6,** 313–320.
Hyde, I. H. (1904). Localisation of the respiratory centre in the skate. *Am. J. Physiol.* **10,** 236–258.
Hynes, H. B. N. (1960). "The Biology of Polluted Waters." Liverpool Univ. Press, Liverpool.
Job, S. V. (1955). The oxygen consumption of *Salvelinus fontinalis. Univ. Toronto Biol. Ser.* **61**; *Publ. Ontario Fisheries Res. Lab.* **73,** 1–39.
Johansen, K., and Hol, R. (1960). A cineradiographic study of respiration in *Myxine glutinosa. J. Exptl. Biol.* **37,** 474–480.
Johansen, K., and Strahan, R. (1963). The respiratory system of *Myxine glutinosa.* In "Biology of Myxine" (A. Brodal and R. Fänge, eds.), pp. 353–371. Oslo Univ. Press, Oslo.
Kahn, N., and Wang, S. C. (1967). Electrophysiologic basis for pontine apneustic centre and its role in integration of the Hering-Breuer Reflex. *J. Neurophysiol.* **30,** 301–318.
Keder-Stepanova, I. A., and Ponomarev, V. A. (1965). Response of the neurones of the region of the respiratory centre to stimulation of the medial zone of the medulla oblongata. *Biofizika* **10,** 324–333.
Kirchhoff, J. (1958). Functionell anatomische Untersuchung des Visceralapparates von *Clupea harengus. Zool. Jahrb., Abt. Anat. Ontog. Tiere* **76,** 461–540.
Kylstra, J., Paganelli, C. V., and Rahn, H. (1967). Some implications of the dynamics of gas transfer in water-breathing dogs. *Ciba Found. Symp., Development of the Lung* pp. 34–63.
Labat, R. (1966). Electrocardiologie chez les poissons Téléostéens: Influence de quelques facteurs écologiques. Ph.D. Dissertation, University of Toulouse, Toulouse.
Laurent, P. (1967). La pseudobranchie des Téléostéens: Preuves électrophysiologiques de ses fonctions chémoréceptrice et baroréceptrice. *Compt. Rend.* **264,** 1879–1882.
Lenfant, C., and Johansen, K. (1966). Respiratory function in the elasmobranch *Squalus suckleyi. Resp. Physiol.* **1,** 13–29.
Lighttoller, G. H. S. (1939). Probable homologues. A study of the comparative anatomy of the mandibular and hyoid arches and their musculature. *Trans. Zool. Soc. London* **24,** 349–444.
Liljestrand, A. (1953). Respiratory reactions elicited from medulla oblongata of the cat. *Acta Physiol. Scand.* **29,** Suppl. 106, 321–393.
Lloyd, R. (1961). Effect of dissolved oxygen concentrations on the toxicity of several poisons to rainbow trout. *J. Exptl. Biol.* **38,** 447–455.
Lutz, B. R. (1930). Respiratory rhythm in the elasmobranch, *Scyllium canicula. Biol. Bull.* **59,** 179–186.
Marvin, D. E., and Heath, A. G. (1968). Cardiac and respiratory responses to gradual hypoxia in three ecologically distinct species of freshwater fish. *Comp. Biochem. Physiol.* **27,** 349–355.
Meyer, H. (1935). Die Atmung von *Uranoscopus scaber* in ihrer Abhängigkeit von Sauerstoffdruck, vom pH und von der Temperatur im Aussenmedium. *Z. Vergleich. Physiol.* **22,** 435–449.
Milic-Emili, G., and Petit, J. M. (1960). Mechanical efficiency of breathing. *J. Appl. Physiol.* **15,** 359–362.

Millen, J. E., Murdaugh, H. V., Hearn, D. C., and Robin, E. D. (1966). Measurement of gill water flow in *Squalus acanthias* using the dye-dilution technique. *Am. J. Physiol.* **211,** 11–14.
Muir, B. S. (1969). Personal communication.
Muir, B. S., and Buckley, R. M. (1967). Gill ventilation in *Remora remora. Copeia* pp. 581–586.
Muir, B. S., and Kendal, J. I. (1968). Structural modifications in the gills of tunas and other fish. *Copeia*, pp. 388–398.
Ogden, E. (1945). Respiratory flow in *Mustelus. Am. J. Physiol.* **145,** 134–139.
Olthof, H. J. (1934). Die Kohlensäure als Atemreiz bei Wassertieren, insbesondere bei den Suszwasserfischen. *Z. Vergleich. Physiol.* **21,** 534–562.
Otis, A. B. (1964). The work of breathing. In "Handbook of Physiology" (Am. Physiol. Soc., J. Field, ed.), Sec. 3, Vol. I, pp. 463–476. Williams & Wilkins, Baltimore, Maryland.
Pasztor, V. M., and Kleerekoper, H. (1962). The role of the gill filament musculature in teleosts. *Can. J. Zool.* **40,** 785–802.
Peyraud, C., and Serfaty, A. (1964). Le rythme respiratoire de la Carpe (*Cyprinus carpio*) et ses relations avec le taux de l'oxygène dissous dans le biotope. *Hydrobiologia* **23,** 165–178.
Piiper, J., and Baumgarten-Schumann, D. (1968a). Transport of O_2 and CO_2 by water and blood in gas exchange of the dogfish (*Scyliorhinus stellaris*). *Resp. Physiol.* **5,** 326–337.
Piiper, J., and Baumgarten-Schumann, D. (1968b). Effectiveness of O_2 and CO_2 exchange in the gills of the dogfish (*Scyliorhinus stellaris*). *Resp. Physiol.* **5,** 338–349.
Piiper, J., and Schumann, D. (1967). Efficiency of oxygen exchange in the gills of the dogfish, *Scyliorhinus stellaris. Resp. Physiol.* **2,** 135–148.
Powers, E. B., and Clark, R. T. (1942). Control of normal breathing in fishes by receptors located in the regions of the gills and innervated by the IXth and Xth cranial nerves. *Am. J. Physiol.* **138,** 104–107.
Prosser, C. L. (1955). Physiological variation in animals. *Biol. Rev.* **30,** 229–262.
Rahn, H. (1966). Aquatic gas exchange: Theory. *Resp. Physiol.* **1,** 1–12.
Randall, D. J. (1966). The nervous control of cardiac activity in the tench (*Tinca tinca*) and the goldfish (*Carassius auratus*). *Physiol. Zool.* **39,** 185–192.
Randall, D. J., and Shelton, G. (1963). The effects of changes in environmental gas concentrations on the breathing and heart rate of a teleost fish. *Comp. Biochem. Physiol.* **9,** 229–239.
Randall, D. J., and Smith, J. C. (1967). The regulation of cardiac activity of fish in a hypoxic environment. *Physiol. Zool.* **40,** 104–113.
Randall, D. J., and Stevens, E. D. (1967). The role of adrenergic receptors in cardiovascular changes associated with exercise in salmon. *Comp. Biochem. Physiol.* **21,** 415–424.
Randall, D. J., Holeton, G. F., and Stevens, E. D. (1967). The exchange of oxygen and carbon dioxide across the gills of rainbow trout. *J. Exptl. Biol.* **6,** 339–348.
Roberts, T. D. M. (1950). The respiratory movements of the lamprey (*Lampetra fluviatilis*). *Proc. Roy. Soc. Edinburgh* B, **64,** 235–252.
Robin, E. D., and Murdaugh, H. V. (1967). Quantitative aspects of vertebrate gas exchange. *Ciba Found. Symp. Development of the Lung* pp. 85–108.
Salmoiraghi, G. C. (1962). Pharmacology of respiratory neurones. *Proc. 1st Intern.*

Pharmacol. Meeting, Stockholm, 1961 Vol. 8, pp. 217–229. Pergamon Press, Oxford.
Salmoiraghi, G. C. (1963). Functional organisation of brain stem respiratory neurones. Ann. N.Y. Acad. Sci. 109, 571–585.
Salmoiraghi, G. C., and Burns, B. D. (1960a). Localisation and patterns of discharge of respiratory neurones in brain stem of cat. J. Neurophysiol. 23, 2–13.
Salmoiraghi, G. C., and Burns, B. D. (1960b). Notes on mechanism of rhythmic respiration. J. Neurophysiol. 23, 14–26.
Salmoiraghi, G. C., and von Baumgarten, R. (1961). Intracellular potentials from respiratory neurones in brain stem of cat and mechanism of rhythmic respiration. J. Neurophysiol. 24, 203–218.
Satchell, G. H. (1959). Respiratory reflexes in the dogfish. J. Exptl. Biol. 36, 62–71.
Satchell, G. H. (1960). The reflex coordination of the heart beat with respiration in the dogfish. J. Exptl. Biol. 37, 719–731.
Satchell, G. H. (1961). The response of the dogfish to anoxia. J. Exptl. Biol. 38, 531–543.
Satchell, G. H., and Way, H. K. (1962). Pharyngeal proprioceptors in the dogfish *Squalus acanthias*. J. Exptl. Biol. 39, 243–250.
Saunders, R. L. (1961). The irrigation of gills in fishes. I. Studies of the mechanism of branchial irrigation. Can. J. Zool. 39, 637–653.
Saunders, R. L. (1962). The irrigation of the gills of fishes. II. Efficiency of oxygen uptake in relation to respiratory flow, activity and concentrations of oxygen and carbon dioxide. Can. J. Zool. 40, 817–862.
Schadé, J. P., and Weiler, I. J. (1959). Electroencephalographic patterns of the goldfish. J. Exptl. Biol. 36, 435–452.
Schumann, D., and Piiper, J. (1966). Der Sauerstoffbedarf der Atmung bei Fischen nach Messungen an der narkotisierten Schlere (*Tinca tinca*). Pflüg. Arch. Ges. Physiol. 288, 15–26.
Sears, T. A. (1964). Some properties and reflex connexions of respiratory motoneurones of the cat's thoracic spinal cord. J. Physiol. (London) 175, 386–403.
Serbenyuk, Ts. V. (1965). The importance of afferentation in the development of rhythmic activity of the respiratory centre in fish. In "Essays on Physiological Evolution" (J. W. S. Pringle, ed.), pp. 262–271. Pergamon Press, Oxford.
Serbenyuk, Ts. V., Shishov, B. A., and Kiprian, T. K. (1959). Relationship between autonomic and reflex processes in the rhythmical activity of the respiratory centre in fish. Biofizika 4, 14–23.
Serfaty, A., and Raynaud, P. (1957). Quelques données sur l'electrocardiogramme de la tanche (*Tinca tinca*) en eau douce. Bull. Soc. Zool. France 82, 49–56.
Serfaty, A., Labat, R., and Bernat, A. (1965). Réactions cardiaque a l'anoxie chez la carpe commune. Rev. Can. Biol. 24, 1–5.
Shelton, G. (1959). The respiratory centre in the tench (*Tinca tinca* L.). I. The effects of brain transection on respiration. J. Exptl. Biol. 36, 191–202.
Shelton, G. (1961). The respiratory centre in the tench (*Tinca tinca*). II. Respiratory neuronal activity in the medulla oblongata. J. Exptl. Biol. 38, 79–92.
Shelton, G. (1966). Unpublished results.
Shelton, G., and Randall, D. J. (1962). The relationship between heart beat and respiration in teleost fish. Comp. Biochem. Physiol. 7, 237–250.
Shelton, G., and Young, S. (1965). Unpublished observations.
Smith, L. S., Brett, J. R., and Davis, J. C. (1967). Cardiovascular dynamics in swimming adult sockeye salmon. J. Fisheries Res. Board Can. 24, 1775–1790.

Springer, M. G. (1928). The nervous mechanisms of respiration in the Selachii. *Arch. Neurol. Psychiat.* **19**, 834–864.

Stevens, E. D., and Randall, D. J. (1967a). Changes in blood pressure, heart rate and breathing rate during moderate swimming activity in rainbow trout. *J. Exptl. Biol.* **46**, 307–315.

Stevens, E. D., and Randall, D. J. (1967b). Changes of gas concentrations in blood and water during moderate swimming activity in rainbow trout. *J. Exptl. Biol.* **46**, 329–337.

Strahan, R. (1958). The velum and respiratory current of Myxine. *Acta. Zool. (Stockholm)* **39**, 227–240.

Sumi, T. (1963). Organisation of spinal respiratory neurones. *Ann. N.Y. Acad. Sci.* **109**, 561–570.

Taylor, W., Houston, A. H., and Horgan, J. D. (1968). Development of a computer model simulating some aspects of the cardiovascular-respiratory dynamics of the salmonid fish. *J. Exptl. Biol.* **49**, 477–493.

Teichmann, H. (1959). Über den Atemmechanismus bei Haifischen und Rochen. *Z. Vergleich. Physiol.* **41**, 449–455.

van Dam, L. (1938). On the utilisation of oxygen and regulation of breathing in some aquatic animals. Ph.D. Dissertation, University of Gröningen, Gröningen.

von Baumgarten, R., and Kanzow, E. (1958). The interaction of two types of inspiratory neurones in the region of the tractus solitarius of the cat. *Arch. Ital. Biol.* **96**, 361–373.

von Baumgarten, R., and Salmoiraghi, G. C. (1962). Respiratory neurones in the goldfish. *Arch. Ital. Biol.* **100**, 31–47.

Wang, S. C., and Ngai, S. H. (1963). Respiration coordinating mechanism of the brain stem—a few controversial points. *Ann. N.Y. Acad. Sci.* **109**, 550–560.

Wang, S. C., and Ngai, S. H. (1964). General organisation of central respiratory mechanisms. *In* "Handbook of Physiology" (Am. Physiol. Soc., J. Field, ed.,), Sec. 3, Vol. I, pp. 487–505. Williams & Wilkins, Baltimore, Maryland.

Wang, S. C., Ngai, S. H., and Frumin, M. J. (1957). Organisation of central respiratory mechanisms in the brain stem of the cat. Genesis of normal respiratory rhythmicity. *Am. J. Physiol.* **190**, 333–342.

Whitmore, C. M., Warren, C. E., and Doudoroff, P. (1960). Avoidance reactions of salmonid and centrachid fishes to low oxygen concentrations. *Trans. Am. Fisheries Soc.* **89**, 17–26.

Willem, V. (1947). Les manoeuvres respiratoires chez les poissons téléostéens. *Bull. Musee Roy. Hist. Nat. Belg. (Brussels)* **23**, No. 29, 1–15.

Woldring, S., and Dirken, M. N. J. (1951). Unit activity in the medulla oblongata of fishes. *J. Exptl. Biol.* **88**, 218–220.

Woskoboinikoff, M. M., and Balabai, P. P. (1936). Vergleichende experimentelle Untersuchung des Atmungsapparates bei Knochenfischen. Vorläufige Mitteilung über die Ergebnisse der Beobachtungen in den Sommern 1933 und 1934. (Ukranian with German summary.) *Trav. Musee Zool. Acad. Sci. Ukr.* **10**, 145–155.

Woskoboinikoff, M. M., and Balabai, P. P. (1937). Vergleichende experimentelle Untersuchung des Atmungsapparates bei Knockenfischen. Beobachtungsergebnisse vom Jahre 1935. (Ukranian with German summary.) *Trav. Musee Zool. Acad. Sci. Ukr.* **16**, 77–127.

Wyss, O. A. M. (1954). Respiratory centre and reflex control of breathing. I. The mode of functioning of the respiratory centre. *Helv. Physiol. Pharmacol. Acta* **12**, Suppl. 10, 5–25.

Yamauchi, A., and Burnstock, G. (1968). An electron miscroscopic study on the innervation of the trout heart. *J. Comp. Neurol.* **132**, 567–588.

Yazdani, G. M., and Alexander, R. McN. (1967). Respiratory currents of flatfish. *Nature* **213**, 96–97.

Young, S. (1969). The activity of respiratory neurones in fish observed with chronically implanted electrodes. *J. Physiol. (London)* **200**, 85P.

9

AIR BREATHING IN FISHES*

KJELL JOHANSEN

I. Occurrence and Bionomics of Air-Breathing Fishes . . . 361
II. Nature of the Structural Adaptations for Air Breathing . . 363
   A. Structural Derivatives of the Mouth and Pharynx as
      Air-Breathing Organs 364
   B. Structural Adaptations of the Gastrointestinal Tract
      for Air Breathing 371
   C. The Air Bladder as a Respiratory Organ 371
III. Physiological Adaptations in Air-Breathing Fishes 375
   A. Respiratory Properties of Blood 375
   B. Gas Exchange in Air-Breathing Fishes 381
   C. Internal Gas Transport in Air-Breathing Fishes . . . 385
   D. Control of Breathing in Air-Breathing Fishes 391
   E. Normal Breathing Behavior 392
   F. Breathing Responses to Changes in External
      Gas Composition 393
   G. Breathing Responses to Mechanical Stimuli 402
   H. Breathing Response to Air Exposure 403
   I. Coupling of Respiratory and Circulatory Events . . . 405
References 408

I. OCCURRENCE AND BIONOMICS OF AIR-BREATHING FISHES

Natural selection of air-breathing habits in fishes has occurred many times and in diverse ways during vertebrate evolution.

The majority of extant fishes showing structural adaptations for direct use of atmospheric oxygen are tropical freshwater or estuarine forms. Most air-breathing fishes are teleosts but typically the dipnoan, chon-

* This chapter was written while the author was supported by grants GB 7166 from the National Science Foundation and HE 12071 from the National Institutes of Health.

drostean, and holostean fishes are dominated by air-breathing forms. Reports of air-breathing habits among elasmobranch fishes have not been adequately confirmed (George, 1953).

A smaller number of air-breathing fishes live in temperate regions, and in exceptional cases such as the bowfin, *Amia calva*, they occupy waters seasonally frozen over.

A shortage of dissolved oxygen constitutes the primary environmental condition which has stimulated the development of air-breathing in fishes (Barrell, 1916; Carter and Beadle, 1931a), although such adaptations may also have been correlated with behavior or activity patterns. A few forms, inhabiting oxygen-deficient tropical swamps, have acquired the faculty of sustained air breathing when entirely removed from water during droughts. Such periods are passsed in an estivating condition. Air breathing has also evolved among fishes living in torrential mountain streams in Asia. Here, the air-breathing habit is an adaptation to drought only since these streams carry well-oxygenated water during the wet season. These fishes either estivate or migrate over land to neighboring streams (Das, 1927, 1940; Saxena, 1963). Many tropical swamps and stagnant tributaries to larger rivers have very acid and CO_2-rich water. Adverse conditions for aquatic breathing with gills also result when water gets excessively turbid with suspended material which may cover the gill surfaces.

The principal causes of oxygen depletion in tropical freshwaters are related to the abundance of organic matter which provides a substrate for rapid oxygen consumption by microorganisms. The prevailing high temperatures of the tropics accelerate oxygen depletion by increasing the O_2 requirements of the microfauna as well as the other inhabitants of the swamps. Diurnal temperature variations are small, and thus little vertical water movement occurs to promote reentry of atmospheric O_2 and the swamps become stagnant. Oxygen release from photosynthetic activity is very slight because of the low intensity of light below the dense foliage of the prolific vegetation in regions of tropical swamps and rain forests. The same foliage cover prevents wind disturbance from effectively stirring and thus aerating the water.

Naturally tropical waters show a vertical gradient in O_2 concentration owing to a more well-oxygenated surface layer. However, as close as 1 cm to the surface, O_2 concentrations can be as low as 2% of saturation value. Several inhabitants of tropical swamps show behavioral adaptations for utilization of the more O_2-rich surface layer by breathing at or near the surface. Often O_2 availability is so limiting that all species not adapted for direct air breathing must utilize the surface water as a source for oxygen (Carter and Beadle, 1931a,b).

Among air-breathing fishes, those occupying brackish and marine habitats are rare because of the well-stirred turbulent conditions of these waters. However, several members of marine gobiid and blennid families show structural adaptations for air breathing (Schöttle, 1931; Oglialoro, 1947). Estuarine waters may show large spatial, diurnal, and seasonal fluctuations in oxygen availability (Todd and Ebeling, 1966). More exceptionally, fishes like *Periopthalmus* have developed air-breathing habits in support of behavioral acts such as feeding and escape reactions.

II. NATURE OF THE STRUCTURAL ADAPTATIONS FOR AIR BREATHING

Structural adaptations for air breathing in fishes are extremely diverse. The extent of modifications from structures typical of an aquatic breather often reflects the relative role of air breathing in overall gas exchange. Some species rely predominantly on water breathing with gills and only supplement gas exchange with air breathing when adverse respiratory qualities in the water make aquatic breathing insufficient or too costly for extraction of the required oxygen. Others are obligate air breathers and succumb if denied access to air for a short time (Rauther, 1910; Johansen, 1968).

All air-breathing fishes have a hollow compartment as a receptacle or holding space for air. The air chamber may have evolved as a modification of an already existing structure or it may be a neomorphic structure. It is richly supplied with blood vessels, and fine capillaries or lacunar spaces invest the membranes bordering the air space. The interface between blood and air is specialized to reduce the diffusion distance and other diffusion barriers between air and blood. Frequently, various forms of structural changes such as foldings, papillations, or arborizations of the gas exchange membranes have resulted in a considerable surface expansion that enhances the diffusion exchange. Such structural specialization is more prominent in species utilizing air breathing as a major means of gas exchange. A concurrent trend is for the primary aquatic gas exchange organs, the gills, to degenerate progressively to few, coarse, and sparsely perfused gill filaments as air breathing assumes a more important role (Carter and Beadle, 1931b; Dubale, 1951). In some species, such as the electric eel, *Electrophorus electricus*, the gills have become vestigial (Evans, 1929; Johansen *et al.*, 1968). Air-breathing fishes also show structural specializations for rhythmic ventilation of the air-breathing organs, while the muscle and skeletal parts involved with purely aquatic breathing have become reduced or modified for the new purpose of breathing air.

Table I lists representative species of air-breathing fishes arranged according to the type of their structural adaptation. The list is not exhaustive, and the interested reader is referred to more specialized morphological texts (Rauther, 1910; Carter and Beadle, 1931b; Das, 1940; Saxena, 1963).

A. Structural Derivatives of the Mouth and Pharynx as Air-Breathing Organs

By far the most common and yet the most diversified structural adjustments for air breathing are those associated with the branchial, opercular, and pharyngeal cavities.

Piscine gills are designed to function in water where lack of net gravitational forces prevents collapse of the fine filaments and lamellae. However, the collapse of these in air makes the gills of most fishes useless as organs of aerial gas exchange. A very few fishes, such as *Symbranchus* and *Hypopomus*, are able to use their gills for air breathing. They fill the buccal cavity with air regularly at the surface, and the used air is voided from the mouth or the opercular openings (Taylor, 1913; Carter and Beadle, 1931b; Johansen, 1966). The gills of these fishes are structurally modified to prevent their collapse in air. Most notably they show a high ratio between the breadth and height of the secondary lamellae. The ratio is more than twice that in purely aquatic fishes or fishes possessing other types of structural adaptations for air breathing (Carter and Beadle, 1931b). The epithelium of the buccal and pharyngeal cavities of these two fishes is richly vascularized and plays a supporting role in aerial gas exchange.

The common eel, *Anguilla anguilla* (Berg and Steen, 1965), and species of *Periopthalmus* (Schöttle, 1931) also employ gill breathing successfully during air exposure.

A further development of aerial mouth breathing is seen in fishes with foldings and papillations of the buccal epithelia projecting into the buccal cavity and forming extensive vascular surfaces for aerial gas exchange. The electric eel, *Electrophorus electricus*, offers a striking example (Evans, 1929; Böker, 1933; Richter, 1935; Carter, 1935; Johansen *et al.*, 1968). In this fish the papillated projections are distributed over both the floor and the roof of the mouth, while smaller prominences are present on the branchial arches and the lateral branchial walls. The system of papillae forms a labyrinth of air passages. Studies on formalin-fixed material revealed a surface area of the mouth respiratory organ about 15% of the total body surface (Johansen et al., 1968). This is considerably less than for either gills in purely aquatic fishes or lungs of higher vertebrates. Yet the mouth organ in *Electrophorus* functions well enough to have permitted the gills of adult fishes to become almost vestigial (Fig. 1).

The next step in a progressive structural adjustment to air breathing includes the formation of an air chamber extending from the buccal or pharyngeal cavities. In the Ophiocephalidae with several air-breathing species in tropical marshes of Asia and Africa, the air chamber lies dorsal to the branchial area and becomes partly enclosed within the cavity of the skull (Munshi, 1962).

In *Amphipnous cuchia*, another air-breathing fish living in India, the paired air chambers referred to as pharyngeal lungs (Das, 1940) are more free and extend several centimeters behind the head. The internal walls of the air chamber are very vascular and may show surface expanding structures in the form of rosettelike arborizations protruding into the lumen. In others, internal trabeculation may give the air chamber a resemblance to an amphibian lung.

Another structural variation is seen in fishes where the opercular cavities have become richly vascularized and serve as receptacles for air. When air is lodged in such chambers they stand out in balloonlike fashion. The opercular walls have become thin and elastic and their bone structures have regressed. The opercular openings are either fused to one small channel ventrally or have otherwise become reduced in size. These modifications and others involving the intrinsic muscles of the opercular walls permit ventilatory movements and renewal of air inside the opercular chambers. Examples of such specialization are seen in *Hypopomus brevirostris, Monopterus* sp., and several species of gobiid fishes (e.g., *Pseudapocryptes lanceolatus*) (Wu and Kung, 1940).

More elaborate use of the opercular cavities in air breathing results when specialized diverticula develop from their dorsal side commonly between the hyoid and the first branchial arches. Two distinctly different types of such opercular air chambers or lungs are found. In some cases they remain free and extend posteriorly. Internally they possess ridges and trabeculations or other protrusions expanding the highly vascular internal walls. Often these air sacs extend half the length of the animal and lie embedded within the myotomes of the body wall, a fact which undoubtedly bears significance in the filling and emptying of the sacs. Indian fishes of the genus *Saccobranchus* are well-known examples of this type of structural adaptation (Rauther, 1910).

The other variety of opercular air chamber results when the opercular diverticulum extends inside the skull to take the form of a labyrinthlike

Table I

A List of Representative Air-Breathing Fishes Arranged According to the Type of Their Structural Adaptation[a]

Type of air breathing	Species	Habitat	Relative importance of air breathing	Remarks	Reference
Gills	*Symbranchus marmoratus*	Swamps and rivers, South America	Essential in O_2-deficient water and during estivation	Gills capable of both aerial and aquatic breathing	Taylor (1913) Johansen (1966)
Pharyngeal and opercular walls	*Hypopomus brevirostris*	Swamps and rivers, South America		Gills capable of both aerial and aquatic breathing	Carter and Beadle (1931b)
Skin	*Anguilla anguilla*	Rivers in Europe, Asia, Africa, North America	Accessory during migrations over land	Rhythmic inflation of branchial chamber with air, when air exposed	Krogh (1904) Berg and Steen (1965)
	Periophthalmus sp.	Marine tropical shores, estuarine and brackish water	Essential during voluntary air exposure	Air breathing appears more related to behavior than to environmental conditions	Schöttle (1931)
	Gillichthys mirabilis	Marine shallow tidal flats	Important in oxygen-deficient water	Remarkable ability to change from aquatic to aerial breathing	Todd and Ebeling (1966)
Pharyngeal epithelium, papillated and vascular	*Electrophorus electricus*	Rivers and swamps of South America	Air breathing obligatory	Air breathing dominates O_2 absorption; CO_2 elimination via vestigial gills and skin	Johansen et al. (1968)

	Species	Habitat	Function	Notes	Reference
Air chambers as diverticula from the opercular and pharyngeal cavities (opercular and pharyngeal lungs)	*Clarias* sp.	Tropical pools, Asia	Important in oxygen-deficient water	Capable of estivation, migrates overland	Rauther (1910) Moussa (1956)
	Anabas scandens	Tropical pools, Asia	Air breathing accessory but necessary in O_2-deficient water	Estivates, migrates out of water	Rauther (1910) Das (1927)
	Amphipnous cuchia	Rivers in northern India and Burma	Facultative air breather	Estivates, gills more reduced than in other air-breathing fishes	Das (1940)
	Ophiocephalus punctatus	Tropical pools, Asia and Africa	Air breathing obligatory	Estivates	Rauther (1910) Das (1927) Qasim *et al.* (1966)
Stomach or intestine modified for air breathing	*Hoplosternum* sp.	Swamps, South America	Obligate air breather	Intestinal breather	Carter and Beadle (1931b)
	Misgurnus fossilis	Temperate and tropical rivers, Europe	Air breathing accessory	Intestinal breather	Calugareanu (1907)
	Ancistrus anisitsi	Swamps, South America	Air breathing accessory	Stomach breather	Carter and Beadle (1931b)
Air bladder functions as lung	*Polypterus* sp.	Lakes and rivers in Africa	Survives on aquatic breathing in well-aerated water. Air breathing obligatory under most natural conditions	Activity markedly increases air breathing	Budgett (1900) Magid (1966)

Table I (*Continued*)

Type of air breathing	Species	Habitat	Relative importance of air breathing	Remarks	Reference
Air bladder functions as lung	*Amia calva*	Rivers and lakes, North America	Air breathing accessory, increased importance at increased temperatures and/or during activity	Capable of estivation	Johansen *et al.* (1970)
	Arapaima gigas	Rivers and swamps, South America	Air breathing obligatory, dies quickly if denied access to air	Very large fish, exceeds 3 meters	Sawaya (1946)
	Neoceratodus forsteri	Rivers, East Australia	Air breathing accessory	Increased importance in O_2-deficient water and during activity	Grigg (1965) Lenfant *et al.* (1966) Johansen *et al.* (1967)
	Lepidosiren paradoxa	Rivers and swamps, South America	Obligatory air breather, succumbs if denied access to air	Structurally most advanced of all air-breathing fishes, estivates	Johansen and Lenfant (1967) Lenfant *et al.* (1970)
	Protopterus sp.	Lakes and swamps, Africa	Air breathing obligatory	Estivates	Lenfant and Johansen (1968) Johansen and Lenfant (1968)

[a] For complete lists, see Rauther (1910), Carter and Beadle (1931b), Das (1940), and Saxena (1963).

9. AIR BREATHING IN FISHES

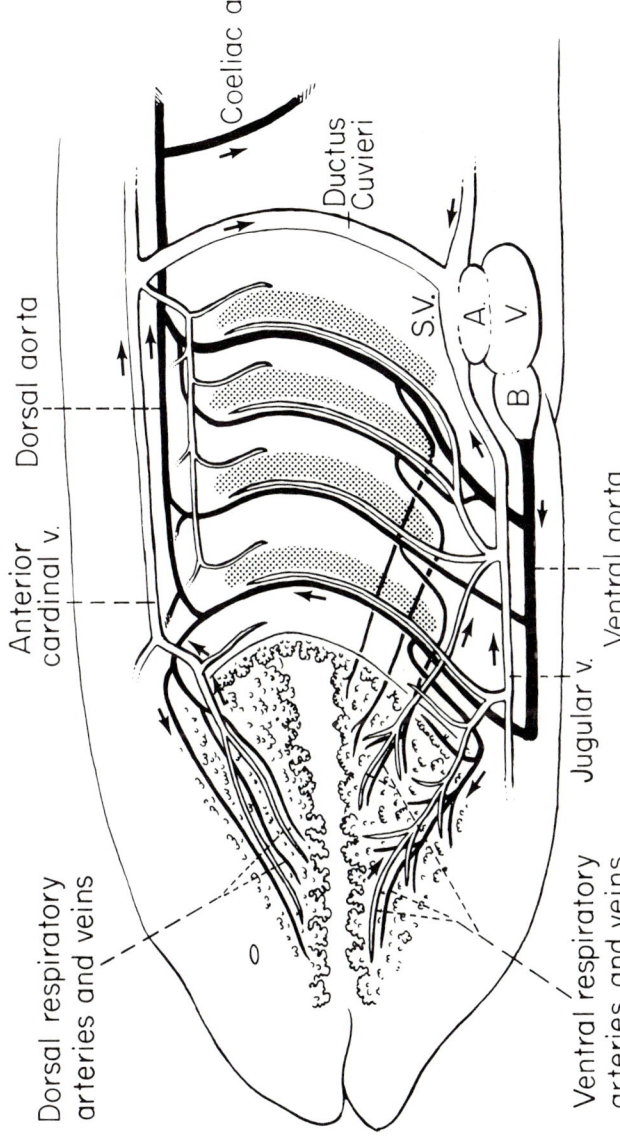

Fig. 1. Schematic drawing of the mouth respiratory organ and its vascular connections with the heart and central circulation in the electric eel. The arrows indicate the direction of the blood flow. From Johansen et al. (1968). A, Atrium; B, bulbusarteriosus; V, ventricle; a, artery; v, vein.

organ (*Anabantidae*) or corallike dendritic formations as seen in species of *Clarias* (Moussa, 1956; Munshi, 1961).

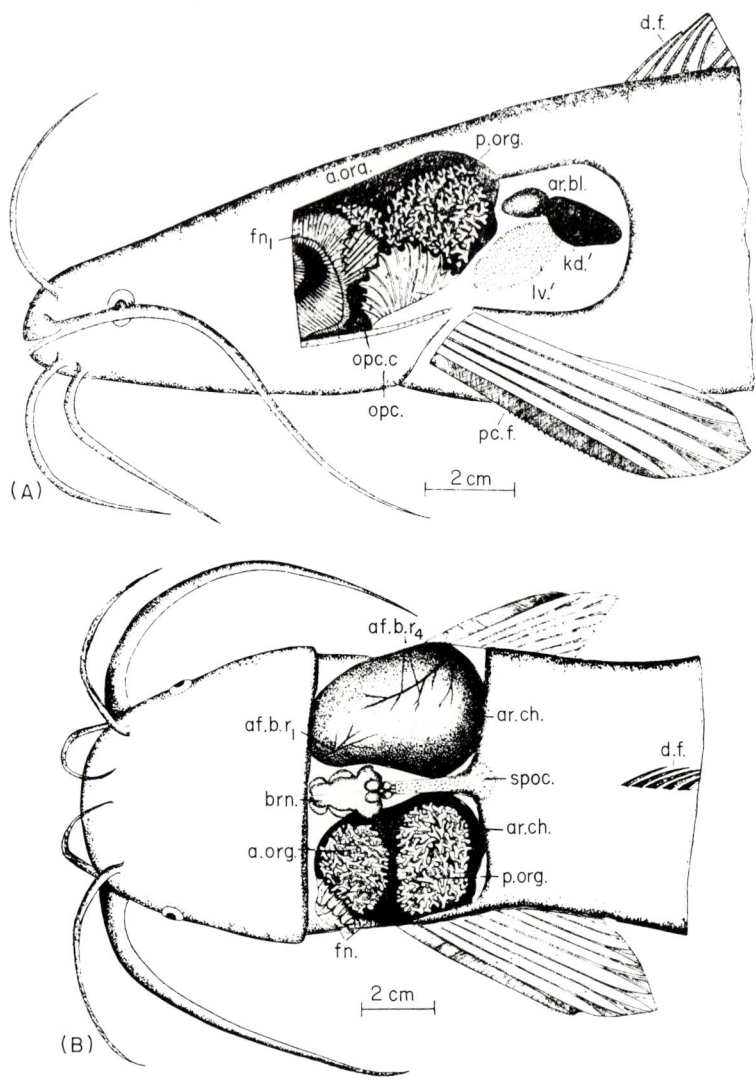

Fig. 2. (A) Lateral and (B) dorsal views showing the gross structure and relative position of the air chamber in *Clarias lazera*. In A and to the left in B the wall of the air chamber has been removed to expose the surface expanding arborescent organs. From Moussa (1956). af.b.r., 1 and 4, ramifications of the first and fourth afferent branchial vessels in the wall of the air-chamber; a.org., anterior arborescent organ; ar.bl., air or swim bladder; ar.ch., air-chamber; brn., brain; d.f., dorsal fin; $fn._{1-4}$, serial order of fans; kd'., part of kidney; lv'., part of liver; opc., operculum; opc.c., opercular cavity; pc.f., pectoral fin; p.org., posterior aborescent organ; spoc., supraoccipital.

Members of the Anabantidae (*Anabas, Macropodus, Trichogaster*, and others) typically show the air labyrinths starting as outgrowths from the epibranchial section of the first branchial arch. Laminated projections contribute to the surface expansion inside the chambers. The detailed structure of the labyrinthine organ is extremely complicated (Munshi, 1968). The diffusion distance from air to blood in *Trichogaster* has been measured as the smallest of all recorded in respiratory organs, showing a minimal value of 600 Å, a medium range of 1300–1700 Å, and a maximum value of 1.2 μ (Schulz, 1960). Based on electron microscopy, Schulz reported that capillaries of some areas of the labyrinthine organ lack both basal membrane and endothelium, thus leaving only a thin epithelium intervening between blood and air.

In *Clarias* the two remarkable vascular rosettelike trees which fill the air chambers on each side receive structural support from gill arches 2 and 4 and their vascular supply from the first and fourth afferent branchial arteries (Moussa, 1956). Figure 2 shows the gross appearance of the labyrinth in *Clarias lazera*.

B. Structural Adaptations of the Gastrointestinal Tract for Air Breathing

In some fishes portions of the gastrointestinal tract, notably the stomach or part of the small intestine, have been modified to serve in gas exchange between the blood and rhythmically swallowed air. The used air is voided at the anus or the mouth. Members of the families Loricariidae, e.g., *Plecostomus plecostomus* (Carter, 1935), and Callichthyidae, e.g., *Callichthys* and *Hoplosternum* (Carter and Beadle, 1931b) offer typical examples of gastrointestinal air breathers. Most fishes employing the gastrointestinal tract for gas exchange use it as an accessory breathing organ when aquatic conditions become severely oxygen deficient. Others depend on it during short migrations over land (*Plecostomus*), and a few are obligate air breathers that need to supplement the aquatic gas exchange with air breathing at all times.

C. The Air Bladder as a Respiratory Organ

It is now generally conceded that the air bladder in fishes originally functioned as an accessory respiratory organ. In most extant actinopterygian fishes the air or swim bladder serves primarily in buoyancy control or sound production or sound detection. In dipnoan fishes the air bladder has retained and further evolved the respiratory function it allegedly possessed in the crossopterygian fishes that also gave rise to amphibians and land vertebrates. Members of the primitive orders

Chondrostei (*Polypterus*) and Holostei (*Amia* and *Lepisosteus*) have also retained a respiratory function of the air bladder but fall short of the refinement in structural adjustment for air breathing seen in the dipnoans (Klika and Lelek, 1967). Among modern teleosts some have secondarily developed the air bladder into a respiratory organ such as seen in the South American fishes *Erythrinus uniteniatus* and *Arapaima gigas* (Carter and Beadle, 1931b; Sawaya, 1946).

Only in the lungfishes, and of these *Protopterus* and *Lepidosiren* in particular, has the air bladder evolved structurally to resemble a lung as it exists in lower vertebrates. Internal septa, ridges, and pillars divide the air space into smaller compartments opening into a median cavity. Further subdivisions of these compartments by progressively smaller reticulated septa terminates in alveoli-like pockets richly covered with blood vessels. *Neoceratodus* has one lung, while *Protopterus* and *Lepidosiren* have two lungs fused in their anterior region where the pneumatic duct and the pneumatic sphincter connect the lungs with the esophagus. The short airway duct, which in no way resembles the air distributing system of the trachea and bronchii in terrestrial vertebrates, is richly supplied with smooth muscle as is the lung parenchyma itself. The smooth musculature is important for the mechanics of breathing and may exert additional influence in distributing the air inside the lung (Grigg, 1965; Johansen and Reite, 1967). Electron microscopy has revealed that the fine structure of the lung in *Protopterus* does not differ basically from that in higher vertebrates (DeGroodt et al., 1960; Schulz, 1960). Klika and Lelek (1967), reporting on lung structure in *Protopterus*, state that the respiratory epithelium adjoins the basement membrane which in turn is separated from the endothelial basement membrane by a thin layer of collagen fibrils. The approximate blood to air diffusion distance in *Protopterus* is about 0.50 μ. The *Protopterus* lung is much more similar to the amphibian lung than that of *Polypterus* which has a smooth inner surface much less vascular than in the lungfishes. Figure 3A shows the gross structure of the *Protopterus* lung.

Structural specializations in the perfusion pattern through the gas exchange organs and the vascular linkage between them and the general circulation is of utmost importance for efficient gas exchange. In air-breathing fishes presumably the highest degree of efficiency will prevail when the air-breathing organ is perfused with systemic venous blood and when the respiratory efferent blood is dispatched directly for systemic arterial distribution. Most air-breathing fishes also employ aquatic breathing and efficient perfusion of the gills and/or skin is important since these areas are the primary sites for CO_2 elimination. The vascular system should hence ensure that both the air- and water-breathing respiratory organs are perfused before the blood is turned over to the nutritive circulation.

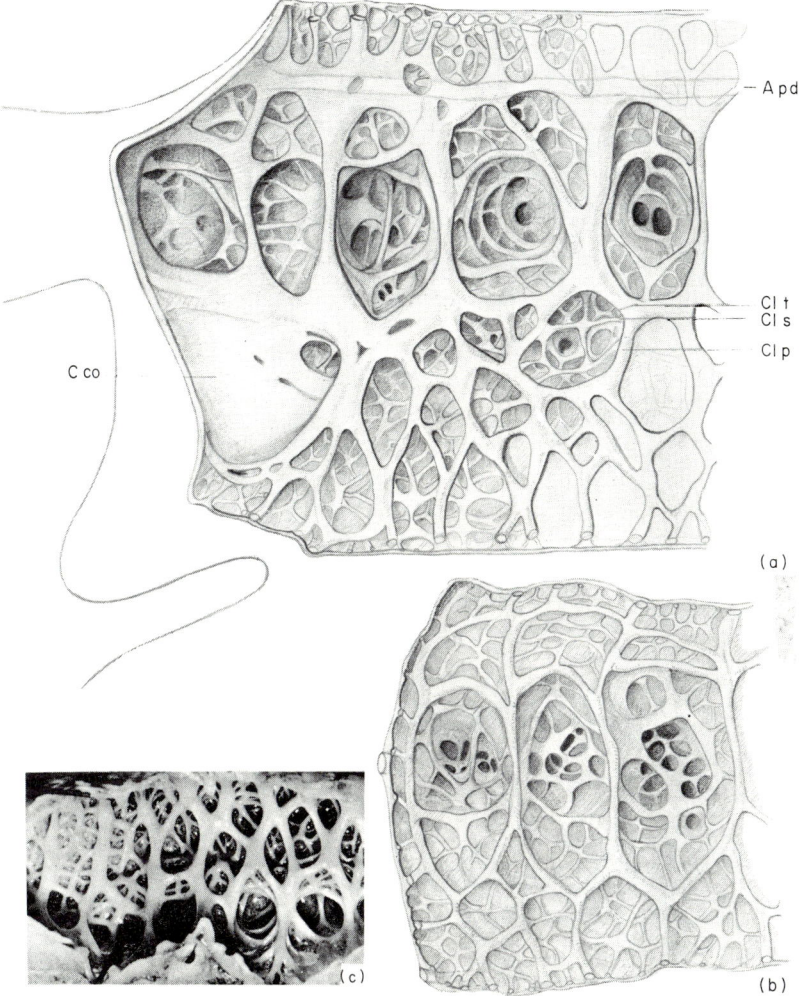

Fig. 3. Gross structure of the anterior region of the lung in *Protopterus aethiopicus* (Heckel). (b and c) Posterior region of the *Protopterus* lung showing extensive trabeculation. From Poll (1962). Apd, dorsal pulmonary artery; C co, communicating cavity between the two lungs; Cl p, primary septum; Cl s, secondary septum; Cl t, tertiary septum.

Figure 4 summarizes the various types of circulatory arrangements seen in air-breathing fishes. All types fall far short of matching the ideal perfusion pattern existing in purely aquatic fishes (type A) or in the highest vertebrates where the respiratory and systemic vascular beds are coupled in direct series. All air-breathing fishes show a parallel linkage between the two types of vascular beds, and only the lungfishes are structurally adapted to minimize shunting between these

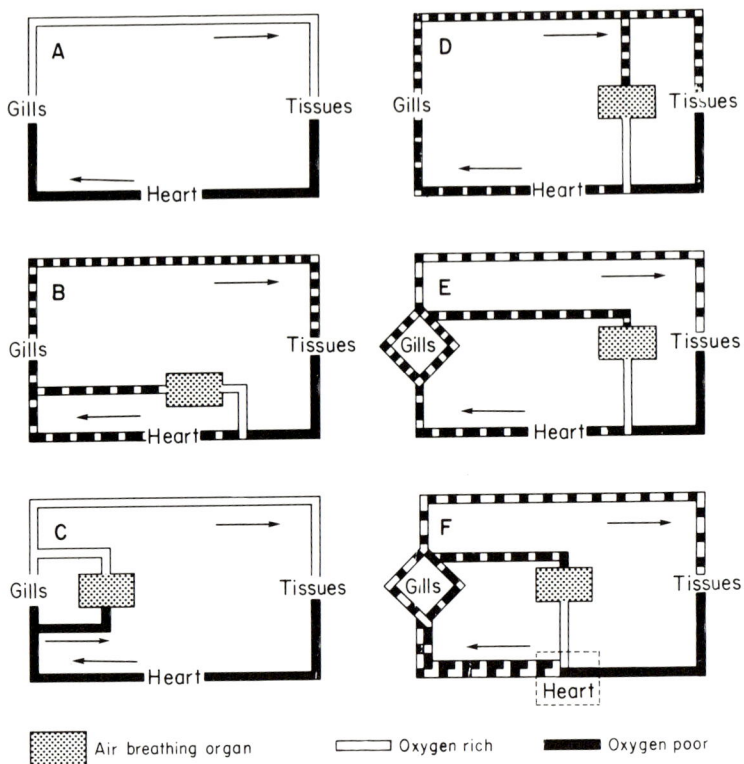

Fig. 4. Schematic representation of the heart and vascular beds in relation to the aquatic (gills) and aerial respiratory organs in air-breathing fishes. The amount of white and black represents only the approximate level of oxygenated and deoxygenated blood carried in the vessels. (A) General piscine arrangement with the branchial (gill) and systemic (tissues) vascular beds in direct series. (B) Air-breathing organ derived from pharyngeal and/or opercular mucosa. Afferent vessels to air-breathing organ arranged in parallel with branchial circulation and derived from afferent branchial vessels. Efferent vessels from air-breathing organ connected to systemic veins. Arrangement typical of *Monopterus*, *Ophiocephalus*, *Electrophorus*, *Amphipnous*, *Periopthalmus*, and *Anabas*. (C) Gills, buccal mucosa, or chambers extending from the opercular cavity serving as air-breathing organs. Afferent vessels to air-breathing organ arranged in parallel with the branchial vascular bed. Efferent circulation largely connected with the efferent branchial circulation. Systemic arterial blood highly oxygenated. *Clarias*, *Saccobranchus*, and in part, *Symbranchus* and *Hypopomus* typical of type C. (D) Air-breathing organ associated with the gastrointestinal tract. Afferent circulation derived from the dorsal aorta (systemic arteries). Efferent circulation connected to systemic veins. Arterial blood mixed. Arrangement typical in *Hoplosternum*, *Plecostomus*, and *Ancistrus*. (E) Air bladder serving as air-breathing organ. Specialized afferent vessel derived from the most posterior epibranchial arteries. Efferent circulation connected to systemic veins. Arterial blood mixed. Arrangement typical of *Polypterus*, *Amia*, and *Lepisosteus*. (F) Air bladder structurally advanced to resemble amphibian lung. Specialized afferent and efferent circulation in parallel with systemic arterial circulation. Partial septation of the

vascular beds. In the electric eel (type B), the entire gain in oxygenation resulting from perfusing the mouth is shunted back to the systemic veins before blood is redistributed to the systemic and respiratory arteries. This deficiency prevails to various degrees in all types except the very few fishes able to use their gills alternately for aerial and aquatic breathing [e.g., *Symbranchus* or the few fishes employing air breathing with specialized air chambers extending from the opercular cavities such as *Clarias* (type C)].

Only in fishes employing the air bladder for aerial gas exchange have specialized vessels developed as connections between the primary branchial circulation and the gas exchange organ (types E and F). These vessels take origin from the sixth aortic arch and are homologue with the pulmonary arteries of higher vertebrates. In addition to the dipnoans, *Polypterus* and *Amia* possess these vessels. The dipnoans are unique among fishes by also having a specialized efferent circulation from the air bladder connected directly with the left side of the heart. *Gymnarchus niloticus*, a teleost using the air bladder for accessory respiration, has an arterial supply to the bladder from the dorsal aorta but a venous drainage connecting directly with the atrium. No septation of the heart is evident. However, the heart in dipnoans is in part structurally adapted for accommodation of two bloodstreams by showing partial atrial and ventricular septa, structures which clearly forecast a further development of the heart and circulation toward conditions in terrestrial vertebrates (Johansen and Hanson, 1968; Johansen and Hol, 1968).

The mechanics of air breathing in lungfishes has been carefully studied (Bishop and Foxon, 1968; McMahon, 1969). The mechanics of inhalation and exhalation of air is derived from a series of basically aquatic breathing cycles typical of fishes modified to serve a specific step in the air-breathing cycle. McMahon (1969) emphasizes the role of the buccal force pump as the main ventilatory mechanism in purely aquatic as well as transitional bimodal breathers among fishes and amphibians.

III. PHYSIOLOGICAL ADAPTATIONS IN AIR-BREATHING FISHES

A. Respiratory Properties of Blood

Studies of respiratory properties of blood from air-breathing fishes have disclosed adaptive features which correlate with environmental

---

heart allows a high degree of selective perfusion in the respiratory and systemic circuits. Arrangement seen in lungfishes and in *Protopterus* and *Lepidosiren* in particular.

Table II
Some Respiratory Properties of Blood in Representative Air-Breathing Fishes[a]

Species	Hemoglobin (%)	Hematocrit (%)	O_2 capacity (vol. %)	P_{50} at pH_a	Bohr effect	Root effect	Haldane effect	Standard bicarbonate	Buffering capacity	Reference
Electrophorus	11.2	41.0	13.90	10.7 (7.57; 28°C)	−0.680	None	4.8 (30)[b]	33.5	4.7	Johansen et al. (1968)
Electrophorus	15.9[c]		19.75	15						Willmer (1934)
Lepidosiren[d] (juvenile)	5.4[c]	19.0	6.80	10.8 (7.6; 23°C)	−0.234	None	3.8 (27)	15.0	1.49	Johansen and Lenfant (1967)
Lepidosiren[d] (adult)	6.6	28.0	8.25		−0.295	None	1.5 (25)	27.2	2.43	Lenfant et al. (1970)
Neoceratodus[d]	7.2[c]	35.0	9.00	12.5 (7.6; 18°C)	−0.620	None	4.0 (10)	11.0	1.33	Lenfant et al. (1966)
Hoplosternum			18.1	15	Slight	None				Willmer (1934)
Myelus setiger			10.7	20	Large	Large				Willmer (1934)
Hoplias malabaricus			6.5	20	Large	Large				Willmer (1934)
Amia calva	5.4	23.0								
Protopterus[d]	7.6	30.5	9.50		−0.470	None	1 (25)	20	1.52	Lenfant and Johansen (1968)
Protopterus				27.3 (7.6; 23°C)	−0.335					Swan and Hall (1966)
Protopterus										Fish (1956)
Symbranchus[d]	13.0	47	17.30	7.2 (7.60; 25°C)	−0.400	None	2.1 (25)	24.1	2.70	Lenfant et al. (1970)
Protopterus[e]	7.8[c]	30.1	9.73	25.0 (7.64; 25°C)	−0.280	None	2.30 (27)	34.2	2.60	Lahari et al. (1968)

[a] *Myelus setiger* and *Hoplias malabaricus* are purely aquatic breathers included for comparison.
[b] Calculated from true plasma.
[c] Derived from O_2 capacity using O_2 combining power of 1.25 ml O_2/g Hb.
[d] Maximum values of Hb, hematocrit, and O_2 capacity.
[e] Average given by authors.

factors and the relative importance of air and water breathing (Willmer, 1934; Fish, 1956; Swan and Hall, 1966; Lenfant and Johansen, 1968).

Table II summarizes available information in representative species. A large variability is apparent in hemoglobin concentration and O_2 capacity, with no adaptive trend evident in air-breathing fishes as a group. However, a comparison of tropical fishes studied by Willmer (1934) disclosed a definite tendency for O_2 capacities to be higher in the air-breathing forms. This difference probably reflects the nature of the fishes' habitats, particularly the degree of O_2 deficiency rather than the habit of air breathing per se. The fast flowing rivers and creeks where the purely aquatic breathers *Myelus setiger* and *Hoplias malabaricus* live are well aerated, whereas the habitat of the air-breathing fishes *Hoplosternum* and *Electrophorus* can be severely O_2 deficient.

In evaluating this type of adaptive adjustment, one must recognize the relative importance of air and aquatic breathing in the total gas exchange of these fishes. If water breathing is the dominate mode and air breathing is only accessory or supplemental, adaptive adjustment of the blood to meet the conditions in the water may be expected. An early documentation by Krogh and Leitch (1919) showed that aquatic species tended to show higher O_2 capacities when living in O_2 deficient water. The obligate air breathers, among fishes, have an ample reservoir of atmospheric oxygen available. However, these fishes may be compared with the diving animals among higher vertebrates which, by means of high oxygen capacities, increase their O_2 stores and consequently extend the time between surfacings. Sluggish or active habits also may influence hemoglobin levels in fishes (Root, 1931). Still another type of selection pressure for development of high O_2 capacities in air-breathing fishes has been suggested and exemplified in the electric eel, *Electrophorus* (Johansen et al., 1968), and applies to most other air-breathing fishes. This relates to the shunting of respiratory efferent blood into systemic veins which results in systemic arterial perfusion with mixed blood (Fig. 4). Thus arterial blood never gets fully saturated, a condition which obviously reduces the efficiency of the gas transport system. The situation in lungfishes differs considerably in that recirculation of the well-oxygenated blood from the lung is largely prevented by a separate pulmonary vascular circuit and a partial division of the heart and its outflow channels.

The general shape of O_2–Hb dissociation curves conform to those of other fishes. Oxygen affinity generally decreases in vertebrates with increasing dependence on aerial gas exchange (McCutcheon and Hall, 1937; Lenfant and Johansen, 1967). Another generalization relates an increased O_2 affinity to the ability of a species to survive in an O_2 poor medium (Krogh and Leitch, 1919). The applicability of these generalizations to air-breathing fishes depends on the relative importance of air breathing in their bimodal gas exchange. Among species emphasizing water breathing, there is a tendency for higher O_2 affinity to correlate with the most O_2-deficient habitats. No clear-cut adaptive trends are apparent from a comparison of the obligate air breathers and those practicing occasional supplemental air breathing. In fishes like *Electrophorus* in which the O_2-rich blood is admixed with O_2-deficient blood before perfusion of the arteries, a high O_2 affinity would improve the efficiency loss caused by the mixed condition of arterial blood.

Several authors have claimed evidence of a relationship between the CO_2 sensitivity of the O_2-Hb dissociation and the habitat and air-breathing habits of fishes. Willmer (1934) offered convincing data to show that tropical fishes from larger rivers and fast flowing creeks with well-

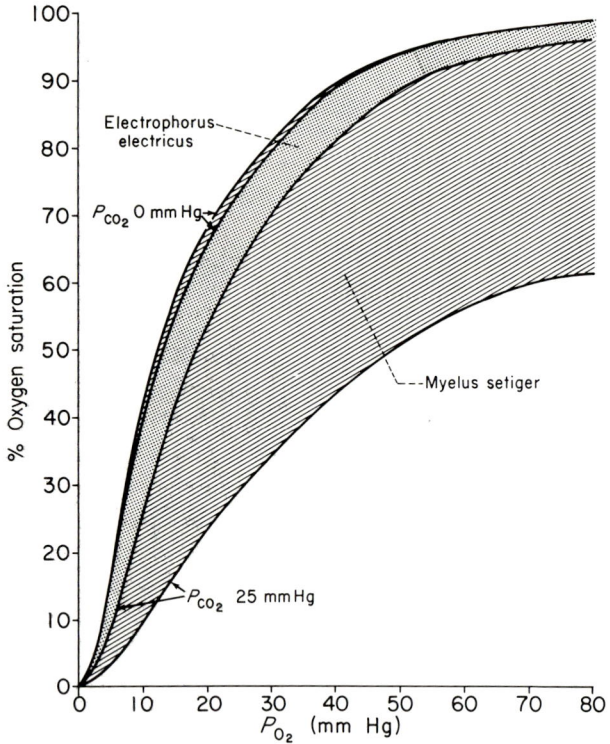

Fig. 5. Oxygen dissociation curves showing a conspicuous Bohr effect in *Myelus* living in well-aerated CO_2-free water. A slight Bohr effect is present in *Electrophorus* living in stagnant oxygen-deficient, CO_2-rich water. Redrawn from Willmer (1934).

aerated water of low CO_2 content have blood which is adversely affected by CO_2 and pH changes. Conversely, blood from fishes living in stagnant swamps and creeks with O_2-deficient, CO_2-rich water were distinctly less sensitive and often even insensitive to CO_2 changes (Fig. 5). For instance, if *Myelus setiger*, normally living in well-aerated CO_2-free water, was transferred to a typical swamp habitat, its blood could never be more than 40% saturated with oxygen because of the large Root and Bohr shifts. It has been argued that the diminishing sensitivity to CO_2

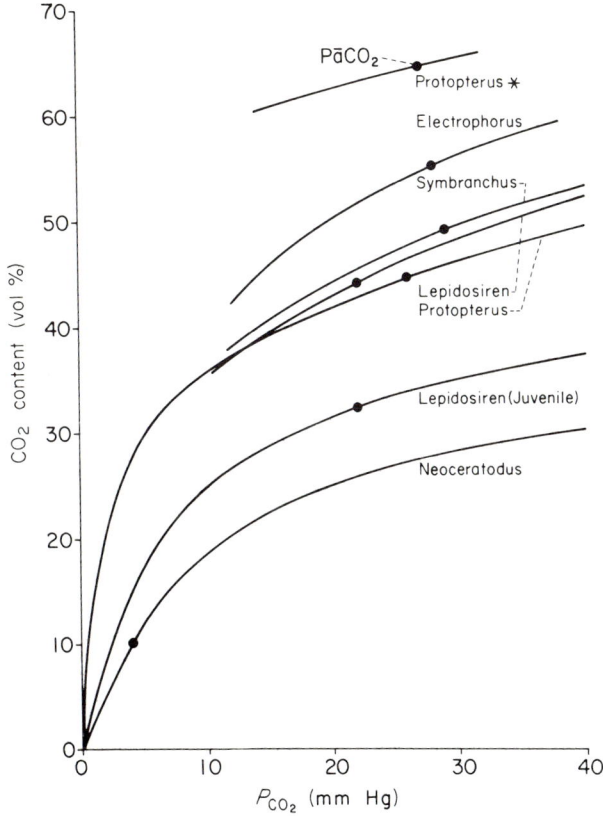

Fig. 6. Comparison CO_2 dissociation curves from representative species of air-breathing fishes. The black dots indicate the normal levels of arterial P_{CO_2}. Information selected from Lahiri et al. (1968), *Protopterus**; Johansen et al. (1968), *Electrophorus;* Lenfant et al. (1970), *Symbranchus;* Lenfant and Johansen (1968), *Lepidosiren* and *Protopterus;* Johansen and Lenfant (1967), *Lepidosiren* (juvenile); and Lenfant et al. (1966), *Neoceratodus*.

in fishes normally living in a CO_2-rich habitat is an important adaptation needed for adoption of air breathing since this change inevitably will entail an elevated concentration of CO_2 at the gas exchange surfaces because of the bidirectional airway of air breathers (Carter, 1951).

The CO_2 combining power of blood shows a clear increase with increased importance of air breathing in overall gas exchange. Figure 6 shows a comparison of CO_2 dissociation curves with normal values of arterial P_{CO_2} indicated for representative air-breathing fishes.

The increase in CO_2 combining power correlates with an increase in the average level of arterial P_{CO_2}, thus meeting the obvious need for a higher capacity in CO_2 transport associated with the air-breathing habit. Figure 7, showing a composite plot of overall buffering capacity in air-breathing fishes, illustrates the same adaptive increase in buffering capacity with the increased importance of air breathing.

Few studies have assessed the influence of temperature on the respiratory properties of blood in air-breathing fishes. A comparison of blood from Neoceratodus and Protopterus shows that the former, which experiences large annual temperature shifts, is notably insensitive to temperature changes, while Protopterus, living in a temperature-stable habitat, has far more temperature-sensitive blood (Lenfant and Johansen, 1968).

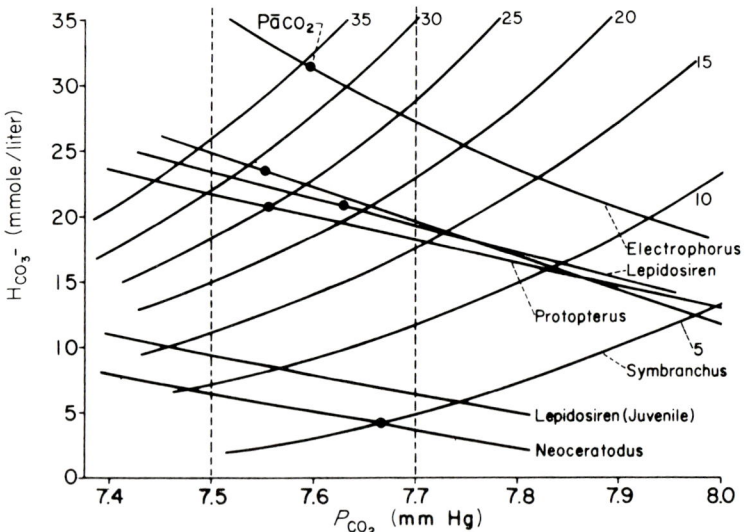

Fig. 7. A comparison of standard bicarbonate and buffering capacities in representative air-breathing fishes. Information selected from Johansen et al. (1968), Electrophorus; Lenfant et al. (1970), Lepidosiren and Symbranchus; Lenfant and Johansen (1968), Protopterus; Johansen and Lenfant (1967), Lepidosiren (juvenile); and Lenfant et al. (1966), Neoceratodus.

B. Gas Exchange in Air-Breathing Fishes

All air-breathing fishes possess some capacity for aquatic breathing. Thus gas exchange is bimodal and depends upon exchange diffusion with the water via gills or skin and with air rhythmically taken into, and expelled, from, the air-breathing organ. The fish may employ both means of gas exchange simultaneously or may resort to one method at a time depending upon external circumstances. A very few fishes undertake voluntary migrations over land in moist or wet grass. Aquatic respiration must then be altogether halted except for some possible exchange via the skin. Estivating air-breathing fishes represent a special case. They must rely almost exclusively on aerial gas exchange although retention of a moist microenvironment in their cocoons or burrows in the substratum is essential for survival. This requisite may be as important for fluid balance and osmotic control as for continued cutaneous gas exchange. The literature is deficient of any attempt to study gas exchange in estivating fishes.

Air-breathing fishes experience two vastly different environments. A low solubility of O_2 in water provides an oxygen availability, in aerated water, about one-thirtieth of that in air in a tropical environment. Hence, aquatic breathers must ventilate much larger volumes than air breathers at similar O_2 uptakes. The greater work of breathing in water is further aggravated by the huge density difference between the two media. The high solubility of CO_2 in water along with the high ventilation of aquatic breathers brings about internal CO_2 tensions which may be one-thirtieth of those usually found in terrestrial vertebrates. Air breathing is advantageous in terms of O_2 availability and high diffusion rates, but it is countered by the hazards of desiccation and by the different requirements for mechanical support under the influence of net gravitational forces in air. Exclusive air breathers also have to meet the need for increased efficiency of internal buffering systems because of elevated internal CO_2 tensions. Air-breathing fishes largely avoid these problems by remaining essentially aquatic, only exploiting the atmosphere as a source of oxygen.

All air-breathing organs in fishes show a low gas exchange ratio indicating that their function is mainly O_2 absorption. Gills and skin conversely play their most important role in CO_2 elimination and show accordingly gas exchange ratios exceeding unity.

Table III offers a compilation of the relative role of aquatic and aerial gas exchange in the few fishes that have been adequately studied.

Among obligate air breathers such as *Lepidosiren* and *Protopterus*, aquatic breathing is of little significance in O_2 absorption and contributes

Table III
Oxygen Uptake, CO_2 Production, and Gas Exchange Ratio for Aquatic and Air Breathing in Representative Air-Breathing Fishes

Species	Oxygen uptake (ml/min/kg)				CO_2 production (ml/min/kg)				Gas exchange ratio			Remarks	Reference
	Aquatic	Air	Total	Aquatic/Air	Aquatic	Air	Total	Aquatic/Air	Aquatic	Air	Total		
Protopterus	0.021	0.169	0.20	0.12	0.101	0.013	0.15	2.34	4.7	0.25	0.75	Fasting, free swimming with access to air, 20°C	Lenfant and Johansen (1968)
Lepidosiren (150 g)	0.90	0.51	1.40	1.80	1.21	0.37	1.58	3.27	1.34	0.72	1.12	Juvenile, 18°C	Johansen and Lenfant (1967)
Amia	0.316	1.20	1.52	0.31	1.54	0.48	2.02	3.2	4.9	0.37	1.33	19°C	Johansen et al. (1970)
Gillichtys	0.34	1.66	2.00	0.205	1.15	0.54	1.69	2.09	3.1	0.33	0.85	27°C	
			1.73							0.28		20°C	Todd and Ebeling (1966)
Neoceratodus	0.25		0.25		0.31							18°C	Lenfant et al. (1966)
Cobitis fossilis	0.24	0.99	1.23	0.24	0.39	0.80	1.19	0.5	1.6	0.2	1.24	25°C, adult	Calugareanu (1907)
Lepidosiren	0.03	0.67	0.70	0.04						0.81	0.98	20°C	Sawaya (1946)
Protopterus			0.87									20°C feeding	Smith (1930)

less than 10% of the total oxygen uptake in adult fish (Johansen and Lenfant, 1967; Lenfant and Johansen, 1968). Aquatic breathing remains very important in CO_2 elimination and is, in *Protopterus* (Lenfant and Johansen, 1968), about 2.5 times more effective than the lungs in removing CO_2. In many fishes the skin plays an important role in gas exchange, particularly in CO_2 elimination. Cunningham (1934) reported a gas exchange ratio in excess of 10 for the skin of *Lepidosiren* in water. During air exposure the skin is also important in gas exchange and marked cutaneous vasodilation has been reported for the common eel, *Anguilla* (Berg and Steen, 1965) and *Lepidosiren* (Johansen and Lenfant, 1967).

In *Cobitis fossilis*, an intestinal air breather (Calugareanu, 1907), and *Gillichtys mirabilis*, a marine gobiid fish practicing accessory air-breathing with the buccopharyngeal cavity (Todd and Ebeling, 1966), O_2 absorption is the dominant feature of air breathing (Table III).

The large variation of total O_2 uptake, apparent in Table III, may be related to feeding. Oxygen uptake in *Protopterus* has been reported to decline rapidly between 24 and 48 hr after the last food intake, and a constant metabolic rate was maintained only as long as the intake of food exceeded the prevailing oxidation rate (Smith, 1930).

A labile oxygen uptake was also apparent during air exposure of juvenile *Lepidosiren* (Johansen and Lenfant, 1967) and adult *Amia*. In *Lepidosiren* the O_2 uptake fell to 20% of the value in water in less than 6

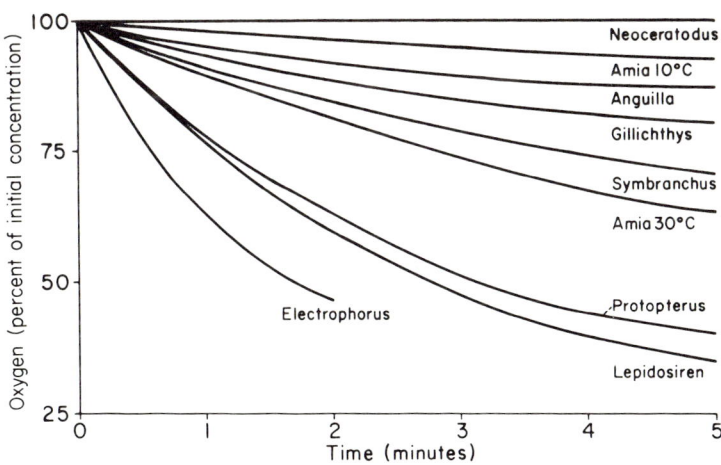

Fig. 8. Rate of O_2 depletion from air-breathing organs in various fishes. The slopes represent average values. Information selected from Lenfant *et al.* (1966), *Neoceratodus*; Johansen *et al.* (1970), *Amia*; Berg and Steen (1965), *Anguilla*; Johansen (1966), *Symbranchus*; Todd and Ebeling (1966), *Gillichtys*; Lenfant and Johansen (1968), *Protopterus*; Johansen and Lenfant (1967), *Lepidosiren*; and Johansen *et al.* (1968), *Electrophorus*.

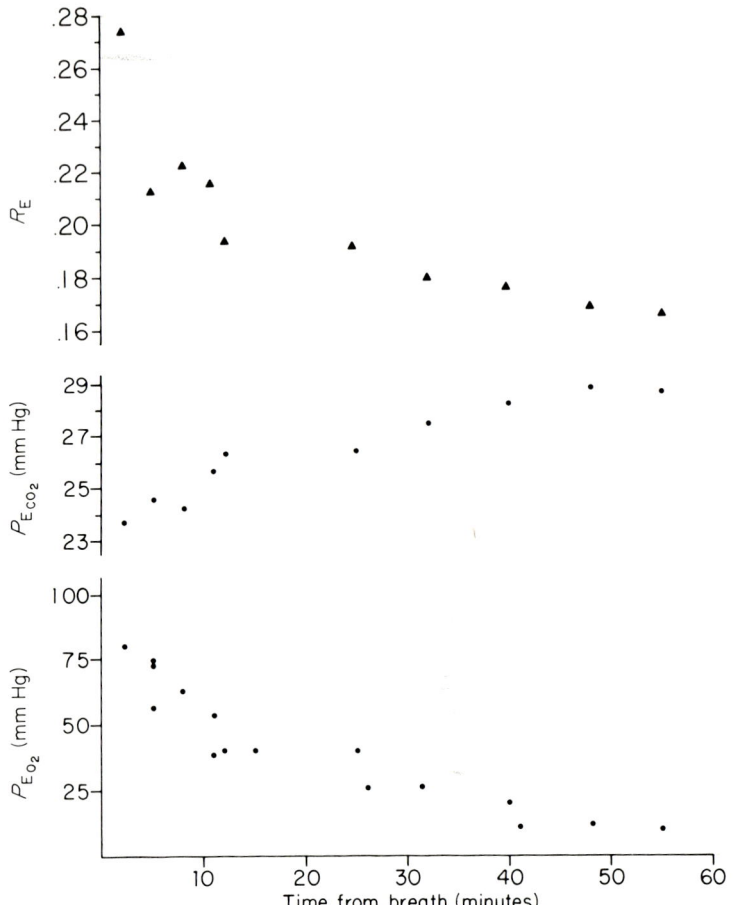

Fig. 9. Time course of changes in O_2 tension, CO_2 tension, and the gas exchange ratio (R_E) inside the lung of *Protopterus* between successive air breaths. From Lenfant and Johansen (1968).

hr. Similarly, Berg and Steen (1965), working on the common eel, reported on O_2 uptake in air half of the value measured in water. A different pattern is evident in *Protopterus* during air exposure. A normal level of O_2 uptake was maintained when rebreathing air until the ambient O_2 tension had fallen to 80 mm Hg and CO_2 tension had risen to 35–40 mm Hg.

Figure 8 shows the rate of change of gas composition inside air-breathing organs of a number of fishes based on samples obtained when the animals were resting in well-aerated water. The rate of O_2 absorption in the obligate air breathers is conspicuously higher than in the fishes employing accessory or supplemental air breathing. Note, also, the marked effect of temperature on the rate of O_2 depletion from the air bladder in *Amia*.

A comparison of the slopes for O_2 depletion and CO_2 accumulation reveals the former to be much steeper, suggesting a low gas exchange ratio. The slopes for CO_2 accumulation also show a more abrupt change to a gentler slope late in the interval between surfacings for air, a fact suggesting that the low gas exchange ratio becomes even lower, later in the interval between air breaths. In the case of *Protopterus* and *Electrophorus*, this change is particularly apparent (Figs. 9 and 12). Thus aquatic CO_2 elimination varies and may increase with time, possibly owing to a vasodilatory effect of increasing CO_2, in gills and skin vessels.

The slopes given for O_2 depletion in Fig. 8 represent average values, and variations likely reflect changes in the blood perfusion through the exchange organ. The average slopes seem related with the breathing habits of the fishes. Thus, in *Electrophorus*, intervals between air breaths rarely exceeded 2 min, while in *Lepidosiren* and *Protopterus* the intervals varied but were usually between 3 and 6 min, whereas, *Symbranchus* and *Amia* both showed infrequent surfacings often exceeding 25–30 min between air breaths.

Few studies have assessed the total volume and tidal volume of air-breathing organs in fishes. Juvenile specimens of *Protopterus* (weight 57–78 g) are reported to show average lung volumes of 1.7 ml (or 25.3 ml/kg) measured by radiological techniques. Tidal volumes measured by spirometry averaged 1.45 ml (or 19.8 ml/kg) and, hence, represent a fraction of total lung volume as large as 80% (Jesse *et al.*, 1967).

C. Internal Gas Transport in Air-Breathing Fishes

Gas transport between the respiratory organs and the metabolizing tissues depends on the metabolic rate, the pattern and rate of blood circulation, and on the respiratory properties of blood. Very few air-breathing fishes have been adequately studied by frequent sampling of blood and gas to allow an evaluation of internal gas transport. Table IV summarizes available information on gas composition in blood and aerial exchange organs.

The Australian lungfish, *Neoceratodus*, is special among the lungfishes by being predominantly a water breather. During rest, in aerated water, the gills are responsible for the entire O_2 absorption which suffices to maintain the arterial blood 95% saturated with O_2. Comparisons of gas tensions in samples from the pulmonary artery, pulmonary vein, and

Table IV
Normal Values of Blood Gas Tensions Sampled from Free Swimming Unrestrained Species of Air-Breathing Fishes

Species	Systemic venous blood		Systemic arterial blood			Afferent blood aerial organ (pulmonary artery)		Efferent blood aerial organ (pulmonary vein)		Temperature (°C)	Reference
	P_{O_2}	P_{CO_2}	P_{O_2}	% sat.	P_{CO_2}	P_{O_2}	P_{CO_2}	P_{O_2}	P_{CO_2}		
Lepidosiren	—	—	31.5	80	—	10.0	—	48–50	90	18–20	Johansen and Lenfant (1967)
Neoceratodus	14.3	6.5	38.9	95	3.60	38.9	3.6	36.3	3.8	20	Lenfant et al. (1966)
Protopterus	2	—	27.0	78	25.7	19.9	25.5	39.8	21.7	20	Lenfant and Johansen (1968)
Electrophorus	—	—	20.8	70	27.7	20.8	27.7	—	—	25–28	Johansen et al. (1968)
Lepidosiren	—	—	38.0	86	22	28	—	69	—	25–28	Lenfant et al. (1970)
Symbranchus	—	—	22	90	29	—	—	—	—	25–28	Lenfant et al. (1970)
Protopterus	—	—	36	—	—	—	—	—	—	25	Lahiri et al. (1968)
Electrophorus	—	—	13	50	29.0	—	—	—	—	—	Garey and Rahn (1968)

the lung disclose a near equilibrium indicating no role of the lung in gas exchange. These samples were obtained between the sporadic breaths for air which often fell more than one hour apart (Lenfant et al., 1966). The two other species of lungfishes emphasize air breathing. In *Lepidosiren paradoxa*, blood from the pulmonary artery was only 40–50% saturated with O_2 while that in the pulmonary vein was almost completely saturated (Table IV). The difference in gas tensions between blood in the dorsal aorta and the pulmonary artery demonstrates a partial selective passage of blood through the heart. This selective passage tends to minimize recirculation of blood from the pulmonary vein back to the lungs (Johansen and Lenfant, 1967; Lenfant et al., 1970).

In *Protopterus aethiopicus* passage of blood through the lung increases the O_2 saturation more than 30% to exceed 90% of full saturation.

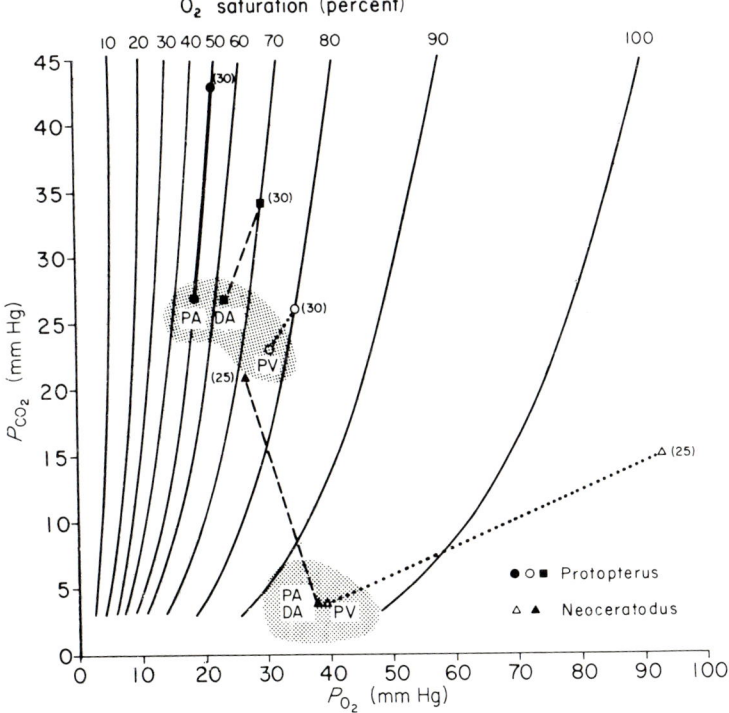

Fig. 10. Nomogram comparing blood gas values in *Neoceratodus* and *Protopterus* during undisturbed breathing in well-aerated water (values on stippled background) and following air exposure. (Duration of air exposure in minutes indicated by numbers in parentheses.) PA, pulmonary artery; DA, dorsal aorta; and PV, pulmonary vein. From Lenfant and Johansen (1968).

There is a concurrent drop in CO_2 tensions but aquatic respiration still dominates the elimination of CO_2 (Table IV).

The marked differences in the importance and efficiency of air breathing in *Neoceratodus* and *Protopterus* can best be discussed when the blood gas data are described in a nomogram (Fig. 10). While resting in aerated water, *Neoceratodus* practices water breathing only, whereas *Protopterus* practices frequent air breathing which results in a sizable gain in O_2 saturation of blood passing the lungs. A significant drop in blood CO_2 tension also accompanies perfusion of the lungs. The much higher CO_2 tensions in *Protopterus* than in *Neoceratodus* is a correlate of the increased importance of air breathing. The behavioral response to air exposure of these two lungfishes differs markedly. While both intensify their air breathing by increased frequency and depth, *Neoceratodus* becomes restless and splashes about in frantic efforts to get back into water. *Protopterus*, on the other hand, normally takes to air exposure with ease. Air exposure of *Neoceratodus* promptly engages the lung in O_2 absorption, and the O_2 tension in blood from the pulmonary vein increases from about 40 mm Hg to more than 90 mm Hg. Yet, it is apparent that this change fails to maintain the systemic arterial O_2 saturation which drops precipitously because of inadequate blood flow through the lung. Note that the changes in oxygenation are accompanied by a conspicuous rise in the blood CO_2 tension. *Protopterus* shows an increased

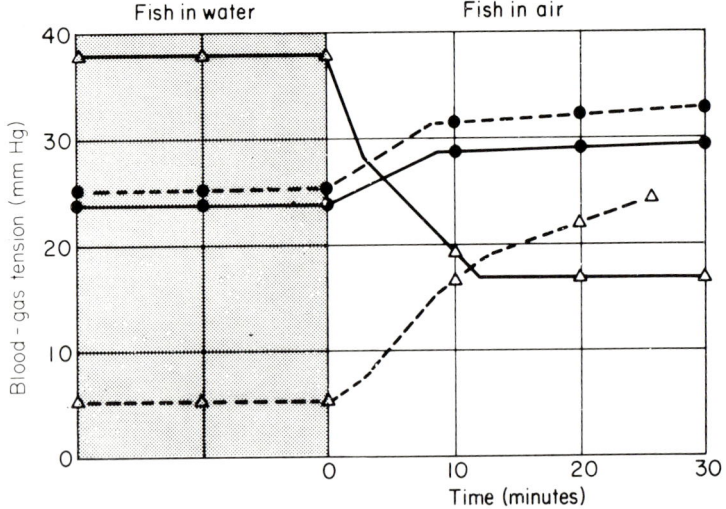

Fig. 11. A comparison of the time course of changes in O_2 and CO_2 tension of arterial blood during air exposure of *Neoceratodus* and *Protopterus*. Broken lines indicate CO_2 tensions. Solid lines indicate O_2 tensions. Adapted from Lenfant and Johansen (1968).

oxygenation in both the pulmonary vein and in systemic arteries. The concurrent increase in CO_2 tension is related to a loss of the important escape route via gills and skin and the inevitable retention of CO_2 resulting from the anatomical dead space of the bidirectional airway. The CO_2 retention is minimized by the high tidal volumes relative to total lung volume. The time course of the blood gas changes during air exposure of the two lungfishes (Fig. 11) also shows the inadequacy of the *Neoceratodus* lung to cope with both O_2 absorption and CO_2 elimination, while the *Protopterus* lung can handle both tasks even though the internal CO_2 levels stabilize at a much higher level in air than in water (Lenfant and Johansen, 1968).

Gas exchange and gas transport in *Electrophorus*, which practices air breathing with the uniquely modified oral mucosa, is illustrated in Fig. 12. The O_2 tension inside the mouth drops from about 140 mm Hg right after a breath to 60 mm Hg at the end of an average 2-min breath

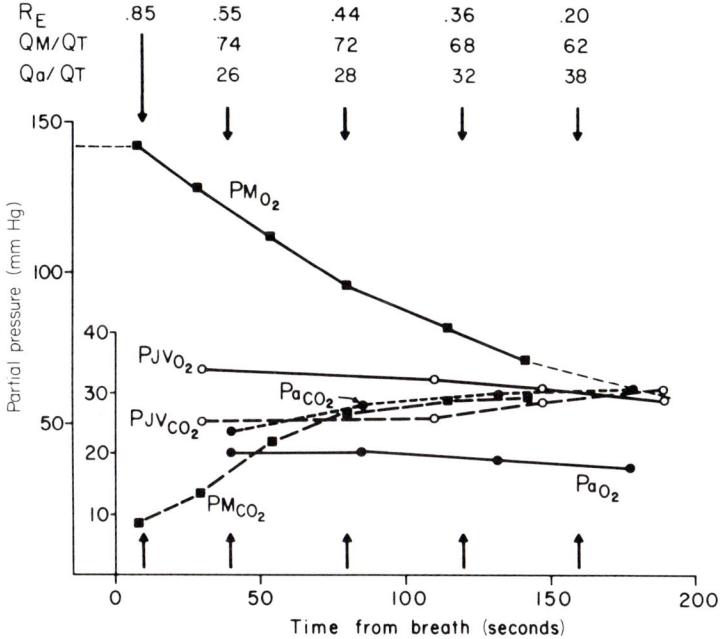

Fig. 12. Changes in O_2 and CO_2 tensions in the mouth gas (M), systemic arterial blood (a), and blood from the jugular vein (JV) in *Electrophorus* between consecutive air breaths. Listed on top are computed values of the gas exchange ratio (R_E), the fraction of cardiac output going to the mouth QM/QT and to the systemic arteries Qa/QT. The vertical arrows indicate the times at which these calculations were made from the corresponding values of gas partial pressures. From Johansen et al. (1968).

interval. Concurrently, the CO_2 tension increases from about 7 to 30 mm Hg inside the mouth, with most of the increase taking place early in the breath interval. The gas exchange ratio drops from about 0.85 just after a breath to 0.20 just before the next breath. Hence, CO_2 is eliminated to a large extent by aquatic gas exchange through the skin or vestigial gills. The low O_2 tension corresponding to about 65% O_2 saturation in arterial blood results from the shunting of blood from the mouth organ back to the systemic veins before arterial distribution. Blood from the jugular vein draining the mouth organ and the anterior systemic veins showed higher O_2 tensions corresponding to about 90% O_2 saturation. This difference in saturation is caused by the additional admixture of systemic venous blood from the posterior end before the blood is distributed from the heart to the arteries (Fig. 1).

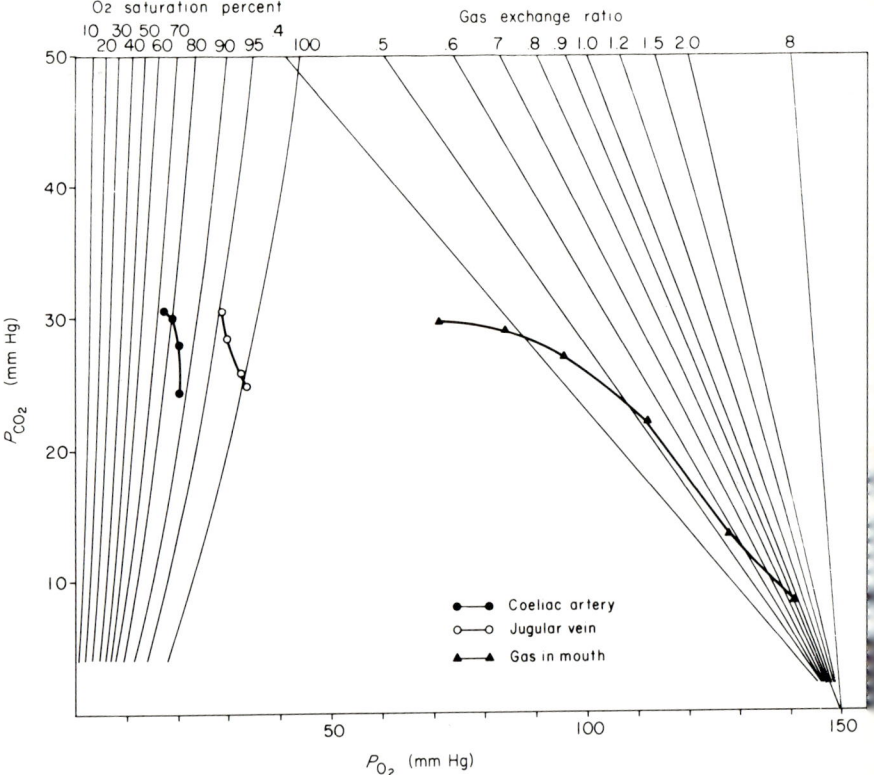

Fig. 13. Nomogram summarizing changes in gas tensions in blood and inside the mouth during an interval between air breaths in *Electrophorus*. From Johansen and Lenfant (1970).

The higher level of CO_2 tension in the jugular vein than in the arterial blood suggests that systemic venous blood returning via the posterior cardinal veins, which receive the major cutaneous drainage, must have the lowest CO_2 tension; otherwise CO_2 must have been largely eliminated during passage through the vestigial gills.

The nomogram (Fig. 13) summarizes results obtained in *Electrophorus* and substantiates the efficiency loss in gas transport caused by the vascular shunt from the mouth organ (Johansen et al., 1968).

D. Control of Breathing in Air-Breathing Fishes

Breathing in vertebrates is, in general, geared to satisfy changing metabolic requirements by maintaining stable internal conditions for gas exchange and gas transport. The factors affecting the control of breathing are markedly different for water breathers and air breathers. The aquatic animal is surrounded by an environment of changing gas composition in which O_2 and CO_2 tensions may change in the same or in opposite directions. Also, in aquatic animals the external and internal environments are in intimate diffusion contact at the gills, and external changes will quickly change the internal environment. Air-breathing terrestrial animals, on the other hand, normally experience an external environment stable in its respiratory qualities; in addition, the presence of an internal gas compartment in the lungs will protect the internal environment against external changes. Control of breathing in the air breather can hence be based on negative feedback from receptors sensing internal parameters important to gas transport such as O_2 and CO_2 tension and pH of arterial blood. The aquatic animal has a more difficult problem in regulation because an internal factor like the arterial O_2 tension may change as a result of external, as well as, internal changes. If, for instance, internal hypoxia occurs when a fish enters severely O_2-deficient water, an increased breathing effort may further aggravate the internal hypoxia. It would be advantageous for the fish to move to water of higher O_2 content. Control of breathing should hence allow the fish to distinguish between changes occurring internally or externally. This complex scheme calls for chemosensitive mechanisms that can screen external as well as internal changes.

It has long been contended that fishes show preference reactions to external gradients in gas composition and that this is linked with an ability to detect the gas tensions in the water by external sensors (Höglund, 1961). Many authors view the orientation of fish to the higher O_2 availability in a gradient as a nondirective escape reaction caused by incipient suffocation. On the other hand, there is little doubt that many fishes living in severely oxygen-deficient swamp water show a clear chemotaxis for the more oxygen-rich surface layers of the swamp. In fact, the existence of many fishes depends on frequent return to the surface layers to satisfy respiratory needs (Carter and Beadle, 1931b; Johansen, 1968).

The apparent variance among fishes in their response to O_2 gradients does not extend to avoidance reactions to P_{CO_2} and pH gradients which reportedly are common and much more acute (Shelford and Allee, 1913; Höglund, 1961). Such reactions are presumably triggered by external gustatory receptors (Scharrer et al., 1947). Some authors consider the breathing responses to CO_2 change in fishes to be connected to chemoreceptors located in the gill region (Dijkstra, 1933; Powers and Clark, 1942; Jesse et al., 1967). Meanwhile, no experimental work has been reported that conclusively identifies the location of and mechanism of chemosensitive reflex control of breathing in fishes.

A consideration of these problems takes on special significance in air-breathing fishes. From their practice of a dual mode of breathing one will expect that air and aquatic breathing are both individually controlled and mutually coordinated.

E. Normal Breathing Behavior

The amplitude and frequency of both air and aquatic breathing in air-breathing fishes are very labile. In fishes employing aquatic as well as air breathing the frequency of branchial pumping generally exceeds the frequency of air breathing. Commonly, the rate of branchial pumping is low just after an air breath, but it increases progressively before the next air breath (Johansen and Lenfant, 1967, 1968). The contribution to gas exchange from air breathing evidently reduces the chemoreceptor drive to branchial breathing. Conversely, this drive will progressively increase with time after an air breath with a resultant higher paced branchial pumping and finally evocation of another air breath. The two types of breathing are apparently under a common or mutual control. In many fishes, the rates of both branchial and air breathing show a notable increase during swimming activity. Young specimens of *Neoceratodus* change from sporadic air breaths during rest in aerated water to a steady higher rate of air breathing during swimming (Grigg, 1965). Similarly, *Polypterus senegalus* and *Amia calva*, both of which use the swim bladder for accessory air breathing, show an augmented rate of branchial and air breathing with the onset of exercise (Magid,

1966). Thus the resting rate of water breathing provides little reserve for increased metabolic activity. The energy expenditure of water ventilation is high, and a definite advantage attends the use of less costly alternative ways of gas exchange during periods of increased demand. Water breathing during low activity, however, permits the fish to exploit its aquatic environment without the need for frequent ascents to the surface for air.

It is generally held that structural adaptations for air breathing in fishes evolved in association with environmental oxygen lack. Yet fishes such as the Australian lungfish, the holostean fishes *Amia* and *Lepisosteus*, and many teleosts breathe air even though their habitats are not oxygen deficient. Thus, internal oxygen deficiencies resulting from activity may have provided a selection pressure for development of air breathing independent of the oxygen availability in the water. Certainly, the exceptional air-breathing habits of marine pelagic fishes such as the tarpon, *Megalops* (Schlaifer and Breder, 1940), support this contention. Recent work comparing the relative importance of water breathing and air breathing in *Amia* at different temperatures has emphasized the importance of air breathing for support of higher metabolic activity (Johansen et al., 1970). The situation resembles the prompt stimulation of breathing in exercising mammals without an apparent initiating role of arterial chemoreceptors.

A few fishes, such as the electric eel, have come so far in the dominance of air breathing that the muscles used for water ventilation and the gills themselves have atrophied to a point where mechanical manifestation of branchial breathing has ceased altogether. However, aquatic gas exchange continues mainly as CO_2 elimination through the skin or the atrophied gills. This important part of overall gas exchange is mainly regulated by changes of blood perfusion at the interfaces between water and blood in the skin or gills.

A marked change in the respiratory behavior of air-breathing fishes likely attend their ontogenetic development. Some species such as the African lungfish are decidedly aquatic breathers during larval and juvenile stages, but air breathing takes on a dominating importance in the adult fish.

F. Breathing Responses to Changes in External Gas Composition

1. Changes of Environmental Oxygenation

Exposure to hypoxic water evokes different breathing responses among air-breathing fishes. The response patterns express an apparent relationship to the relative role of aquatic and air breathing in overall gas exchange.

The Australian lungfish is an almost exclusive water breather when at rest in aerated water. It reacts to hypoxic water by promptly increasing its branchial breathing efforts. Water ventilation increased as much as five times when the O_2 tension in the water was reduced to about 80 mm Hg. Figure 14 shows the result of a typical experiment. Ventilation doubled, while the extraction of O_2 decreased as did the overall O_2 uptake. The increased ventilation was caused by an increase of the

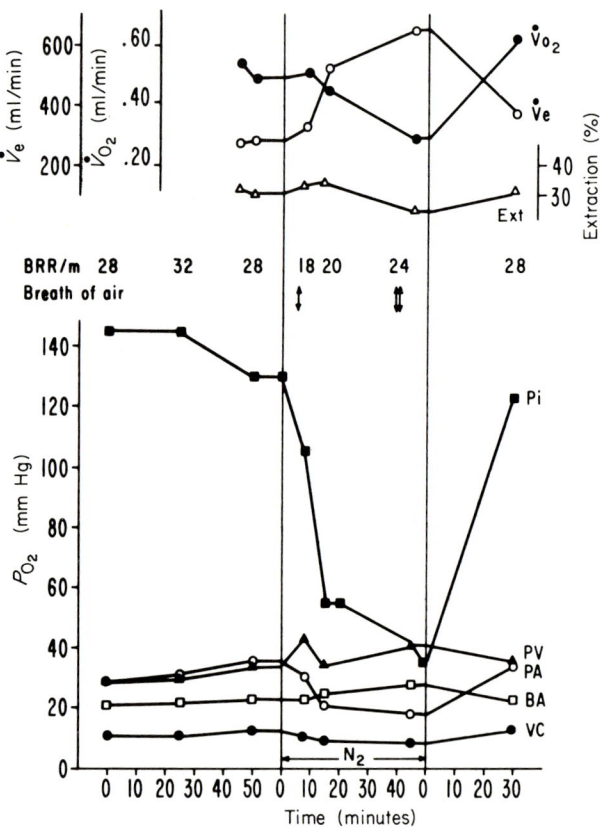

Fig. 14. Responses to breathing deoxygenated water in *Neoceratodus*. $\dot{V}e$, ventilation of water; $\dot{V}_{O_2}$, oxygen uptake from the water; Ext, percent extraction of oxygen from the water; and BRR, branchial respiratory rate. Arrows mark breaths of air. Oxygen tensions followed in: Pi, inspired water; PV, pulmonary vein; Pa, pulmonary artery; BA, an afferent branchial artery; and VC, vena cava. From Johansen et al. (1967).

volume propelled by each branchial pumping movement and not by an increased frequency of the branchial pumping. The pattern of air breathing changed from no air breaths in the hour preceding the hypoxia to more than three air breaths in the first half-hour of hypoxia. The stimulation of air breathing was much slower in onset than the increase of branchial breathing. The commencement of air breathing, however, initiated important changes in the pattern of blood perfusion. Most notably, the O_2 tensions increased in blood drawn from those branchial arteries, which supply the systemic circulation, while blood in the pulmonary arteries decreased in O_2 tension. The O_2 tension of blood in the pulmonary vein also increased. Thus the lung had assumed a role in O_2 uptake, and important circulatory changes attended this change (Lenfant et al., 1966; Johansen et al., 1967).

The yarrow, Erythrinus sp., is another fish utilizing the swim bladder for accessory air breathing (Carter and Beadle, 1931b; Willmer, 1934). Figure 15 shows how the breathing behavior is related to the external gas composition for Erythrinus. Depending on the CO_2 content of the water (see later discussion) aquatic respiration suffices to satisfy the

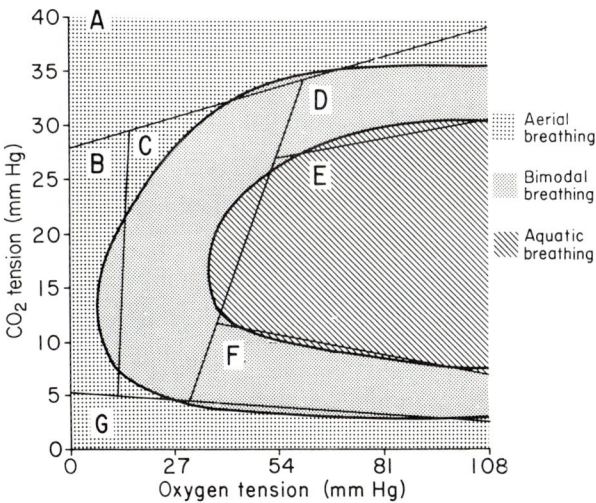

Fig. 15. Relationship between the mode of respiration in the yarrow Erythrinus unitaeniatus, and the O_2 and CO_2 tensions in the external water. Suggested reasons for the respiratory behavior: In area A, CO_2 tensions too high. Gill opercula actively closed. Area B, oxygen lost to water if gills remain active. Area C, gills inadequate to saturate the blood with oxygen. Area D, CO_2 too high, but kept within bounds by aerial respiration. Area E, aquatic breathing sufficient. Area F, gill movements insufficient to saturate blood. Area G, CO_2 content too low to stimulate gill movements. Figure and comments from Willmer (1934).

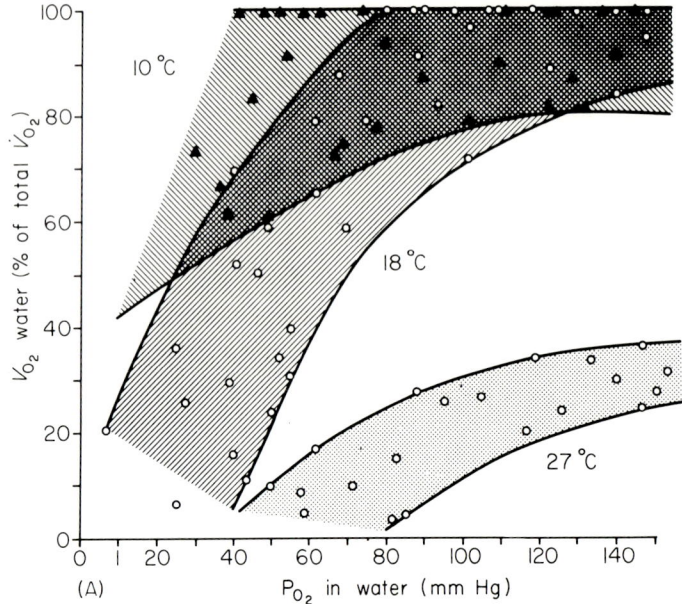

Fig. 16. (A) Oxygen uptake (V_{O_2}) of *Amia calva* from water in percent of the total oxygen uptake, related to the oxygen tension and temperature of the water. (B) Rate of air breathing in *Amia calva* (top) at different temperatures and rates of branchial breathing (temp. range, 14°–26°C) both as functions of oxygen tensions in the ambient water. From Johansen et al. (1970).

respiratory needs at decreasing O_2 tension down to about 40 mm Hg.

The bowfin, *Amia calva*, also utilizes the air bladder as an accessory gas exchanger. The relative importance of air breathing in this fish depends in a large measure on the temperature of the water (Fig. 16A). The fish occupies temperate regions of North America and experiences large annual temperature variations. At water temperatures of 20°–25°C the fish regularly ascends for air even in well-aerated water. Upon deoxygenation of the water the fish initially increases the branchial breathing rate moderately. Further deoxygenation elicits an increased rate of air breathing. When the latter occurs, both the rate and amplitude of branchial breathing drop below the level prevailing before deoxygenation. This depression of branchial breathing may be an adaptive response which reduces the loss of oxygen from the air bladder to the oxygen-deficient water via the gills (Fig. 16B).

Among obligate air breathers such as the African and South American lungfishes or the electric eel, the breathing responses to external changes are different. In these fishes the gills are variously degenerated, and the

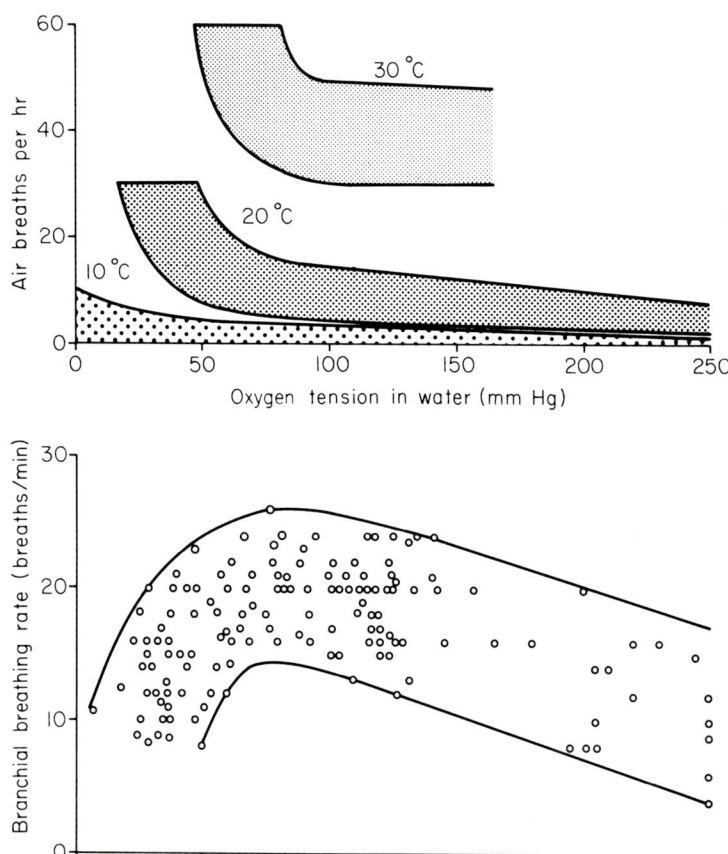

Fig. 16B.

gill surface area is reduced. Thus there is a gradual reduction of the diffusion link between the external water and the blood as the fish shifts emphasis from water to air breathing. Deoxygenation of the water elicits no change in the breathing pattern of these fishes. In a sense they have become relieved of the environmental deoxygenation stress which originally provided the selection pressure favoring the air-breathing habit. Blood gas measurements have substantiated that no change in internal O_2 tension occurs in response to deoxygenation of the water, nor do the fishes exhibit any restlessness suggesting the presence of external chemosensation of O_2 that could guide the fish in avoidance or attraction to the degree of external oxygenation (Johansen and Lenfant, 1968).

Experiments on juvenile specimens of the African lungfish have produced data in conflict with those cited above (Jesse et al., 1967). Deoxygenated water elicited an increase of both branchial and air breathing in the juveniles. However, these specimens survived on water breathing alone for several days, which contrasts markedly with other observations on adult fish. In addition, the experimental procedure evoked hypoxia simultaneously in water and in the air phase and consequently could stimulate the fish via the lung, as well as the gills. The interesting possibility emerges from these studies that control of breathing changes markedly during the ontogenetic development of the African lungfish.

A logical and simple experiment used to test the importance of air breathing in fish is to deprive them of access to air. Depending on the temperature and gas composition of the external water, such experiments have produced variable results, but they show that the degree of internal oxygenation clearly influences the breathing efforts. The attempts to breathe air increase in all air-breathing fishes, while branchial breathing appears to increase only in species normally dependent on branchial breathing for an important part of their oxygen uptake.

Experiments in which the fish have surfaced into atmospheres of controlled gas composition, while the conditions in the water have remained stable, have been more conclusive. Only in a few experiments have internal blood gas tensions been correlated with such procedures. In the African lungfish, introduction of hypoxic gas into the lung causes a prompt and sharp increase in the rate of air breathing. Simultaneous measurements of arterial O_2 tension show that air breathing is stimulated at much higher values than those recorded when the fish practices undisturbed breathing in normal air (Johansen and Lenfant, 1968). This implies that the chemosensitive areas which trigger the response to inhalation of hypoxic gas are more sensitive to the rate of change than to the actual level of oxygen tension in the hypoxic condition. Experiments on *Electrophorus* support such a contention by showing remarkably prompt responses to changes in composition of inhaled air. Figure 17a illustrates how the rate of air breathing was changed after a single breath of air following hypoxic breathing. This implies a location of the chemoreceptors in the buccal mucosa or in the blood path in intimate contact with the gas in the mouth. A location in the systemic arteries is ruled out because of the complete shunt of blood from the gas exchange organ to the systemic veins. The resulting changes in systemic arterial blood are much too slow to represent an important feedback stimulus. Figures 17b and 18 show that inhalation of oxygen-enriched atmospheres causes a depression of air breathing and is important by suggesting the removal of a tonic P_{O_2}-dependent stimulus. This indicates that normal spontaneous

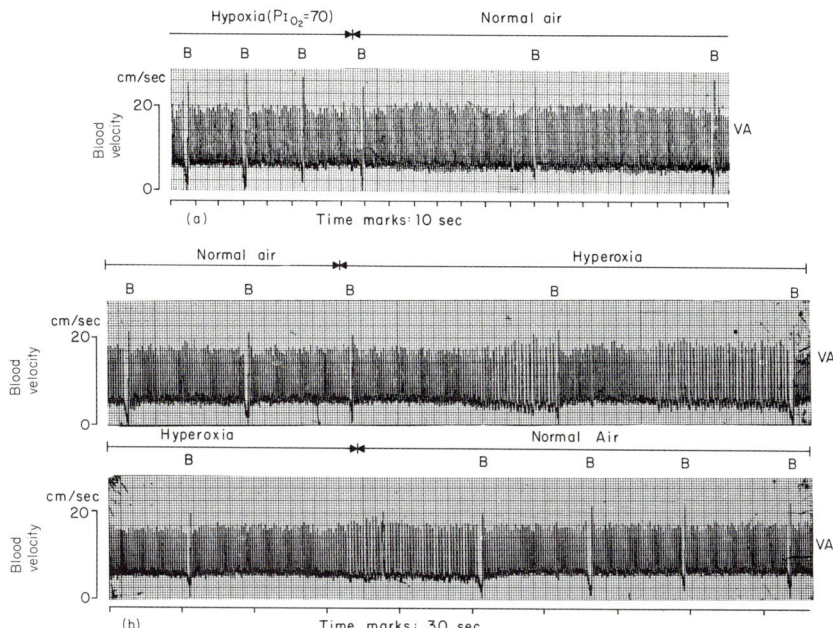

Fig. 17. Continuous tracings of blood velocity in the ventral aorta (VA) of *Electrophorus*. Each air breath shows as a distinct larger excursion (B). (a) shows the very prompt change in the rate of air breathing following the first breath of air after hypoxic breathing. (b) Inhalation of hyperoxic air immediately prolongs the interval between air breaths. From Johansen et al. (1968).

breathing is governed by changes of oxygen tension in the air-breathing organ. The situation corresponds to conditions in higher vertebrates where oxygen inhalation is effective in removing the tonic activity of chemoreceptor cells stimulated by the normally prevailing levels of P_{O_2}. Hyperoxic breathing also reduces the rate of air breathing in *Protopterus* and other air-breathing fishes. Species depending largely on aquatic breathing show a depression of branchial breathing in hyperoxic water or following inhalation of hyperoxic air.

2. Changes of Environmental CO_2 Tensions

Internal CO_2 tensions are very low among purely water-breathing fishes living in well-aerated water (Lenfant and Johansen, 1966; Piiper and Schumann, 1967). A negative feedback control of breathing based on the level of metabolically produced CO_2 appears to be of little consequence in controlling breathing in such fishes. However, in stagnant tropical swamps, the CO_2 levels can build up high enough to adversely

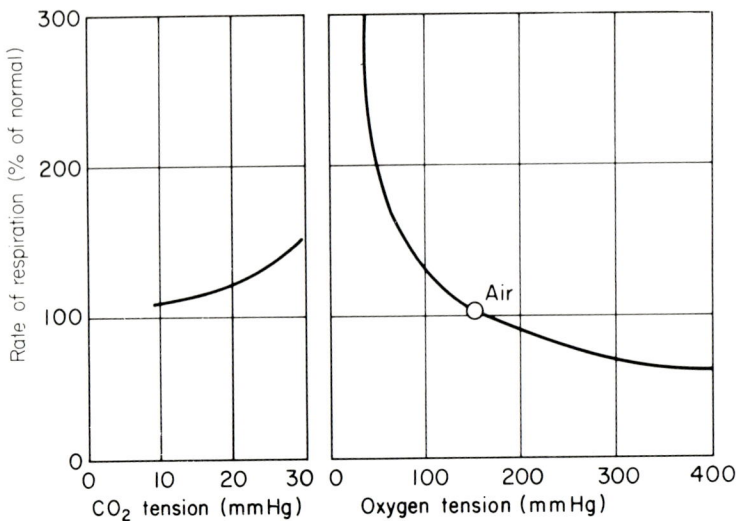

Fig. 18. Percent change in the rate of air breathing of *Electrophorus* in relation to the gas composition of the inhaled air. From Johansen et al. (1968).

affect the dissociation of oxy-hemoglobin and to disrupt internal acid–base balance. Adoption of air breathing also inevitably leads to a retention and increase of internal CO_2 tensions. For tropical air-breathing fishes the external and internal levels of CO_2 may hence be important factors in the control of breathing.

Exposure to increasing CO_2 tensions in the external water evokes extremely varied breathing responses in different species of air-breathing fishes. This variety and the paucity of controlled experiments complicate the evaluation of the role of CO_2 in respiratory control. It appears that CO_2 at moderately high tensions stimulates both branchial and air breathing. At increasing tensions, however, a depression of branchial breathing occurs while air breathing continues at increased rates. The response pattern also seems to depend on the relative importance of air breathing and water breathing in the normal respiratory behavior of the fish.

In a few fishes studied adequately, such as the teleost *Symbranchus* and the Australian lungfish *Neoceratodus*, exposure to CO_2 concentration as low as 1–2% caused a prompt depression of branchial breathing while the rate of air breathing was accelerated; CO_2 concentrations below 1% were not tested (Johansen, 1966; Johansen et al., 1967). In *Amia calva*, which employs accessory air breathing using the air bladder, CO_2 concentrations up to 3% stimulated both branchial and air breathing while higher concentrations depressed branchial breathing. Figure 19 demonstrates the rapid depression of branchial breathing in *Neoceratodus* following exposure to 2% CO_2 in the water. Upon removal of the hypercarbic stimulus, the changes were promptly reversed. Atropinization prevented the CO_2 response. This suggests a cholinergic link in the reflex chain eliciting the response. It is of considerable interest that filling the lung in *Neoceratodus* with a CO_2-rich gas mixture, which greatly elevates the CO_2 level in the pulmonary vein and in blood afferent to the gills, evokes no breathing response whatever. However, blood sampled distal to the gills was very low in CO_2 and attested to a remarkable efficiency of the gills in removing that gas. These data prompted the suggestion that the CO_2-sensitive receptors eliciting the depression of branchial breathing must be external sensors like gustatory or olfactory receptors or they must be located in the blood path efferent to the branchial circulation. Strong corroborative evidence was gained from studies of *Symbranchus*, which is capable of both air and aquatic breathing with

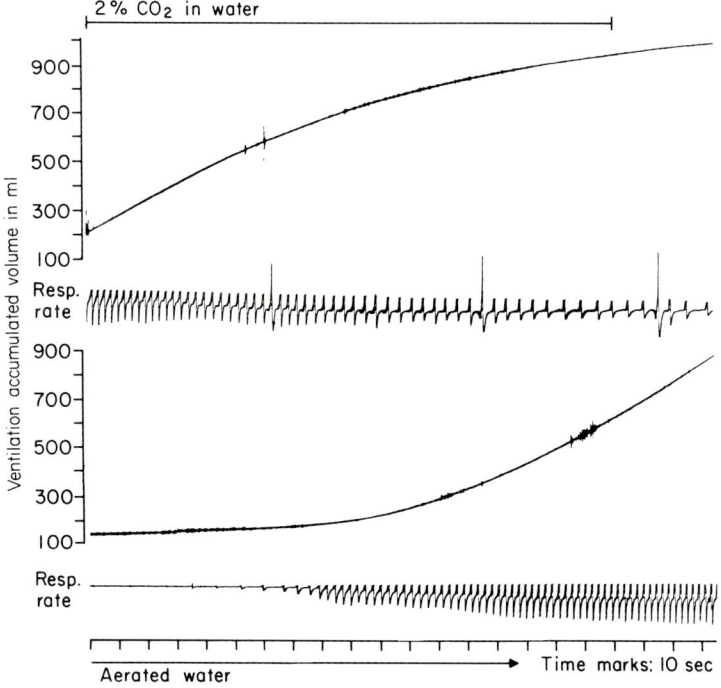

Fig. 19. Continuous recording of water ventilation and branchial respiratory rate in *Neoceratodus* during (top) and after (bottom) an increase in P_{CO_2} of the ambient water. From Johansen *et al.* (1967).

its gills (Johansen, 1966). The fish shows inhibition of branchial breathing in CO_2-rich water which persists as CO_2 leaks into the bloodstream. However, when the fish is transferred to CO_2-free water, a high frequency of forceful branchial movements is established. Hence, the fish appears to possess a sensory mechanism that can differentiate between the external and internal CO_2 level. Similarly, in experiments with *Protopterus*, bubbling of 5% CO_2 in the water stopped branchial breathing before there were appreciable changes in CO_2 tension of the arterial blood.

The stimulation of air breathing by CO_2 may be a direct effect, or it may be secondary to the reduced oxygen uptake caused by the depression of branchial breathing. The response of *Protopterus* and *Electrophorus*, in which branchial breathing contributes little or nothing to the oxygen uptake, suggests a direct effect. Also, inhalation of CO_2-enriched gas stimulates air breathing in these fish while internal oxygen levels remain the same or increase. However, in predominantly water-breathing fishes such as *Neoceratodus* or *Amia*, the influence of CO_2 on air breathing may be both direct and indirect.

The variability found in the few studies on CO_2 response in air-breathing fishes was also apparent in the important work of Willmer (1934). Figure 15 summarizes his work on the factors controlling breathing in the yarrow, *Erythrinus*, a teleost which uses its air bladder for accessory breathing. Willmer contended that a certain CO_2 concentration in the blood is necessary to drive the branchial breathing. At very low O_2 contents of the water, branchial movements ceased, possibly to prevent loss to the water of O_2 gained by air breathing. The opercula also remain actively closed at higher external CO_2 concentrations, allegedly to prevent internal acidosis which would have resulted if gill breathing persisted. The figure also explains the conditions under which respiratory needs are satisfied by aquatic respiration alone and similarly the factors in the environment which cause the fish to employ bimodal breathing.

G. Breathing Responses to Mechanical Stimuli

A possible role of external mechanosensitive receptors in control of breathing rhythms is suggested by recent experiments showing a marked acceleration of branchial breathing rate in *Lepidosiren* and *Protopterus* when the water was stirred or agitated (Johansen and Lenfant, 1967, 1968). Such a role of mechanoreceptors in fish takes on special significance in the normally O_2-deficient stagnant tropical swamps, where stirring of surface water by wind movement or other mechanical agitation should improve the respiratory quality of the water and make branchial breathing more efficient.

Inflation and deflation of the air bladder in *Amia* definitely influence the breathing pattern. A deflation causes the fish to promptly ascend for air, while fishes approaching the surface for air breathing interrupt their ascent and settle to the bottom again if the bladder is artificially inflated. This response occurs whether air, oxygen, or nitrogen is injected. This inflation response shows striking similarity with the inflation reflex described long ago and known as the Hering-Breuer reflex in mammals. In higher vertebrates the inhibition of breathing caused by stretch in the lungs is no longer held to exert any important influence in control of breathing since such changes do not supply information to the respiratory center about the respiratory quality of the blood. In air-breathing fishes a stretch reflex may prove to play an important role since the volume and thus the distention of the air bladder will change during the long intervals between air breaths because of the low gas exchange ratio for the air-breathing organ. Since the air bladder is primarily an O_2 absorber, the change in volume with time provides a measure of the rate of O_2 depletion. If the reflex described in fishes such as *Amia* proves to be homologous with the vagus-mediated inflation reflex in higher vertebrates, the case may exemplify a basic reflex type that exerted a more important regulatory role early in its phylogenetic history.

An important role of the air which a fish takes below the surface is its effect on buoyancy and, indeed, in most teleosts the air bladder is engaged mainly in buoyancy control. In fishes which use the air bladder primarily as a respiratory organ, changes of buoyancy clearly influence breathing behavior. Ascent for air in such fishes may be provoked simply by tying a weight around them (Spurway and Haldane, 1963). The air-breathing fishes consequently face the additional task of coordinating the mechanisms for buoyancy control with those engaged in satisfying the respiratory needs.

H. Breathing Response to Air Exposure

Only a few of the fishes adapted for air breathing make voluntary excursions onto dry land. Estivating fishes are, of course, compelled to subsist on air breathing when entrapped in the substratum during drought.

It seems typical that fishes engaged in voluntary air exposure under natural conditions promptly commence frequent ventilatory movements and show no signs of distress and erratic behavior such as seen in purely aquatic breathers when air-exposed in the laboratory (Berg and Steen, 1965; Johansen and Lenfant, 1968). Only in the South American teleost *Symbranchus* and in the lungfishes have blood gas values been correlated with breathing behavior during air exposure (Johansen, 1966; Lenfant and Johansen, 1968). In *Symbranchus*, arterial P_{O_2} and P_{CO_2} both increased during air exposure. In *Protopterus*, the onset of the accelerated air breathing was too rapid to have been elicited by internal changes in blood gas composition. The sensory input evoking the response must have been the physical act of air exposure itself. It is of interest that *Protopterus*, while recovering from anesthesia, will attempt to breathe air when air-exposed, even while still unresponsive to tactile or painful stimuli. The blood gas values show an increase in both P_{O_2} and P_{CO_2} (Fig. 20); the latter obviously results from the interruption of aquatic CO_2 exchange. The sustained rate of air breathing during air exposure could be driven by the increased level of internal P_{CO_2}, although this stimulus could not play a role in the initial elicitation of the response.

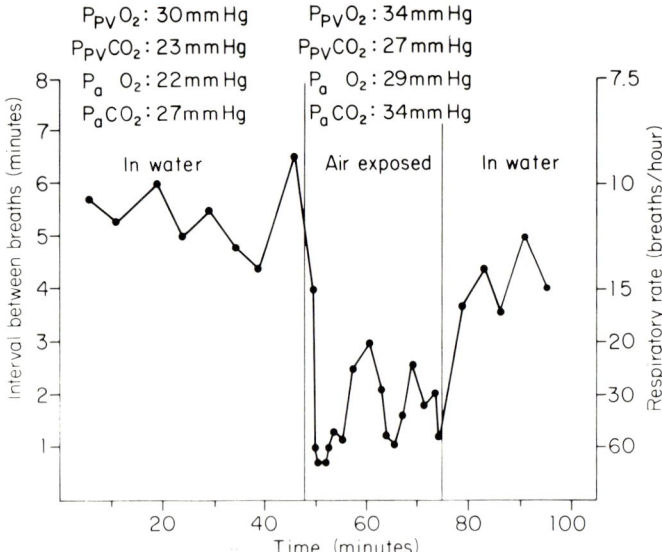

Fig. 20. Changes in rate of air breathing and blood gas tensions following air exposure in *Protopterus*. PV, pulmonary vein; PA, pulmonary artery; and a, systemic artery. From Johansen and Lenfant (1968).

I. Coupling of Respiratory and Circulatory Events

Cardiovascular and respiratory events are strongly interrelated in fishes. Cardiac output determination by the Fick principle in *Neoceratodus* indicated a marked increase during exposure to hypoxic water. The increase in blood perfusion exceeded the relative increase in the ventilation of water. Consequently, the ventilation perfusion ratio of the gills changed from about 12 before hypoxia to about 4, after 45 min exposure to hypoxic water, then climbed to 17 during recovery. The ratio of pulmonary blood flow to total blood flow rose with increased rate of air breathing (Johansen *et al.*, 1967). In *Protopterus*, direct blood flow measurements have shown that spontaneous air breathing causes distinct changes in heart rate, total cardiac output, and regional blood flow. Cardiac output increased sharply with each air breath and declined slowly during the subsequent breath interval (Fig. 21A) (Johansen *et al.*, 1968). Branchial vascular resistance tended to decrease in conjunction with an air breath. The most consistent change was a marked increase of pulmonary blood flow which exceeded that in the vena cava and thus indicated a regional flow shift from the systemic to the pulmonary circuit. Thus, pulmonary flow could vary from less than 20% to more than 70% of the cardiac output.

In *Electrophorus* there is an equally strong interrelationship between circulatory and respiratory events. When the intervals between air breaths were relatively long (exceeding 1 min), cardiac output and heart rate decreased progressively in the late phase of the interval. At the next breath, both heart rate and blood velocity promptly increased and were reestablished at the levels prevailing just after the preceding breath (Fig. 21B). This cyclic phenomenon occurred as a normal event when the intervals between breaths were long, whereas no time-dependent changes were recorded at short intervals. The sudden changes in blood perfusion elicited by air breaths are at least, in part, controlled by the degree of mechanical distension of the air-breathing organ inside the mouth. Artificial inflation of the mouth via a catheter elicited an increase in heart rate and cardiac output whether nitrogen, air, or oxygen were injected; thus, the chemical composition of the gas in the mouth does not constitute the actual stimulus. A similar interrelation between cardiovascular events and distention of the air-breathing organ has been reported for *Symbranchus* (Johansen, 1966). Based on blood gas values the fractional distribution of cardiac output has been calculated in *Electrophorus*. Just after a breath the blood flow fraction

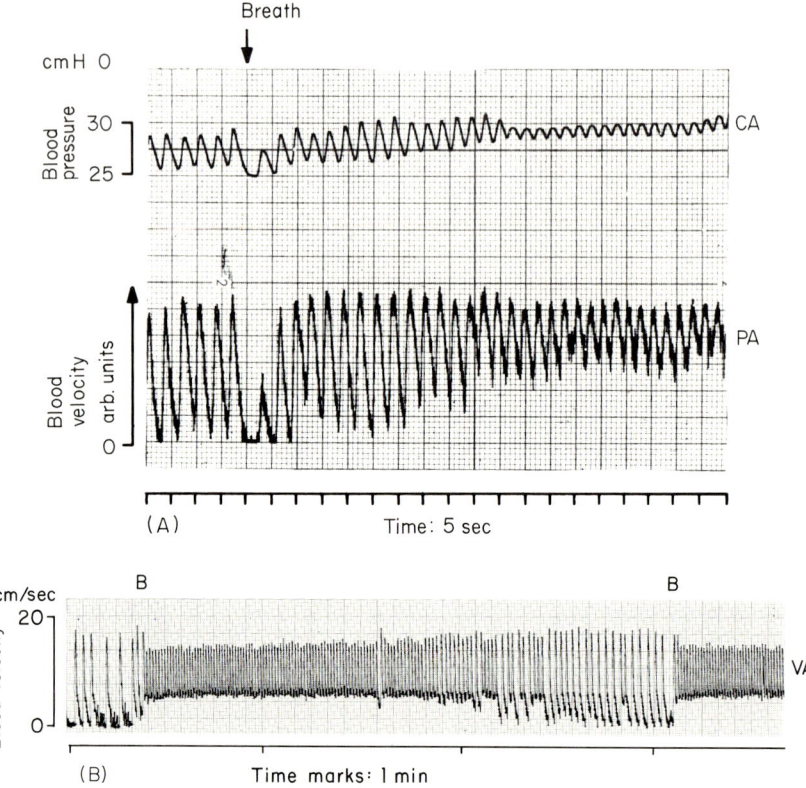

Fig. 21. (A) Change in blood velocity in the pulmonary artery associated with an air breath in *Protopterus*. CA, coeliac artery; and PA, pulmonary artery. From Johansen et al. (1968). (B) Changes in ventral aortic blood velocity related to the phase of the interval between air breaths in *Electrophorus*. VA, ventral aorta; and B, air breath. From Johansen et al. (1968).

going to the air-breathing organ is at its highest, and it declines steadily later in the breathing interval. Similar experiments during hypoxic and hyperoxic breathing showed that blood will shift toward the respiratory organ when the O_2 tension in the mouth is high (Fig. 22). This adjustment has the obvious rationale of promoting the matching process between blood and gas in the air-breathing organ.

Increased perfusion through skin and mucosal linings of the buccal cavity has been frequently reported in air-exposed air-breathing fishes (Johansen and Lenfant, 1967; Berg and Steen, 1965). The marine gobiid fish, *Gillichtys*, displays some remarkable vascular changes when placed in hypoxic water. Under such conditions this fish habitually visits the surface and gulps air which is placed as a bubble between the tongue and the buccal roof. When stimulated to breath air by low O_2 tension in the water, the buccal epithelial capillaries promptly dilate and become engorged with blood, giving the epithelium a corrugated lunglike appearance, bright red from a proliferous circulation. This change in perfusion occurs within a few minutes and is totally absent in related fishes which are purely aquatic breathers (Todd and Ebeling, 1966). The close coupling between mechanisms controlling breathing and vasomotor responses seems basic to very diverse types of respiratory organs ranging from the gills in fishes to the lungs of both lower and higher vertebrates.

Only in the lungfishes among air-breathing fishes has the vascular circuit to the air-breathing organ gained enough anatomical separation from the systemic circuit to allow separate selective perfusion of the two circuits.

During rest in aerated water, the lung in *Neoceratodus* is of no

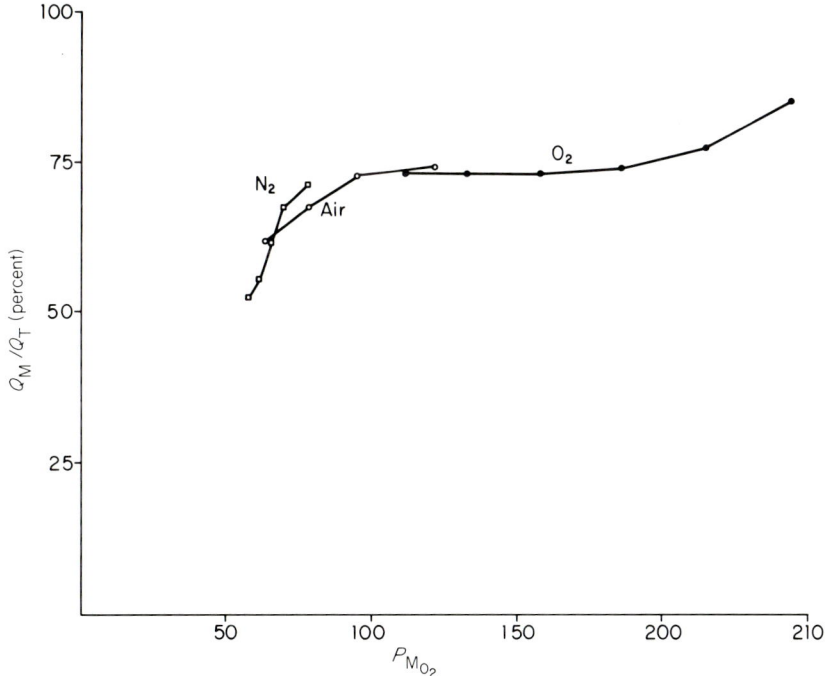

Fig. 22. Relationship between the fractional distribution of blood flow to the mouth air-breathing organ and the oxygen tension of the air inside the mouth of *Electrophorus*. The three curves show the relationship when breathing a hypoxic and a hyperoxic atmosphere and normal air, respectively. From Johansen et al. (1968). Q_M, Blood flow to mouth; Q_A, total cardiac output.

consequence to gas exchange, and no apparent tendency for a preferential distribution of blood from the pulmonary vein to the systemic arteries is apparent. Exposure to hypoxic water evokes an active use of the lung and results in a clear preferential passage of blood from the lung to the anterior branchial arteries which carry the major portion of the blood to the dorsal aorta. One analysis showed that more than 83% blood from the pulmonary vein passes this way, whereas recirculation of systemic venous blood to the anterior arches was reduced to about 16%.

Lepidosiren and *Protopterus* are much more dependent on air breathing and show a consistent gradient of O_2 tension between blood in systemic and pulmonary arteries. This gradient reflects the degree of preferential perfusion which increases with an increased rate of air breathing (Table III). In *Protopterus* (Johansen et al., 1968) the proportion of blood conveyed to the systemic arteries from the pulmonary circuit could be as high as 95% shortly after a breath and decline to about 65% just prior to the next air breath. The factors important in the control of the preferential passage of blood are hence also coupled to the breathing rhythm.

REFERENCES

Barrell, J. (1916). Influence of silurian-devonian climates on the rise of air-breathing vertebrates. *Bull. Geol. Soc. Am.* **27**, 387–436.

Berg, T., and Steen, J. B. (1965). Physiological mechanisms for aerial respiration in the eel. *Comp. Biochem. Physiol.* **15**, 469–484.

Bishop, I. R., and Foxon, G. E. H. (1968). The mechanism of breathing in the South American lungfish, *Lepidosiren paradoxa;* a radiological study. *J. Zool., Lond.* **154**, 263–271.

Boker, H. (1933). Uber einige neue Organe bei luftatmenden Fischen und im Uterus der Anakonda. *Anat. Anz.* **76**, 118–155.

Budgett, J. S. (1900). Observations on *Polypterus* and *Protopterus*. *Proc. Cambridge Phil. Soc.* **10**, 236.

Calugareanu, D. (1907). Die Darmatumung von *Cobitis fossilis*. II. Mitteilung: Über den Gaswechsel. *Arch. Ges. Physiol.* **120**, 425–450.

Carter, G. S. (1935). Reports of the Cambridge expedition to British Guiana 1933. Respiratory adaptations of the fishes of the forest waters, with descriptions of the accessory respiratory organs of Electrophorus electricus L. and Plecostomus plecostomus L. *J. Linnean Soc. London* **39**, 219–233.

Carter, G. S. (1951). "Animal Evolution." Sidgwick & Jackson, London.

Carter, G. S., and Beadle, L. C. (1931a). The fauna of the swamps of the Paraguayan Chaco in relation to its environment. I. Physico-chemical nature of the environment. *J. Linnean Soc. London* **37**, 205–258.

Carter, G. S., and Beadle, L. C. (1931b). The fauna of the swamps of the Paraguayan Chaco in relation to its environment. II. Respiratory adaptations in the fishes. *J. Linnean Soc. London* **37**, 327–366.

Cunningham, J. T. (1934). Experiments on the interchange of oxygen and carbon dioxide between the skin of *Lepidosiren* and the surrounding water, and the probable emission of oxygen by the male *Symbranchus*. *Proc. Zool. Soc. London* 875–887.

Das, B. K. (1927). The bionomics of certain airbreathing fishes of India, together with an account of the development of their airbreathing organs. *Phil. Trans. Roy. Soc. London* **216**, 183–219.
Das, B. K. (1940). Nature and causes of evolution and adaptation of the air-breathing fishes. *Proc. 27th Indian Sci. Congr.* Presidential address, No. 2, pp. 215–260.
DeGroodt, M., Lagasse A., and Sebruyns, M. (1960). Elektronenmikroskopische morphologie der lungenalveolen des *Protopterus* und *Amblystoma*. *Proc. 4th Intern. Conf. Electron Microscopy, Berlin, 1958* Vol. 1, pp. 418–421. Springer, Berlin.
Dijkstra, S. J. (1933). Über Wesen und Ursache der Notatmung. *Z. Vergleich Physiol.* **19**, 666–672.
Dubale, M. S. (1951). A comparative study of the extent of gill-surface in some representative Indian fishes, and its bearing on the origin of the airbreathing habit. *J. Univ. Bombay* [N.S.] **19**, 90–101.
Evans, M. (1929). Some notes on the anatomy of the electric eel, *Gymnotus electrophorus*, with special reference to a mouthbreathing organ and the swimbladder. *Proc. Zool. Soc. London* 17–23.
Fish, G. R. (1956). Some aspects of the respiration of six species of fish from Uganda. *J. Exptl. Biol.* **33**, 186–195.
Garey, W. F., and Rahn, H. (1968). Unpublished data.
George, J. C. (1953). Observations on the air-breathing habit in *Chiloscyllium*. *J. Univ. Bombay* [N.S.] **21**, 80.
Grigg, G. C. (1965). Studies of the Queensland lungfish, *Neoceratodus forsteri* (Krefft). *Australian J. Zool.* **13**, 243–253.
Höglund, L. B. (1961). The reactions of fish in concentration gradients. A comparative study based on fluviarium experiments with special reference to oxygen, acidity, carbon dioxide, and sulphite waste liquor (SWL). Rept. No. 43, pp. 1–147. Inst. Freshwater Res. Fish Board, Sweden.
Jesse, M. T., Shub, C., and Fishman, A. P. (1967). Lung and gill ventilation of the African lungfish. *Respiration Physiol.* **3**, 267–287.
Johansen, K. (1966). Airbreathing in the teleost *Symbranchus marmoratus*. *Comp. Biochem. Physiol.* **18**, 383–395.
Johansen, K. (1968). Airbreathing fishes. *Sci. Am.* **219**, 102–111.
Johansen, K., and Hanson, D. (1968). Functional anatomy of the hearts of lungfishes and amphibians. *Am. Zoologist* **8**, 191–210.
Johansen, K., and Hol, R. (1968). A radiological study of the central circulation in the African lungfish, *Protopterus aethiopicus*. *J. Morphol.* **126**, 333–438.
Johansen, K., and Lenfant, C. (1967). Respiratory function in the South American lungfish, *Lepidosiren paradoxa* (Fitz). *J. Exptl. Biol.* **46**, 205–218.
Johansen, K., and Lenfant, C. (1968). Respiration in the African lungfish. II. Control breathing. *J. Exptl. Biol.* **49**, 453–468.
Johansen, K., and Lenfant, C. (1970). Unpublished data.
Johansen, K., and Reite, O. B. (1967). Influence of acetylcholine and biogenic amines on pulmonary smooth muscle in the African lungfish, *Protopterus aethiopicus*. *Acta Physiol. Scand.* **71**, 248–252.
Johansen, K., Lenfant, C., and Grigg, G. C. (1967). Respiratory control in the lungfish, *Neoceratodus forsteri* (Krefft). *Comp. Biochem. Physiol.* **20**, 835–854.
Johansen, K. Lenfant, C., Schmidt-Nielsen, K., and Petersen, J. A. (1968). Gas exchange and control of breathing in the electric eel, *Electrophorus electricus*. *Z. Vergleich. Physiol.* **61**, 137–163.

Johansen, K., Hanson, D., and Lenfant, C. (1970). Respiration in a primitive air breather, *Amia calva. Respiration Physiol.* (in press).
Klika, E., and Lelek, A. (1967). A contribution to the study of the lungs of the *Protopterus annectens* and *Polypterus senegalensis. Folia Morphol.* **15,** 168–175.
Krogh, A. (1904). Some experiments on the cutaneous respiration of vertebrate animals. *Skand. Arch. Physiol.* **16,** 348–357.
Krogh, A., and Leitch, I. (1919). The respiratory function of the blood of fishes. *J. Physiol. London* **52,** 288–300.
Lahiri, S., Shub, C., and Fishman, A. P. (1968). Respiratory properties of blood of the African lungfish. (*Protopterus*) *Proc. 24th Intern. Union Physiol. Sci. Washington, D.C.* Vol. **7,** 251.
Lenfant, C., and Johansen, K. (1966). Respiratory function in the elasmobranch *Squalus suckleyi. Respiration Physiol.* **1,** 13–29.
Lenfant, C., and Johansen, K. (1967). Respiratory adaptations in selected amphibians. *Respiration Physiol.* **2,** 247–260.
Lenfant, C., and Johansen, K. (1968). Respiration in the African lungfish, *Protopterus aethiopicus.* I. Respiratory properties of blood and normal patterns of breathing and gas exchange. *J. Exptl. Biol.* **49,** 437–452.
Lenfant, C., Johansen, K., and Grigg, G. C. (1966). Respiratory properties of blood and pattern of gas exchange in the lungfish *Neoceratodus forsteri* (Krefft). *Respiration Physiol.* **2,** 1–21.
Lenfant, C., Johansen, K., Petersen, J. A., and Schmidt-Nielsen, K. (1970). Gas exchange in the airbreathing fishes; *Lepidosiren paradoxa* and *Symbranchus marmoratus.* In preparation.
McCutcheon, F. H., and Hall, C. G. (1937). Hemoglobin in the Amphibia. *J. Cellular Comp. Physiol.* **9,** 191–197.
McMahon, B. R. (1969). A functional analysis of the aquatic and aerial respiratory movements of an African lungfish, *Protopterus aethiopicus,* with reference to the evolution of the lung-ventilation mechanism in vertebrates. *J. Exptl. Biol.* **51,** 407–430.
Magid, A. M. A. (1966). Breathing and function of the spiracles in *Polypterus senegalus. Animal Behaviour* **14,** 530–533.
Moussa, T. A. (1956). Morphology of the accessory airbreathing organs of the teleost, *Clarias lazera* (C & V). *J. Morph.* **98,** 125–160.
Munshi, J. S. D. (1961). The accessory respiratory organs of *Clarias batrachus* (Linn). *J. Morph.* **109,** 115–140.
Munshi, J. S. D. (1962). On the accessory respiratory organs of *Ophiocephalus punctatus* (bloch) and *Ophiocephalus striatus* (bloch). *J. Linnean Soc. London, Zool.* **44,** 616–624.
Munshi, J. S. D. (1968). The accessory respiratory organs of *Anabas testudineus* (Bloch) (Anabantidae, Pisces). *Proc. Linnean Soc. London* **179,** 107–126.
Oglialoro, C. M. (1947). Vascolarizzazione della mucosa bucco-faringea e respirazione accessoria attraverso tale regione in alcuni *Teleosti marini. Boll. Soc. Ital. Biol. Sper.* **23,** 990–992.
Piiper, J., and Schumann, D. (1967). Efficiency of O_2 exchange in the gills of the dogfish, *Scyliorhinus stellaris. Respiration Physiol.* **2,** 135–148.
Poll, M. (1962). Etude sur la structure adulte et la formation des sacs pulmonaires des Protopteres. *Ann. Reeks Zool. Wetenschap.* **108,** 131–172.
Powers, E. B., and Clark, R. T. (1942). Control of normal breathing in fishes by receptors located in the regions of the gills and innervated by the Xth and Xth cranial nerves. *Am. J. Physiol.* **138,** 104–107.

Qasim, S. Z., Qayyum, A., and Garg, R. K. (1966). The measurement of carbon dioxide produced by air-breathing fishes and evidence of the respiratory function of the accessory respiratory organs. *Proc. Indian Acad. Sci.* **52B**, 19–26.
Rauther, M. (1910). Die akzessorischen Atmungsorgane der Knochenfische. *Ergeb. Zool.* **2**, 517–585.
Richter, G. (1935). Die Luftatmung und die akzessorischen Atmungsorgane von *Gymnotus electricus* L. *Morphol. Jahrb.* **75**, 469–475.
Root, R. W. (1931). The respiratory function of the blood of marine fishes. *Biol. Bull.* **61**, 427–456.
Sawaya, P. (1946). Sobre abiologia de alguns peixes de respiracao aerea. *Univ. Sao Paulo, Fac. Filosof., Cienc. Letras, Zool.* **11**, 255–286.
Saxena, D. B. (1963). A review on ecological studies and their importance in the physiology of air-breathing fishes. *Ichthyologica* **2**, 116–128.
Scharrer, E., Smith, S. W., and Palay, S. L. (1947). Chemical sense and taste in the fishes, *Prionotus* and *Trichogaster. J. Comp. Physiol.* **86**, 183–198.
Schlaifer, A., and Breder, C. M. (1940). Social and respiratory behaviour of small tarpon. *Zoologica* **25**, 493–512.
Schöttle, E. (1931). Morphologie und Physiologie der Atmung bei wasser, schlamm und landlebenden Gobiiformes. *Z. Wiss. Zool.* **140**, 1–114.
Schulz, H. (1960). Die submikroskopische Morphologie des Kiemenepithels. *Proc. 4th Intern. Conf. Electron Microscopy, Berlin, 1958* Vol. 2, pp. 421–426. Springer, Berlin.
Shelford, V. E., and Allee, W. C. (1913). The reactions of fishes to gradients of dissolved atmospheric gases. *J. Exptl. Zool.* **14**, 207–266.
Smith, H. W. (1930). Metabolism of the lungfish, *Protopterus aethiopicus. J. Biol. Chem.* **88**, 97–130.
Spurway, H., and Haldane, J. B. S. (1963). The regulation of breathing in a fish, *Anabas testudineus*. In "The Regulation of Human Respiration" (D. J. C. Cunningham and B. B. Lloyd, eds.), pp. 431–434. Blackwell, Oxford.
Swan, H., and Hall, F. G. (1966). Oxy-hemoglobin dissociation in *Protopterus aethiopicus. Am. J. Physiol.* **210**, 487–489.
Taylor, M. (1913). The development of *Symbranchus marmoratus. Quart. J. Microscop. Sci.* **59**, 1–51.
Todd, E. S., and Ebeling, A. W. (1966). Aerial respiration in the longjaw mudsucker *Gillichtys mirabilis* (Teleostei: Gobiidae). *Biol. Bull.* **130**, 265–288.
Willmer, E. N. (1934). Some observations on the respiration of certain tropical fresh water fish. *J. Exptl. Biol.* **11**, 283–306.
Wu, H. W., and Kung, C. C. (1940). On the accessory respiratory organ of *Monopterus. Sinensia* **11**, 59–67.

10

THE SWIM BLADDER
AS A HYDROSTATIC ORGAN

JOHAN B. STEEN

I. Introduction 414
II. The Biological Significance of the Swim Bladder . . . 414
III. The Architecture of the Swim Bladder 415
   A. General Organization and Cytology 415
   B. Circulation 419
   C. The Rete Mirabile 420
IV. The Performance of the Swim Bladder 423
   A. The Behavior of the Swim Bladder during
      Vertical Displacements 423
   B. The Swim Bladder as a Float 424
   C. Energy Saved by Neutral Buoyancy 425
   D. The Cost of Depositing Gas 425
   E. The Ability to Concentrate Gases—The Composition
      of the Swim Bladder Gas 426
V. The Mechanisms of Gas Transport 428
   A. The Removal of Gas 428
   B. The Maintenance of a Gas-Filled Swim Bladder . . . 429
   C. The Concentrating Ability of the Rete 430
VI. The Realization of Countercurrent Multiplication
   in the Swim Bladder 432
   A. Release of O_2 from Blood 432
   B. Release of Inert Gases from Blood 433
   C. Multiplication by the Rete 433
   D. Present Concept of Countercurrent Multiplication
      in the Rete 436
   E. Gas Deposition in Fishes with Poorly Developed Rete . . 438
VII. Nervous Control of the Hydrostatic Function
   of the Swim Bladder 438
References 440

I. INTRODUCTION

The swim bladder acts as a hydrostatic organ by replacing part of the fish body, which is heavier than water, with gas. The specific weight of the fish as a whole is thereby brought toward that of the ambient water, and the fish becomes neutrally buoyant.

The function of the swim bladder rests on its ability to maintain a gas space inside a fish and to vary this amount of gas according to changing hydrostatic demands. The problems connected with this function are amplified by the fact that while most natural waters, and consequently the arterial blood of fishes, have a P_{O_2} and P_{N_2} of about 0.2 and 0.8 atm, respectively, the partial pressure of O_2 and N_2 in the swim bladder may be 100 and 20 atm, respectively. This ability of the bladder to concentrate O_2 and N_2 some 500 and 30 times, respectively, is the unique property of the organ. This aspect will therefore be the focus of our interest throughout this chapter. In addition we shall discuss the problems of hormonal and nervous regulation of its activities, as well as the "biological importance" of the organ.

In the swim bladder an intimate relationship between anatomy and function is unusually evident. This holds true for the microstructures but is particularly well displayed in the macroanatomy or, more properly, the architecture of the organ. Since so much of the swim bladder problem hinges on the anatomy, we shall describe this aspect in some detail.

Various aspects of the swim bladder problem have been reviewed at intervals. The articles by Fänge (1953, 1966), Denton (1961), Marshall (1962), Kuhn *et al.* (1963), Steen (1963a,b,c), and Alexander (1966) are of particular interest.

II. THE BIOLOGICAL SIGNIFICANCE OF THE SWIM BLADDER

Marshall (1962) gives an excellent account of the distribution of those marine fishes that have a swim bladder. The bladder is found most frequently among fishes in the upper 200 meters and is less frequently encountered down to 1000 meters. The organ is not found in fishes living in free water below this depth, but it is found regularly in those living near the bottom down to 2000 meters. The report by Nielsen and Munk

10. THE SWIM BLADDER AS A HYDROSTATIC ORGAN   415

(1964) of a swim bladder in a fish caught at about 7000 meters depth seems to represent the present depth record. Except for typical bottom fishes most freshwater teleosts possess a swim bladder.

The fact that typical bottom fishes do not possess a swim bladder is reasonable since they presumably rest on the bottom and thus may benefit by a slight overweight. The arguments presented to justify the lack of a bladder in elasmobranchs and in many active pelagic fishes are less convincing. These arguments often imply that since these fishes swim all or most of their lives anyway, the additional energy needed to counteract sinking is insignificant. While the relative energy expenditure for this purpose will be less in active than in passive fishes, it is hardly less expensive in absolute terms. It is more tempting to suggest that certain disadvantages are associated with the presence of a bladder or certain advantages with its absence.

A fish with a certain amount of gas in the bladder is in buoyancy equilibrium at one depth only. Like a Cartesian diver the slightest vertical displacement will bring it out of equilibrium and the force pushing upward or downward will increase with the displacement. Volume changes are checked by an interplay between removal and deposition of gas. However, these processes are slow as compared to the movements of the fish. Few fishes are able to refill their bladder at surface levels in less than a few hours. Thus while a swim bladder saves locomotory energy for the fish, it inflicts upon it an instability with regard to vertical position. This may be particularly inconvenient for active fishes which change depth rapidly, e.g., to chase prey. These difficulties connected with having a bladder will obviously be less marked the deeper the fish lives since the volume change is proportional to the relative change in hydrostatic pressure.

III. THE ARCHITECTURE OF THE SWIM BLADDER

The anatomy of the swim bladder is, from a functional point of view, very similar in all fishes. It is most instructively displayed in the common eel, *Anguilla vulgaris*, which will therefore serve as an example throughout this section (Fig. 1).

A. General Organization and Cytology

Embryologically the swim bladder develops as a pouch protruding from the foregut, and in some fishes the connection between the

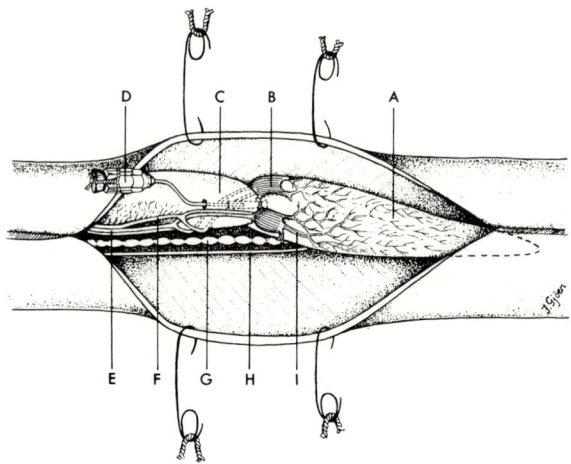

Fig. 1. A full-scale drawing of the swim bladder of an eel. A modified cannula is inserted into the secretory bladder. Intestines and gonads removed. A, secretory bladder; B, rete; C, reabsorbent bladder; D, modified cannula; E, prerete artery; F, postrete vein; G, dorsal artery; H, dorsal vein; and I, postrete artery and prerete vein. From Steen (1963c).

bladder and the digestive tract is retained in adult life. These fishes are called physostomes, and the gas content of their bladder is thus potentially in direct communication with the exterior. This avenue, the pneumatic duct is, however, always guarded by a muscular valve (Fig. 2).

In most fishes the proximal part of the pneumatic duct disappears early in their embryological development, whereas the distal part often develops into a well-vascularized, thin-walled resorbent part. The rest of the bladder becomes the site of gas deposition (Fig. 2). In accordance with earlier nomenclature this is still termed the secretory area or bladder, and the tissue from which gas emanates is called the gas gland. As we shall see later there is no secretory or glandular function in the strict sense of the terms in this tissue.

Fishes with a closed swim bladder are called physoclists. In the literature one will also find the terms paraphysoclist and euphysoclist. The former refers to fishes where the resorbent and secretory part of the bladder is not sharply separated from one another. In the euphysoclists the two areas are separate.

The swim bladder in the eel (Fig. 1) displays to some degree the early embryological "physostome stage" of the physoclist bladder. But although the eel is a physostome, its swim bladder has in all essential

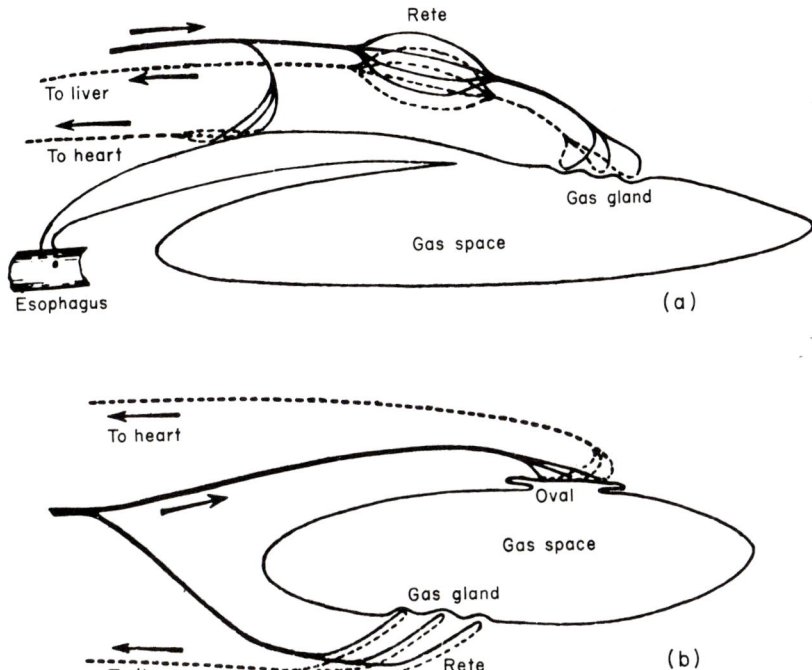

Fig. 2. The two main types of swim bladders: (a) a physostome bladder (from the eel, *Anguilla vulgaris*), and (b) a physoclist bladder (from the perch, *Perca fluviatilis*). Reprinted with permission from E. J. Denton (1961), The buoyancy of fish and cephalopods, *Progr. Biophys.* **11**. Pergamon Press Ltd.

features the same organization as that found in physoclists (Rauther, 1937; Fänge, 1953).

The variability in organization of the different parts of the swim bladder is illustrated in Fig. 3 (see also Woodland, 1911; Marshall, 1962). In some euphysoclists the resorbent part forms a posterior chamber separated from the secretory part by a transverse membrane with a central adjustable opening. This membrane is called the diaphragm. In other euphysoclists the resorbent part is not a bladder but only a well-vascularized dorsal region of the swim bladder mucosa, separated from the rest of the mucosa by a muscular sheath. This is termed the oval. There also exist species provided with swim bladders of undoubtedly physoclist type, but in which neither oval nor diaphragm can be distinguished (Rauther, 1937; Fänge, 1953).

The wall of the secretory bladder consists of an outer silvery layer which is rich in collagen fibers, a middle vascular layer which frequently

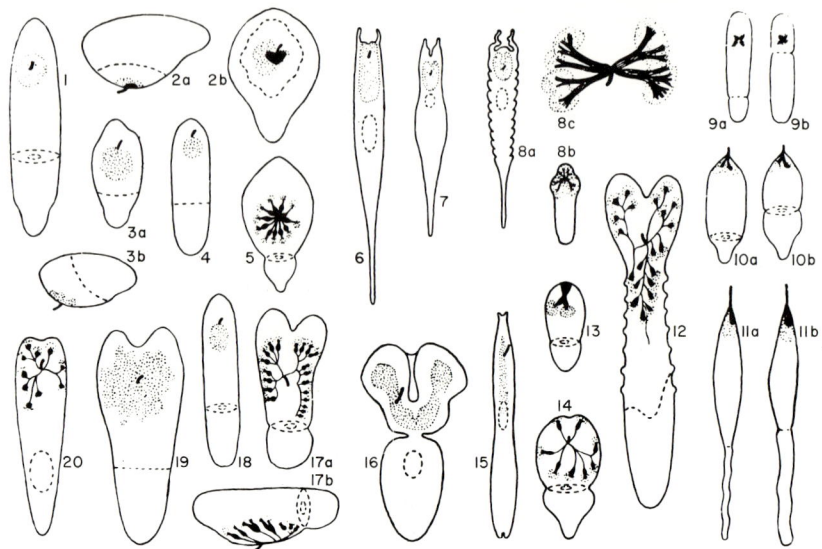

Fig. 3. The variability in architecture of physoclist swim bladders. Dark areas show the retial structure(s). Dotted areas indicate the area from where gas appears. The dotted line indicates the boundary between areas of gas deposition and gas reabsorption. Species names are found in the original paper. From Fänge (1953). 1. *Labrus ossigragus*. Diaphragm. 2. *Gobius niger*. Oval: the secretory bladder is maximally contracted. (a) Left side view; (b) ventral view. 3. *Crenilabrus melops*: (a) ventral view; (b) left side view. 4. *Centrolabrus exoletus*. 5. *Sebastes viviparus*. Diaphragm. 6. *Gadus pollachius*. Nearly closed oval. 7. *Gadus minutus*. Closed oval. 8. *Gadus callarias*: (a) closed oval; (b) embryo. Gas gland and rete mirabile. 9. *Gasterosteus aculeatus*: (a) normal position of the diaphragm; (b) diaphragm in a more cranial position owing to contraction of the secretory bladder. 10. *Spinachia vulgaris*: (a) normal position of the diaphragm; (b) the diaphragm translocated in a cranial direction owing to contraction of the secretory bladder. 11. *Syngnathus acus*: (a) secretory bladder relaxed; (b) secretory bladder contracted. The diaphragm has moved only slightly. The thin-walled caudal part of the bladder is distended. 12. *Lota vulgaris*. The diaphragm has an oblique position. The secretory bladder is halfway contracted. 13. *Onos cimbrius*. Diaphragm. 14. *Onos mustela*. Diaphragm. 15. *Caranx trachurus*. Nearly closed oval. 16. *Raniceps raninus*. Nearly closed oval. 17. *Trigla gurnardus*. Diaphragm: (a) ventral view; (b) left side view. 18. *Ctenolabrus rupestris*. Diaphragm. 19. *Labrus berggylta*. 20. *Perca fluviatilis*. Oval.

contains muscle fibers, and an inner epithelial layer (for details, see Fänge, 1953).

The internal surface of the resorbent part is usually covered by flat epithelial cells, while the secretory epithelium is cuboidal. Dorn (1961) has done an electron microscopic study of the swim bladder of the eel. The resorbent epithelium is reminiscent of the respiratory epithelium

10. THE SWIM BLADDER AS A HYDROSTATIC ORGAN

of primitive lungs. The secretory epithelium on the other hand has a microstructure similar to secretory tissues, with an abundance of reticulum, inclusions, and mitochondria. Thus the cytology of the two parts of the bladder clearly indicates two functionally different elements: a metabolically passive one and a metabolically active one.

B. Circulation

Figure 4 shows the two main types of circulatory arrangements in the swim bladder. In most fishes the swim bladder receives its blood supply from a branch of the coeliac (pneumogastric) artery. This branch first gives off one or more branches to the capillary network of

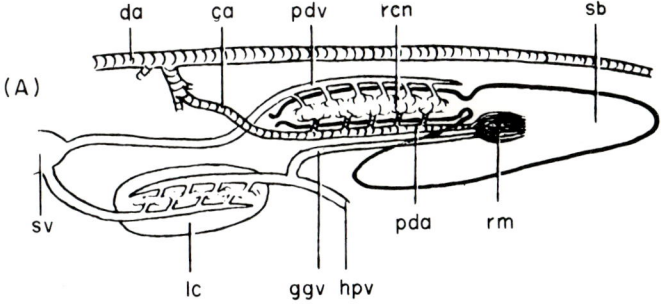

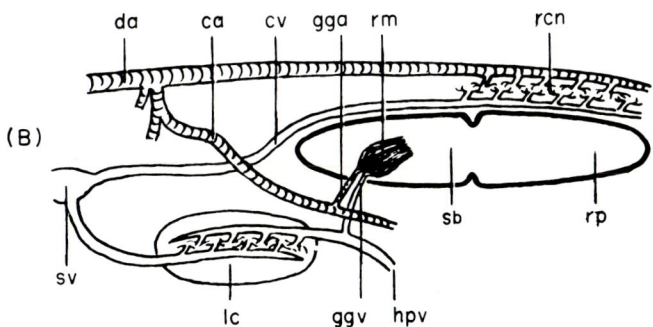

Fig. 4. Diagram of vascular supply to the (A) physostome type of swim bladder (the eel) and (B) euphysoclist type. Abbreviations: ca, coeliac artery; cv, cardial vein; da, dorsal aorta; gga, gas gland artery; ggv, gas gland vein; hpv, hepatic portal vein; lc, liver circulation; pda, pneumatic duct artery; pdv, pneumatic duct vein; rcn, resorbent capillary network; rm, rete mirabile; rp, resorbent part; sb, secretory bladder; and sv, sinus venosus. From Fänge (1953).

the resorbent part, then forms one or more retia and next the capillary network of the secretory bladder is formed. Whereas the arterial supply to the two parts of the bladder comes from one single source, the venous drainage is separate. The blood from the secretory part flows back through the gas gland vein to the hepatic portal vein. Thus it passes the liver before entering the sinus venosus of the heart. The blood from the reabsorbent bladder, on the other hand, runs directly to the sinus venosus through a pneumatic duct vein.

In some physoclists, the arterial supply emanates from two sources (Fig. 4B). The secretory part receives blood from the coeliac artery, while the resorbent part is supplied from intercostal arteries.

It is important to note that all the blood to and from the secretory tissue passes through a rete, while that to the reabsorbent area does not.

C. The Rete Mirabile

Nowhere in biology is the countercurrent principle realized more perfectly than in the rete mirabile of the swim bladder. It is well developed in most physoclists, while in many physostomes it is only weakly developed. The herring is one of the rare species that has a bladder but no rete (Fahlén, 1967).

Usually the rete consists of one or more distinctly discernible bundles of capillaries (Figs. 1 and 3). The relationship between the bundle and the epithelium of the bladder varies. In all euphysoclist bladders the rete and the secretory epithelium have positions resembling the stem and hat of a mushroom (Marshall, 1962). In these cases (Fig. 2b) the rete is termed unipolar since the retial capillaries continue directly into the secretory epithelium without first running together in larger vessels. The whole structure may properly be called "the gas gland." In the bladder of all stomiatoids and in the eel, *Anguilla*, the retia are bipolar (Fig. 2a), that is, with its capillaries distinct from the epithelial ones. In these cases the term gas gland is less meaningful since neither the rete nor the epithelium alone has "glandular" capacity. If used, the gland term should include both structures.

In the case of a bipolar rete there are two capillary beds in parallel. This makes the vascular nomenclature somewhat complex. Steen (1963c) put "pre rete" or "post rete" in front of the artery or vein to distinguish between vessels before and after they had passed the rete (Fig. 5).

From a functional point of view the most important aspect of this structure is that the arterial blood to the bladder and the venous blood

10. THE SWIM BLADDER AS A HYDROSTATIC ORGAN

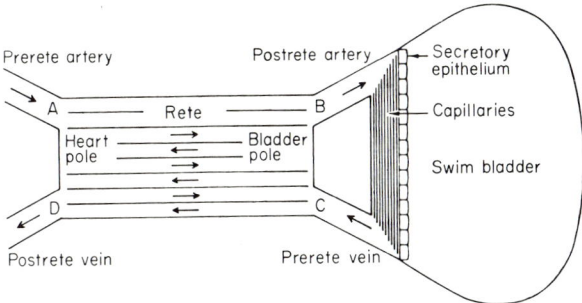

Fig. 5. A simplified drawing of the gas gland with rete and swim bladder. Also given are the anatomical designations used in the text. From Steen (1963c).

from the bladder are in intimate diffusion contact with each other. The arrangement is called a hairpin countercurrent system. This term is not strictly an anatomical one, but it expresses the loop nature of the circulation (Fig. 6). Krogh (1929, pp. 42–45) counted 88,000 venous and 116,000 arterial capillaries in the two retia of an eel. These had an aggregated length of 352 and 464 meters, respectively. The total capillary surface was 106 meters^2 for the venous and 105 meters^2 for the arterial vessels. All this surface was contained in a volume of 0.064 cm^3 of which two-thirds were occupied by blood. The ratio of the total diffusion area to the rete volume is therefore some 1700 cm^2/cm^3. For the long-nosed eel, *Synaphobranchus pinnatus*, the ratio is 1200 (calculated from the data of Scholander, 1954). By comparison the ratio between the alveolar diffusion area and the volume of the human lung is about 100 cm^2/cm^3 (calculated from Forster, 1957). As shown in Fig. 6 the arterial and venous capillaries are arranged in a checkerboard or hexagonal pattern to provide maximum contact area (Scholander, 1954). The distance between the two bloodstreams in the rete of some species may average 1.5 μ (Scholander, 1954; Dorn, 1961) or approximately the same as the distance between air and blood in the human lung alveoli.

The capillary cells of the rete of the eel, *Anguilla*, contain only few and small mitochondria (Dorn, 1961). This observation indicates, when compared to the abundance of large mitochondria in tissues known to have a high aerobic metabolism, that the functioning of the rete does not involve a high expenditure of energy. The electron microscopic appearance of rete capillaries resembles, on the other hand, that of diffusion tissues like lungs. Its abundance of round vesicles and the sparse reticulum gives it a spongy appearance. The microstructure of the rete capillaries thus fits in with the view that their main function is to support passive diffusion between arterial and venous blood. This is also sup-

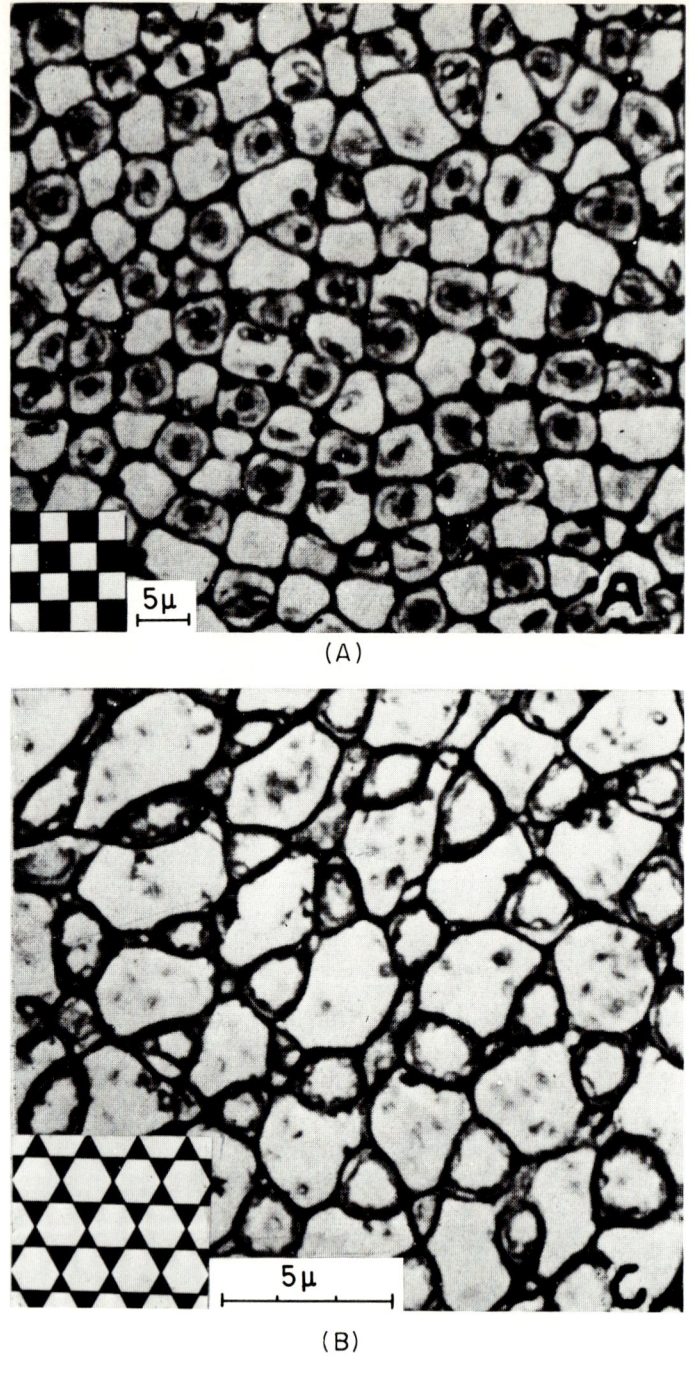

Fig. 6. An example of checkerboard and hexagonal arrangements of arterial and venous capillaries in the rete: (A) from Synaphobranchus and (B) from Coryphoenoides. Note also the intracapillary distance. From Scholander (1954), by permission of Managing Editor, Biol. Bull.

10. THE SWIM BLADDER AS A HYDROSTATIC ORGAN

ported by the observation that a high capacity to deposit gases is correlated with the presence of a rete system which gives the two bloodstreams a large, but thin, common exchange area. This aspect will be discussed in quantitative terms together with the functional aspects of the rete.

IV. THE PERFORMANCE OF THE SWIM BLADDER

A. The Behavior of the Swim Bladder during Vertical Displacements

The behavior of the swim bladder during vertical displacements is shown in Fig. 7. As alluded to earlier, a fish with a swim bladder is in buoyancy equilibrium at one depth only. This seemingly inconvenient situation led earlier workers to suppose that the muscles around it must control its volume. Borelli (1680) believed that the swim bladder was also an organ of locomotion, the muscles being actively constricted when the fish wished to descend and relaxed when the fish wished to rise. Dispute later arose between the supporters of Borelli's views and those who thought that the muscles around the swim bladder were used only to hold its volume constant. These very rational differences of opinion as

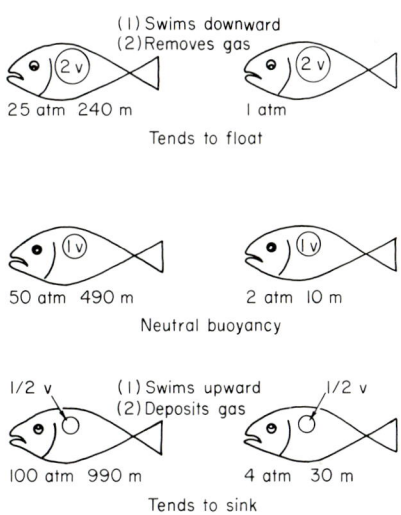

Fig. 7. The behavior of the swim bladder volume of a fish during vertical displacements. The response of the fish is also indicated.

to just how the muscles control the volume of the swim bladder vanished when Moreau (1876) showed that they exert no control whatsoever. Surprisingly enough, fishes are in fact in a state of unstable equilibrium, as we are when standing upright.

Fishes compensate for *transient changes* of pressure by swimming. Moreau showed that when a fish voluntarily changes depth, even by a very small distance, the volume of gas within the swim bladder simply changes in accordance with Boyle's law. Even the gurnards, which have particularly well-developed muscles around the bladder, do not use them primarily for controlling the volume but for making sounds (Moreau, 1876).

Further research has added details to Moreau's conclusions. Alexander (1959a–d) showed that only in Cyprinidae does the swim bladder restrict, although passively, the expansion of the gas when the external pressure falls. For 22 other species he found that the swim bladder obeyed Boyle's law remarkably well. The sea horse, *Hippocampus brevirostris*, has been shown to use the muscles of its swim bladder wall to control not its density but its posture (Peters, 1951).

Moreau also demonstrated that in response to *persistent changes* in pressure the fish can restore the original bladder volume by changing the amount of gas in the bladder (the pressure of the gas is always equal to the external hydrostatic pressure). It thus became easy to see why codfish caught at 100-meter depth had very distended swim bladders when brought to the surface while cod caught in shallow waters had no such difficulties.

These early investigations thus drew up the contours of the hydrostatic function of the swim bladder. When a fish ascends the bladder expands and gas is removed; when it descends the bladder is compressed and gas is deposited. The crucial point is that the fish is able to vary its gas content in such a way that the gas volume is almost constant regardless of the hydrostatic pressure.

B. The Swim Bladder as a Float

In most common marine fishes the swim bladder occupies about 5% of the total volume. This brings the density of the fish very close to that of seawater, which is about 1.025. The density of muscle tissue is about 1.05 and that of the skeleton 2–3 times that of water. The density of fat, on the other hand, is about 0.9. In comparison the density of air at 1 atm is about 0.00125.

The efficiency of the swim bladder as a float is reduced with depth. Thus, while water and the organic components of animals are compressed

10. THE SWIM BLADDER AS A HYDROSTATIC ORGAN   425

to half the volume by 20,000 atm, gases obey Boyle's law fairly well. Thus their volume is halved when the pressure is doubled, for example, on going from the surface to 10-meter depth of the sea. According to Alexander (1966) the specific gravity of O_2 at 700 atm is 0.7, while the density of water is changed insignificantly. Thus even at 7000-meter depth a gas-filled swim bladder gives some buoyancy.

C. Energy Saved by Neutral Buoyancy

The possession of neutral buoyancy enables a fish to save energy in two ways: (1) It can remain motionless in midwater, and (2) the energy expenditure to swim horizontally is reduced.

(1) Since a fish without a bladder is about 5% heavier than water, it will have to exert a force of 5% of its body weight downward on the water if it wishes to stay at one level. A few percent of the body weight may seem quite a small load, but a weight easily supported on land can only be sustained in water by continuous movement. Even an active pelagic fish very infrequently exerts a swimming force corresponding to more than 25% of its weight (Gray, 1953). To exert continuously a force of a few percent of its body weight seems therefore to be a formidable task (see Denton and Shaw, quoted by Denton, 1961).

(2) Alexander (1966) has made an interesting calculation of the amount of energy saved by the presence of a swim bladder during horizontal swimming. When a fish moves at a velocity of one body length per second, the power needed to overcome the sinking tendency is 60% of the total power of movement. When it moves at 10 lengths per second, only 5% of the total power is used to keep it at the same vertical position. At normal cruising speeds, which according to Bainbridge (1960) is 3–4 lengths/sec, about 20% of the total power is expended to overcome sinking. Since we do not know the efficiency of horizontal versus vertical movements, we can only guess that the energy partition must also be of this magnitude.

At present we are not in a position to make a quantitative comparison between the energy needed to stay at a certain depth by swimming as compared to the cost of maintaining a gas-filled swim bladder. It appears, however, that the swim bladder solution is by far the least expensive.

D. The Cost of Depositing Gas

Kanwisher and Ebeling (1957) investigated the metabolic cost connected with vertical migrations. On the assumption that the fish must keep a constant bladder volume during vertical migrations,

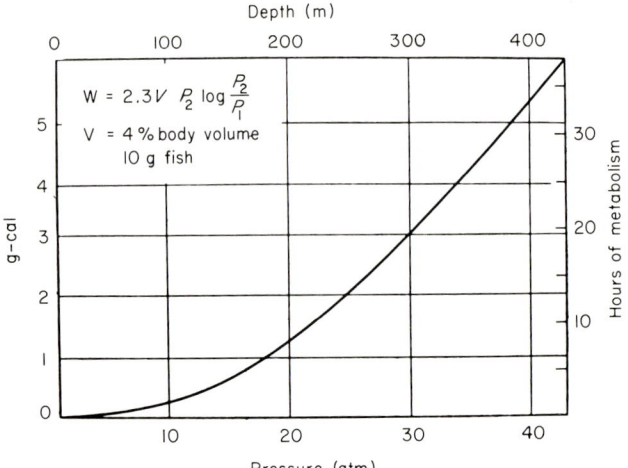

Fig. 8. The accumulated work necessary to migrate from the surface to various depths for a 10-g fish with a 0.4-ml swim bladder. The scale at the right indicates the number of hours necessary to attain neutral buoyancy at the various depths under condition specified in the text. From Kanwisher and Ebeling (1957).

the work performed by the bladder in migrating from pressure P_1 to P_2 is given by the formula:

$$W = 2.3vP_2 \log P_2/P_1$$

where v is the volume.

Figure 8 shows the accumulated work necessary to migrate from the surface to various depths. This work can be compared to the total metabolism of a fish. From data of Scholander and van Dam (1953) a 10-g fish should use about 0.4 ml of O_2 per hour under conditions of aerobic metabolism. This is equal to about 2 g cal. If one-third of this can be used by the swim bladder and the process of gas deposition is 25% efficient, then 0.15 g cal could be used per hour to deposit gas. On this assumption a fish would need 5 hr to go from the surface to a depth of 150 meters all the while being neutrally buoyant. Direct observations on the rate of gas deposition (Fänge, 1953) indicate, however, that an appreciably longer time is needed.

E. The Ability to Concentrate Gases—The Composition of the Swim Bladder Gas

Analysis of the swim bladder gas has been an important method for assessing the capacity of the bladder to concentrate gases. To take full

10. THE SWIM BLADDER AS A HYDROSTATIC ORGAN

advantage of such information two more parameters must be known. One is the total pressure at which the fish was caught, the other is at what pressure the swim bladder gas would have made the fish neutrally buoyant. The first of these values is used to calculate the partial pressure of each gas, the second gives the minimum pressure against which the fish is able to deposit gas.

Biot (1807) was the first to analyze the gas of the swim bladder. He constructed a gas analyzer where he mixed the unknown gas with hydrogen, ignited the mixture, and weighed the resultant water. He was primarily interested in variations of the composition of the atmosphere; and possibly bored by its constancy, he introduced a sample of swim bladder gas into his analyzer. Much to his delight, it is presumed, the gas analyzer exploded. Biot reconstructed his gadget and confirmed the high O_2 content of the gas. More extensive analysis of swim bladder gases from deep sea fishes was first reported by Schloesing and Richard in 1898. A typical example of their results in a specimen caught at 900-meter depth was: oxygen 75.1%, nitrogen 20.5%, carbon dioxide 3.1%, and argon 0.4%. Scholander and van Dam (1953) studied gas taken from the swim bladders of fishes living at depths down to 1400 meters. The general picture emerging from these and other studies is that in fishes living near the surface the swim bladder contains a gas which is much like air, while at greater depths O_2 becomes an increasingly dominant component (Fig. 9). In deep sea fishes the P_{O_2} and P_{N_2} in the bladder may be 100 and 25 atm, respectively, while the P_{O_2} and P_{N_2} of the ambient water, and consequently of the arterial blood, does not exceed 0.2 and 0.8 atm, respectively. The swim bladder has thus the ability to concentrate O_2 by a factor of at least 500 and N_2 by a factor of 30.

Almost a century after Biot's experiments, Hüfner (1892) published the observation that the swim bladder of whitefish, *Coregonus acronius*, from the bottom of Lake Constance at a depth of 60–80 meters contained 99% N_2. Accordingly, the P_{N_2} in the bladder at the depth at which the fish was caught must have been 7–9 atm. Similar findings on the swim bladder gas of several other species of freshwater physostomes from Lake Huron and Lake Michigan were later reported by Saunders (1953), Tait (1956), Scholander *et al.* (1956a), and Sundnes *et al.* (1958). Sundnes (1963) added a significant piece of information when he observed that high P_{N_2} in several species of physostomes was found only in those fishes that had stayed for several weeks at the same depth. In the same species he found high P_{O_2} values in samples taken from fishes that had recently migrated to greater depths.

From the point of view of the mechanism of gas deposition the composition of freshly deposited gas has special interest. Unfortunately, such

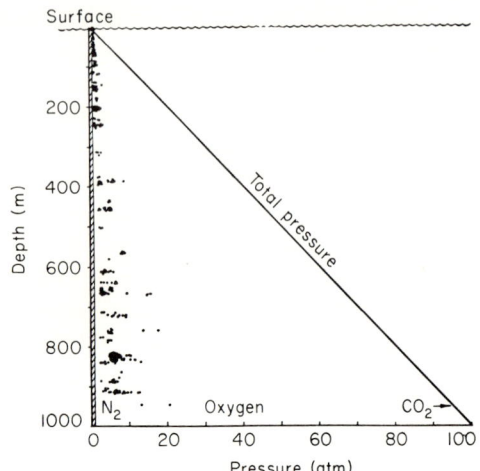

Fig. 9. The P_{N_2}, P_{O_2}, and P_{CO_2} in the swim bladder of fishes at increasing depth. The P_{N_2} is represented by the distance from the black dots to the ordinate, the P_{O_2} by the distance from the dots to the diagonal line. The P_{CO_2} is less than that represented by the thickness of the diagonal line. The shaded area along the ordinate represents the P_{N_2} in the water. From Scholander and van Dam (1953), by permission of Managing Editor, *Biol. Bull.*

data are available only from experiments at surface pressures. Scholander (1956) covered the exposed gas gland of codfish and barracuda (physoclists) with a transparent material and analyzed gas bubbles as they appeared. They contained 5–15% CO_2, 30–81% O_2, and 10–55% N_2. Wittenberg (1958) determined the compositions of gases in swim bladders from physostomes 12 hr to a week after they had been emptied. The collected gas always had a higher P_{O_2} and lower inert gases than the gas mixture with which the water was in equilibrium.

The information on the partial pressure of the gases in the swim bladder seems to indicate, therefore, that O_2 is the prominent swim bladder gas in both physoclists and in physostomes.

V. THE MECHANISMS OF GAS TRANSPORT

A. The Removal of Gas

The removal of gas occurs in two ways: direct release of gas into the water or reabsorption of gas into the bloodstream.

Direct release of gas occurs in physostomes as a consequence of decreased hydrostatic pressure. This is well demonstrated when a herring filled purse seine is hauled up through the water. An ascending school of herring can also be identified on an echo sounder screen by a "cloud" above it. The cloud is reflections from rising gas bubbles, the school itself is detected by reflection from their bladders.

In physoclists, and also in physostomes exposed to mild reduction in external pressure, gas is removed by reabsorption into the blood circulating in the resorbent area. The mechanism in this case is identical to that in respiratory organs: Gases diffuse according to their partial pressure, and the amount of reabsorption is equal to the arteriovenous difference in gas content times the blood flow (Steen, 1963b). Reabsorption is regulated partly by varying the proportion of this area exposed to the gas phase and partly by variation of the blood flow through the resorbent area. In a few physostome fishes (the common eel, Anguilla vulgaris, is an example) the pneumatic duct acts as an organ of gas reabsorption in addition to being a direct route for gas release.

The oxygen reabsorbed for hydrostatic purposes is most likely utilized metabolically. However, since the blood from the reabsorbent region—be it a pneumatic duct or an oval—drains directly to the gills, the possibility exists that some of the reabsorbed O_2, under certain conditions, may be lost to the ambient water.

B. The Maintenance of a Gas-Filled Swim Bladder

The swim bladder is in principle in the same situation as a gas bubble submerged in water which is equilibrated with air: The gas is compressed, and diffusion will ceaselessly cause reduction in the amount of gas. This is directly limited by a supposedly low gas permeability of the swim bladder wall. However, the most important mechanism to reduce gas loss is the barrier function created by arteriovenous exchange in the rete. Since this organ also plays the crucial role in the deposition of gas, we shall discuss the general features of countercurrent exchange.

COUNTERCURRENT EXCHANGE

Consider Fig. 10. The arterial blood has a P_{O_2} of 0.2 atm, and the bladder is filled with gas containing O_2 at 100 atm. Let us assume that blood from the bladder enters the rete with a P_{O_2} of 100 atm. In the rete O_2 will therefore diffuse into the arterial blood, thereby reducing the venous P_{O_2} and increasing the arterial P_{O_2} in the two flow directions. Thus, as arterial blood enters the bladder, its P_{O_2} will approach that of the venous blood leaving the rete. In the same way the venous P_{O_2} will

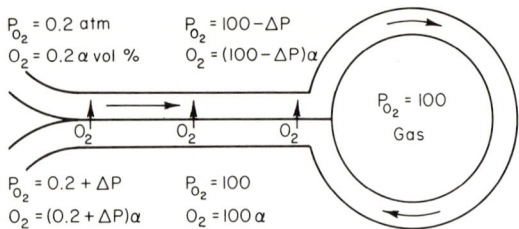

Fig. 10. The barrier function of the rete. See text for details.

approach the arterial P_{O_2} at the heart pole of the rete. The closeness of approach (i.e., the efficiency of the rete as a gas exchanger) is determined by the area, the thickness, and the gas permeability of the membrane shared by arterial and venous blood, as well as by the length of time the two types of blood are in contact.

We should be aware of the importance of the number of capillaries in the rete. In a rete with many capillaries the intrarete diffusion will be large in comparison to diffusion between the rete and the surrounding tissue. In deep sea fishes with a P_{O_2} gradient of 1:1000 between the bladder and the tissue, this may be a factor of some concern. The number of rete capillaries will, on the other hand, not influence the efficiency of the rete as an exchanger but only its circulatory and thereby functional capacity.

Scholander (1954) developed a mathematical formula which expressed the efficiency of the rete as a barrier, its barrier function. He based his calculations on anatomical dimensions from the rete and on reasonable values for velocity of blood flow. He found that the barrier function of the rete was so great that only when the P_{O_2} in the bladder exceeded 3000 atm (compared to an arterial P_{O_2} of 0.2 atm) would the O_2 loss via the bloodstream exceed that which one would accept as a reasonable amount of deposited gas. There is thus good reason to believe that the rete can reduce the gas leak from the bladder under natural conditions.

C. The Concentrating Ability of the Rete

Compared to the barrier function of the rete (see Section V, B, 1) the multiplier function is distinguished by a reduction of the gas solubility, i.e., in the ratio of gas content to its partial pressure as blood circulates in the bladder. This results in an increasing P_{O_2} in the blood circulating in the bladder with consequent diffusion of O_2 into the bladder lumen. It also creates a higher P_{O_2} in prerete venous blood than in postrete arterial blood. This will cause diffusion of O_2 from venous to arterial blood of the rete and consequently an increased P_{O_2} in the arterial blood entering the bladder. When this blood in turn experiences a reduced solubility, the P_{O_2} will increase further and still more O_2 will enter the bladder. Thus the small increase in P_{O_2} caused by reduced solubility, the primary effect, will be accumulated by circulation through the rete. The process is cumulative and is termed "hairpin countercurrent multiplication" (ccm).

Haldane (1922) and, in particular, Koch (1934) were the first to suggest how the rete might function to concentrate O_2 and inert gases subsequent to release from the blood. The potential properties of such a system were, however, first clearly realized by Kuhn and Rüffel (1942), who introduced to biology the concept of countercurrent multiplication. Hargitay and Kuhn (1951) developed an extensive model of such a mechanism in the case of urine concentration in the loop of Henle.

Kuhn and Kuhn (1961) expounded the ccm of gases in the rete mathematically. [Helpful reviews are given by Kuhn et al. (1963), Alexander (1966), and Enns et al. (1967).] They assumed the rete tissue to be *impermeable* to salts and acids and treated separately the importance of a salting out effect, the Bohr effect, and the Root effect on the concentrating ability. (The Bohr effect denotes the pH sensitive O_2 affinity of blood; the Root effect denotes the pH sensitive O_2 capacity.) When they inserted experimentally determined values for the permeability of O_2, blood flow, lactic acid production, and dimensions of the rete in the final formula they reached maximum values for P_{O_2} and P_{N_2} produced. These are shown in Table I.

These calculations indicate that a 1-cm long rete can produce gas pressures well above those ever recorded in deep sea fishes. Even without a Root effect the system can produce a P_{O_2} of 300 atm. The retia of *Anguilla* (Kuhn et al., 1963) and of *Pomatomus* (Wittenberg et al.,

Table I

Estimated P_{N_2} and P_{O_2} in Atmospheres That Can Be Developed by Retia of Various Lengths

	Length of rete (mm)		
	0	5	10
N_2 (by salting out)	0.8	4	23
O_2			
(1) By salting out; Bohr effect and Root effect	0.2	300	2000
(2) By salting out and Bohr effect only	0.2	60	300

1964) are about 1 cm long, but retia 2 cm long or more occur in benthic fish (Marshall, 1962). Kuhn and Kuhn (1961) also showed that the rate of gas deposition depends on the extent of the reduction in solubility. Thus for O_2 it is higher when both the Bohr and the Root effect participates, than when only the salting out effect causes the primary reduction in solubility.

VI. THE REALIZATION OF COUNTERCURRENT MULTIPLICATION IN THE SWIM BLADDER

Much information has accumulated which supports the view that O_2 in particular, and probably also inert gases, are in fact concentrated by countercurrent multiplication of a primary gradient caused by acidification of the blood.

A. Release of O_2 from Blood

Hall (1924) found that pH of a dialysate of homogenate of gas glands from the yellow perch, *Perca flavescens*, during gas deposition was 6.4 against 7.1 for inactive glands. Ball et al. (1955) found that gas gland tissue from several species converts most of its glucose to lactic acid even in an atmosphere of pure O_2. Fänge (1953) found higher lactate content in homogenates of active glands than in those homogenates of passive ones in codfish. Steen (1963c) found that the postrete venous blood contained more lactic acid than the prerete arterial blood and that the pH of the blood was drastically reduced as it circulated the secretory bladder. These results indicate that the blood is acidified by lactic acid as it circulates in the bladder.

The notion that this plays an essential role in the swim bladder function is supported through the demonstration by Root (1931) that fish blood apparently suffers not only a reduced O_2 affinity upon acidification but also a reduced O_2 capacity (see Chapter 6 by Riggs, this volume, for details). This characteristic of fish blood is termed a "Root effect."[*] Scholander and van Dam (1954) found a pronounced Root effect in the blood of several species of deep sea fishes where the hemoglobin at low pH attained full saturation only at P_{O_2} between 20 and 130 atm.

[*] The Root shift can occur in two directions in that pH can either increase or decrease. The phenomenon connected with acidification is termed a "Root off-shift," while the opposite is termed a "Root on-shift," "off" and "on" referring to the direction of O_2 movement relative to Hb (or to O_2 affinity).

10. THE SWIM BLADDER AS A HYDROSTATIC ORGAN   433

In some fishes a P_{O_2} of up to 140 atm did not increase the hemoglobin saturation above that recorded at 20 atm, indicating strongly that the acid blood suffers a permanently reduced O_2 capacity (Root effect). In response to reversal of the pH the blood regained normal O_2 capacity, thus showing that the Root effect is reversible. On the other hand, a Root effect is not found in fishes lacking a bladder.

The proposal that the O_2 present in the swim bladder is derived from molecular O_2 present in the water and transported in this form by Hb is directly supported by the experiments of Scholander et al. (1956b). Mass spectrometer analyses of gas from the bladder of codfish from aquaria to which $H_2^{18}O$ had been added revealed no trace of the heavy isotope. It has further been shown by tracer experiments (Wittenberg, 1961) that there is no exchange of O atoms between O_2 molecules as these pass from the water via the blood to the bladder during gas deposition in the toadfish, Opsanus tau.

Acidification alone is, however, not sufficient to explain the high P_{O_2} values found in deep sea fishes. Even if all the O_2 of fish blood were split off it would not raise the P_{O_2} to above some 5 atm.

B. Release of Inert Gases from Blood

As we have noticed in an earlier section the P_{N_2} in the bladder may also exceed that in the water. The mechanism of inert gas deposition is thought to be based on release of some of the gas present in physical solution in the blood. Koch (1934) pointed out that an acid added to the blood would not only dissociate HbO_2, it would also increase the ionic strength of the blood and reduce its gas solubility. This type of gas release will of course affect all gases and may be brought about by any ion added to blood. It is commonly termed the "salting out" effect. If half of the dissolved gas were salted out, the partial pressure would double. Such an increase would, however, require an increase of ionic strength far above that which is compatible with normal physiology.

C. Multiplication by the Rete

The notion that the primary gradient caused by acidification is multiplied by ccm in the rete receives support from the very existence of the rete as well as from its unique architecture. This, together with the pH sensitivity of HbO_2, led Haldane (1922) to suggest that O_2 is deposited as a consequence to acidification and that the effect is possibly amplified by circulation through the rete.

The central position of the rete in the swim bladder function is clearly demonstrated by a comparison between the extent of its development and the depth at which the fish lives. Unfortunately, the only parameters of the rete structure that have been sufficiently studied are the length of the rete and the number of retial capillaries (Marshall, 1962). Deep sea fishes have longer retia than shallow water forms. This indicates that the rete plays an essential role in gas deposition.

Before accepting the ccm explanation, three basic assumptions have to be discussed. (1) The effect of the added substance must be present at the pressure against which gas is to be deposited. (2) During gas deposition the gas pressure must be higher in postrete venous blood than in prerete arterial blood. (3) The rete must be impermeable to the substance causing the increase in gas pressure.

1. PRESSURE-DEPENDENT SOLUBILITY

The most likely mechanism for a reduction in solubility is the addition of ions. The specific effect of H^+ ions on the O_2 capacity of blood (Root effect) has been mentioned earlier. In many fishes the effect does not persist to a P_{O_2} which is known to be present in their swim bladders. The effect of ions on the physical solubility will not, however, be influenced by the partial pressure of the gas.

It seems, therefore, that an acid like lactic acid will fulfill the desired requirements by salting out effect alone or, depending on the species, in combination with a Bohr or Root effect.

2. ARTERIOVENOUS GAS GRADIENT

The P_{O_2} in arterial and venous blood at both poles of the rete of the eel was estimated by Steen (1963c) from O_2 content, pH, and hematocrit of appropriate blood samples. During established deposition of gas the postrete venous P_{O_2} was found to be *lower* than the prerete arterial. Similar results were obtained by Steen and Iversen (1966) who measured P_{O_2} polarographically.

3. IMPERMEABILITY FOR THE RETE TO ACIDS

In circulating blood CO_2 will be in equilibrium with H^+ ions. Thus a change in pH will cause a change in P_{CO_2}. Carbon dioxide is known to pass biological membranes faster than O_2 (about 20 times). If this is the case also for rete tissue, then the ccm will be short-circuited. On the other hand, ions are known to penetrate slower than water (about 10 times). Thus lactic acid deposited in the swim bladder capillaries

10. THE SWIM BLADDER AS A HYDROSTATIC ORGAN

may be instrumental in a concentration mechanism based on salting out, but hardly on one based on its acid effects. Steen (1963c) experimentally showed that the rete is in fact permeable to acids. Thus a considerable pH change is detectable in both arterial and venous blood during circulation of the rete.

It thus appeared that the experimental evidence on two of the three crucial points clearly contradicted the requirements of the model. For a while it appeared that we had in the swim bladder a system which was perfectly engineered for gas concentration by ccm but that the fishes did not use it that way.

The anomalous evidence was based on measurements done on blood samples. Thus, transient changes in pH or P_{O_2} might have been overlooked. Manwell (1960) urged students of the swim bladder to be aware of such transients. How right he was was demonstrated by Berg

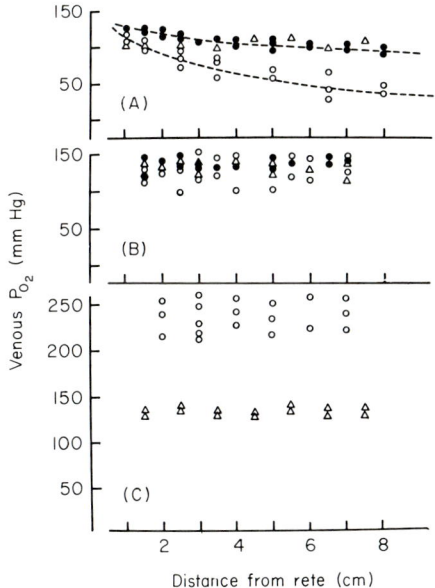

Fig. 11. (A) P_{O_2} of prerete arterial blood and of postrete venous blood from an actively secreting swim bladder. (△) arterial P_{O_2}, (●) venous P_{O_2} at high flow where each centimeter from the rete corresponds to 2 sec, and (○) venous P_{O_2} at low flow where each centimeter corresponds to 4 sec. (B) P_{O_2} of arterial (△) and venous blood (●) at various distances from the rete of a nonsecreting swim bladder. The symbol (○) shows venous P_{O_2} where swim bladder was replaced by loop of plastic tubing. (C) P_{O_2} of prerete arterial (△) and postrete (○) venous blood in experiments where countercurrent exchange had been abolished by appropriate cannulation. From Berg and Steen (1968).

and Steen (1968). In venous blood leaving the rete they recorded a lower P_{O_2} the farther away from the exit of the rete it was recorded (Fig. 11). The half-time of the change was estimated to be 10–20 sec. This implies that the earlier estimates of venous P_{O_2} represented the stable end value and not the P_{O_2} actually present at the exit of the rete. In fact the P_{O_2} at that point seemed equal to or even slightly higher than the arterial P_{O_2} which was shown not to vary with distance from the rete (Berg and Steen, 1968).

The matter was further studied by measurements on the rate of the Root shifts in suspensions of eel red cells. At 23°C, the Root off-shift (acidification and subsequent splitting off of O_2) showed a half-time of 0.05 sec, while the reverse process, the Root on-shift, had a $t_{1/2}$ of 10–20 sec (Forster and Steen, 1969).

D. Present Concept of Countercurrent Multiplication in the Rete

Blood enters the rete with arterial pH and P_{O_2} values. As it flows through the rete it receives O_2 and acid (CO_2 and H^+) from the venous blood draining the bladder. The rate of the concomitant Root off-shift has a $t_{1/2}$ of 50 msec at 23°C (Forster and Steen, 1969).

Upon entering the bladder epithelium the arterial blood has thus become enriched in O_2 and acidified. During its further circulation through the bladder epithelium the blood receives still more lactic acid whereby its P_{O_2} is further increased causing O_2 diffusion into the bladder. The addition of acid probably also increases the osmotic pressure of the plasma, causing a reduction in the physical solubility of all gases.

When the blood returns to the rete it has a lower pH and O_2 content, but a higher P_{O_2} than when it left the rete to enter the bladder. It may also have a lower physical gas solubility. As this blood flows through the rete, O_2 will diffuse according to the gradient in P_{O_2} from venous to arterial blood. At the same time, lactic acid and CO_2 pass in the same direction thus causing increasing pH in venous blood (Steen, 1963c). The concomitant decrease of venous P_{O_2} (Root on-shift) has a $t_{1/2}$ of 10–20 sec. The degree to which the venous P_{O_2} changes are completed when the blood leaves the rete will depend on the rate of blood flow. Final equilibrium between H^+ ions, Hb, and O_2 is established as the blood flows away from the rete.

Thus, as suggested by the Kuhns, O_2 is concentrated by countercurrent multiplication of a P_{O_2} gradient caused by acid; however, this gradient is not based on impermeability of the rete to acid but instead on the slow response of the O_2 capacity of blood to decreasing H^+ ion concentration. The slow Root on-shift thus explains how the rete can work in spite of the fact that it is permeable to acid.

The rate constants hereby introduced into the system will also make the system dependent upon blood flow in another and biologically more attractive manner than previously assumed. The Kuhn model implies that at a low flow rate the extraction of O_2 from the blood is maximum. Introduction of rate constants indicates that when the flow is low the arteriovenous acid flux will have sufficient time to influence the arteriovenous P_{O_2} gradient and thus reduce the degree of extraction. It is thus quite possible to circulate a bladder at a low perfusion rate without significant deposition of O_2. The full consequence of these kinetic parameters will be evident only when incorporated into a mathematical model of the countercurrent multiplying system.

It was shown by Fänge (1953) that acetozolamide (Diamox), which inhibits the catalytic action of carbonic anhydrase on the reaction of H_2O with CO_2, will abolish gas deposition. In a recent investigation, Forster and Steen (1969) showed that this inhibitor lengthened the half-time of the Root off-shift from a normal value of 50 msec at 20°C to at least 30 sec. It is very likely that this explains Fänge's observation. The acid secreted by the bladder epithelium would act too slowly to cause significant O_2 release from the blood during the period of time it circulates in the swim bladder.

The model for inert gas concentrations based on salting out effect (Kuhn and Kuhn, 1961) should also take into account the diffusion constants of inert gases relative to that of water and ions. The half-time for water penetration of eel red cells during osmotic swelling has been measured by Blum (personal communication) to be 2–3 min, as compared to less than 20 msec for O_2 diffusion out of the red cell (Forster and Steen, 1969). N_2 will diffuse through water roughly at half the speed of O_2 and is thus apparently so much faster than H_2O diffusion that countercurrent multiplication of inert gases via the salting out effect can take place.

CONCLUSION

At present it seems feasible to explain the concentration of O_2 and of inert gases in fishes with a rete by countercurrent multiplication of a primary effect caused by lactic acid. Other ions, yet unidentified, may possibly be involved.

E. Gas Deposition in Fishes with Poorly Developed Rete

The situation becomes more difficult when we attempt to explain the high concentration of inert gases found in many physostomes, the above-mentioned whitefish of Hüfner is an example. The swim bladder of these fishes possesses only a very poorly developed rete; in many cases, it consists of only a few arteries and veins (Fänge, 1958; Fahlén, 1967). One possibility is that inert gases in these fishes are concentrated by an entirely different process. However, Wittenberg (1958) showed that even in these fishes the freshly deposited gas contained predominantly O_2. This indicates that the final high tension of inert gases may result from secondary reabsorption of O_2. The presence of Hb would certainly increase the rate of O_2 reabsorption above that of inert gases. It is most likely, as concluded by Wittenberg (1958) and substantiated by Sundnes (1963), that the high partial pressures of inert gases found in these fishes result from the interaction between a countercurrent mechanism of lower capacity than that found in physoclists and a reabsorptive process with a high capacity.

Gas deposition in the herring, which lacks a rete altogether (Fahlén, 1967), has not been investigated. Most likely it depends on acidification alone. Variation in gas composition will be accomplished by the interplay between the rate of deposition and that of reabsorption. The problem is obviously in need of experimental data.

VII. NERVOUS CONTROL OF THE HYDROSTATIC FUNCTION OF THE SWIM BLADDER

The volume of the swim bladder may be considered to be steadily controlled by two reflex mechanisms: an inflatory reflex (gas deposition) and a deflatory reflex (gas reabsorption). Both these reflexes seem to be complex processes involving reactions of several categories of effectors in the swim bladder.

Treatments that increase the specific gravity of the fish act as stimuli initiating the inflatory reflex. Thus gas deposition (inflation) can be produced experimentally by an increase of the external hydrostatic pressure, by removal of a part of its gas content, or by attachment of small weights to the fish (see Fänge, 1953). All these treatments reduce the degree of stretch in the bladder wall.

Treatments that lower the specific gravity of the fish, or increase the stretch in the bladder wall, act as stimuli for the deflatory reflex. Thus von Ledebur (1937) reports that reabsorption of gases can be provoked by a decrease of the hydrostatic pressure of the surrounding water, by attachment of air-filled balloons to the fish, or by injection of gases into the swim bladder. It is also known that during asphyxiation oxygen is reabsorbed from the bladder (Berg and Steen, 1966).

The swim bladder is innervated by branches from the vagi and from coeliac ganglia (Fig. 12). Nerve endings have been observed in the reabsorbent region, the oval, the rete, and in the secretory epithelium. Near the heart pole of the rete there are numerous large ganglion cells. The muscular layer of the swim bladder wall is also well supplied by nerves.

The functioning of this nerve supply is not yet clearly understood. For example, we do not know what part of the swim bladder function is nervously regulated. Most likely the quantitative extent of gas reabsorption and gas deposition is accomplished by regulation of the blood flow to the appropriate areas. Gas reabsorption, being a pure diffusion process, will increase when the blood flow through the reabsorbent area increases. Deposition of gas, being dependent on interaction between acid deposition and circulation rate, appears at first sight more complex. However, as shown by Ball et al. (1955), the production of lactic acid decreases with increasing H^+ concentration. This implies that lactic acid production will be increased at increased blood flow, since under these conditions H^+ ions are more quickly removed. In addition, the retention of H^+ ions in the bladder circulation will increase at low flow owing to enhanced arteriovenous exchange in the rete.

Fänge (1953) showed that electric stimulation of the vagus will

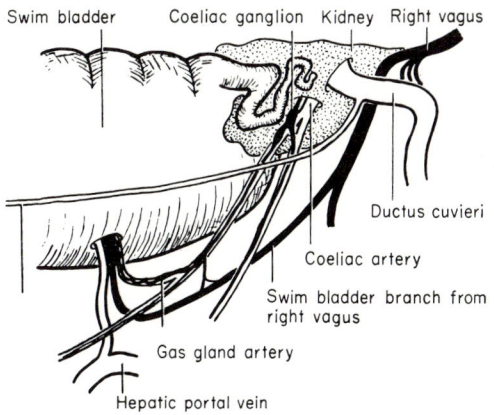

Fig. 12. Blood vessels and nerves to the swim bladder of the cod. From Fänge (1953).

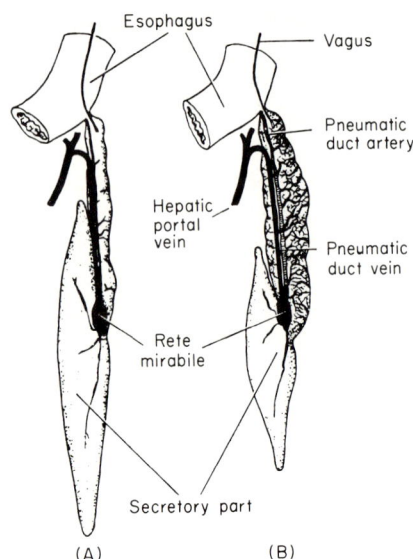

Fig. 13. (A) Normal appearance of the eel swim bladder. (B) Its appearance after stimulation for 5 min of the left vagus. The secretory part is contracted and the resorbent part relaxed. From Fänge (1953).

relax the resorbent areas (Fig. 13) or cause an opening of the oval. Adrenaline had the same effect on the resorbent part but caused the secretory bladder to contract. Acetylcholine has no effect on the secretory bladder, but it has a contracting influence on the resorbent bladder.

Stray-Pedersen (1969) has shown that the blood flow through the secretory bladder of the eel is especially reduced by adrenaline and to some extent by acetylcholine. The vascular resistance of the rete was unaffected by acetylcholine, but it increased on adrenaline injections. The reabsorbent bladder showed increased circulation when adrenaline was added to the perfusate.

It may appear, therefore, that reabsorption of gas is stimulated by catecholamines, while deposition is increased by the absence of such stimuli. It is difficult to find support for the presence of "secretory cholinergic stimuli."

REFERENCES

Alexander, R. McN. (1959a). The physical properties of the swimbladder in intact Cypriniformes. *J. Exptl. Biol.* **36**, 315–332.

Alexander, R. McN. (1959b). The densities of Cyprinidae. *J. Exptl. Biol.* **36**, 333–340.

Alexander, R. McN. (1959c). The physical properties of the isolated swimbladder in Cyprinidae. *J. Exptl. Biol.* **36**, 341–346.
Alexander, R. McN. (1959d). The physical properties of the swimbladders of fish other than Cypriniformes. *J. Exptl. Biol.* **36**, 347–356.
Alexander, R. McN. (1966). Physical aspects of swimbladder function. *Biol. Rev.* **41**, 141–176.
Bainbridge, R. (1960). Speed and stamina in three fish. *J. Exptl. Biol.* **37**, 129–153.
Ball, E., Strittmatter, F., and Cooper, O. (1955). Metabolic studies on the gas gland of the swimbladder. *Biol. Bull.* **108**, 1–17.
Berg, T., and Steen, J. B. (1966). Regulation of ventilation in eels exposed to air. *Comp. Biochem. Physiol.* **18**, 511–516.
Berg, T. and Steen, J. B. (1968). The mechanism of oxygen concentration in the swimbladder of the eel. *J. Physiol. (London)* **195**, 631–638.
Biot, M. (1807). Sur la nature de l'air contenue dans la vessie natatoire des poissons. *Mem. Phys. Chem. Soc. Arcuiel* **1**, 252.
Blum, R. Personal communication.
Borelli, I. A. (1680). "De motu animalium," Pars. I. Rome.
Denton, E. J. (1961). The bouyancy of fish and cephalopods. *Progr. Biophys.* **11**, 178–234.
Dorn, E. (1961). Uber den Feinbau der Schwimmblase von *Anguilla vulgaris* L. *Z. Zellforsch. Mikroskop. Anat.* **55**, 849–912.
Enns, T., Douglas, E., and Scholander, P. F. (1967). Role of the swimbladder rete of fish in secretion of inert gas and oxygen. *Advan. Biol. Med. Phys.* **11**, 231–244.
Fahlén, G. (1967). Morphological aspects on the hydrostatic function of the gas bladder of Clupea harengus. *Acta Univ. Lund. Sect. II* **1**, 1–49.
Fänge, R. (1953). The mechanism of gas transport in the euphysoclist swimbladder. *Acta Physiol. Scand.* **23**, Suppl. 110, 1–133.
Fänge, R. (1958). The structure and function of the gas bladder in *Argentina silus*. *Quart. J. Microscop. Sci.* **99**, 95–102.
Fänge, R. (1966). Physiology of the Swimbladder. *Physiol. Rev.* **46**, No. 2, 299–322.
Forster, R. E. (1957). Exchange of gases between alveolar air and pulmonary capillary blood: Pulmonary diffusing capacity. *Physiol. Rev.* **37**, 391–453.
Forster, R. E., and Steen, J. B. (1969). The rate of the Root shift of eel red cells and haemoglobulin solution. *J. Physiol. (London)* **204**, 259–282.
Gray, J. (1953). The locomotion of fishes. "Essays in Marine Biology," pp. 1–16. Oliver & Boyd, Edinburgh and London.
Haldane, J. S. (1922). "Respiration," pp. 250–257. Yale Univ. Press, New Haven, Connecticut.
Hall, F. G. (1924). The functions of the swimbladder of fishes. *Biol. Bull.* **47**, 79–126.
Hargitay, B., and Kuhn, W. (1951). Das Multiplikationsprinzip als Grundlage der Harnkonzentrierung in der Niere. *Z. Elecktróchem.* **55**, 539–558.
Hüfner, G. (1892). Zur physikalischen Chemie der Schwimmblasengase. *Arch. Anat. Physiol. (Leipzig)* pp. 54–80.
Kanwisher, J., and Ebeling, A. (1957). Composition of the swimbladder gas in bathypelagic fishes. *Deep-Sea Res.* **4**, 211–217.
Koch, H. (1934). L'emission de gaz dans la vésicule gazeuse des poissons. *Rev. Questions Sci.* **26**, 385–409.
Krogh, A. (1929). "The Anatomy and Physiology of Capillaries." Yale Univ. Press, New Haven, Connecticut (reprinted, Hafner, New York, 1959).

Kuhn, W., and Kuhn, H. J. (1961). Multiplikation von Aussaltz- und anderen Einzeleffekten für die Bereitung hoher Gasdrüke in der Schwimmblase. *Z. Elektrochem.* **65**, 426–439.

Kuhn, W., and Rüffel, K. (1942). Herstellung konzentrierter Lösungen aus verdünnten durch blasse Membranwirkungen. Ein Modell versuch zur Funktion der Niere. *Z. Physiol. Chem.* **276**, 145–178.

Kuhn, W., Ramel, A., Kuhn, H. J., and Marti, E. (1963). The filling mechanism of the swimbladder. *Experientia* **19**, 497–511.

Manwell, C. (1960). Comparative physiology: Blood pigments. *Ann. Rev. Physiol.* **22**, 191–244.

Marshall, N. B. (1962). Swimbladder structure of deep-sea fishes in relation to their systematics and biology. *'Discovery' Rept.* **31**, 1–122.

Moreau, A. (1876). Recherches expérimentales sur les fonctions de la vessie natatoire. *Ann. Sci. Nat. Zool. Palaentol.* **4**, No. 8, 1–85.

Nielsen, J. G., and Munk, O. (1964). A hadal fish (*Bassogigas profundissimus*) with a functional swimbladder. *Nature* **204**, 594–595.

Peters, H. M. (1951). Beiträge zur ökologischen Physiologies des Seepferdes (*Hippocampus brevirostris*). *Z. Vergleich. Physiol.* **33**, 207–265.

Rauther, M. (1937). Echte Fische. Die pneumatischen Darmanhänge. *Bronn's Klassen* **6**, Sect. I, Book 2, 759–826.

Root, R. W. (1931). The respiratory function of the blood of marine fishes. *Biol. Bull.* **61**, 427–456.

Saunders, R. L. (1953). The swimbladder gas content of some freshwater fish with particular reference to the Physostomes. *Can. J. Zool.* **31**, 547–560.

Schloesing, T., and Richard, J. (1898). Recherche de l'argon dans les gaz de la vessie natatoire des poissons et des physalies. *Compt. Rend.* **122**, 615–617.

Scholander, P. F. (1954). Secretion of gases against high pressures in the swimbladder of deep sea fishes. II. The rete mirabile. *Biol. Bull.* **107**, 260–277.

Scholander, P. F. (1956). Observations on the gas gland in living fish. *J. Cellular Comp. Physiol.* **48**, 523–528.

Scholander, P. F., and van Dam, L. (1953). Composition of the swimbladder gas in deep sea fishes. *Biol. Bull.* **104**, 75–86.

Scholander, P. F., and van Dam, L. (1954). Secretion of gases against high pressures in the swimbladder of deep sea fishes. I. Oxygen dissociation in blood. *Biol. Bull.* **107**, 247–259.

Scholander, P. F., van Dam, L., and Enns, T. (1956a). Nitrogen secretion in the swimbladder of whitefish. *Science* **123**, 59–60.

Scholander, P. F., van Dam, L., and Enns, T. (1956b). The source of oxygen secreted into the swimbladder of the cod. *J. Cellular Comp. Physiol.* **48**, 517–522.

Steen, J. B. (1963a). The physiology of the swimbladder of the eel *Aguilla vulgaris*. I. The solubility of gases and the buffer capacity of the blood. *Acta Physiol. Scand.* **58**, 124–137.

Steen, J. B. (1963b). The physiology of the swimbladder of the eel *Anguilla vulgaris*. II. The reabsorption of gases. *Acta Physiol. Scand.* **58**, 138–149.

Steen, J. B. (1963c). The physiology of the swimbladder in the eel *Anguilla vulgaris*. III. The mechanism of gas secretion. *Acta Physiol. Scand.* **59**, 221–241.

Steen, J. B., and Iversen, O. (1966). Oxygen secretion in the isolated blood perfused swimbladder. *Hvalradets Skrifter* **48**, 101–109.

Stray-Pedersen, S. (1970). Vascular responses induced by drugs and by vagal stimulation in the swimbladder of the eel, *Anguilla vulgaris*. *Comp. Gen. Pharm.* (in press).

Sundnes, G. (1963). Studies on the high nitrogen content in the physostome swimbladder. *Rept. Norweg. Fishery Invest.* **13**, 1–8.
Sundnes, G., Enns, T., and Scholander, P. F. (1958). Gas secretion in fishes lacking rete mirabile. *J. Exptl. Biol.* **35**, 671–676.
Tait, J. S. (1956). Nitrogen and argon in salmonoid swimbladders. *Can. J. Zool.* **34**, 58–68.
von Ledebur, J. F. (1937). Über die Sekretion und Resorption von Gasen in der Fischschwimmblase. *Biol. Rev.* **12**, 217–244.
Wittenberg, J. B. (1958). The secretion of inert gas into the swimbladder of fish. *J. Gen. Physiol.* **41**, 783–804.
Wittenberg, J. B. (1961). The secretion of oxygen into the swimbladder of fish. I. The transport of molecular oxygen. *J. Gen. Physiol.* **44**, 521–526.
Wittenberg, J. B., Schwend, M. J., and Wittenberg, B. A. (1964). The secretion of oxygen into swimbladder of fish. III. The role of carbon dioxide. *J. Gen. Physiol.* **48**, 337–355.
Woodland, W. N. F. (1911). On the structure and function of the gas glands and retia mirabila associated with the gas bladder of some teleosteam fishes, with notes on the teleost pancreas. *Proc. Zool. Soc. London* **1**, 183–248.

11

HYDROSTATIC PRESSURE

*MALCOLM S. GORDON**

I. Introduction 445
   A. History 445
   B. Pressure in Natural Environments 446
   C. Physicochemical Effects of Pressure 447
   D. Prospects 449
II. Fishes Living under High Pressures 450
   A. General Remarks 450
   B. Freshwater Forms 451
   C. Marine Forms 451
III. Experimental Studies 454
   A. General Remarks 454
   B. Low Pressures 454
   C. High Pressures 456
   D. Explosions 460
References 460

I. INTRODUCTION

A. History

The effects of elevated hydrostatic pressures on living systems have attracted considerable scientific attention during the past century. Experimental interest in the subject appears to have been awakened primarily by the discovery of organisms living in the deep sea. These organisms were first collected under conditions which precluded serious argument about their origin by a series of oceanographic expeditions during the

---

* This work has been partially supported by National Science Foundation Research Grant GB 5661.

1860's and 1870's. The famous Challenger expedition (1873–1876) was one of the later of these efforts (Murray and Hjort, 1912).

Fishes were among the earliest subjects for the study of pressure effects. However, experimental efforts involving fishes have since occurred only sporadically. The reviews of Cattell (1936), Johnson et al. (1954), Knight-Jones and Morgan (1966), Morita (1967), and Fenn (1967) document the fact that studies of fishes began with Regnard in France in the 1880's, then were completely neglected for 40 years until taken up by Fontaine, again in France, in the 1920's. The 1930's saw more activity in this area, especially in the United States and Germany. The 1940's and 1950's were again quiet (the author found only two papers). The present decade, however, has produced a burst of activity. Omitted from this survey and from this chapter are papers dealing with swim bladder function and with pressure-sensitive sense organs, since these topics are discussed elsewhere in these volumes.

This pattern of historical development has produced a picture of pressure effects on fishes which is in many respects vague, fragmentary, and probably inaccurate. As is also the case in many other fields of physiology, much of what has been done has been directed at laboratory phenomena having only limited relevance to conditions encountered by intact fishes in nature. Our understanding of the effects of hydrostatic pressure on the lives of fishes is in a primitive state.

B. Pressure in Natural Environments

Steady hydrostatic pressures encountered by fishes in natural environments vary from about 0.5 atm to no more than 1100 atm. Fishes living in streams and shallow lakes at high altitudes in mountain areas such as the Andes and Himalayas live at the lower end of the range. Fishes living in the hadal zones (below 6000 meters) of the great ocean trenches live near the upper end.

Depths of water (in meters) can be readily converted to equivalent pressures (in atmospheres) with an error of only a few percent. Ten meters water depth (either fresh or salt) approximates one atmosphere. Other conversions frequently used include: 1 atm = 14.7 lb/in^2 (psi), 1.03 kg/cm^2, 1.01 bars, 101 dynes/cm^2, and 760 mm Hg at 0°C.

Very few freshwater fishes encounter pressures higher than 40 atm. Hutchinson (1957) lists only 18 lakes worldwide known to reach depths greater than 400 meters. (He points out that there may be a few more, as yet unmeasured, in southern South America.) Most of these very deep lakes are vertically stratified and essentially permanently anaerobic (hence devoid of fish life) at depths below about 100–250 meters. The major exception to this situation is Lake Baikal in Siberia, which is 1620 meters deep. Baikal is the deepest known lake, is aerobic throughout its volume, and supports an abundant and diverse fish fauna down to its greatest depths (Kozhov, 1963).

A much larger proportion of marine fishes encounters high pressures. Considerable numbers of bottom-dwelling fishes have been collected by many expeditions at depths in excess of 3000 meters (Grey, 1956). Recent observations with underwater cameras indicate such fishes are more abundant, and they are often of far larger size than had been suspected (Hersey, 1967). Water depths between 3000 and 6000 meters occupy about 83% of the area of the oceans (Sverdrup et al., 1942).

Very few fishes have been collected from depths greater than 6000 meters. Only three genera belonging to two families are known. These are *Grimaldichthys* (family Brotulidae), maximum collection depth 6035 meters in the Atlantic Ocean, *Bassogigas* (family Brotulidae), maximum collection depth 7160 meters in the Sunda Trench, and *Careproctus* (family Liparidae), maximum collection depth 7579 meters in the Japan Trench (Rass, 1958; Wolff, 1960). Water depths greater than 6000 meters occupy about 1.3% of the area of the oceans. The greatest depth known is near 11,000 meters, in the Marianas Trench. Our ideas on maximum depth of fish occurrence, also variety and abundance of hadal fishes, will almost certainly change dramatically as collecting and observing techniques are improved.

A variety of transitory and/or fluctuating pressure fields are superimposed upon the steady pressures just described. These include fields from wave phenomena of many kinds, ranging from wind driven surface waves to tides and including sound waves. Most of these fields are of relatively small amplitude compared with the ambient steady pressure field. Their physiological effects lie primarily in the area of sensory physiology; hence, they will not be considered here.

Exceptions to this situation are the transitory pressure fields associated with violent events such as explosions. Explosions affect fishes often fatally, both as a result of the very high compressive pressures often associated with their primary shock waves (rising to hundreds of atmospheres above ambient in substantial water volumes around even quite small explosions) and because of the rarefaction waves which usually follow the shock waves (Hubbs and Rechnitzer, 1952).

C. Physicochemical Effects of Pressure

Pressure has multiple effects on both nonliving and living systems. Reviews of basic principles for nonliving systems include those by

Bradley and Munro (1965), Horne (1965), and Weale (1967). Whalley (1967) surveys recent developments in studies of nonliving systems. Chapters 9 and 10 in Johnson et al. (1954) remain the most penetrating and comprehensive reviews of pressure effects on biological systems. Knight-Jones and Morgan (1966) and Morita (1967) review much of the more recent biological literature. A brief summary of the major effects of pressure follows.

Pure water and complex electrolyte solutions such as seawater up to salinities of at least 41‰ (presumably also, therefore, the body fluids of fishes) all have comparable compressibilities. Over the temperature range from 0° to 25°C, increase in pressure from 1 to 1000 atm produces a decrease in specific volume of 3–4% (Newton and Kennedy, 1965).

The compression of solutions like seawater appears to decrease the average size of hydrogen-bonded aggregations of water molecules in the solutions and to increase the number of such aggregations. This change in solvent structure produces viscosity versus pressure curves which initially decrease, go through a minimum, then increase. The pressures at which the minima occur vary with temperature, as does the magnitude of the maximum viscosity decrease. At 0°C the viscosity decrease reaches a minimum about 4% below viscosity at 1 atm at a pressure of about 700 atm. The decrease is smaller at higher temperatures (Horne, 1965).

The pressure-induced changes in water structure combine with pressure-induced changes in degree of hydration of dissolved ionic substances to produce changes in both the electrical and the colligative properties of the solutions. For example, the specific conductance of 35‰ salinity seawater at 25°C increases about 6% between pressures of 1 and 1000 atm. The activation energy for electrical conduction in seawater decreases by a comparable amount over the same pressure range (Horne, 1965). The freezing point depression of pure water, fish body fluids, and seawater increases 9°C between 1 and 1000 atm (Hodgman, 1962–1963; Scholander et al., 1957).

Increased pressure produces a multiplicity of effects on the properties and chemical reactivities of solutes present in aqueous solutions. Essentially all of the drastic effects of pressure on solutes, such as denaturation of proteins, occur at pressures far higher than the 1100-atm limit present in the ocean. Less drastic effects which do occur within this range include changes in solubility, changes in rate constants for reactions, and changes in equilibrium constants. Most of these changes appear to be in large part attributable to the occurrence of volume changes associated with the processes. Processes involving volume decreases are accelerated by high pressures, while processes involving volume increases are slowed.

Pressure effects on solubilities depend on the ionization state of the dissolved substance, its partial specific volume, and the density (therefore the compressibility) of the solution. For most salts, solubility increases with increased pressure. Gas solubilities vary more widely. For example, oxygen solubility is relatively insensitive to pressure, while nitrogen solubility decreases substantially (Klotz, 1963a,b).

The details of reaction mechanisms critically determine the direction and magnitude of pressure effects on chemical reaction rates. Theoretical analyses of these effects have been attempted using both collision theory and transition state theory. Most recent discussions use the latter approach, in which pressure effects are interpreted as being largely determined by the direction and magnitude of changes in volumes of activation of the reacting substances (Whalley, 1967).

Changes in reaction rates may result from changes in the concentrations of the reacting substances or from changes in the reaction rate constant. For reactions in solution, the former effect is usually the most important at pressures below about 1000 atm. Polymerization reactions are often particularly sensitive to pressure (Bradley and Munro, 1965).

Activated states in many biologically important reactions are ionized states. Ionic dissociation in aqueous solutions generally produces a contraction of the surrounding water structure (electrostriction). Thus most ionization reactions are speeded by pressure and, in many cases, their equilibrium constants are increased. The dissociation constants of carbonic acid in seawater at 22°C, for example, are increased by factors of 1.84 for K_1' and 1.48 for K_2' at 654 atm as compared with 1 atm (Culberson et al., 1967). Dissociation constants for most small ions increase 2–6 times between 1 and 1000 atm (Bradley and Munro, 1965).

D. Prospects

The variety of possible physicochemical effects of pressure, combined with the diversity of the living world, make it improbable that many areas of substantial uniformity will be uncovered by studies of the biological effects of pressure. The fundamental nature of pressure effects simultaneously make it probable that important effects occur at every level of biological organization, from the molecular to the populational. Microorganisms have been most thoroughly studied at the former level, with results clearly supporting this position (Fenn, 1967; Morita, 1967). The most extensive studies of higher organisms, including fishes, have been at the behavioral level. Similar diversity exists there (Knight-Jones and Morgan, 1966).

The vast majority of experimental studies of pressure effects on higher organisms carried out to date has dealt with surface dwelling or relatively shallow water forms. It has only been within the past few years that physiological experiments have begun on some of the most accessible of the deep-sea creatures, namely, members of the vertically migratory deep scattering layer communities (Napora, 1964; Teal, 1966; Teal and Carey, 1967; Pearcy and Small, 1968; Gordon, 1970). The technical problems associated with such work are considerable. However, the prospective gains in understanding which should come from studies of organisms which have evolved adaptations to high pressures over very long periods of time should make these efforts amply worthwhile.

II. FISHES LIVING UNDER HIGH PRESSURES

A. General Remarks

Although this discussion is intended to be primarily physiological, some comments about other subjects seem necessary. The kinds of fishes that live under high pressures are rarely seen by most workers in fish physiology, and information on aspects of their structure and natural history is widely scattered in the technical literature. The author knows of no general discussion of the biology of deep water fishes which is directed at professional biologists. Several semi-popular discussions are available, including Marshall (1954, 1966) and Idyll (1964). A few topics have been given thorough scientific review (Marshall, 1960; Denton, 1961; Mead et al., 1964; parts of chapters by Munz, Steen, and Nicol in this treatise). Current ideas on the phylogenetic and systematic relationships of the teleosts among these fishes are summarized by Greenwood et al. (1966).

Easily 95% of what we know about deep water fishes derives from studies of dead specimens. The recent development of research submersibles has permitted a beginning at *in situ* observations of behavior, ecological associations, etc., but only the first few steps toward full understanding have been taken. No one has succeeded for more than a few days in maintaining in captivity, alive and in good condition, any really deep water fish (i.e., a species which normally occurs at depths greater than 500 meters). At the present time we have no way of deciding unequivocally to what extent high hydrostatic pressures have

11. HYDROSTATIC PRESSURE   451

played a significant selective role in the evolutionary development of the many bizarre and specialized patterns of structure and natural history shown by deep water fishes.

B. Freshwater Forms

The only freshwater lake which is of sufficient geological age, of sufficient depth and area, and sufficiently aerobic in its deeper parts to have evolved a truly deep water fish fauna (both pelagic and benthic) is Lake Baikal. Kozhov (1963) provides the most thorough description of Baikal available in English. Baikal presently supports 50 species of fishes representing nine families. One of these families, the cottoid Comephoridae, is endemic.

Kozhov (1963) calls abyssal those parts of Baikal with depths of 250 meters or more. The two species of Comephoridae are both pelagic and apparently frequently enter the abyssal zone, reaching depths well in excess of 500 meters. Both species attain lengths of 18–20 cm, are scaleless and dull glassy and translucent in appearance. They have large heads and well-developed lateral lines.

The benthic abyssal fish fauna of Baikal is composed entirely of cottids. The upper abyssal subzone (250–500 meters) is occupied by at least 10 species belonging to five genera. The lower abyssal subzone (500 meters and below) contains three species, representing two genera, which appear not to occur in shallower areas. Several of the upper abyssal species occasionally enter the lower subzone.

The abyssal cottids generally have flabby bodies covered with loose, thin skin. They have reduced eyes and well-developed lateral lines. Their bodies are generally whitish or light pale yellow in color, without spots. Adults of these species generally attain maximum sizes of 10–20 cm.

The abyssal zone of Baikal has temperatures of 3.4°–3.6°C all year round. Light is practically absent. None of the abyssal fishes is known to be luminescent. The oxygen content of the deepest water layers is high, usually 75–90% of saturation. As is the case in the deep oceans, the biomass of abyssal benthic organisms is only a small fraction of that in shallower zones.

C. Marine Forms

Deep-sea fishes occur in all oceans and most major adjacent seas. Grey (1956) and Wolff (1960) list most species known to occur at depths greater than 2000 meters. They are an extremely diverse group, including representatives or, in many cases, all known species of over 60 families of teleost fishes (Greenwood et al., 1966). There are also several families of chondrichthyan fishes represented (Bigelow and Schroeder, 1948, 1953).

Most types of deep-sea fishes are small. The most abundant species are usually less than 5 cm in length. However, species reaching 20 cm or longer are not rare, and some forms regularly attain lengths of many meters.

The deep-sea fishes most often seen in pictures or described in the popular literature are members of the abundant mesopelagic, diurnally vertically migratory, deep scattering layer communities (Taylor, 1968). These fishes are typically black and silver in color, have luminescent photophores of various types, and often have large mouths with long, sharp teeth. About 50% of the species have gas-filled swim bladders (Marshall, 1960). Their vertical migrations take them toward the surface at night, back into the depths during the day. Some species appear to migrate more than 1000 meters vertically in each direction. Different species have different ranges for migration. Some occur at the surface at night, while others approach no nearer the surface than 100–200 meters in depth. Daytime depths also vary between species. Scattering layer communities occur in all the oceans, most often between latitudes 40°N and 40°S.

The scattering layer fishes are not only the shallowest of the true deep-sea fishes, but also are the deep-sea groups encountering the greatest environmental variability. In all areas their migrations take them twice daily between pressures of 50–100 atm and pressures of only one or a few atmospheres. In thermally stratified regions, like most of the tropics, they undergo twice daily changes in temperature, which often are in the range of 10°–15°C. In substantial regions (e.g., much of the eastern North Pacific) their migrations also either take them through, or their daytime depths lie within, well-developed oxygen minimum layers. Dissolved oxygen concentrations in these layers are often well below 0.5 ml/liter (5–10% saturation; Childress, 1968; Lavenberg and Ebeling, 1967; Longhurst, 1967). Vertical stratification of salinity is usually small, variations generally being less than 0.5‰.

Vertical migrations also appear to occur among truly bathypelagic fishes, but the frequency and extent of these migrations are very poorly known. The environmental changes encountered in a move between, say, 4000 and 3000 meters are, in any case, much smaller than the changes just described for mesopelagic species. In this really deep-sea situation, pressure would change by only about 25%, temperature by no more than a fraction of 1°C, oxygen probably by less than 0.1 ml/liter, and salinity probably not at all (Menzies, 1965).

11. HYDROSTATIC PRESSURE   453

Bathypelagic fishes are often more uniformly dark in color than mesopelagic forms. Many groups are not known to be luminescent. Species restricted to depths below 1000 meters universally lack a gas-filled swim bladder (Marshall, 1960).

Not all meso- and bathypelagic fishes are vertically migratory, at least as adults. However, these groups include many forms which have epipelagic eggs and larvae; hence, they must normally make at least one extended migration all the way up to the surface layers (Mead et al., 1964).

Many kinds of meso- and bathypelagic fishes, mostly predatory species, have greatly reduced skeletons and trunk musculature, as compared with epipelagic fishes. Marshall (1960) and Denton (1961) argue that these reductions are primarily adaptations of buoyancy control mechanisms in response to restricted food supplies in the deep-sea midwater environment. The systematic distribution of these reductions correlates well with the absence of gas-filled swim bladders, the maintenance of which both authors consider to be energetically expensive. If this hypothesis is correct, most of the major structural differences between epi-, meso-, and bathypelagic fishes would appear to be primarily responses to environmental factors other than hydrostatic pressure.

Much the same situation may hold for abyssal and hadal benthic fishes as well. These latter fishes are generally larger and more solidly constructed than the pelagic forms. Associated with this circumstance is the occurrence of gas-filled swim bladders in more than half the species, including some which occur to depths of 5000 meters (Marshall, 1960). The food supply on and near the bottom appears to be much greater than in midwater (although still much less than in shallow water environments). Many other morphological features of these fishes seem explicable in terms of requirements for food and mate location, species recognition, and reproduction (Mead et al., 1964).

Except for hydrostatic pressure, the physical features of the abyssal environment are very uniform (Menzies, 1965). In all oceans, at latitudes between 50°N and 58°S, the temperature range at depths below 2000 meters is only from —0.6° to 3.6°C. Salinities are universally near 34‰, with variations of less than 0.4‰ from this figure. Sunlight does not penetrate below 1000 meters, but measurable bioluminescent light intensities occur at most greater depths. Oxygen concentrations are generally high, anaerobic conditions being restricted to a few enclosed or semi-enclosed basins.

The abyssal environment has generally been considered to have been as uniform in its properties over time as it spatially is now. Russian ichthyologists especially have based theories of the evolutionary history of the deep-sea fauna upon this presumption (Rass, 1959). However, recent geological findings have led some workers to doubt this. They suggest that the environment has undergone substantial changes within geologically relatively recent times (Menzies et al., 1961).

III. EXPERIMENTAL STUDIES

A. General Remarks

The following survey of the experimental literature on pressure effects on fishes emphasizes work done since the thorough review of Cattell (1936). Behavioral studies are de-emphasized, since they have recently been summarized by Knight-Jones and Morgan (1966). Morita (1967) provides a useful listing of various types of apparatus which have been used in studies of biological effects of pressure.

B. Low Pressures

Studies of the effects of low pressures (up to 5 atm) on fishes have been directed solely toward intact organisms, with most effort expended on behavioral studies. A variety of freshwater and marine species have been tested with respect to their ability to detect and behaviorally respond to small changes in pressure. All species tested respond, those having swim bladders showing lower response thresholds than those lacking swim bladders. Fish with swim bladders generally have thresholds in the range of 5–10 mbars, those without bladders at about 25 mbars (Knight-Jones and Morgan, 1966).

The tolerance of developing fish eggs, larvae, and various ages of juvenile fishes for changes in pressure has been studied by several workers. The majority of the eggs and larvae of several species of shallow water, freshwater, and marine fishes, at their normal environmental temperatures, can tolerate both rapid and gradual changes in pressure. Developing eggs of bream and pike perch showed no significant effects of being maintained for 5 days at 1.5–2.0 atm pressure instead of more normal pressures near 1 atm (Belyi, 1963). Larvae of these two species (1–6 days old) subjected to 3.5–5.0 atm pressure for 5–6 days also were unaffected as long as the release of pressure at the end of the experiment was relatively slow (reduced over 65 min). More rapid pressure release (25 min) raised mortality to three times control levels. Larvae of both species developed normal gas-filled swim bladders while under high pressures. Semenov (1954, cited by Belyi, 1963) reported that sturgeon eggs and larvae developed better under increased pressures of unspecified magnitude than they did at 1 atm.

Bishai (1961b) found that herring, *Clupea harengus*, larvae survived and behaved normally at 2 atm pressure during exposures to pressure lasting 12 days. Indeed, larvae under pressure survived somewhat better than controls maintained at 1 atm. Similar results were obtained with other larvae subjected to the following regime: 1 day at 2 atm, 1 day at 3 atm, and 3 days at 3–4 atm. Young plaice, *Pleuronectes platessa*, were similarly unaffected by pressures of 2 atm maintained for up to 8 days. All experiments involved rapid changes in pressure (5–30 min) for both compression and decompression. The swim bladders had not yet developed in the herring larvae used; the plaice does not have a swim bladder.

Similar experiments carried out on larvae and young of salmonid fishes (*Salmo salar; Salmo trutta*, both sea and brown trout types) showed a distinct dependence of pressure tolerance on age (Bishai, 1961b). Alevins can tolerate pressures up to 5 atm until the disappearance of the yolk sac. Older alevins, to at least 56 days old and having partly and fully developed swim bladders, survived exposures to 2–5 atm pressure for up to 8 days, although many of them experienced periods of acute disequilibrium following decompression at fairly slow rates. Some of these fish died, apparently from asphyxia resulting from gill blockage by gas bubbles. Young fish 98–178 days old all died within 24 hr when subjected to pressure increases of more than 0.2 atm. Fish older than 272 days once again survived high pressures, in these cases up to at least 4 atm. Bishai interpreted these results as reflections of the gradual development of greater ability to control swim bladder gas volumes in older fish. No suggestion was offered as to why the 98–178-day old fish die following even very small pressure increases.

A more extensive survey of low pressure tolerances of a variety of freshwater fishes all having well-developed swim bladders, but some being physostomous, others physoclistous, has also been made by Bishai (1961a, 1963). In these experiments pressures of only 2 atm were used, with rapid compressions (1–2 min) and variable rates of decompression (1–35 min). Durations at high pressure were 1–5 days, with observations of experimental fishes often continued for weeks after decompression. All species studied (5 physoclists and 5 physostomes) survived the experiments undamaged. Some mortality occurred, but it was attributed to gas disease rather than pressure effects. The physoclists had more difficulty in adjusting, and they took longer to adjust to both pressure increase and fall than did the physostomes.

Kulikov (1964) has made some observations on the effects of sudden decompressions on Bering Sea sablefish, *Anoplopoma fimbria*. These fish generally live at considerable depths but can perform long vertical migrations. Substantial sudden decompressions (magnitude unspecified) had no adverse effects.

A group of Russian workers has made a series of metabolic studies on various Siberian freshwater fishes subjected for periods of a few hours to pressures of 2–5 atm (Lopukhova, 1964; Pegel' *et al.*, 1964, 1966). Unavailability of the original papers makes the following summary entirely dependent upon English translations of the authors' abstracts.

Experiments were carried out with ide, dace, gudgeon, crucian carp, and perch. Measurements made included oxygen consumption, carbon dioxide excretion, oxygen content of blood hemoglobin, blood sugar level, electrocardiogram, and conditioned reflex responses. There is no indication that the oxygen consumption and carbon dioxide excretion rate measurements took into account possible changes in activity levels of the fishes. Experimental temperatures varied seasonally from 9° to 11°C in winter to 18°C in summer.

The overall conclusion was that during spring–summer periods of increased activity increased pressures stimulated the fishes' physiological processes. During fall–winter periods of reduced activity increased pressures inhibited these same processes. There were variations in detail between species, but results were broadly as follows.

Pressure increases of 2–4 atm in spring and summer produced increases of 20% in oxygen consumption, increased blood levels of oxyhemoglobin, decreased blood sugar levels, and no change in conditioned reflex responses. Similar pressure increases in fall and winter produced a 20% decrease in oxygen uptake, 60–80% increase in carbon dioxide excretion, higher blood sugar levels, lower blood oxyhemoglobin levels, lowered R waves in the electrocardiograms (interpreted as decreased conductivity of heart muscle), and increased lag times for conditioned reflexes. The reflexes apparently eventually disappeared altogether. The experiments with dace were apparently particularly variable and sometimes inconsistent (Pegel' *et al.*, 1964). The abstract of the most recent summary of the work (Pegel' *et al.*, 1966) is in several ways inconsistent with the abstract of what appears to be the major experimental paper (Lopukhova, 1964).

C. High Pressures

Work on the effects on fishes of high pressures (above 5 atm) has been a bit more diversified than the work at low pressures. Some consideration has been given to phenomena at the biochemical, cellular, and tissue levels, as well as to whole organisms. However, compared with what might be done, very little has been done.

Biochemical comparisons between deep-sea and shallow water fishes to date have not uncovered any differences which can be unequivocally attributed to the effects of increased pressure. Only fragmentary data are available on tissue metabolic rates and pathways for carbohydrate metabolism in a few mesopelagic species (Soprunov, 1968).

Lewis (1962, 1967) has studied the abundances of fatty acids of different chain lengths and degrees of unsaturation in the lipids of a variety of species of epi-, meso-, and bathypelagic fishes, also several benthic species from various depths. His results indicate a general trend of decreasing abundance of medium-chain-length saturated and long-chain-length polyunsaturated acids with increasing depth, while oleic acid greatly increases.

Similar results have been obtained by Nevenzel et al. (1965, 1966, 1969) on many species of mesopelagic fishes. Nevenzel et al. have shown that the unusual chain-length distribution in these species results from the occurrence of large proportions of wax esters, rather than glyceryl esters, in the lipids. Reasons for this difference are not apparent, although there may be small differences between the specific gravities of two types of lipids which may be of some significance in buoyancy control. However, similar fatty acid chain-length distributions occur in deep-sea benthic fishes and crustaceans (Lewis, 1967). Lewis argues that the wax esters in these latter forms may derive from feeding on bathypelagic species. The causative factors underlying this fundamental biochemical difference remain to be determined.

The author has made a similar survey of a variety of littoral, epipelagic, and mesopelagic fishes with respect to the spectrum of abundances of free amino acids and similar substances in the muscles. This survey has found no statistically reliable differences in terms of either absolute or relative abundance of 18 ninhydrin-positive compounds between the various groups. Variables taken into account in the statistical analysis of the results of about 120 analyses carried out on a Beckman Model 120B amino acid analyzer included systematic position of the species, water depth, water temperature, and geographical area. It is possible that variability between duplicate samples resulting from technical difficulties in sample preparation under field (oftentimes shipboard) conditions may have obliterated some small differences between the various groups. However, differences comparable in magnitude to those occurring among the lipids should have been clearly detected (Gordon and Yuen, 1969).

All other biochemical studies relating to pressure and involving materials deriving from fishes, likewise all studies of pressure effects at the cell and tissue levels which involve preparations coming from fishes, have been directed primarily toward understanding of general physiological questions rather than toward understanding the nature of pressure effects on fishes. However, even these papers are few in number.

Lebedeva et al. (1965) included carp muscle actomyosin as one of four types of vertebrate skeletal muscle actomyosin (the others were rabbit, pigeon, and frog proteins) on which they studied the effects of pressures of 500–5000 atm. Their primary interest was in determining whether or not across-the-board differences exist in the properties of this important protein between ectothermous and endothermous animals. Among other things, they found that adenosinetriphosphatase (ATPase) activity of this protein was more sensitive to pressure in the ectotherm preparations than it was in the endotherm samples. They concluded that there appear to be differences between the two groups, and that carp preparations, in particular, are not very resistant to high pressures.

Brown (1934–1935) studied pressure effects on isolated pectoral fin muscles of red groupers, *Epinephelus morio*. This work was part of a more extensive study of the mechanism of striated muscle contraction. It is placed in its proper perspective in the review of Johnson et al. (1954).

Ebbecke (1935a) studied the effects of rapidly applied and released short duration compressions up to 600 atm on the beating of isolated hearts of cat sharks, *Scyllium canicula*. He did only a few experiments with shark hearts, with results generally similar to those obtained in much more extensive studies on isolated frog hearts. Pressure pulses were able to stop the beating of active hearts or start up inactive hearts. Pressures to about 400 atm generally increased both the frequency and the amplitude of the beat. Higher pressures produced irreversible damage to the preparations. Cattell (1936) and Johnson et al. (1954) put this work into perspective in terms of understanding of mechanisms of cardiac function.

Marsland (1944), as part of a series of studies on pressure effects on protoplasmic movements within single cells, studied the effects of pressures up to 500 atm on the melanophores of isolated scales of the killifish, *Fundulus heteroclitus*. He found that increased pressure inhibited the concentration of melanophore pigment, that this effect was independent of the innervation of the cells, and that low temperature (6°C) enhanced the effect while high temperature (30°C) reduced it. Johnson et al. (1954) and Knight-Jones and Morgan (1966) discuss these and related results.

The studies of Regnard (1885) and of Draper and Edwards (1932) remain the only work the author knows of on the effects of high pressures on the developing eggs of fishes. Regnard exposed salmonid eggs to pressures of 100–650 atm for periods of 6 hr. He then followed the course of their development. Control eggs and eggs exposed to 100 and 200 atm hatched in 21 days. Eggs exposed to 300 atm hatched normally, but 2 days later. All other eggs died, those exposed to 400 and 500 atm after 6 days, and those exposed to 650 atm after 2 days.

Draper and Edwards exposed killifish, *Fundulus* sp., eggs for periods of up to 3 hr to pressures up to 130 atm. No observable effects on cell constituents occurred. Cleavage rates were slowed, and cases of abnormal development occurred. The hearts of young embryos beat more slowly under pressure, sometimes stopping altogether if pressure was continued. This stoppage was usually reversed within a few minutes after pressure release. Pressure also induced cardiac arrhythmia, types of local block, and isolated fibrillary activity. The cardiac effects of pressure apparently resulted from direct influences. Similar effects occurred in still non-innervated hearts of young embryos and in de-innervated hearts of older embryos.

High pressure effects on intact fishes have received very little attention since Cattell's review (1936). The author knows of only one brief, entirely behavioral paper (Nishiyama, 1965). Cattell himself cites only three substantial papers (Regnard, 1891; Fontaine, 1930; Ebbecke, 1935b). The papers by Regnard and Fontaine both completely summarize previously published series of shorter papers by the same authors.

Nishiyama (1965) observed the behavior of rainbow trout, goldfish, bitterling, and loach during exposures of up to 5 days' duration to pressures up to about 25 atm. Abnormal behavior patterns were observed in all species (e.g., unusual swimming postures or cessation of swimming, irregular respiratory rhythms, and "convulsions" during pressure changes). Paling of body color because of melanophore contraction also occurred in the trout.

A brief description of the major features of the results previously reviewed by Cattell seems useful here as well. Regnard, Fontaine, and Ebbecke each observed behavioral changes and tolerances for high pressures (to 1000 atm for Regnard, about 700 atm for Fontaine, and 500 atm for Ebbecke) in a variety of both freshwater and marine shallow water fishes, including forms with and without swim bladders. Details varied somewhat between species, but the basic pattern of response to quickly applied and released pressures, usually over durations of a few minutes to at most a few hours, was an increase in physical activity under moderate pressures, then depression of activity, ultimately leading to cessation at higher pressures. The region of stimulation extended to about 100–150 atm. The severity of depression during and following exposures to higher pressures depended to considerable extents upon both the duration of the compression and the speed of pressure buildup and release. Almost immediate death resulted from exposure to pressures higher than 300–400 atm.

Fontaine also measured oxygen consumption by small individuals of several species of shallow water marine fishes at pressure up to 150 atm. The rate–pressure curves agreed well with the behavioral observations.

D. Explosions

The literature on the pressure-related effects of explosions upon fishes is probably substantial, but a large part of it must be buried in internal reports to government agencies. The only published experimental study the author knows of is that by Hubbs and Rechnitzer (1952).

Hubbs and Rechnitzer investigated the ability of conventional explosives (dynamite, Hercomite, black powder) to kill a variety of marine shallow water fishes when detonations of varying sizes were set off under varying conditions. The major interest was in trying to develop recommendations for procedures to be used in geophysical explorations which would minimize mortality to fishes. Results indicated that slow burning explosives such as black powder are much less destructive than the other types. The reasons for this appeared to be a combination of the slower rate of rise to peak pressures in the compressive shock waves of black powder explosions and the lesser development of negative pressure waves by these explosions.

ACKNOWLEDGMENTS

The author wishes to thank Mr. John C. Marr, Director, and the staff of the Biological Laboratory, U.S. Bureau of Commercial Fisheries, Honolulu, Hawaii for facilities and assistance with the preparation of this chapter.

REFERENCES

Belyi, N. D. (1963). Development of bream and pike-perch larvae at great depths. *Dokl. Akad. Nauk SSSR* **149**, 373–374 (English Trans.).

Bigelow, H. B., and Schroeder, W. C. (1948). Sharks. In "Fishes of the Western North Atlantic" (J. Tee-Van *et al.*, eds.), Mem. 1, No. 1, pp. 59–546. Sears Found. Marine Res., Yale Univ., New Haven, Connecticut.

Bigelow, H. B., and Schroeder, W. C. (1953). Sawfishes, guitarfishes, skates and rays. Chimaeroids. In "Fishes of the Western North Atlantic" (J. Tee-Van *et al.*, eds.), Mem. 1, No. 2, pp. 1–588. Sears Found. Marine Res. Yale Univ., New Haven, Connecticut.

Bishai, H. M. (1961a). The effect of pressure on the distribution of some Nile fish. *J. Exptl. Zool.* **147**, 113–124.
Bishai, H. M. (1961b). The effect of pressure on the survival and distribution of larval and young fish. *J. Conseil, Conseil Perm. Intern. Exploration Mer* **26**, 292–311.
Bishai, H. M. (1963). Effect of pressure on young *Mugil cephalus* L., *Mugil chelo* C.V., and *Atherina* sp. *Proc. Zool. Soc. UAR* **1**, 197–206.
Bradley, R. S., and Munro, D. C. (1965). "High Pressure Chemistry." Pergamon Press, Oxford.
Brown, D. E. (1934–1935). "Cellular Reactions to High Hydrostatic Pressure," Ann. Rept. Tortugas Lab., pp. 76–77. Carnegie Inst., Washington, D.C.
Cattell, M. (1936). The physiological effects of pressure. *Biol. Rev.* **11**, 441–476.
Childress, J. J. (1968). Oxygen minimum layer: Vertical distribution and respiration of the mysid. *Gnathophausia ingens. Science* **160**, 1242–1243.
Culberson, C., Kester, D. R., and Pytkowicz, R. M. (1967). High-pressure dissociation of carbonic and boric acids in seawater. *Science* **157**, 59–61.
Denton, E. J. (1961). The buoyancy of fish and cephalopods. *Progr. Biophys. Biophys. Chem.* **11**, 177–234.
Draper, J. W., and Edwards, D. J. (1932). Some effects of high pressure on developing marine forms. *Biol. Bull.* **63**, 99–107.
Ebbecke, U. (1935a). Über die Wirkung hoher Drucke auf Herzschlag und Elektrokardiogram. *Arch. Ges. Physiol.* **236**, 416–426.
Ebbecke, U. (1935b). Über die Wirkungen hoher Drucke auf marine Lebewesen. *Arch. Ges. Physiol.* **236**, 648–657.
Fenn, W. O. (1967). Possible role of hydrostatic pressure in diving. *In* "Proceedings of the Third Symposium of Underwater Physiology" (C. J. Lambertsen, ed.), pp. 395–403. Williams & Wilkins, Baltimore, Maryland.
Fontaine, M. (1930). Recherches expérimentales sur les réactions des êtres vivants aux fortes pressions. *Ann. Inst. Oceanog. (Paris)* [N.S.] **8**, 1–99.
Gordon, M. S. (1970). Unpublished study.
Gordon, M. S., and Yuen K. (1969). Unpublished data.
Greenwood, P. H., Rosen, D. E., Weitzman, S. H., and Meyers, G. S. (1966). Phyletic studies of teleostean fishes, with a provisional classification of living forms. *Bull. Am. Museum Nat. Hist.* **131**, 339–456.
Grey, M. (1956). The distribution of fishes found below a depth of 2000 meters. *Fieldiana, Zool.* **36**, 75–337.
Hersey, J. B. (1967). "Deep-Sea Photography." Johns Hopkins Press, Baltimore, Maryland.
Hodgman, C. D., ed. (1962–1963). "Handbook of Chemistry and Physics," 44th ed. Chem. Rubber Publ. Co., Cleveland, Ohio.
Horne, R. A. (1965). The physical chemistry and structure of sea water. *Water Resources Res.* **1**, 263–276.
Hubbs, C. L., and Rechnitzer, A. B. (1952). Report on experiments designed to determine effects of underwater explosions on fish life. *Calif. Fish Game* **38**, 334–366.
Hutchinson, G. E. (1957). "A Treatise on Limnology," Vol. 1, Geography, Physics and Chemistry. Wiley, New York.
Idyll, C. P. (1964). "Abyss: The Deep Sea and the Creatures that Live in It." Crowell, New York.

Johnson, F. H., Eyring, H., and Polissar, M. J. (1954). "The Kinetic Basis of Molecular Biology." Wiley, New York.

Klotz, I. M. (1963a). Comment on the variations of solubility with depth. *Limnol. Oceanog.* **8**, 486.

Klotz, I. M. (1963b). Variation of solubility with depth in the ocean: A thermodynamic analysis. *Limnol. Oceanog.* **8**, 149–151.

Knight-Jones, E. W., and Morgan, E. (1966). Responses of marine animals to changes in hydrostatic pressure. *Oceanog. Marine Biol. Ann. Rev.* **4**, 267–299.

Kozhov, M. M. (1963). Lake Baikal and its life. *Monographiae Biol.* **11**, 1–344.

Kulikov, M. Y. (1964). O prisposoblennosti ugol'noi ryby *Anoplopoma fimbria* (Pallas) k rezkim izmeneniyam davleniya. (On the adaptation of sablefish *Anoplopoma fimbria* (Pallas) to sudden changes in pressure.) *Izv. Tikhookeans. Nauchn.-Issled. Inst. Rybn. Khoz. i Okeanogr.* **55**, 247–248; *Biol. Abstr.* **48**, 552 (1967).

Lavenberg, R. J., and Ebeling, A. W. (1967). Distribution of midwater fishes among deep water basins of the southern California shelf. *In* "Proceedings of the Symposium on the Biology of the California Islands" (R. N. Philbrick, ed.), pp. 185–201.

Lebedeva, N. A., Skvortsova, N. V., and Ivanov, I. I. (1965). Vliyanie vysokogo davleniya na svoistva aktomiozina teplokrovnykh i kholodnokrovnykh zhivotnykh. (Effect of high pressures on the properties of [muscle] actomyosin in warm- and cold-blooded animals.) *Zh. Evolyutsionnoi Biokhim. i Fiziol.* **1**, 133–137.

Lewis, R. W. (1962). Temperature and pressure effects on the fatty acids of some marine ectotherms. *Comp. Biochem. Physiol.* **6**, 75–89.

Lewis, R. W. (1967). Fatty acid composition of some marine animals from various depths. *J. Fisheries Res. Board Can.* **24**, 1101–1115.

Longhurst, A. R. (1967). Vertical distribution of zooplankton in relation to the eastern Pacific oxygen minimum. *Deep-Sea Res.* **14**, 51–63.

Lopukhova, V. V. (1964). Vliyanie povyshennogo davleniya vody na gazoobmen, pokazateli krovi i uslovno-reflektornuyu deyatel'nost' u nekotorykh presnovodnykh ryb. (The effect of elevated water pressure on the gaseous exchange, blood indices and conditioned-reflex response of freshwater fishes.) *Uch. Zap. Tomsk. Gos. Univ.* **49**, 215–225; *Biol. Abstr.* **47**, 11144 (1966).

Marshall, N. B. (1954). "Aspects of Deep Sea Biology." Hutchinson, London.

Marshall, N. B. (1960). Swimbladder structure of deep-sea fishes in relation to their systematics and biology. *'Discovery' Rept.* **31**, 1–122.

Marshall, N. B. (1966). "The Life of Fishes." World Publ. Co., Cleveland, Ohio.

Marsland, D. A. (1944). Mechanism of pigment displacement in unicellular chromatophores. *Biol. Bull.* **87**, 252–261.

Mead, G. W., Bertelsen, E., and Cohen, D. M. (1964). Reproduction among deep-sea fishes. *Deep-Sea Res.* **11**, 569–596.

Menzies, R. J. (1965). Conditions for the existence of life on the abyssal sea floor. *Oceanog. Marine Biol. Ann. Rev.* **3**, 195–210.

Menzies, R. J., Imbrie, J., and Heezen, B. C. (1961). Further considerations regarding the antiquity of the abyssal fauna with evidence for a changing abyssal environment. *Deep-Sea Res.* **8**, 79–94.

Morita, R. Y. (1967). Effects of hydrostatic pressure on marine microorganisms. *Oceanog. Marine Biol. Ann. Rev.* **5**, 187–203.

Murray, J., and Hjort, J. (1912). "The Depths of the Ocean." Macmillan, New York.

Napora, T. A. (1964). "The Effect of Hydrostatic Pressure on the Prawn, *Systellaspis*

debilis," Proc. Symp. Exptl. Marine Ecol., Occasional Publ. No. 2, pp. 92–94. Grad. School Oceanog., University of Rhode Island.
Nevenzel, J. C., Rodegker, W., and Mead, J. F. (1965). The lipids of *Ruvettus pretiosus* muscle and liver. *Biochemistry* 4, 1589–1594.
Nevenzel, J. C., Rodegker, W., Mead., J. F., and Gordon, M. S. (1966). Lipids of the living coelacanth, *Latimeria chalumnae*. *Science* 152, 1753–1755.
Nevenzel, J. C., Rodegker, W., Robinson, J. S., and Kayama, M. (1969). The lipids of some lantern fishes (family Myctophidae). *Comp. Biochem. Physiol.* 31, 25–36.
Newton, M. S., and Kennedy, G. C. (1965). An experimental study of the P-V-T-S relations of seawater. *J. Marine Res.* 23, 88–103.
Nishiyama, T. (1965). A preliminary note on the effect of hydrostatic pressure on the behavior of some fish. *Bull. Fac. Fisheries, Hokkaido Univ.* 15, 213–214.
Pearcy, W. G., and Small, L. F. (1968). Effects of pressure on the respiration of vertically migrating crustaceans. *J. Fisheries Res. Board Can.* 25, 1311–1316.
Pegel', V. A., Remorov, V. A., and Lopukhova, V. V. (1964). Vliyanie izmeneniya davleniya vody na gazoobmen u ryb. (The effect of changes in hydrostatic pressure on the gaseous exchange in fish.) *Nauchn. Dokl. Vysshei Shkoly, Biol. Nauki* 1, 62–64; *Biol. Abstr.* 47, 11674 (1966).
Pegel', V. A., Remorov, V. A., and Lopukhova, V. V. (1966). Reaktsiya obmena veshchestv u raznykh vidov presnovodnykh kostistykh ryb na izmenenie davleniya vody. (The metabolic response of various freshwater teleosts to changes in water pressure.) *In* "Tezisy dokladov Vsesoyusnogo soveshchaniya po ekologii i fiziologii ryb 1966" ("Report Summaries of the All-Union Conference on the Ecology and Physiology of Fishes, 1966"), pp. 34–35. *Biol. Abstr.* 48, 120847 (1967).
Rass, T. S. (1958). Ryby samykh bolshikh glubin. (Fish from the deepest depths.) *Priroda* 7, 107–108.
Rass, T. S. (1959). Ryby bolshikh glubin. (Deep-sea fishes.) *Itogi Nauki: Dostijeniya Okeanol.* 1, 285–315.
Regnard, P. (1885). Influence des hautes pressions sur l'éclosion des oeufs des poissons. *Compt. Rend. Soc. Biol.* 37, 48.
Regnard, P. (1891). "Recherches expérimentales sur les conditions physiques de la vie dans les eaux." Masson, Paris.
Scholander, P. F., van Dam, L., Kanwisher, J., Hammel, H. T., and Gordon, M. S. (1957). Supercooling and osmoregulation in Arctic fishes. *J. Cellular Comp. Physiol.* 49, 5–24.
Soprunov, F. F. (1968). Nekotorie osobennosti tkanevogo dikhaniya i obmena uglevodov i predstavitelei ultrabissalnoi Faune Kurilo-Kamchatskoi vpadine. (Some peculiarities of tissue respiration and carbohydrate metabolism in representatives of deep-water fauna of the Kurile-Kamchatka abyss.) *Zh. Evolyutsionnoi Biokhim. i Fisiol.* 4, 24–31.
Sverdrup, H. U., Johnson, M. W., and Fleming, R. H. (1942). "The Oceans, Their Physics, Chemistry and General Biology." Prentice-Hall, Englewood Cliffs, New Jersey.
Taylor, F. H. C. (1968). The relationship of midwater trawl catches to sound scattering layers off the coast of northern British Columbia. *J. Fisheries Res. Board Can.* 25, 457–472.
Teal, J. M. (1966). The effects of pressure on the respiration of *Euphausia* from

the scattering layer. *Abstr. 2nd Intern. Oceanog. Congr., Moscow, 1966* pp. 361–362.
Teal, J. M., and Carey, F. G. (1967). Effects of pressure and temperature on the respiration of euphausiids. *Deep-Sea Res.* **14**, 725–733.
Weale, K. E. (1967). "Chemical Reactions at High Pressures." Spon, London.
Whalley, E. (1967). High pressure. *Ann. Rev. Phys. Chem.* **18**, 205–232.
Wolff, T. (1960). The hadal community, an introduction. *Deep-Sea Res.* **6**, 95–124.

12

IMMUNOLOGY OF FISH

JOHN E. CUSHING

I. Introduction 465
   A. Historical 466
   B. Methods 467
   C. General References 468
II. Antibodies 468
   A. The Immunoglobulins 469
   B. Allotypes 470
   C. The γM Molecule 472
   D. The Primary Amino Acid Sequences of L and H Chains . 472
   E. The Immunoglobulins of Fish 474
   F. The Specificity of Fish Antibodies 477
III. The Cellular Basis of the Immunological Response . . . 479
   A. Phylogeny of the Lymphoid System 481
   B. Transplantation Studies in Fish 482
IV. Complement 483
   Comparative Immunology of Fish Complement 485
V. Blood Groups 486
VI. Cyclostomes 490
VII. Final Considerations 491
References 494

I. INTRODUCTION

The problem of understanding the mechanism of antibody synthesis is not only of extreme interest to immunologists and geneticists but also to biologists in general. One major reason for this is that the mechanism as far as it is understood has fundamental paradoxical aspects that have not yet been experimentally integrated with present concepts of genetics and protein synthesis. A second major reason is that the immunological response of higher vertebrate is a magnificient example of a highly evolved adaptation, the phylogeny of which is still being worked out.

The purpose of this chapter is to acquaint persons interested in fishes with the basic aspects of the biology and evolution of antibody synthesis and associated phenomena and to show in some detail how the study of fish relates to this area. Since the point of view is that of comparative immunology, a brief history of this subject is considered first.

A. Historical

The history of the comparative study of the immunological responses of vertebrates can be divided into an early period extending from the initiation of immunological research to 1966, and a recent period extending from this date to the present. The dividing date, while obviously quite arbitrary, is conveniently taken as the year of publication of "Phylogeny of Immunity" (R. T. Smith et al., 1966). This book was the first to be published on the subject and presents an assemblage of papers, most of which are relevant to this chapter. The book in turn was inspired by papers on the phylogenetic development of immunological responsiveness in vertebrates (cf. Good and Papermaster, 1964). These papers were produced by Robert Good and his associates and, together with a symposium on immunological phenomena in cold-blooded vertebrates (Sigel, 1963), served to trigger and solidify the intensification of interest in comparative immunology among the very heterogeneous population of immunologists. A general survey of the field of comparative immunology as it existed in 1957 (Cushing and Campbell, 1957) presents a further basis for discovering earlier literature and evaluating the advances made in the intervening years. Prior to 1966, Hildemann and his associates securely established the area of comparative transplantation immunology by their work on the lower vertebrates (cf. Hildemann, 1962; Hildemann and Cooper, 1963), and Fujino, Ridgway, Sprague and others (cf. Symposium, 1961) each independently did extensive work in relating blood grouping concepts and techniques to the identification of subpopulations of marine animals. The researches of these and other workers were reviewed in 1964 (Cushing, 1964) and are an outgrowth of the idea that blood group polymorphism could provide useful, genetically controlled criteria for the analysis of the isolating effects of nonheritable imprinted homing specificities in subpopulations of Pacific salmonids (and other migratory forms) (Cushing, 1941, 1952).

In addition to studies on vertebrates, the defense mechanism of invertebrates has also long been under investigation, but it was not until 1967 that the first symposium devoted solely to this subject was organized (Bang, 1967). Evans and his associates form a group that has been active in studying the comparative immunology of both vertebrates and invertebrates for some time (cf. the several references in this chapter).

The sketch and references just given provide a background against which the consideration of the immunology of fish can be developed. Speaking very broadly, this background of research shows that the immunological responses of fishes and related phenomena such as blood group polymorphism and the nature of fish complement are much like those of other classes of vertebrates. However, variations continue to be discovered within this broad likeness. These are not only interesting in their own right, but their study contributes to fundamental knowledge concerning the immunological response, its evolution, ontogenesis, population ecology, and medical significance. That fish have numerous advantages as experimental organisms is brought out in these volumes, and it is hoped that this chapter provides a basis for further research into the immunological facet of fish physiology. Emphasis will be placed on those aspects of the physiology of comparative immunology referred to above rather than on the problems relating to the identification of subpopulations (see Section V on blood groups).

B. Methods

The methods of immunology have multiplied enormously over the past decade as an integral part of the "immunology explosion." It is not feasible to review these in any detail, but some general discussion will be of value.

Several recent books dealing with the vast array of methods are now available. Among these, the series entitled "Methods in Immunology and Immunochemistry" edited by Williams and Chase (1967, 1970), the book "Experimental Immunology" edited by Weir (1967), and the book by Campbell *et al.* entitled "Methods in Immunology" (1970) are outstanding. References to relatively simple applications of blood grouping methods to the study of fish are reviewed in Cushing (1964). The most extensive and recent book on blood grouping is Race and Sanger's "Blood Groups in Man" (1968).

Obviously, the methods used by individual workers on fish show the most specific application to problems of direct concern to this chapter, and it is recommended that they be followed in detail through the literature citations. It is worthy of note that gel diffusion methods (also called "immunodiffusion") of demonstrating antigen–antibody reactions are in their variety of forms among the most useful, yet simple, of those in current use. Their value is further enhanced by the coupling of simple diffusion with electrophoresis to produce immunoelectrophoretic systems capable of resolutions of extreme specificity.

As in all methods, initial difficulties are invariably encountered in the development of specific applications, but generally speaking, the careful and persistent worker with general laboratory know-how will find that with a reasonable expenditure of efforts, he will be rewarded.

The sections to follow are written with the assumption that the reader initially will be concerned with observations and concepts and will follow through the literature those citations of methods that are of specific interest to him.

C. General References

Those who would like to read broadly on immunology will find a number of books that have appeared recently. The following are suggested among the several that this writer considers useful: Kabat's "Structural Concepts in Immunology and Immunochemistry" (1968), Haurowitz's "Immunochemistry and the Biosynthesis of Antibodies" (1968), Hildemann's "Immunogenetics" (1970), Pressman and Grossberg's "The Structural Basis of Antibody Specificity" (1968), the annual series" Advances in Immunology (1961 to date), the Cold Spring Harbor Symposium, Volume XXXII, "Antibodies" (1968), and the Oak Ridge National Laboratory Symposium (1966).

II. ANTIBODIES

As noted in the Introduction, the study of the nature and synthesis of antibodies, such as are found in the serums of humans and readily available laboratory mammals, currently occupies the attention of so many persons as to have become one of the major areas of interest in biology. The principal reasons for this follow. First, the study of antibody molecules forms the heart not only of immunochemistry but of immunology in general. Second, the synthesis of antibodies cannot yet be experimentally related to the so-called dogma of the genetic code and therefore presents an outstanding challenge to those who would understand the mechanism of protein synthesis in general. Third, the differentiation during morphogenesis of the specialized antibody synthesizing cells provides the embryologist with challenging material for consideration. Fourth, the phylogeny and progressive evolution of this highly evolved response are still far from understood.

The purpose of this section is to provide a general concept from the work on mammalian antibodies, especially human, and to show how the study of fishes is assisting in enlarging this concept which is in a state of rapid evolution.

A. The Immunoglobulins

Intensive work by the family of researchers, whose names and literature citations are to be found through the references cited below and in the Introduction, has established the following general picture of the nature of antibodies found in human and other serums. This picture is well presented in considerable detail and with particular concern for evolutionary considerations in a recent review by Putnam (1969).

The antibodies are distinguished as a class from other proteins found in blood, such as the esterases, transferins, haptoglobulins, albumins, and hemoglobins, through the physicochemical characteristics that associate them with the immunoglobulins, as is recognized by the name of this class of proteins. Writers generally note that while antibodies are all immunoglobulins (often abbreviated as Ig), it is difficult to prove that all Ig are antibodies; however, this point does not present significant difficulties in the considerations here. It will become apparent as we proceed that the outstanding and unique feature of Ig is the great heterogeneity of structure that is superimposed on the class characteristics of these molecules. These characteristics include such attributes as relative electrophoretic mobility and precipitability in contrast with other proteins in blood.

Early confusions in the nomenclature of the Ig and their molecular subunits have been largely resolved through international agreement (cf. Putnam, 1969) and this nomenclature will be used here. Four major classes (isotypes) of Ig are distinguishable by their molecular weights, antigenic specificities, biological and other properties. These are termed, respectively, IgG or γG, IgA or γA, IgM or γM, and IgD or γD. Of these, γG and γM are so much the best known that attention will be confined to them; γG is a protein with a molecular weight around 160,000, in contrast with γM with a molecular weight of about 10^6. The differential sedimentation of these two molecules in the ultracentrifuge has often led to their being referred to as the 7 S and 19 S antibodies, respectively. Physicochemical and physiological dissections (through the study of Bence-

Jones proteins and other Ig fragments produced by individuals with pathological defects) are leading to a profound understanding of the subunit structure of these molecules. Both are composed of four chains of polypeptides, one pair of shorter, "light" (L) chains and one pair of longer, "heavy" (H) chains. These chains are relatively similar in both molecules, the major contrasts between γG and γM being that while γG is a single molecular entity with two combining sites per antibody the γM molecule consists of a complex of five repeating molecular entities, each grossly similar to a γG molecule with a total combining valence of ten. The γM molecule is readily inactivated by mercaptoethanol, whereas the γG molecule is not, a simple but far from comprehensive example of a contrasting characteristic of the molecules. An individual serum has all four classes of Ig in variable amounts depending upon physiological conditions such as location and length of immunization.

When the L and H chains of γG are separated from each other, they are found to be polypeptides consisting of from 210 to 220 amino acid residues for the L chains, and around 440 for the H chains. Injection of separate chains into rabbits produces antibodies that show that L chains are of two types with respect to antigenic specificity; that is, a single L chain has either a specificity designated kappa (κ) or one designated lambda (λ). Both L chains in any single molecule of γG are always the same, but any one individual's serum contains a mixture of molecules showing either the κ or λ specificity. This reflects the concept that single cells among those that manufacture anitbodies only synthesize a single type of L chain.

In contrast to L chains, γG H chains are found to have a different set of antigenic specificities existing as four alternate possibilities. These are termed γG1, γG2, γG3, and γG4, respectively. As with the L chains, both H chains of a single antibody molecule are of the same specificity, but any single human produces a mixture of varying percentages of all four kinds of H-chain molecules, each with a κ or λ L-chain dimer. Again, this reflects the capacity of synthesis of single Ig-producing cells, some making one kind of γG, some another. Data are accumulating to show that both L- and H- antigenic specificities reflect differences in the primary amino acid sequences of the chains involved.

B. Allotypes

Still additional antigenic heterogeneities are imposed upon the γG molecule. These include a series of genetically controlled polymorphic variations termed "allotypes" which appear on both L and H chains in addition to the subclass variations referred to above. The interrelations of the allotypes and the other γG specificities are complex to the point where further discussion is not warranted here. However, it should be kept in mind that these specificities not only are proving very useful in understanding the nature of the genetic control of γG synthesis but also in providing data of value in subpopulation studies. The reader is referred to the paper of Natvig et al. (Cold Spring Harbor Symposium, 1968) for a lucid introduction to this phenomenon.

Allotype variation is not only found in humans but also is well known in such diverse forms as the mouse and rabbit. It has yet to be demonstrated in fish; however, observations by Krauel and Ridgway (1963) point in this general direction. Studies on the genetics of allotypes strengthen the concept that the L chains are produced by two different cistrons and the H chains by four other cistrons all present in every individual with mutations giving rise to allotypic alleles in the various cistrons. New allotypic variations continue to be discovered, and other antigenic variations exist among γG molecules, but the above information should be sufficient to show the complex nature of the variations in antigenic specificities that occur in this class of molecule.

As Putnam (1969) shows in some detail, it is possible to "slice" the γG molecule in various ways, using appropriate enzymes and other reagents, and to obtain fragments that include or exclude the combining site areas and a carbohydrate moiety that is attached to the C-terminal portion of the H chains. Figure 1 presents a schematic presentation currently in vogue of some of the salient features of the γG molecule.

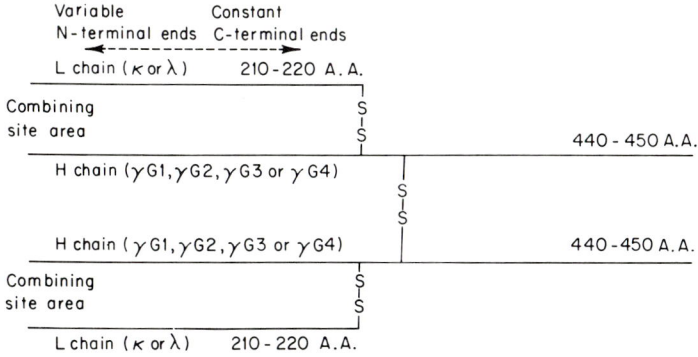

Fig. 1. A widely accepted concept of the relations of several of the features of the γG antibody molecule in its primary structure. The reader is referred to the text for further explanation and to Putnam (1969) for a more detailed figure. (A.A., amino acids.)

C. The γM Molecule

It will be recalled that this molecule consists of a complex of five subunits, and it is therefore considerably larger than γG. Each single subunit is constructed very much like γG, consisting of two L and two H chains of the same relative lengths. The "same" L chains that occur in γG are used in constructing this molecule even to the point of showing either the κ or λ specificities. However, while of comparable length, the H chains used show a single unique antigenic specificity (designated μ), in contrast to the four possible γG, which helps to characterize this class of globulin. Allotypes have not yet been demonstrated in the γM molecules.

This class of Ig appears first during the ontogenetic development of immunological competence, and also first, in subsequent life, during a course of immunizations; γM antibodies also have a more efficient hemolytic relationship with complement than do γG and exhibit other differences in immunological relations.

D. The Primary Amino Acid Sequences of L and H Chains

It is well known that only a small percentage of the surface of the intact antibody molecule is devoted to forming a combining site and that the γG molecule has two such sites, both of which appear to be identical in any one molecule with respect to their specific affinity for antigen. It is also well known that a population of antibody molecules having affinity for a given antigenic specificity shows great heterogeneity, not only of the sort already reviewed in preceding sections but also at the site itself as measured by degree of affinity and specificity. This variability is imposed upon the more general heterogeneity to the point that essentially an "infinite" number of antigenic specificities can be recognized by the antibody synthesizing machinery. This capacity for recognition is the basis for considering the synthesis of antibody as the most unique and remarkable protein synthesis known. Immunochemists are devoting much effort to the determination of the molecular architecture of Ig molecules with the aim of learning how the remarkable properties of the combining site are contrived. These studies are still in progress and rapidly evolving. However, some concepts are emerging that will be considered here, again largely with reference to the γG molecule.

A combining site area appears to involve N-terminal portions of both the H and L chains (cf. Pressman *et al.*, Cold Spring Harbor Symposium, 1968). The specific nature of any one site is apparently determined by variants in the primary amino acid sequence of the regions involved. These sequences in turn determine the ultimate three-dimensional tertiary structure of the site which must "fit" intimately into contact with the antigenic determinant for which it is contrived. (Note that the growth of polypeptide chains starts at the N-terminal end and finishes at the C-terminal.) From 40 to 73 of the 100 or so amino acid positions along the N-terminal side of L chains appear to be variable with respect to the amino acids that can occupy them, and the N-terminal fragment appears to be equally variable in H chains. In contrast, the remaining positions (C terminating) along both chains appear to be invariant except as stated below.

The invariant lengths of the chains, while interesting in their own right, do not appear to be particularly remarkable, but rather they seem to be synthesized according to the relations known to occur among genes and the polypeptides such as those of hemoglobin. These relations include the demonstration that the invariant portions of antibody molecules are much more alike than different among the individuals of several mammalian species so far investigated. Differences that do occur seem readily explainable as the result of mutations that substitute one amino acid for another and that are reflected in the antigenic variants discussed earlier. Considering the "invariant" regions, the story seems to be that a good case can be made for their common genetic ancestry, modified through evolution by gene duplication and the relatively minor evolutionary disturbances introduced by codon mutation (Jukes, 1966).

Not so, however, for the "variable" regions of antibody molecules. Here, as shown above, it is found that while some positions along the chain are comparable to those of the invariant areas and remain constant from individual to individual and species to species many other positions are extremely variable with respect to the kind of amino acid that occurs in them. The magnitude of this variability is indicated by the statement of Haurowitz (1968, p. 105) that in human L chains sequenced at the time of his writing, approximately 22 of the 105 N-terminal amino acids have been observed to be replaced by other amino acids. As he points out, if it is assumed that each amino acid is replaced by only one other amino acid, the number of combinations possible could produce four million different L chains. Considering the same degree of variability in the H chains and that L and H chains are synthesized separately and then assembled, there would seem to be a physical basis for providing for the "infinite" heterogeneity that is observed at Ig combining sites. The nonvariant positions of the N-terminal sequences of both L and H chains appear to be similar among widely diverse organisms, again suggesting a common origin from an ancestral gene (cf. Wikler et al., 1969; Putnam, 1969).

The extreme variability of the numerous positions in the N-terminal region cannot yet be understood in terms of genetic and evolutionary concepts. If this variability reflects the occurrence of gene mutation, it is not possible to conceive of the antigenic determinant as a mutator capable of specifically and simultaneously inducing so many mutations at once and in a way that predetermines how the polypeptides will finally fold in order to achieve a "fit" complementary to the three-dimensional structure of the antigen. Similarly, it is difficult to conceive of hereditary or somatic mutations occurring in advance of contact with an antigen and waiting for it to come along and select them out of the tremendously large pool of combinations of such mutations that would have to be stored and available. Finally, no knowledge exists of any mechanism whereby amino acid sequences are determined by information introduced (as would be the case with antigen) at levels intermediate to the nucleic acids and their final polypeptide product or between the finished protein and its polypeptide subunits. Several ingenious hypotheses suggest ways in which one or another of the above difficulties might be overcome, but to date no experimental evidence is available that will decide among them. The biologist, therefore, has a first class puzzle involving a remarkable evolutionary adaptation that is being painstakingly resolved. There is every reason to believe that this resolution will provide some very profound insights into the mechanisms of genetic, embryological, and evolutionary biology. The next section will show how the study of fish relates to this problem.

E. The Immunoglobulins of Fish

The preceding section has outlined the complex nature of the immunoglobulins. Obivously, the mechanism for the synthesis of these molecules is highly evolved in mammals, and it is of great interest to understand its evolutionary origins and phylogeny. Comparative studies on lower vertebrates and invertebrates are now in progress in several laboratories. It is not the purpose of this chapter to review these in detail. However, some general comments will help place research on fish Ig in proper perspective. References for background reading should include "Phylogeny of Immunity" (R. T. Smith et al., 1966), Good and Papermaster (1964), and the symposium "Defense Reactions in Invertebrates" (Bang, 1967). It must be emphasized that research is already modifying some of the concepts and observations cited (see, for example, Section VI). One major concept that emerges from these studies is that vertebrates generally have a basic capacity for Ig synthesis and its associated phenomena [the total complex of these reactions have been defined by Good and Papermaster (1964) as adaptive immunity]. This capacity has yet to be demonstrated in any invertebrates, and it is probable that "plants and microorganisms" do not make Ig (although some proteins with hemaglutinating properties are found in these groups). This puts fish as a class in a very interesting position, for they include a diversity of forms that can provide information on the details of the evolutionary progress of the complication of the Ig mechanism.

The best introduction to the early literature of the immunological response in teleosts is that of Ridgway *et al.* (1966) and the following statements are based on this paper, supplemented by notes from the review of Sindermann (1966). The capacity of teleosts to respond to antigenic stimulation was demonstrated early in the 1900's, so that by 1950 it was known that teleosts could produce specific bacterial agglutinins and develop protective immunity to a variety of antigens. These responses were found to be depressed or suppressed at lower temperatures, as might be expected in cold-blooded animals (see, for example, Ridgway's discussion of sablefish that normally live below 10°C).

Early investigations of the plasma proteins of fish extend back at least as far as 1929. Electrophoretic techniques were introduced in 1945 using human plasma as a standard against which to demonstrate general similarities in numbers of lower vertebrates including species of fish (cf. Cushing and Campbell, 1957, for references). These electrophoretic studies were further extended by Engle and Woods (cf. Good and Papermaster, 1964) to include the demonstration that while some teleosts apparently lacked Ig, several species of elasmobranchs all had bands that behaved like those of human immunoglobulins. Evans and his associates (cf. Evans, 1963) and Elek *et al.* (1962) demonstrated immunoglobulins in amphibians and reptiles, and Baril *et al.* (1961) showed similar proteins in the alligator. The stage was thus set by the early 1960's for a more sophisticated look at the phylogeny of antibody synthesis. Research moved rapidly in the hands of several groups so that by 1964 it was possible for Good and Papermaster to state on review that a comparison of a diverse sample of species of fish produced antibodies like γM in general aspects and to suggest that the ontogenetic and immunological shift from the production of 19 S-like antibodies to that of 7 S-like antibodies observed in higher vertebrates was in the process of various stages of evolutionary expression among the fishes. They also noted the great potential interest inherent in subunit analysis of the Ig involved, and the need for clarifying the "limits of resolving power" of lower vertebrate antibody with respect to the recognition of antigenic specificities (see Section II, F).

Progressive studies since 1964 have demonstrated that modern elasmobranchs and several other species of fish produce both 19 S-like and 7 S-like Ig and may possess some capacity for immunological memory, although this point is still equivocal, being difficult to demonstrate (Sigel and Clem, 1966; Clem and Small, 1967; Marchalonis and Edelman, 1965, 1966). Elasmobranch antibodies are similar to those of mammals in being composed of light and heavy chain subunits. They differ, however, in that the H chains of both kinds of antibodies appear to be identical with respect to such characteristics as amino acid composition, size, carbohydrate content, and antigenicity. Acrylamide-gel disc electrophoresis shows that elasmobranch light chains probably exists as two classes and are heterogeneous to the same degree as those of humans. Chain sequence studies by Suran et al. (1967) show that the N-terminal pentapeptide amino acids sequence of most L and H chains of leopard shark was identical to human κ chains. This chain type was predominant (80–100%) in both L and H leopard shark chains. The remaining chains are presumed to have blocked N-terminal amino acid groups similar to the λ and H chains of human chains. The unblocked H chains were very different in sequence from the blocked rabbit and human H chains.

Pollara et al. (1968) report a study on the H and L chains of paddlefish, *Polyodon spathula*, 19 S Ig (this species apparently does not make a 7 S antibody). The intact molecule had a molecular weight of 870,000 ± 7,300, its H chain that of 75,000 ± 750, and its L chain 23,500 ± 560. These values correspond to those of human and rabbit γM H chains, shark H chains, and the L-chain molecular weights of other species of fish and mammals. N-terminal pentapeptides (ASP-Ile-Val-Ile(Leu)-Thr-) are not only identical for H and L chains but also homologous to those for sharks and mouse and human κ L chains. The demonstration of an unblocked N-terminal amino acid in H chains from this lower vertebrate is the second such, the other being reported in the leopard shark (Suran et al., 1967). This is in contrast to mammals where cyclized N (glutamine pyrollidone carboxylic acid) appears as the N terminus of many 7 S H chains. A low yield of N-terminal aspartic acid from L chains suggests that as in elasmobranchs there may be two classes of L chains, a κ plus a λ type with blocked sequence.

The homologies just reviewed strengthen the concept of a common origin of the genetic apparatus for antibody synthesis from a primitive ancestral gene (cf. Marchalonis and Edelman, 1965; Kabat, 1967a,b; Putnam, 1969). They also support the idea that the conserved amino acid positions within the variable portion of L chains may have the function of acting in the establishment of the general teritary structure of antibodies. Further information on fish Ig has been obtained by Trump (1968) who used the passive hemagglutination technique to study antibodies for bovine serum albumin produced by goldfish. Two populations of Ig containing antibodies were obtained and characterized. Both of these sediment at 16 S, and both had a common antigen determinant. One population was electrophoretically faster than the other and had a unique second determinant. Whether or not these represent antibody classes similar to those discussed above is still uncertain, but their occurrence is a good indication of the need for additional comparative studies on fish antibodies.

This section shows that sufficient studies on the molecular architecture of fish antibodies now exist to verify earlier assumptions that research on this class will contribute profoundly to the devolpment of our understanding of the immunoglobulins, particularly in view of the observations reported on the differences between fish and mammalian 7 S antibodies with respect to their relations to homologous γM, and the observations of Trump just referred to (see also Uhr et al., 1962).

F. The Specificity of Fish Antibodies

As stated earlier, there is good evidence that fish in general make phage neutralizing and agglutinating antibodies following antigenic stimulation. Precipitating antibodies are another story, however. Good and Papermaster, in 1964, considered the nature of the evolution of the specificity (avidity) of the antibody combining site. At this time, only the holostean bowfin and a few other species had been demonstrated to be capable of precipitin production, while several failures had been reported (cf. also Ridgway et al., 1966).

The successful cases had not been pursued in depth, and one of them (Sirotinin, 1959) even led to the conclusion that fish antibody against the serum of other species reacted indiscriminately with heterologous antigen. This stimulated Everhart and Shefner (1966) to a further investigation of the specificity of goldfish antibodies, using bovine serum (BSA) as homologous antigen. The serum albumins of the human, sheep, and horse were used as heterologous antigen, with rabbit anti-BSA as a comparative control. The standard quantitative precipitin test was used to determine equivalence points and cross-reactivity. In addition, gel diffusion and immunoelectrophoretic comparisons were made, as were characterizations obtained from antibodies eluted from an immunoabsorbent column. The results showed that fish anti-BSA was at least as specific as rabbit anti-BSA and possibly more so (in this sample), since fish antibody did not cross-react with sheep serum albumin whereas rabbit did. Differences observed between the two kinds of antibodies included the following: (1) The goldfish antibodies had three equivalence points with homologous antigen, whereas the rabbit antibodies used had only one; (2) the fish serum had a larger amount of low avidity antibody than did the rabbit; and (3) the fish antibodies migrated electrophoretically more slowly than did those of the rabbit. The observation concerning low avidity antibody agrees with that of Suran et al. (1967) on that of the leopard shark. Here both phage-neutralizing and hemocyanin-precipitating antibodies were less effective than those of rabbits.

Clem and Sigel (1966) have shown that two species of shark, the margate, *Halmulon album*, the gray snapper, *Lutjanus griseus*, as well as the gar, *Lepisoteus platyrhincus*, are capable of production precipitating antibodies for bovine serum albumin. In contrast to the sharks and the goldfish (referred to above), the marine teleosts appeared to produce only γM-like antibodies as far as could be detected by electrophoretic and other methods. Gar serum included two protein fractions of γG type. However, these did not increase during prolonged immunization of this fish with BSA.

Chuba et al. (1968) have shown that the brown bullhead, *Ictalurus nebulosus*, immunized with appropriate human secretor saliva, responds by producing potent fractions of agglutinating anibodies specific for human A, B, and O blood types. While precipitating antibodies are not yet under investigation, these studies demonstrate that this species is excellent material for further work on the specific competence of the combining sites of fish antibodies. The value of the brown bullheads is enhanced because of the ease with which they are cultured, and the fact that they are known to have blood grouping antigens of their own (Cushing and Durall, 1957) that should make specific study of iso-antibodies possible. Further studies on the responses of this species to human antigens are discussed in Section V on blood groups.

Consideration of the studies reviewed in this section shows that fish offer favorable material for the investigation of the evolution of the specific competence of the antibody combining site. It also shows, however, that much work will need to be done before generalities will emerge in this area. Finally, it shows that great care will be necessary to distinguish between variations induced by methodological vicissitudes and those that represent real evolutionary differences.

III. THE CELLULAR BASIS OF THE IMMUNOLOGICAL RESPONSE

Much ingenious research has been devoted to discovering the cells that produce humoral antibodies and that are responsible for the class of reactions resulting in cellular immunity, such as are observed in delayed hypersensitivities and transplant rejections. This is proving to be a very complex field indeed, and one in which many questions remain unanswered. The reader is referred to the Oak Ridge Symposium (1966) and the Cold Spring Harbor volume entitled "Antibodies" (1968) for papers and references upon which to base further reading. Some of the concepts that are emerging relevant to this section are as follows.

Specialized cells are involved in the immunological reactions of mammals termed "lymphocytes" or the "lymphoid series." These originate in the bone marrow along with other hemopoietic cells and appear to differentiate into two populations, one concerned with the production of immunoglobulins and the other with cellular immunity. As maturation proceeds, both kinds of cells are to be found, along with other cell types, in the lymph nodes and spleen and also circulating in the blood. The thymus gland, while not the site of immunological reactions, appears to serve during youth as a reservoir of small lymphocytes that seed the other lymphoid tissues with cells competent to react in immunity. The thymus also appears to produce a hormone that "enhances" the immunological capabilities of lymphoidal cells. The lymphocytes exist in several morphological forms including a series ranging from small through intermediate to plasma cells, some of which proliferate markedly, with the exception of notably mature plasma cells, following stimulation with antigen. This series colonizes the germinal centers of the lymphoid system and produces immunoglobulins, while the cellular immunity series colonize the deeper regions of these organs.

The above cell types are also to be found in the circulating blood and immunologically active sites, such as those of infection or graft rejection. Remarkably, none of these cells, responsible as they are for specific immunity, are phagocytic. This responsibility is that of other leukocytes, also derived from hematopoietic tissue in the bone marrow, which are widely distributed throughout the body, along the linings of the circulatory system, and in the blood itself. These also gather at foci of activity.

There is, therefore, a complex relation between phagocytic cells which injest and digest antigen and those mediating specific immunity. This relationship is not yet understood, but evidence is accumulating to suggest that the phagocytes, after initial processing of antigen, transmit specific molecular information to the immunological system that leads to the reactions of adaptive immunity.

A final point is that not all the immunologically competent cells react following a given antigenic stimulus. The extent of reaction is determined in part by the geography of the stimulus in relation to the cells involved, and in any case only a minority of available cells becomes committed to a given response. There is a general concept that any single cell and its descendents can produce only one kind of several antibody types genetically available to it, but exceptions to this are still being collated. While this point may not always be true, it does make it easy to visualize the immunological system as consisting of a coexisting array of clones of lymphoidal cells capable of recognizing and reacting to different ones among the total antigens that an individual will be exposed to and of making one or another of the heterogeneous immunglobulins reactive to the antigen involved.

The reader should be aware that the above discussion presents only a very brief sketch of a very difficult research area. It is, for example, easy to distinguish an array of cell types by their morphology, but it is very difficult to determine their ontogenetic and physiological relationships and also to be certain that cells of the same appearance in one species are homologous with those in another. The reader is recommended to read descriptions of experiments that are being performed in this area, notably those showing the ingenuity of transplantation researchers, who are able, for example, to "empty" mice of lymphocytes of their own, replace them with genetically marked cells of other individuals in an array of combinations, and apply a battery of methods available to the immunologist and histochemist in interpreting the results. One basically simple experiment, easily visualized, takes advantage of an evolutionary "dissection" occurring in the chicken. This has been the development of the distinctive organ called the "bursa of Fabricius" along the lower gut of this species. Removal and transplantation of this organ in various combinations with the thymus has contributed greatly to the idea of two basic lymphoidal populations of cells, one producing immunoglobulins mediated by the bursa of Fabricius, and one responsible for delayed hypersensitivity and transplantation reactions represented by small lymphocytes originating in the thymus.

A. Phylogeny of the Lymphoid System

Good and several of his associates present a paper (1966) devoted to "morphologic studies on the evolution of the lymphoid tissues among the lower vertebrates." This integrates present immunological knowledge with the rather extensive and sometimes obscure literature of comparative histology and anatomy that extends back into times well before the advent of modern comparative immunology. With respect to fish, it is sufficient to say here that this system is found to be well established, at least as homologues of the thymus and spleen, in primitive sharks and rays, and that plasma cells are "first" found in the paddlefish. From this point on up the evolutionary line, emergence, consolidation, and reshuffling of other kinds of lymphoidal tissues occurs within patterns too detailed to present here. As the authors note, their paper serves to sketch a scheme for the evolution of the lymphoid system that provides a basis for experiment and further clarification.

One tantalizing observation among the many awaiting such clarification is the nature of the lymphoidal system of chimaeroids (Fänge, 1966). These fish, unlike other elasmobranchs, lack lymphoid tissue in their excretory system, but they do have a thymus and uniquely prominent lymphoidal-like structures in the cranial region. A second observation worthy of further research, also introduced by Fänge, is that a production of lymphoid cells takes place in the excretory system of several kinds of fishes (as well as in the cyclostomes, see below), suggesting a common origin from this type of tissue (see also A. M. Smith et al., 1967). It is also possible to speculate from morphological data that the thymus evolved from the gill region.

As to experimentation, none seems to have been done on fishes that is directly aimed at elucidating the activities of their lymphoidal system except the papers of A. M. Smith et al. (1967) and Chiller et al. (1968) discussed in the complement section. As Hildemann and Cooper (1963) have noted, such experiments could readily include manipulation of the thymus. They also might well consider the transplantation of other lymphoidal organs into "empty" fish. The attention of the reader who considers such studies is called to the excellent work of Hildemann and his associates, Cohen, Haas, and Cooper, on both fish and amphibians since this sets standards of methods and interpretations that have yet to be excelled among researchers on lower vertebrates (citations in Hildemann and Cooper, 1963, and also below).

B. Transplantation Studies in Fish

The transplantation of tissues in fish is conveniently reviewed in connection with consideration of their lymphoid system, for the rejection reactions observed are known to be those of delayed hypersensitivity (cellular immunity). Transplant rejections were first demonstrated by Goodrich and Nichols (1933), using a technique for scale transplantation developed by Mori (1931). This technique is very simple and effective, consisting merely of removing scales with a forceps from their "pocket" and exchanging them for other scales. This is easily done with anesthetized fish since the transplanted scales slip readily into their host "pockets" and start to grow (sometimes a bit of trimming to fit is in order). Goodrich and Nichols observed that autotransplants of goldfish scales succeeded without any tissue loss, while varying degrees of tissue incompatibility occurred with all homografts (those among individuals). This varied from only the destruction of chromatophores to inflammation and destruction of the whole scale. Two other workers (Sauter, 1934; Nardi, 1935) confirmed these observations with scales and other tissues, but none recognized the immunological significance of their results.

Hildemann reopened these investigations in the 1950's and performed an extensive series of experiments firmly establishing the immunological basis of scale rejection (Hildemann, 1957, 1958). Specific results include the development of quantitative methods for evaluating scale rejection times, and demonstrations of rejection specificity, enhanced secondary reactions, the influence of temperature, the cellular basis of the rejection action, and the indepedence of this reaction from circulating antibodies. The histocompatibility genetics of these scale transplantations are considered by Hildemann and Owen (1956) in a paper that shows a general method for this purpose. Hildemann and Haas (1960) extended the generality of observations on goldfish to include the carp, *Cyprinus carpo*, the black-spotted barb, *Barbus filamentosus*, the blue acara, *Aequidens latifrons*, and the guppy, *Lebistes reticulatus*. Additional studies on the kinetics of homograft rejection and the effects of antimetabolites and X-radiation on *Fundulus* are reported by Hildemann and Cooper (1963). The effects of a variety of metabolites, antimetabolites, and antibiotics on transplantations in *Fund-*

ulus have also been studied extensively by Goss and his associates (references in Hildemann and Cooper, 1963).

Triplett and Barrymore (1960) have investigated the ontogenic appearance of transplant rejection in the shiner sea perch, *Cymatogaster aggregata*, a viviparous embiotocid. They have shown that embryos accept scale transplants, but reject homografts, starting one to three days following birth. The potential experimental value of this group of fishes demonstrated by their work has yet to be taken advantage of by other investigators. Reid and Triplett (1968) report observations on some initial autograft destruction in scale transplants in *Micropterus salmoides* and relate their recovery to a lag period observed in homograft destruction.

Finally, it is to be noted that fin transplants have been studied by Kallman and Gordon (1958) in the viviparous platyfish, *Xiphophorus maculatus*, taking advantage of crosses obtained from inbred strains. The results of this work show that histocompatibility differences occur according to the same genetic relationships observed in mammals. That is, isografts and those from parent to F_1 hybrid offspring were completely accepted, while reciprocal grafts or those among parents and F_2 offspring were rejected as expected.

Kallman has shown the potential usefulness of fin transplants in evaluating the degree of genetic homogeneity of small isolated fish populations (1961), and Hildemann (1962) has pointed out the potential usefulness of scale transplants in subpopulations analogous to those conducted on amphibians (Hildemann and Haas, 1961).

Ball and Kallman (1962) also report the success of whole pituitary transplants into the tail musculature of adults of the all-female, gynogenetic teleost, *Molliensia formosa*. This experiment is noted because it is representative of the area of transplantation so well exploited, especially in amphibians, by embryologists, and it is concerned with discovering aspects of physiology other than those that are the subject of this chapter.

IV. COMPLEMENT

Early immunologists soon discovered that the fresh normal serum of humans and laboratory animals contained a system of molecules that worked together to lyse red blood cells (and some bacteria) to which specific antibodies were attached. The system has come to be called "complement" (symbolized C'). Some of its components are inactivated at temperatures well below those which destroy antibodies, a property that has been found most useful in demonstrating its distinctive nature. Other general properties of the system are that it reacts with "all" antigen–antibody complexes, it is not influenced by immunization, it maintains relatively constant *in vivo* levels, it acts with antibody both as a bactericidin and cytotoxin, it enhances phagocytosis, and it is much more efficient in action with γM than with γG antibodies. Many additional attributes of complement reveal it to be an essential and complex part of the immunological machinery, the activities of which are still only partially understood. Comparative immunologists therefore are paying increasing attention to it in conjunction with other aspects of the immunological system.

The complexity of the complement system was long indicated by the knowledge that it could be inactivated in different ways and reactivated by recombining the inactivated aliquots. For example, heat inactivated serum regains its activity when combined with zymosan (prepared from yeast) inactivated serum or with ammonia inactivated serum. In turn, the latter two serums regain activity when combined. Classically, four separate components have been recognized through the inactivation and recombination patterns that were studied. It was also recognized that the methods involved left much to be desired in terms of specifying the molecular nature and interactions of components. Some of these reagents, useful today, are wonderfully bizarre, such as cobra, *Naja naja*, venom and that of the brown recluse spider, *Loxoceles reclusa*.

The complement story has changed radically over the past few years. Müller-Eberhard, one of the leading investigators in this area, has noted that 85% of the references listed in his comprehensive review (1968) were published within the past 6 years, and 33% of these within the last 15 months. Present knowledge shows that most of the nine components now known can be characterized in such biochemical detail that it is predicted that the essential features of the molecular basis of complement function will be known within the next few years and that this information will lead to accelerated understanding of the less well known biological parameters of the system. The details concerning the complement system are too complex to be presented here, and the reader is referred to Müller-Eberhard's review quoted above, or for less comprehensive treatments, to the texts of Haurowitz (1968) and Kabat (1968).

An interesting aspect of the complement story are the inherited complement deficiencies now known in guinea pigs, mice, rabbits, and humans. These show that direct and "simple" genetic relationships exist with respect to the biosynthesis of several different components and provide a firm basis for evolutionary considerations.

Comparative Immunology of Fish Complement

It is now recognized that the complement system is an attribute of all classes of vertebrates including the fish but that the system has yet to be demonstrated in the invertebrates (see Evans *et al.*, 1968a). Gewurz *et al.* (1966) review the literature on this subject, discuss various papers demonstrating the classic four components in some higher vertebrates, and give reasons for believing that these components are present in all species examined.

The hemolytic nature of fish serum and its heat lability has been known for many years, as has its relationship to hemolytic antibody. Three of the four classic components in guinea pig serum were demonstrated in carp in 1945 (Cushing, 1945), as was the reactivation of some heterospecific combinations of inactive carp, frog, and guinea pig serums. Subsequent investigation of fish complement remained essentially static until very recently when members of Good's laboratory turned their attention to that of the chondrostean paddlefish, *Polyodon spathula*, and several species of elasmobranchs (Gerwurz *et al.*, 1966). The complement system of these species was demonstrated in several ways, including potentiation with Carbowax, reversible blockage by EDTA (ethylenediaminetetraacetate), heat inactivation and interaction, and potentiation with some guinea components. Legler *et al.* (1967) and Evans *et al.* (1968a) have also demonstrated the occurrence and potentiation of complement in teleost fishes.

More recently, Jensen *et al.* (1968) have investigated the complement of the nurse shark, *Ginglymostoma cirratum*, using natural antibody of this species for sensitized sheep erythrocytes. They were able to partially purify the first component of shark complement and to demonstrate its capacity for several kinds of interactions with different components of guinea pig complement. The complement of species of teleosts including a catfish has also been examined in some detail and found to be comparable in activity to that of the guinea pig (Legler *et al.*, 1967).

Another kind of study of fish complement has been made by A. M. Smith *et al.* (1967). These workers developed a modification of the Jerne plaque technique for detecting antibody production by fish spleen and pronephros. Classically, this technique is carried out by preparing agar plates in which numerous intact sheep or other erythrocytes are embedded together with separated cells, such as lymphocytes from animals previously stimulated with sheep erythrocytes, being tested for antibody production. Active cells diffuse antibodies into their vicinity during a period of incubation. At the end of this period the agar plate is flooded with guinea pig complement with the result that hemolytic plaques appear around antibody producing cells. Smith and his associates found that the guinea pig complement was unsatisfactory in the demonstration of antibody production by the bluegill, *Lepomis macrohirus*. However, effective results were obtained using either isologous complement or that of the pumpkinseed, *Lepomis gibbosus*, or the marine white perch, *Morene americana*. Hemolytic antibodies were produced by cells from both spleen and pronephros, which contained large numbers of active cells. The latter observation is most interesting when, as these authors discuss, one considers that there are marked similarities between the cellular structure of the pronephros and mammalian lymph nodes that suggest a common concern with the filtration of fluids in both organs.

Chiller *et al.* (1968) found, using the Jerne method, that the demonstration of antibody formation by rainbow trout cells was dependent on isologous complement or that of closely related species, and, in their case, incubation within a range of temperature between 4° and 20°C. Sephadex G-200 fractionation showed that plaque activating complement occurred in a different fraction from the natural hemolytic activity of trout serum, a point as yet unexplained. Two types of active cells were observed. The first, forming small plaques, appeared in the anterior kidney and then in the spleen. The second, forming large plaques, appeared in the same sequence, but occurred later during immunization. The largest number of active cells appeared in the spleen.

The above two papers have great significance not only to the study of fish complement but also because of their potential value in bringing the relatively simple and effective Jerne method to bear on problems concerned with the evolution of the cellular basis of antibody production in fish. A final point to note is that the differences observed relative to incubation temperatures for the two species reported above further confirm that different species have different temperature optimums of immunological activity (Cushing, 1945; Gewurz *et al.*, 1966; and especially Ridgway *et al.*, 1966).

V. BLOOD GROUPS

Blood group polymorphism has been found in all species of vertebrates so far investigated and some of the antigens involved, notably those of the ABO system, also occur among diverse forms of invertebrates and microorganisms (cf. Cushing et al., 1963). Of course, many species have yet to be studied, including the cyclostomes and the fish lacking erythrocyte nuclei, but it is a reasonable prediction that the occurrence of polymorphic blood group antigens (if not on erythrocytes, then on tissues or in solution) is ubiquitous, at least as far as the vertebrates are concerned. The study of blood groups relates to immunology in a variety of ways, the most fundamental of which is that the variations involved are antigenic polymorphism that are detected through the use of specific antibodies. These variations are determined in a simple and direct manner by a number of different genetic loci exhibiting multiple allelism. So many combinations of specificities are known in extensively studied species such as humans and cattle that it is possible to demonstrate that each individual potentially has an overall blood type individuality all his own. Additionally, evolution has produced differences among the frequencies of gene controlling antigens in different populations. Comparison of these can, under suitable circumstances, provide information useful as an aid in evaluating the extent to which a species is divided into subpopulations (demes, races). As noted in the Introduction, the subject of the blood groups of marine animals has been reviewed with particular attention to the problem of subpopulation identification (Cushing, 1964), and W. de Ligny (1969) includes additional information, together with a very comprehensive consideration of other molecular polymorphisms in fish. In addition to references cited in Cushing (1964), the reader's attention is called to two texts useful in providing background material. The first of these is Hildemann's "Immunogenetics" (1970), and the second, the fifth edition of Race and Sanger's "Blood Groups in Man" (1968).

The most exciting blood group studies on fish currently underway are those of Fujino (1967) and Fujino and Kazama (1968) on the Pacific skipjack, *Katsuwonis pelamis*. Blood group antibodies and antigenic polymorphism in tuna were first reported by Cushing (1952, 1956). Sprague and his associates, working at the U.S. Bureau of Commercial Fisheries Biological Laboratory, Honolulu, subsequently did extensive work to establish reagents revealing a large series of blood groups in skipjack and to show that these provided evidence of at least three reproductively isolated populations distributed over a wide expanse of the Central Pacific (Sprague and Holloway, 1962), as well as the possibility that two different subpopulations associated with changes in water masses occur around Hawaii (Sprague, 1964). A new blood group system termed Y, as distinct from the B system of Sprague and Holloway, has been discovered by Fujino and Kazama in skipjack in Hawaiian waters. This system, independent of sex and size, is postulated to be based on six codominant alleles determining 15 phenotypes. These are differentiated by seven reagents. Rigorous statistical analysis of frequencies of the various phenotyes among almost 1000 fish, taken over the summer of 1965, showed that the frequencies of the hypothetical genes determining these conformed to the expectation for a population in Hardy-Weinberg equilibrium. The development of the understanding of this system provides an excellent basis for subpopulation investigation in this and other species of tuna. Such population studies will assume added significance as they continue to be combined with those on serum esterase (Fujino and Kang, 1968a) and transferrin (Fujino and Kang, 1968b) groups in tuna. In fact, when one considers the additional parameters provided by all other researches on subpopulations of tunas, it looks as though the next few years should provide a fascinating story, concerning as it does the vast reaches of the major ocean systems and the biological magnificence of the tunas with all their associated fisheries (cf. Cushing, 1964, for additional references on tuna blood group research).

Another recent involvement of fish with blood group research is that concerned with the production of antibodies for human red cells in the brown bullhead catfish, *Ictalurus nebulosus* (Chuba *et al.*, 1968; Wiener *et al.*, 1969). As noted in the section on antibody specificity, antibodies were produced by the injection of saliva from human secretors. The most striking of the results obtained was a reagent, anti-Z, that detected a new specificity on human cells. This reacted equally with all human cells irrespective of type except that of Bombay. It also reacted with the cells of anthropoid apes but failed to react with those of monkeys. Additional observations are reported that lead the authors to postulate a basic blood group substance A on which the genes A and B and the gene H superimpose their specificities.

Of interest in connection with the above studies is the fact that the brown bullhead is the first species in which a system of erythrocyte antigens and natural agglutinins analogous to that of the human ABO system was detected (Cushing and Durall, 1957). Additional intraspecific agglutinations between the channel catfish, *Ictalurus p. punctatus*, and the brown bullhead were also observed at this time, as was the occurrence of rare isoagglutinations among the white croakers, *Genyonemus lineatus*. An isoagglutinin in the skipjack, *Katsuwonis pelamis*, was also found in this period (cf. Cushing, 1956). Recently, Kuhns and Chuba (1968) have reported on other intrageneric blood group differences among freshwater catfish (Ictalurids). These include a Forssman-like antigen present not only on red cells but also on epithelial cells and in the gastric mucus, and a second non-Forssman-like antigen recognized by anti-H-like antibodies in the serum of *Ictalurus nebulosus* and eel serum. The catfish agglutinins behave electrophoretically like human γM, and in the ultracentrifuge they have light and heavy components, both of which are inactivated by mercaptoethanol. Chuba (1968) has found Forssman-like antigens in several other species of teleosts and demonstrated that this antigen can be concentrated from goldfish, *Carassius auratus*, aquarium water.

Vann (1966; Vann and Cushing, 1966) has studied the reaction of *Dolicos biflorus* lectin with the red cells of the California bonito, *Sarda chiliensis*. [Lectins are proteins extracted from plants and invertebrates that are not antibodies but have the property of agglutinating erythrocytes; cf. Cushing *et al.* (1963) and Brown *et al.* (1968) for references, especially to the work of Boyd's group.] This lectin distinguishes between two kinds of individuals in this species, those that react positively (D-positive) and those that react negatively. Analysis of population data showed that D-positive fish were more frequent among larger sized individuals in samples even when taken at the same time and place. Unfortunately, it was not possible to maintain bonito in aquaria so that the difference observed could not be followed during the growth of individual fish. [de Ligny (1969) observed a similar phenomenon in plaice, *Hippoglossoides platessoides* (Fabr.), of the North Sea. This species appears to form subpopulations in that area and can be bred in captivity so that it should provide especially favorable material for further study.] A number of alternative explanations are considered by Vann, ranging from the possibility of subpopulation differences to changes in phenotype during the life of individual fish. This last phenomenon has potentially related precedence in humans, chickens, and fish (Sanders and Wright, 1962) and, most pertinently here, in the Atlantic herring, *Clupea harengus*. This species has a blood group system described as the C system (Sindermann, 1962) that consists of a single positive antigen or its absence. Considerable variation was discovered among herring samples with respect to the frequency of occurrence of this antigen, and it looked as though it would be extremely useful in subpopulation studies of this species. However, Ridgway (1968) has discovered that the situation is complicated in that the number of C-negative fish increase markedly in fish held at relatively high temperatures and has produced considerable evidence that herring erythrocytes become inagglutinable by any reagents in fish living under temperature stress, probably as a reflection of anemia and the relatively high output of immature erythrocytes. Obviously, this and the observations of Vann are among the "unusual deviations" that might be anticipated in the study of fish (Cushing, 1964, p. 92) and are a warning that care must be taken in interpreting the results of field observations in the absence of genetic data. This care should include the use of a number of reagents and consideration of sex, size, and gene frequency relationship, as well, of course, as the practice of a general alertness for the unexpected appearance of ecological and other complications [cf. further discussion in de Ligny (1969)].

The most ideal situation for the study of blood groups in fish is, of course, where genetic data can be obtained directly. A number of investigators have or are taking advantage of the possibilities offered in such species [see, for example, Sanders and Wright (1962), Ridgway *et al.* (1966), the ictalurid (catfish) research and that on plaice of de Ligny (1969) referred to above]. The potential usefulness for genetic studies of some wild populations that cannot be cultured is illustrated by the demonstration of Sindermann and Mairs (1961) that the spiny dogfish shark, *Squalus acanthias*, like the brown bullhead, has a blood group system with isoantibody correlates similar to the human ABO system. This species is particularly valuable since it is viviparous and also may be indicative of similar systems to be found in other elasmobranchs. Attention is called, in closing, to some additional studies on fish blood groups recently published. One shows the potential value of blood groups for evolutionary and distributional studies of golden trout (Calaprice and Cushing, 1967), the other, an application of a useful and simplified technique (Ridgway *et al.*, 1958) for comparing erythrocyte antigenic variability among different fish populations (Utter *et al.*, 1966). The technique was employed in this case in the evaluation of the relative degree of inbreeding in Lahontan trout, *Salmo clarkii henshawi*, populations. The paper of Kilambi *et al.* (1965) applies this technique in the differentiation of spawning populations of surf smelt, *Hypomesus pretiosus*.

VI. CYCLOSTOMES

This class of vertebrates requires special consideration for the reason that Good and his associates have developed the hypothesis that adaptive immunity evolved within this phylogenetic level of organization. Reasons for believing this are discussed by them in "Phylogeny of Immunity" (Good *et al.*, 1966) and elsewhere. These are based on the demonstration that the lamprey, *Petromyzon marinus* (cf. also Pollara *et al.*, 1966; Boffa *et al.*, 1967; Marchalonis and Edelman, 1968), possesses a set of immunological responses that fulfills the criteria for adaptive immunity and leads them to the conclusion that lampreys have the most primitive such responses studied in contrast to the more primitive Pacific hagfish, *Eptatretus stoutii*, where a number of immunological reactions were not detected.

More recently, however, Hildemann has pointed out that it seems improbable that a physiological system as complex as that of adaptive immunity would have evolved so rapidly (1966, p. 240) between the phylogenetic stages represented by these two forms and, further, Hildemann and Thoenes (1969) have shown that hagfish "animal husbandry" has important consequences for results obtained. Their work demonstrates in considerable detail that hagfish can reject transplants in the same manner, including immunological memory, as higher forms, and also produce circulating antibodies against sheep erythrocytes. Further, Evans and Hildemann and their associates (1969) have shown that hagfish are also capable of producing bactericidins (although the latter have not yet been characterized as necessarily being antibodies). Hildemann and Thoenes conclude that a "closer look at thymus-lymphoid cell-immunoglobulin relationships" among lower vertebrates is needed. Since invertebrates have so far not been demonstrated to produce antibodies (cf. Cushing, 1967), the origins of adaptive immunity remain to be discovered. The work of Cooper (1968) on earthworm transplants, of Stevens on induced bactericidins in insects (1967), of Bang on induced lysins in sipunculids (1967), of Seaman and Robert on cockroaches (1968), and of Evans *et al.* on induced bactericidins in spiny lobsters (1968b, and 1969a) and sipunculids (1969b) offer exciting material for further research in this area.

VII. FINAL CONSIDERATIONS

This section considers briefly some relevant matters not covered above. The first is the point, which should not be overlooked, that the immunological mechanism obviously was evolved as a defense against parasitic infection. The reader is referred to Sindermann's excellent review of the parasites of marine fishes (1966) and also to Davis's book entitled "Culture and Diseases of Game Fishes" (1967). The relative newness of the relation between the immunology and parasitology of fishes is indicated in that, while the first publication has an extensive section and other comments dealing with immunity, the second, although an authoritative treatise, does not even have subject references to this field in the index.

Sindermann's review adds a number of references that complement those of Ridgway et al. (1966) concerned with aspects of the immunization of fishes. These reviews both include those on efforts, sometimes partially successful, to protect fish such as salmonids from hatchery diseases and to show that oral and other forms of immunization hold considerable promise (cf. Ross and Klontz, 1965). Additionally, Sindermann presents reasons for believing that some diseases of marine fishes may come to be controllable, not, probably, by giving them "shots," but by manipulating their ecology so that they can develop their own immunities, either active or ecological. Among the many interesting observations discussed by Sindermann is that of Nigrelli and Breeder (cf. Nigrelli, 1947) that trematode infections are in part combated by immunological reactions in fish mucus. This substance, so abundantly produced, apparently can acquire protective properties and offers opportunities for research. Attention is called to the paper of Hildemann (1959) showing that the newly hatched fry of the cichlid fish, *Symphysodon discus*, feed exclusively on an abundant mucus secretion produced over the skin of both parents. It is possible that this highly specialized diet may have immunological as well as nutritional attributes, and other kinds of mucoid relations between parents, eggs, and young might be investigated from such a point of view.

The so-called natural antibodies of fish present an additional area of interest. The antibodies are to be found in essentially all "normal" serums investigated and are detected by their ability to agglutinate erythrocytes that can come from a variety of species, ranging from those of different individuals in the same species to other fishes and on to birds and mammals. These are often highly specific and include antibodies that recognize human antigens of ABO and other systems. One such antibody, long known, is that found in serums of marine eels that has anti-H activity (reacting most strongly with O- and A_2-type cells, cf. Race and Sanger, 1968). These antibodies appear to be generated by infections or by normal intestinal flora. Support for this concept derives from seasonal fluctuations observed among them, for example, as found in the brown bullhead (Cushing and Durall, 1957) and the winter skate (Sindermann and Honey, 1964). These fluctuations have been partially correlated with bacterial contamination in some freshwater species by Bisset (1948) and emphasize, as has been pointed out by Noble (1957, 1960), the continued need to maintain an ecological point of view in the study of fish parasites and their effects. Medical significance is attached to fish natural agglutinins in the paper of Janssen and Meyers (1968) who find that the serums of white perch, *Roccus americanus*, taken adjacent to heavily populated areas of Chesapeake

Bay, contain specific antibodies for several microbes pathogenic for humans, while those from less populated shorelines are free of these antibodies. [*N.B.* The antibody response of fish to injected dead bacteria is not entirely unknown, the authors having overlooked the paper of Ross and Klontz (1965) and research preceding it.] As the authors comment, alternative explanations are possible, but their study presents further evidence that it is better to fish in "still" rather than polluted waters.

Two incredible ichthyological observations show us that much is to be learned with respect to fish in the area of transplantation. The first of these is the well-known occurrence of parasitic males among the deep-sea family ceratioidea. Here, males (sometimes more than one) attach themselves orally to females, form continuous tissue connections, and become degenerated parasitic forms, nourished by female blood and capable only of reproduction. Hopefully, the growing competence in marine research will enable these fish to be brought under experimental control so that we can learn the nature of the tolerance involved.

The second incredible observation is concerned with parasitic copepods of the group Lernaeoceriformes, where adult gravid females partially bury their head and anterior body in the fish host, the ultimate being achieved by members of the genus *Cardiodectes* where the head is buried in the heart of the host. Here, without apparent "damage" to either party, intimate contact is maintained with the host blood supply (Ho, 1966; Shino, 1958; Sindermann, 1966, p. 33). Again, tantalizing questions are raised about the nature of the tolerances involved. (The author is indebted to Dr. Sneed Collard, University of West Florida, Pensacola, for calling the above phenomenon to his attention and for showing him examples of the parasitism involved.)

Ridgway and his associates have made a unique and useful contribution by demonstrating that it is not only possible to show subpopulation differences in serum proteins between salmons from the eastern and western North Pacific but also to demonstrate that mature female salmon have a protein of unique antigenic specificity in their serum (Ridgway *et al.*, 1962; Olivereau and Ridgway, 1962; Krauel and Ridgway, 1963). Their extensive work shows that the serological detection of this protein in salmonids and also English sole, *Parophys vetulus*, and Pacific halibut, *Hippoglyossus stenolopis* (Utter and Ridgway, 1967), has potential usefulness in anticipating the appearance of maturation.

Additional relations of fish to immunology could be further elaborated such as the effects of cortical hormones reported by Bisset (1949), the potential usefulness of eggs and antibodies for the identification of larvae with adult forms (Ridgway, 1962), the further elaboration of very restricted studies on fish anaphylactic and delayed hypersensitivity phenomena (cf. Good and Papermaster, 1964), and the report (Barrow, 1955) that not only temperature, *but social behavior*, influenced the production of lytic and agglutinating antibodies against trypanosomes by tench, goldfish, and the perch, *Perca fluviatilis*. However, the author believes that the present chapter has demonstrated the nature of the essential contributions that the study of fish is capable of making to the advancement of comparative immunology.

ACKNOWLEDGMENTS

The author gratefully acknowledges the partial support of the various researches of his referred to in this chapter by the U.S. Office of Naval Research, Biology Branch, and the U.S. Public Health Service, National Institutes of Health, Allergy and Immunology Branch.

REFERENCES

"Advances in Immunology." (1961–1969). Vols. 1–9. Academic Press, New York.
Ball, J. N., and Kallman, K. D. (1962). Functional pituitary transplants in the all-female, gynogenetic teleost, *Mollienesia formosa* (Girard). *Am. Zoologist* **2**, 264.
Bang, F. K. (1967). Defense reactions in invertebrates. *Federation Proc.* **26**, 1664–1715 (Symp.).
Baril, E. F., Palmer, J. L., and Bartel, A. H. (1961). Electrophoretic analysis of young alligator serum. *Science* **133**, 278.
Barrow, J. H., Jr. (1955). Social behavior in freshwater fish and its effect on resistance to trypanosomes. *Proc. Natl. Acad. Sci. U.S.* **41**, 676–679.
Bisset, K. A. (1948). Natural antibodies in the blood serum of freshwater fish. *J. Hyg.* **46**, 267–270.
Bisset, K. A. (1949). The influence of adrenal cortical hormones upon immunity in cold-blooded vertebrates. *J. Endocrinol.* **6**, 99–104.
Boffa, G. A., Fine, J. M., Drillon, A., and Amouch, P. (1967). Immunoglobulins and transferrin in marine lamprey sera. *Nature* **214**, 700.
Brown, R., Almodovar, L. R., Bhatia, H. M., and Boyd, W. C. (1968). Blood group specific agglutinins in invertebrates. *J. Immunol.* **100**, 214–216.
Calaprice, J. R., and Cushing, J. E. (1967). Serological analysis of three populations of golden trout, *Salmo aguabonita* Jordan. *Calif. Fish Game* **53**, 273–281.
Campbell, D. H., Garvey, J. S., Cremer, N. E., and Sussdorf, D. H. (1970). "Methods in Immunology" 2nd ed. Benjamin, New York (in press).
Chiller, J. M., Hodgins, H. O., and Weiser, R. S. (1968). Characteristics and application of the Jerne test in studies of the immune response of rainbow trout (*Salmo gairdneri*). *Federation Proc.* **27**, 492 (abstr.).
Chuba, J. V. (1968). Personal communication.
Chuba, J. V., Kuhns, W. J., and Nigrelli, R. F. (1968). The use of catfish, *Ictalurus nebulosus* (Le Sueur), as experimental animals for immunization with human secretor and other antigenic materials **101**, 1–5.
Clem, L. W., and Sigel, M. M. (1966). Immunological and immunochemical studies on holostean and marine teleost fishes immunized with bovine serum albumin. *In* "Phylogeny of Immunity" (R. T. Smith *et al.*, eds.), pp. 209–217. Univ. of Florida Press, Gainesville, Florida.

Clem, L. W., and Small, P. A. (1967). Phylogeny of immunoglobulin structure and function. I. Immunoglobulins of the lemon shark. *J. Exptl. Med.* **125**, 893–920.
Cold Spring Harbor Symposia on Quantitative Biology. (1968). "Antibodies," Vol. XXXII. Cold Spring Harbor Lab. Quant. Biol., Cold Spring Harbor, Long Island, New York.
Cooper, E. L. (1968). Transplantation immunity in annelids. I. Rejection xenografts exchanged between *Lumbricus terrestris* and *Eisenia foetid. Transplantation* **6**, 322–337.
Cushing, J. E. (1941). An experiment on olfactory conditioning in *Drosophila guttifera. Proc. Natl. Acad. Sci. U.S.*, **27**, 496–499.
Cushing, J. E. (1945). A comparative study of complement. I. The specific inactivation of the components. II. The interaction of components of different species. *J. Immunol.* **50**, 61–89.
Cushing, J. E. (1952). Serological differentiation of fish bloods. *Science* **115**, 404–405.
Cushing, J. E. (1956). Serology of tuna. *U.S. Fish Wildlife Serv., Spec. Sci. Rept., Fisheries* **183**, 1–14.
Cushing, J. E. (1964). The blood groups of marine animals. *Advan. Marine Biol.* **2**, 85–131.
Cushing, J. E. (1967). Invertebrates, immunology and evolution. *Federation Proc.* **26**, 1666–1679.
Cushing, J. E., and Campbell, D. H. (1957). "Principles of Immunology." McGraw-Hill, New York.
Cushing, J. E., and Durall, G. L. (1957). Isoagglutination in fish. *Am. Naturalist* **91**, 121–126.
Cushing, J. E., Calaprice, N. L., and Trump, G. (1963). Blood group reactive substances in some marine invertebrates. *Biol. Bull.* **125**, 69–80.
Davis, H. S. (1967). "Culture and Diseases of Game Fishes." Univ. of California Press, Berkeley, California.
de Ligny, W. (1969). Serological and biochemical studies on fish populations. *Oceanogr. Mar. Biol. Ann. Rev.* **7**, 411–513.
Elek, S. D., Rees, T. A., and Gowing, N. F. C. (1962). Studies on the immune response in a poikilothermic species (*Xenopus laevis daudin*). *Comp. Biochem. Physiol.* **7**, 255–267.
Evans, E. E. (1963). Antibody response in amphibia and reptilia. *Federation Proc.* **22**, 1132–1137 (Symp.).
Evans, E. E., and Hildemann, W. H. (1969). Personal communication.
Evans, E. E., Legler, D. W., Painter, B., Acton, R. T., and Attleberger, M. (1968a). Complement-dependent systems and the phylogeny of immunity. *In Vitro* **3**, 146–153.
Evans, E. E., Painter, B., Evans, M. L., Weinheimer, P., and Acton, R. T. (1968b). An induced bactericidin in the spiny lobster, *Panulirus argus. Proc. Soc. Exptl. Biol. Med.* **128**, 394–398 (additional papers on this species in press).
Evans, E. E., Cushing, J. E., Sawyer, S., Weinheimer, P., Acton, R. T., and McNeeley, J. L. (1969a). An induced bactericidin in the California spiny lobster, *Panulirus interruptus. Proc. Soc. Exptl. Biol. Med.* **132** No. 1, 111–114.
Evans, E. E., Weinheimer, P. F., Acton, R. T., and Cushing, J. E. (1969b). Induced bactericidal response in a sipunculid worm. *Nature* **222**, 695.
Everhart, D. L., and Shefner, A. M. (1966). Specificity of fish antibody. *J. Immunol.* **97**, 231–234.
Fänge, R. (1966). Comparative aspects of excretory and lymphoid tissue. *In*

"Phylogeny of Immunity (R. T. Smith, *et al.*, eds.), pp. 141–147. Univ. of Florida Press, Gainesville, Florida.
Fujino, K. (1967). Review of subpopulation studies on skipjack tuna. *Proc. 47th Ann. Conf. West Assoc. State Game Fish Comm., Honolulu, Hawaii* pp. 349–371. Hawaiian State Game Fish Comm.
Fujino, K., and Kang, T. (1968a). Serum esterase groups of Pacific and Atlantic tunas. *Copeia* No. 1, 383–395.
Fujino, K., and Kang, T. (1968b). Transferrin groups of tunas. *Genetics* **59**, 79–91.
Fujino, K., and Kazama, T. K. (1968). The Y system of skipjack tuna blood groups. *Vox Sanguinis* [N.S.] **14**, 383–395.
Gewurz, H., Finstad, J., Muschel, L. H., and Good, R. A. (1966). Phylogenetic inquiry into the orgins of the complement system. *In* "Pylogeny of Immunity" (R. T. Smith *et al.*, eds.), pp. 105–177. Univ. of Florida Press, Gainesville, Florida.
Good, R. A., and Papermaster, B. W. (1964). Ontogeny and phylogeny of adaptive immunity. *Advan. Immunol.* **4**, 1–115.
Good, R. A., Finstad, J., Pollara, B., and Gabrielsen, A. E. (1966). Morphologic studies on the evolution of the lymphoid tissues among the lower vertebrates. *In* "Phylogeny of Immunity" (R. T. Smith *et al.*, eds.), pp. 149–170. Univ. of Florida Press, Gainesville, Florida.
Goodrich, H. B., and Nichols, R. (1933). Scale transplantation in the goldfish, *Carassius auratus*. Effects of chromatophores. II. Tissue reactions. *Biol. Bull.* **56**, 253–265.
Haurowitz, F. (1968). "Immunochemistry and the Biosynthesis of Antibodies." Wiley (Interscience), New York.
Hildemann, W. H. (1957). Scale homotransplantation in goldfish (*Carassius auratus*). *Ann. N.Y. Acad. Sci.* **64**, 775–791.
Hildemann, W. H. (1958). Tissue transplantation immunity in goldfish. *Immunology* **1**, 46–53.
Hildemann, W. H. (1959). A cichlid fish, *Symphysodon discus*, with unique nuture habits. *Am. Naturalist* **93**, 27–34.
Hildemann, W. H. (1962). Immunogenetic studies of poikilothermic animals. *Am. Naturalist* **96**, 195–204.
Hildemann, W. H. (1966). Immune responsiveness—Some development comparisons from bullfrog to mice. *In* "Phylogeny of Immunity" (R. T. Smith *et al.*, eds.), pp. 236–242. Univ. of Florida Press, Gainesville, Florida.
Hildemann, W. H. (1970). "Immunogenetics." Holden-Day, San Francisco, California.
Hildemann, W. H., and Cooper, E. L. (1963). Immunogenesis of homograft reactions in fishes and amphibians. *Federation Proc.* **22**, 1145–1151 (Symp.).
Hildemann, W. H., and Haas, R. (1960). Comparative studies of homotransplantation in fishes. *J. Cellular Comp. Physiol.* **55**, 227–233.
Hildemann, W. H., and Haas, R. (1961). Histocompatibility genetics of bullfrog populations. *Evolution* **15**, 276–281.
Hildemann, W. H., and Owen, R. D. (1956). Histocompatibility genetics of scale transplantation. *Transplant. Bull.* **4**, 132–134.
Hildemann, W. H., and Thoenes, G. H. (1969). Immunological responses of Pacific hagfish. I. Skin transplantation immunity. *Transplantation* **7**, 506–521.
Ho, Ju-Shey (1966). Larval stages of Cardiodectes sp. (Caligoida: Lernaeoceriformes), a copepod parasitic on fishes. *Bull. Marine Sci.* **16**, 159–199.

Janssen, W. A., and Meyers, C. D. (1968). Fish: Serologic evidence of infection with human pathogens. *Science* **159**, 547–548.
Jensen, J. A., Sigel, M. M., and Ross, G. D. (1968). Natural antibody (A_N) and complement (C'_N) of the nurse shark. *Federation Proc.* **27**, 491 (abstr.).
Jukes, T. H. (1966). "Molecules and Evolution." Columbia Univ. Press, New York.
Kabat, E. A. (1967a). The paucity of species-specific amino acid residues in the variable regions of human and mouse Bence-Jones proteins and its evolutionary and genetic implications. *Proc. Natl. Acad. Sci. U.S.* **57**, 1345–1349.
Kabat, E. A. (1967b). A comparison in invariant residues in the variable and constant regions of human K, human λ and mouse κ Bence-Jones proteins. *Proc. Natl. Acad. Sci. U.S.*, **58**, 229–233.
Kabat, E. A. (1968). "Structural Concepts in Immunology and Immunochemistry." Holt, New York.
Kallman, K. D. (1961). Genetic homogeneity of a small isolated population of viviparous fish as revealed by tissue transplantation. *Am. Zoologist* **1**, No. 4, 204.
Kallman, K. D., and Gordon, M. (1958). Genetics of fin transplantation in xiphophorin fishes. *Ann. N.Y. Acad. Sci.* **73**, 599–610.
Kilambi, R. V., Utter, F. M., and de Lacy, A. C. (1965). Differentiation of spawning populations of surf smelt, *Hypomesus pretiosus* (Girard) by serological methods. *J. Marine Biol. Assoc. India* **7** No. 2, 364–368.
Kuhns, W. J., and Chuba, J. (1968). Intrageneric blood group differences between Ictalurids (fresh water catfishes). *Federation Proc.* **27**, 491 (abstr.).
Krauel, K. K., and Ridgway, G. J. (1963). Immunoelectrophoretic studies of red salmon serum. *Intern. Arch. Allergy Appl. Immunol.* **23**, 246–255.
Legler, D. W., Evans, E. E., and Dupree, H. K. (1967). Comparative Immunology: Serum complement of freshwater fish. *Trans. Am. Fisheries Soc.* **96**, 237.
Marchalonis, J., and Edelman, G. M. (1965). Phylogenetic origins of antibody structure. I. Immunoglobulins in smooth dogfish (*Mustelus canis*). *J. Exptl. Med.* **122**, 601–618.
Marchalonis, J., and Edelman, G. M. (1966). Polypeptide chains of immunoglobulins from the smooth dogfish (*Mustelus canis*). *Science* **154**, 1567–1568.
Marchalonis, J., and Edelman, G. M. (1968). Phylogenetic origins of antibody structure. III. Antibodies in the primary immune response of the sea lamprey, *Petromyzon marinus*. *J. Exptl. Med.* **127**, 891–914.
Mori, Y. (1931). On the transformation of ordinary scales into lateral-line scales and lateral-line organs in the goldfish. *J. Fac. Sci., Imp. Univ. Tokyo, Sect. IV* **2**, 185–194.
Müller-Eberhard, H. J. (1968). Chemistry and reaction mechanism of complement. *Advan. Immunol.* **8**, 1–80.
Nardi, F. (1935). Das verhalten der schuppen erwachsener Fishe bei regenerations und transplantationsversuchen. *Arch. Entwicklungsmech. Organ.* **133**, 621–663.
Nigrelli, R. F. (1947). Susceptibility and immunity of marine fishes to *Benedenia* (=Epidella) *melleni* (MacCallum), a monogenetic trematode. III. Natural hosts in the West Indies. *J. Parasitol.* **33**, Suppl., 25.
Noble, E. R. (1957). Seasonal variations in host-parasite relations between fish and and their protozoa. *J. Marine Biol. Assoc. U.K.* **36**, 143–155.
Noble, E. R. (1960). Fishes and their parasite-mix as objects of ecological studies. *Ecology* **41**, 593–596.
Oak Ridge National Laboratory, Biology Division. (1966). Symposium on differentia-

tion and growth of hemoglobin- and immunoglobulin-synthesizing cells. *J. Cellular Physiol.* **67**, Suppl. 1, 1–224.

Olivereau, M., and Ridgway, G. J. (1962). Cytologie hypophysaire et antigéne sérique en relation avec la maturation sexuelle chez *Oncorhynchus* species. *Compt. Rend.* **4**, 753–755.

Pollara, B. J., Finstad, J., Good, R. A., and Bridges, R. A. (1966). Immunoglobulins of the sea lamprey (*Petromyzon marinus*). *Federation Proc.* **25**, 1390.

Pollara, B., Suran, A., Finstad, J., and Good, R. A. (1968). N-terminal amino acid sequences of immunoglobulin chains in polydon spatula. *Proc. Natl. Acad. Sci. U.S.* **59**, 1307–1312.

Pressman, D., and Grossberg, A. L. (1968). "The Structural Basis of Antibody Specificity." Benjamin, New York.

Putnam, F. W. (1969). Immunoglobulin structure: Variability and homology. *Science* **163**, 633–644.

Race, R. R., and Sanger, R. (1968). "Blood Groups in Man," 5th ed. Blackwell, Oxford.

Reid, P., and Triplett, E. L. (1968). Observations on the immune system of *Micropterus salmoides*. *Transplantation* **6**, 338–341.

Ridgway, G. J. (1962). The application of some special immunological methods to marine population problems. *Am. Naturalist* **96**, 219–224.

Ridgway, G. J. (1968). Personal communication.

Ridgway, G. J., Cushing, J. E., and Durall, G. L. (1958). Serological differentiation of populations of sockeye salmon, *Oncorhynchus nerka*. *U.S. Fish Wildlife Serv., Spec. Sci. Rept. Fisheries* **257**, 1–9; republished *Bull. Intern. N. Pacific Fishery Comm.* **3**, 5–10 (1961).

Ridgway, G. J., Klontz, G. W., and Matsumoto, C. (1962). Intraspecific differences in serum antigens of red salmon demonstrated by immunochemical methods. *Bull. Intern. N. Pacific Fishery Comm.* **8**, 1–13.

Ridgway, G. J., Hodgins, H. O., and Klontz, G. W. (1966). The immune response in teleosts. *In* "Phylogeny of Immunity" (R. T. Smith *et al.*, eds.), pp. 199–208. Univ. of Florida Press, Gainesville, Florida.

Ross, A. J., and Klontz, G. W. (1965). Oral immunization of rainbow trout (*Salmo gairdneri*) against an etiologic agent of "redmouth disease." *J. Fisheries Res. Board Can.* **22**, No. 3, 713–719.

Sanders, B. G., and Wright, J. E. (1962). Immunogenetic studies in two trout species of the genus *Salmo*. *Ann. N.Y. Acad. Sci.* **97**, 116–130.

Sauter, V. (1934). Regeneration und Transplantation bei erwachsenen Fischen. *Arch. Entwicklungmech. Organ.* **132**, 1–41.

Seaman, G. R., and Robert, N. L. (1968). Immunological response of male cockroaches to injection of tetrahymena pyriformis. *Science* **161**, 1359–1361.

Shino, S. M. (1958). Copepods parasitic on Japanese fishes. *Rept. Fac. Fisheries, Perfect. Univ. Mie.* **3**, No. 1, 75–100.

Sigel, M. M. (1963). Symposium on immunologic phenomena in cold-blooded vertebrates. *Federation Proc.* **22**, 1131–1155 (Symp.).

Sigel, M. M., and Clem, L. W. (1966). Immunologic anamnesis in elasmobranchs. *In* "Phylogeny of Immunity" (R. T. Smith *et al.*, eds.), pp. 190–198. Univ. of Florida Press, Gainesville, Florida.

Sindermann, C. J. (1962). Serology of Atlantic clupeiod fishes. *Am. Naturalist* **96**, 225–231.

Sindermann, C. J. (1966). Diseases of marine fishes. *Advan. Marine Biol.* **4**, 1–89.

Sinderman, C. J., and Honey, K. A. (1964). Serum hemagglutinins of winter skates (*Raja ocellata*) from the Western North Atlantic. *Copeia* No. 1, 139–144.
Sindermann, C. J., and Mairs, D. F. (1961). A blood group system for spiny dogfish, *Squalus acanthias* L. *Biol. Bull.* **120**, 401–410.
Sirotinin, N. M. (1959). "Mechanism of Antibody Formation," Proc. Symp. Czech. Acad. Sci., Prague.
Smith, A. M., Potter, M., and Mechant, B. (1967). Antibody-forming cells in the pronephros of the teleost *Lepomis marochirus*. *J. Immunol.* **99**, 876–882.
Smith, R. T., Miescher, P. A., and Good, R. A., eds. (1966). "Phylogeny of Immunity," Univ. of Florida Press, Gainesville, Florida.
Sprague, L. M. (1964). Personal communication.
Sprague, L. M., and Holloway, J. R. (1962). Studies of the erythrocyte antigens of the skipjack tuna (*Katsuwonus pelamis*). *Am. Naturalist* **96**, 233–238.
Stevens, J. M. (1967). Serological responses of insects. *Federation Proc.* **26**, 1675–1679.
Suran, A. A., Tarail, M. H., and Papermaster, B. W. (1967). Immunoglobulins of the leopard shark. I. Isolation and characterization of 17 S and 7 S immunoglobulins with precipitating activity. *J. Immunol.* **99**, 679–686.
Symposium on Immunogenetic Concepts in Marine Population Research. (1961). *Am. Naturalist.* **96**, 193–246.
Triplett, E. L., and Barrymore, S. (1960). Tissue specificity in embryonic and adult Cymatogaster aggregata studies by scale transplantation. *Biol. Bull.* **118**, 463–471.
Trump, G. N. (1968). Immunoglobulins of goldfish: Antibodies to bovine serum alubumin. *Federation Proc.* **27**, 491, (abstr.); Ph.D. Dissertation, Dept. of Microbiol. and Immunol., University of California, Los Angeles, 1968.
Uhr, J. W., Finkelstein, M. S., and Franklin, E. C. (1962). Antibody response to bacteriophage X174 in non-mammalian vertebrates. *Proc. Soc. Exptl. Biol. Med.* **111**, 13–15.
Utter, F. M., and Ridgway, G. J. (1967). A serologically detected serum factor associated with maturity in English sole, *Parophys vetulus*, and Pacific halibut, *Hippoglossus stenolopis*. *U.S. Fish Wildlife Serv., Fishery Bull.* **66**, 47–58.
Utter, F. M., Ridgway, G. J., and Warren, J. W. (1966). Serological evidence for inbreeding in Lahontan cutthroat trout, *Salmo clarkii henshawi*, in Summit Lake, Nevada. *Calif. Fish Game* **52**, 180–184.
Vann, D. C. (1966). Studies on an unusual blood factor in the California bonito and the lectin with which it reacts. P.D. Dissertation, Dept. of Biol. Sci., University of California, Santa Barbara, California.
Vann, D. C., and Cushing, J. E. (1966). Reactions of the lectin from *Dolichos bifloris* with erythrocytes from California bonito, *Sardis chilensis*. *Federation Proc.* **25**, 437 (abstr.).
Weir, D. M. (1967). "Handbook of Experimental Immunology." Davis, Philadelphia, Pennsylvania.
Wiener, A. S., Chuba, J. V., Gordon, E. B., and Kuhns, W. J. (1969). Hemagglutinins in the plasma of catfish (*Ictalurus nebulosus*) injected with saliva from human secretors of various A-B-O blood groups. *Transfusion* (in press).
Wikler, M., Köhler, H., Tomataka, S. and Putnam, F. W. (1969). Macroglobulin structure: Homology of mu and gamma heavy chains of human immunoglobulins. *Science* **163**, 75–78.

Williams, C. A., and Chase, M. W., eds. (1967–1970). "Methods in Immunology and Immunochemistry," Vols. 1–4. Academic Press, New York.

REFERENCES ADDED IN PROOF

Evans, E. E., Acton, R. T., Bennett, J. C., and Weinheimer, P. F. (1969). Evolution of the immune response. In "Protides of the Biological Fluids," (H. Peetsrs, ed.) Vol. 17, pp. 29–38. Pergamon Press, Oxford.

Fletcher, T. C., and Grant, P. T. (1969). Immunoglobulins in the serum and mucus of the Plaice (*Pleuronectes platessa*). *Biochem. J.* **115**, No. 5, 65.

Hildemann, W. H., and Cooper, E. L. (1970). Phylogeny of Transplantation Immunity. *Transplantation Proc.* (in press).

Hood, L., and Talmage, D. W. (1970). Mechanism of antibody diversity: germ line basis for variability. *Science.* **168**, 325–334.

International Council for the Exploration of the Sea (1970). "The Biochemical and Serological Identification of Fish Stocks" (W. de Ligny, A. Jamieson, Ph. Serène, and N. P. Wilkins, eds.). Rapportes and Procès Verbaux des Reunions, Høstand Søns Forlag, 1260 Copenhagen (in press).

AUTHOR INDEX

Numbers in italics refer to the pages on which the complete references are listed.

A

Abe, Y., 141, *170*
Acton, R. T., 485, 491, *495*, *500*
Adam, H., 98, *104*
Adams-Ray, J., 112, *128*, 155, 166, 167, *169*
Adinolfi, M., 213, *246*
Adrian, E. D., 322, *352*
Afzelius, B. J., 71, *78*
Albers, C., 181, 187, 188, 194, 195, 197, 198, 200, 201, *205*, 239, *246*
Albers, R. W., 101, *104*
Alexander, R. McN., 62, *78*, 300, 305, 306, *352*, *359*, 414, 424, 425, 431, *440*, *441*
Aljure, E., 7, 69, *79*
Allee, W. C., 392, *411*
Allison, A., 211, *248*
Almodovar, L. R., 489, *494*
Altman, P. L., 256, *286*
Altner, H., 26, *78*, 94, 100, *104*
Altschule, M. D., 91, 102, *106*
Amouch, P., 490, *494*
Andersen, R., 67, *82*
Anderson, H., 278, *289*
Anderson, S., 226, *246*
Anthony, E. H., 271, *286*, 342, *352*
Antonini, E., 210, 213, 221, 226, 230, 232, 233, 234, 237, 238, *246*, *248*, *251*, 273, *291*
Ariëns Kappers, C. U., 12, 13, 28, 33, 36, 37, 41, 43, 54, 68, 70, 78
Ariëns Kappers, J., 91, 94, 95, 96, 98, 99, 103, *104*
Armand, J., 205, *206*, 255, 275, 279, 287, 344, *353*
Arnold, G. P., 300, *352*
Aronson, L. R., 12, 18, 21, 22, 26, 27, 33, 50, 51, 54, *78*, *84*

Arora, H. L., 41, 44, 45, 47, 76, 77, 78, *79*
Attardi, D. G., 44, 45, 46, 75, *79*
Attleberger, M., 485, *495*
Atz, J. W., 28, *87*
Augustinsson, K. B., 110, 111, 112, *128*, 150, *168*
Axelrod, J., 101, 103, *104*, *108*

B

Babkin, B. P., 119, *128*
Baglioni, S., 297, *352*
Bainbridge, R., 298, *352*, 425, *441*
Bakay, L., 7, *79*
Baker, C. A., 213, 244, *250*
Balabai, P. P., 295, *358*
Ball, E., 432, 439, *441*
Ball, J. N., 32, *79*, *87*, 259, 277, *286*, 483, *494*
Ballintijn, C. M., 306, 308, 309, 313, 314, 323, 340, *352*, *354*
Bang, F. K., 467, 475, 491, *494*
Barcroft, J., 215, 227, *247*
Bardach, J. E., 58, 59, 60, 61, 62, *79*, *82*
Bargmann, W., 91, *104*
Baril, E. F., 475, *494*
Barrell, J., 362, *408*
Barrett, I., 271, *289*
Barrington, E. J. W., 117, 119, *128*
Barrow, J. H., Jr., 494, *494*
Barrymore, S., 483, *499*
Bartel, A. H., 475, *494*
Bartels, H., *206*
Bastos, J. R., 268, *286*
Basu, S. P., 329, 331, *352*
Bateman, J. B., 116, *130*, 165, 166, *170*, 267, *289*
Bauer, C., 240, *247*
Baumgarten, D., 160, *168*, 260, 261, *286*, 339, 342, *352*

501

Baumgarten, H. G., 124, 125, *128*
Baumgarten-Schumann, D., 200, *206*, 255, 261, 266, 267, 268, 271, 274, 275, 278, 280, 281, 282, 285, 286, *286*, *290*, 310, 334, 335, 336, 339, 344, 348, *352*, *356*
Beadle, L. C., 362, 363, 364, 366, 367, 368, 371, 372, 392, 395, *408*
Beamish, F. W. H., 304, 328, 329, 342, 343, 344, *352*, *353*
Beaton, B., 160, *171*
Beauman, W., 216, 226, *247*
Beaumont, C., 255, 272, 273, *286*
Becker, E. L., 184, *206*
Belekova, M. G., 34, *84*
Bellelle, L., 221, 226, 237, 238, *246*
Belyi, N. D., 454, 455, *460*
Benesch, R., 215, 226, *247*
Benesch, R. E., 215, 226, *247*
Bennett, J. C., *500*
Bennett, M. V. L., 7, 31, 69, 72, 73, 79, *86*
Bennion, G. R., 142, 144, 145, 151, 152, 153, 154, 155, 156, 157, 166, *169*, *172*
Berg, J., 277, *286*
Berg, T., 257, 274, *286*, *292*, 364, 366, 383, 384, 404, 406, *408*, 435, 436, 439, *441*
Bergstrom, E., 213, *249*
Berkowitz, E. C., 8, 9, *89*
Bern, H. A., 7, 68, 70, 71, 72, 73, 74, 75, 76, 79, *80*, *82*, *90*
Bernat, A., 350, *357*
Bernstein, J. J., 19, 20, 23, 25, 26, 35, 38, 40, 41, 42, 75, *80*, *81*, *85*
Bernstein, M. E., 75, *80*
Bertelsen, E., 450, 453, *462*
Bertolini, B., 8, 26, *80*
Bhatia, H. M., 489, *494*
Bianki, V. L., 43, 44, 52, *80*
Bigelow, H. B., 452, *460*
Biot, M., 427, *441*
Bird, R., 184, *206*
Bishai, H. M., 455, *461*
Bishop, I. R., 375, *408*
Bisset, K. A., 492, 493, *494*
Bitjel, J. H., 264, *286*
Black, E. C., 162, *172*, 189, 193, 194, 195, 196, 197, 198, 199, *206*, 235,
247, 254, 255, 268, 273, *286*, *287*, *289*, *291*, 294, 331, 342, *353*
Black, V. S., 273, *286*, 342, *353*
Blazka, P., 331, *353*
Bloom, G., 112, *128*, 155, 166, 167, *169*
Blum, R., 437, *441*
Bodenheimer, T. S., 6, 7, 8, 9, 26, 27, *85*, *87*, *88*
Boffa, G. A., 490, *494*
Bohr, C., 119, *128*, 229, *247*
Boker, H., 364, *408*
Bonaventura, J., 220, 221, 222, 227, 228, 237, 245, 246, *247*, *249*
Bone, Q., 50, 68, *80*, 114, *128*, 162, *169*
Boon, A. A., 97, *104*
Borelli, I. A., 423, *441*
Bottazzi, F., 126, *128*
Bourguet, J., 71, *85*
Boyd, E. S., 24, *80*
Boyd, W. C., 489, *494*
Bradley, A. F., 180, *208*
Bradley, R. S., 448, 449, *461*
Bradley, S. E., 124, 125, *128*, 136, *169*
Brandt, W., 110, *128*
Braunitzer, G., 210, 215, 217, 218, 221, *247*, *249*
Brecht, K., 116, *128*
Breckenridge, C. G., 318, *354*
Breder, C. M., 99, *104*, *105*, 393, *411*
Brett, J. R., 44, *80*, 136, *171*, *172*, 283, 284, 285, *287*, 297, 298, 328, 346, *353*, *357*
Breucker, H., 26, 27, *80*, 98, *105*
Brewin, E. G., 201, *206*
Bricker, N., 216, 226, *247*
Bridges, C. D. B., 28, *80*
Bridges, R. A., 490, *498*
Briehl, R., 236, 237, 238, *247*
Brown, C. E., 299, *353*
Brown, D. E., 458, *461*
Brown, F. A., *207*, 210, 236, *250*
Brown, R., 489, *494*
Brownell, K. A., 184, *207*
Brunori, M., 230, *247*
Buckley, R. M., 261, *290*, 299, 300, *356*
Budgett, J. S., 367, *408*
Buhler, D., 212, 222, *247*
Burger, J. W., 124, 125, *128*, 136, *169*
Burke, J. D., 273, *287*
Burns, B. D., 325, *353*, *357*

AUTHOR INDEX

Burnstock, G., 112, 117, 119, 123, 124, 125, *128*, *129*, *130*, *132*, 141, 150, *169*, *172*, 352, *359*
Butler, T. C., 203, 204, *206*, *208*
Buytendijk, F. J. J., 322, *352*
Byrne, J., *105*

C

Calaprice, J. R., 490, *494*
Calaprice, N. L., 487, 489, *495*
Callegarini, C., 212, *247*
Calugareanu, D., 367, 382, 383, *408*
Campbell, D. H., 466, 467, 475, *494*, *495*
Campbell, G., 112, 117, 118, 119, 125, *129*
Cann, J., 216, *247*
Caputo, A., 210, *251*
Carey, F. G., 163, *164*, 165, *169*, 233, *247*, 450, *464*
Carlson, A. J., 110, 111, 116, *129*, *130*
Carter, G. S., 259, 287, 362, 363, 364, 366, 367, 368, 371, 372, 380, 392, 395, *408*
Case, J., 58, 59, 60, *79*
Cattell, M., 446, 454, 458, 459, *461*
Chan, D. K. O., 166, 167, 168, *169*, 286, *290*
Chanutin, A., 215, *247*, *248*
Chapman, C. B., 71, *83*, 110, *129*, 137, 138, 140, 141, 151, 155, 158, 167, *169*, 255, *287*
Chase, M. W., 467, *500*
Chaston, I., *105*
Chester Jones, I., 286, *290*
Chiancone, E., 213, *248*
Chiba, K., 271, *287*
Chieffi, G., 213, *246*
Childers, W., 213, 244, *250*
Childress, J. J., 452, *461*
Chiller, J. M., 481, 486, *494*
Chipperfield, J., 234, *251*
Chiu, W., 193, 196, 199, *206*, 331, *353*
Christomanos, A., 222, *248*
Chuba, J. V., 478, 488, 489, *494*, *497*, *499*
Chung, M. Y., 24, 53, 54, *81*
Clark, E., 41, 63, *80*, *89*
Clark, M. R., 24, 53, 54, *81*
Clark, R. T., 327, *356*, 392, *410*
Clark, S. L., 24, 53, 54, *81*

Clegg, M. T., 94, *105*
Clem, L. W., 476, 478, *494*, *495*, *498*
Clemente, C. D., 74, *81*
Cohen, D. M., 450, 453, *462*
Cohen, M. J., 67, *81*, 323, *353*
Cokelet, G., 227, *248*
Connor, A. R., 331, *353*
Conte, F. P., 160, *169*
Conway, E. J., 203, *206*
Cooper, E. L., 466, 481, 482, 483, 491, *495*, *496*, *500*
Cooper, O., 432, 439, *441*
Couteaux, R., 352, *353*
Cox, J., 211, 219, 220, *250*
Craig, L., 212, 213, *248*
Crain, S. M., 69, *79*
Cremer, N. E., 467, *494*
Cronly-Dillon, J. R., 37, 43, *81*
Crosby, E. C., 12, 13, 28, 33, 36, 37, 41, 43, 54, 68, 70, 78
Crowther, R. A., 218, 230, *250*
Cucchi, C., 212, *247*
Culberson, C., 449, *461*
Cunningham, J. T., 383, *408*
Curnish, R., 215, *247*, *248*
Cushing, J. E., 466, 467, 475, 478, 485, 486, 487, 488, 489, 490, 491, 492, *494*, *495*, *498*, *499*

D

Daniel, J. F., 255, *287*
Das, B. K., 362, 364, 365, 367, 368, *409*
Datta Munshi, J. S., 255, 259, 266, *287*, *288*
Davidson, T. M., 38, 42, 43, *86*
Davies, D., 211, *249*
Davis, H. S., 491, *495*
Davis, J. C., 136, 156, *169*, *172*, 283, *287*, *297*, *357*
Davis, R., 218, *248*
Dawson, W., 35, 38, 40, *81*
Day, F., 277, *283*
Dayhoff, M., 210, 218, *248*
De Groodt, M., 372, 373, *409*
Dejours, P., 205, *206*, 255, 275, 279, *287*, 344, 347, *353*
de Kock, L. L., 158, *169*, 341, *353*
de Lacy, A. C., 490, *497*
De laMotte, I., 98, *105*
de Ligny, W., 487, 489, 490, *495*

Demina, G. A., 43, 44, 52, *80*
Dendy, A., 96, 98, *105*
Denton, E. J., 414, 417, 425, *441*, 450, 453, *461*
de Reuck, A. V. S., 260, 270, *291*
Dewilde, M. A., 272, *287*
Dewsbury, D. A., 23, *81*
Dickerson, R., 211, *249*
Dijkgraaf, S., 63, *81*
Dijkstra, S. J., 392, *409*
Dill, D. B., 189, 198, 200, *206*, 273, *287*
Dirken, M. N. J., 319, 320, *358*
Dittmer, D. S., 256, *286*, *287*
Dixit, V. P., 31, *81*
Dixon G., 210, *248*
Dodt, E., 28, *81*, 96, 98, *105*
Döving, K. B., 16, 17, *81*
Donaldson, E. D., 286, *287*
Donaldson, E. M., 159, *170*
Dorn, E., 418, 421, *441*
Dornesco, G. J., 158, *169*
Doudoroff, P., 328, *358*
Douglas, E., 431, *441*
Draper, J. W., 458, *461*
Drewry, W. F., 159, *171*
Drillon, A., 490, *494*
Droogleever Fortuyn, J., 11, 12, 13, *81*
Dubale, M. S., 258, 259, *287*, 363, *409*
Dupree, H. K., 485, *497*
Durall, G. L., 478, 488, 490, 492, *495*, *498*

E

Eakin, R. M., 26, *81*, 98, *105*
Eakins, K. E., 123, *129*
Ebbecke, U., 458, 459, *461*
Ebeling, A., 363, 366, 382, 383, 407, *411*, 425, 426, *441* 452, *462*
Eccles, R. M., 320, 321, *353*
Eddy, F. B., *287*
Edelman, G. M., 476, 477, 490, *497*
Edström, A., 50, *81*
Edwards, D. J., 458, *461*
Edwards, H. T., 189, 198, 200, *206*, 273, *287*
Eguchi, H., 222, *248*
Eigenmann, C. H., *105*
Elek, S. D., 475, *495*
Elvehjem, C. A., 184, *206*
Enami, M., 71, 72, *81*, 126, *129*

Enger, P. S., 18, 34, 55, 62, 63, 64, 65, 66, 67, 68, *82*, 110, 112, 113, *129*
English, E., 216, *249*
Enns, T., 427, 431, 433, *441*, *442*, *443*
Enoki, Y., 226, *248*
Epp, O., 211, *249*
Erichsen Jones, J. R., 328, 338, *353*
Evans, E. E., 475, 485, 491, *495*, *497*, *500*
Evans, J., 213, *249*
Evans, M., 363, 364, *409*
Evans, M. L., 491, *495*
Evenhuis, H. H., 268, *288*
Everhart, D. L., 477, *495*
Eyring, H., 446, 448, 458, *462*

F

Fänge, R., 110, 111, 112, 113, 116, 118, 119, 121, 125, *128*, *129*, *132*, 150, 155, 166, *168*, *171*, *172*, 231, *248*, 267, *290*, 414, 417, 418, 419, 426, 432, 437, 438, 439, 440, *441*, 481, *495*
Fahlén, G., 121, 124, 126, *129*, 420, 438, *441*
Fain, W. B., 99, *105*
Fair, E., 111, *131*
Fairey, R., 261, *287*
Falck, B., 111, 121, 123, 124, 126, *129*, 150, 155, 156, *169*
Fanelli, E. M., Jr., 159, *169*
Farrell, G., 102, *105*
Faura, J., 216, *249*
Fearon, P. J., 203, *206*
Fenn, W. O., 279, *291*, 446, 449, *461*
Fenwick, J. C., 97, 99, 100, 101, 102, 103, *105*, *108*
Ferguson, J. K. W., 188, 189, 193, 194, 195, 197, 198, *206*, 268, 273, *287*
Field, J. B., 184, *206*
Finch, C., 216, *249*
Fine, J. M., 490, *494*
Fingerman, M., 126, *129*
Finkelstein, M. S., 477, *499*
Finstad, J., 476, 481, 485, 486, 490, *496*, *498*
Fish, G. R., 273, *287*, 376, 377, *409*
Fishman, A. P., 376, 379, 385, 386, 392, 398, *409*, *410*

Fleming, R. H., 447, 463
Fleming, W. R., 28, 75, 76, 82, 86
Fletcher, T. C., 500
Flock, A., 63, 82
Florkin, M., 189, 198, 200, 206, 273, 287
Fontaine, M., 459, 461
Forbes, F. D., 193, 196, 199, 206
Ford, P., 26, 27, 82, 94, 95, 96, 97, 98, 99, 100, 101, 103, 105
Forlani, L., 213, 248
Formanek, H., 211, 249
Forster, M., 286, 290
Forster, R. E., 274, 287, 421, 436, 437, 441
Forster, R. P., 159, 169
Fox, S., 72, 73, 79
Foxon, G. E. H., 140, 169, 375, 408
Franklin, D. L., 148, 150, 159, 166, 170
Franklin, E. C., 477, 499
Fridberg, G., 7, 68, 70, 71, 72, 75, 76, 78, 82
Frieden, E., 214, 248
Friedman, M. H. F., 119, 127, 128, 129
Friedrich-Freksa, H., 93, 95, 100, 101, 105
Fromm, P. O., 160, 172, 191, 207, 276, 279, 288
Frumin, M. J., 325, 358
Fry, F. E. J., 189, 206, 254, 255, 262, 273, 278, 283, 284, 287, 294, 297, 328, 329, 342, 353
Fujiki, H., 210, 217, 247
Fujino, K., 487, 488, 496
Fujiya, M., 59, 79, 82

G

Gabrielsen, A. E., 481, 490, 496
Gameson, A. L. H., 256, 288, 292
Gannon, B. J., 123, 124, 129, 150, 169
Ganong, W. F., 94, 105
Garcia Romeu, F., 205, 207, 275, 289
Gardner, L. C., 24, 80
Garey, W. F., 136, 159, 169, 254, 255, 260, 261, 266, 273, 274, 278, 282, 284, 288, 348, 353, 386, 409
Garg, R. K., 367, 411
Garvey, J. S., 467, 494
Gaze, R. M., 35, 36, 37, 38, 43, 45, 46, 47, 48, 82, 83

Gelderd, J. B., 80
Gemne G., 16, 17, 81
George, J. C., 362, 409
Gewurz, H., 485, 486, 496
Gibson, Q., 231, 232, 233, 250
Gilbert, P. W., 17, 83
Gill, K. P., 318, 353
Gimenez, M., 7, 31, 69, 79
Glauser, E. M., 304, 354
Glauser, S. C., 304, 354
Goad, W., 216, 247
Goaman, L., 211, 219, 220, 250
Goldstein, L., 159, 169
Good, R. A., 466, 474, 475, 476, 477, 481, 485, 486, 490, 494, 496, 498, 499
Goodrich, H. B., 482, 496
Gordman, A., 15, 16, 17, 30, 31, 32, 83, 88, 90, 97, 106, 107
Gordon, E. B., 488, 499
Gordon, M. S., 4, 87, 448, 450, 457, 461, 463, 483, 497
Goto, M., 141, 170
Gould, R. P., 201, 206
Govyrin, V. A., 123, 129
Gowing, N. F. C., 475, 495
Grafstein, B., 49, 82
Grant, P. T., 500
Gratzer, W., 211, 248
Gray, I. E., 255, 257, 288
Gray, J., 425, 441
Grebe, R. M., 256, 287
Green, A., 231, 248
Greene, C. W., 110, 130
Greenwood, P. H., 450, 452, 461
Greer, J., 218, 230, 250
Grey, M., 447, 451, 461
Grigg, G. C., 245, 249, 273, 274, 275, 288, 289, 368, 372, 376, 379, 380, 382, 383, 386, 387, 392, 394, 395, 400, 401, 405, 409, 410
Grillo, M. A., 4, 87
Grimm, R. J., 23, 24, 82
Grimstone, A. V., 255, 266, 288
Grossberg, A. L., 468, 498
Grundfest, H., 69, 79
Grunewald-Lowenstein, M., 98, 100, 105
Guerra, L., 216, 226, 247
Guest, G., 216, 250
Guidotti, G., 212, 213, 248

H

Haas, R., 482, 483, *496*
Hadley, M. E., 99, *105*
Hafeez, M. A., 26, 27, *82*, 94, 95, 96, 97, 98, 99, 100, 101, 103, *105*
Hagadorn, I. R., 70, 71, *79*, *80*
Hagiwara, S., 69, *82*
Hainsworth, F. R., 20, 21, *82*
Haldane, J. B. S., 227, *248*, 403, *411*, 431, 433, *441*
Hall, C. G., 377, *410*
Hall, F. G., 261, 273, *288*, 292, 299, 329, *354*, 376, 377, *411*, 432, *441*
Hamada, K., 222, 243, 244, *252*
Hamana, K., 70, *82*
Hammel, H. T., 448, *463*
Hane, S., 136, *171*
Haning, Q. C., 203, *207*, 278, *288*
Hanslip, A., 193, 196, 199, *206*
Hanson, A., 134, 135, 136, 137, 138, 139, 140, 145, *170*
Hanson, D., 116, *130*, 136, 145, 146, 147, 148, 156, 158, 159, 166, *169*, *170*, 254, 267, 282, *288*, 289, 310, *354*, 368, 375, 382, 383, 393, 396, *409*, *410*
Hara, T. J., 15, 16, 17, 30, *83*, 90, 97, *106*
Hargitay, B., 431, *441*
Harms, H., 200, *207*
Harris, T., 160, *169*
Hart, J. S., 283, *287*
Hart, R., 211, *249*
Hartline, D. K., 319, *354*
Hartman, F. A., 184, *207*
Hartmann, J. F., 7, 8, 9, *85*
Harvey, H. W., 185, *207*
Hashimoto, K., 213, 222, 231, 241, 242, 243, *248*, *252*
Hasler, A. D., 14, *83*
Hasselbalch, K., 229, *247*
Haurowitz, F., 468, 473, 484, *496*
Hayashi, K., 199, *206*
Hazelhoff, E. H., 268, *288*
Healey, E. G., 14, 23, 26, 44, 47, 51, 62, 74, *83*
Hearn, D. C., 333, *356*
Heath, A. G., 331, 373, *354*, *355*
Hedenius, A., 213, *251*

Heezen, B. C., 454, *462*
Henderson, L. J., 188, 196, *207*
Henschel, J., 295, 300, 306, *354*
Herberman, R., 22, *78*
Herner, A., 214, *248*
Herrick, C. J., 26, *83*
Hersey, J. B., 447, *461*
Hewer, H. R., 99, *105*, 126, *130*
Hibbard, E., 74, *83*
Hibiya, T., 96, 97, 101, *107*
Hidaka, I., 59, *83*
Hildemann, W. H., 466, 468, 481, 482, 483, 487, 491, 492, *495*, *496*, *500*
Hill, C., 92, 93, 95, 96, *105*
Hill R., 218, *248*
Hilschmann, N., 210, *247*
Hilse, K., 210, 217, 218, *247*, *249*
Hjort, J., 446, *462*
Ho, Ju-Shey., 493, *496*
Hoar, W. S., 99, 103, *105*
Hochachka, P. W., 198, *207*
Hodgins, H. O., 475, 477, 481, 486, 490, 492, *494*, *498*
Hodgman, C. D., 256, *288*, 448, *461*
Hodgson, E. S., 17, *83*
Höglund, L. B., 391, 392, *409*
Hoff, H. E., 318, *354*
Hoffert, J. R., 191, *207*, 276, 279, *288*
Hoffman, R. A., 97, *106*
Hoffmeister, H., 110, 112, *130*
Hol, R., 315, *355*, 375, *409*
Holeton, G. F., 116, *130*, 136, 137, 159, *169*, 255, 260, 261, 268, 271, 278, 280, 281, 282, 284, 285, *288*, *291*, 299, 304, 305, 332, 333, 334, 336, 338, 344, 346, 347, 348, 350, 351, *354*, *356*
Holl, A., 59, *79*
Holloway, J. R., 487, *499*
Holmes, W. N., 159, *170*
Holmgren, N., 94, 96, 98, 100, *106*
Holmgren, U., 27, 71, *83*, 92, 93, 94, 95, 96, 98, 100, 101, *106*
Holt, E. W., 96, *106*
Honey, K. A., 492, *499*
Hood, L., *500*
Horgan, J. D., 347, 348, *358*
Hori, T., 318, *354*
Horne, R. A., 448, *461*
Horstmann, E., 26, 27, *80*, 98, *105*

AUTHOR INDEX

Horvath, S. M., 188, 189, 193, 194, 198, 206
Houston, A. H., 272, 287, 341, 347, 358
Hubbs, C. L., 447, 460, 461
Huber, G. C., 12, 13, 28, 33, 36, 37, 41, 43, 54, 68, 70, 78
Huber, R., 211, 249
Huckauf, H., 223, 249
Hüfner, G., 427, 441
Hughes, G. M., 254, 255, 257, 259, 260, 262, 263, 264, 266, 268, 281, 282, 288, 290, 294, 295, 296, 297, 299, 300, 301, 303, 304, 306, 308, 309, 310, 311, 312, 313, 314, 317, 320, 329, 331, 332, 333, 334, 335, 339, 340, 341, 343, 345, 346, 347, 348, 349, 350, 352, 354, 355
Hukuhara, T., 320, 322, 355
Hunn, J. B., 134, 170, 274, 288
Hutchinson, G. E., 446, 461
Hutten, H., 223, 249
Hyde, I. H., 321, 355
Hynes, H. B. N., 345, 355

I

Idll, C. P., 450, 461
Iles, T., 212, 213, 214, 252
Imai, K., 71, 75, 76, 81, 83
Imbrie, J., 454, 462
Ingle, D. J., 22, 42, 83
Ingram, V., 210, 249
Irisawa, H., 141, 172
Irving, L., 150, 170, 231, 251, 273, 286, 289, 291
Ishibashi, T., 72, 73, 83, 86
Itano, H., 212, 249
Itina, N. A., 111, 130
Ivanov, I. I., 458, 462
Iversen, O., 434, 442

J

Jacobson, M., 35, 36, 37, 38, 40, 41, 43, 45, 46, 47, 48, 82, 83
Jaeger, R., 141, 170
Jakubowki, M., 277, 289
Janssen, W. A., 492, 497
Janzen, W., 106
Jasinski, A., 30, 31, 83, 97, 106
Jensen, D., 110, 129, 137, 138, 140, 141, 151, 155, 158, 167, 169, 170, 255, 287
Jensen, J. A., 485, 497
Jesse, M. T., 385, 392, 398, 409
Job, S. V., 328, 329, 331, 355
Johansen, K., 116, 117, 123, 130, 134, 135, 136, 137, 138, 139, 140, 145, 148, 150, 155, 158, 159, 166, 170, 196, 198, 207, 239, 240, 244, 245, 249, 254, 255, 267, 268, 269, 273, 274, 275, 289, 315, 336, 344, 355, 363, 364, 365, 366, 368, 369, 372, 375, 376, 377, 379, 380, 382, 383, 384, 386, 387, 388, 389, 390, 391, 392, 393, 394, 395, 396, 397, 398, 399, 400, 401, 402, 404, 405, 406, 407, 409, 410
Johnels, A. G., 110, 111, 112, 113, 128, 129, 130, 150, 168
Johnson, F. W., 446, 448, 458, 462
Johnson, M. W., 447, 463
Jones, D. R., 136, 172
Jones, M. P., 136, 137, 145, 147, 148, 159, 171
Jones, R., 213, 217, 251, 252
Juday, C., 184, 206
Jukes, T. H., 473, 497
Julien, M., 28, 85
Jullien, A., 117, 118, 130

K

Kabat, E. A., 468, 477, 484, 497
Kahling, J., 98, 108
Kahn, N., 318, 325, 355
Kajita, A., 222, 243, 244, 252
Kallman, K. D., 32, 79, 483, 494, 497
Kamrin, R. P., 18, 84
Kandel, E. R., 29, 30, 31, 84
Kang, T., 488, 496
Kanwisher, J., 425, 426, 441, 448, 463
Kanzow, E., 323, 358
Kaplan, H., 21, 78, 84
Karamyan, A. I., 18, 22, 44, 51, 52, 84
Katz, R. L., 123, 129
Kawamoto, M., 71, 88
Kawamoto, N. Y., 72, 84
Kayama, M., 457, 463
Kaye, G. W. C., 256, 289
Kazama, T. K., 487, 496
Keder-Stepanova, I. A., 318, 355

Kelley, J., 235, 249
Kelly, D. E., 93, 98, 106
Kelly, J. W., 184, 206
Kendall, J. I., 260, 261, 264, 290, 301, 356
Kendrew, J., 211, 249
Kennedy, G. C., 448, 463
Kester, D. R., 449, 461
Keys, A., 116, 130, 165, 166, 170, 267, 289
Kholodov, Y. A., 22, 84
Kilambi, R. V., 490, 497
Kilmartin, J. V., 218, 230, 249, 250
Kingsbury, B. F., 93, 106
Kiprian, T. K., 327, 357
Kirby, S., 124, 130
Kirchhoff, J., 306, 355
Kirkpatrick, D., 189, 206, 255, 273, 286
Kirsche, K., 48, 84
Kirsche, W., 8, 9, 25, 48, 74, 75, 84
Kirschstein, H., 26, 27, 87, 96, 98, 107
Kirtzler, H., 63, 84
Kisch, B., 141, 170
Kitay, J., 91, 102, 106
Kitto, G., 237, 249
Klahr, S., 216, 226, 247
Klatzo, I., 7, 8, 9, 12, 84, 85
Klawe, B. M. H., 271, 289
Klawe, W. L., 271, 289
Kleerekoper, H., 14, 62, 84, 264, 290, 301, 356
Klika, E., 372, 410
Klontz, G. W., 475, 477, 486, 490, 492, 493, 498
Klotz, I. M., 449, 462
Knight-Jones, E. W., 446, 448, 450, 454, 458, 462
Knights, B., 333, 354
Knoop, A., 71, 88
Koch, H., 213, 249, 431, 433, 441
Kochiyama, Y., 243, 252
Köhler, H., 474, 499
Kohn, P. G., 256, 289
Konigsberg, W., 212, 213, 248
Konishi, J., 35, 59, 60, 61, 84, 85
Koppanyi, T., 74, 85
Kozhov, M. M., 447, 451, 462
Krabbe, K. H., 97, 106
Krauel, K. K., 471, 493, 497
Krawkow, N. P., 116, 130

Krockert, G., 101, 102, 106
Krogh, A., 229, 236, 247, 249, 254, 256, 262, 263, 273, 277, 289, 366, 377, 378, 410, 421, 441
Kruger, L., 8, 9, 35, 37, 38, 85, 89
Krupp, M. A., 136, 171
Kruysse, A., 165, 166, 172, 259, 266, 267, 268, 274, 292
Kuhn, H. J., 119, 130, 414, 431, 432, 437, 442
Kuhn, W., 119, 130, 414, 431, 432, 437, 441, 442
Kuhns, W. J., 477, 488, 494, 497, 499
Kulaev, B. S., 123, 130
Kulikov, M. Y., 456, 462
Kung, C. C., 365, 411
Kuno, M., 318, 353
Kuriyama, H. A., 141, 170
Kylstra, J. A., 254, 264, 265, 289, 344, 355

L

Labat, R., 155, 170, 283, 289, 350, 355, 357
Laby, T. H., 256, 289
Laffont, J., 155, 170
Lagasse, A., 372, 373, 409
Lahiri, S., 376, 379, 386, 410
Lahlou, B., 71, 85
Lam, K., 199, 206, 331, 353
Lam, T., 259, 289
Larimer, J. L., 191, 207
Larsell, O., 50, 85
Laurent, P., 150, 170, 341, 346, 352, 353, 355
Lavenberg, R. J., 452, 462
Lebedeva, N. A., 458, 462
Lederis, K., 28, 85
Lee, J. C., 7, 79
Legler, D. W., 485, 495, 497
LeGros, Clarke, W. E., 96, 97, 106
Leitch, I., 236, 249, 273, 289, 377, 378, 410
Leivestad, H., 278, 289
Lelek, A., 372, 410
Lenfant, C., 116, 130, 196, 198, 207, 216, 239, 240, 244, 245, 249, 254, 268, 269, 273, 274, 275, 289, 336, 344, 355, 363, 364, 365, 366, 368, 369, 376, 377, 379, 380, 382, 383,

AUTHOR INDEX 509

384, 386, 387, 388, 389, 390, 391, 392, 393, 394, 395, 396, 397, 398, 399, 400, 401, 402, 404, 405, 406, 407, 409, 410
Leont'eva, G. R., 114, 123, 124, 129, 130
Lewis, L. A., 184, 207
Lewis, R. W., 457, 462
Li, S. L., 221, 249
Lickfield, K., 110, 112, 130
Lighttoller, G. H. S., 313, 355
Liljestrand, A., 318, 355
Lindström, T., 110, 130
Lipmann, F., 215, 250
Lishajko, F., 112, 128, 155, 166, 167, 169
Lloyd, R., 334, 355
Loewenstein, W. R., 29, 87
Long, D. M., 7, 8, 9, 85
Longhurst, A. R., 452, 462
Lopukhova, V. V., 456, 462, 463
Love, W., 211, 221, 236, 237, 238, 250, 251
Lowenstein, O., 62, 63, 85
Lubinska, L., 49, 85
Luckhardt, A. B., 116, 130
Lutz, B. R., 124, 131, 326, 355

M

McBridge, J. R., 286, 287
McCleary, R. A., 19, 41, 85
McCutcheon, F. H., 239, 250, 273, 288, 377, 410
MacDuff, G., 226, 247
McGee-Russell, S. M., 4, 87
MacKay, M. E., 124, 128, 131
MacKay-Sawyer, M. E., 119, 128
Mackintosh, J., 37, 85
MacLean, P. D., 101, 104
McMahon, B. R., 375, 410
Macnab, H. C., 160, 171
McNeeley, J. L., 491, 495
MacNichol, E. F., Jr., 35, 38, 85, 90
McWilliam, J. A., 141, 170
Maeno, T., 141, 170
Maetz, J., 28, 71, 85, 275, 286, 289, 290
Magid, A. M. A., 367, 392, 410
Mahn, R., 121, 131
Mairs, D. F., 490, 499
Malyusz, M., 160, 168, 285, 286, 286
Mangia, F., 26, 80
Manning, G. T., 199, 206

Manwell, C., 210, 213, 228, 236, 238, 239, 240, 244, 250, 435, 442
Marchalonis, J. J., 476, 477, 490, 497
Marchena, C., 216, 226, 247
Marchis-Mouren, G., 215, 250
Marcus, H., 112, 131
Maren, T. H., 190, 191, 207, 275, 276, 279, 289
Margoliash, E., 210, 250
Mark, R. F., 38, 42, 43, 77, 85, 86
Marks, W. B., 40, 86
Marón, K., 25, 74, 86
Marshall, N. B., 414, 417, 420, 432, 434, 442, 450, 452, 453, 462
Marsland, D. A., 458, 462
Marti, E., 119, 130, 414, 431, 442
Martin, A. W., 117, 123, 130, 134, 145, 170, 310, 354
Martinson, J., 112, 118, 131
Marvin, D. E., 337, 355
Mathewson, R. F., 17, 83
Matioli, G., 223, 250
Matsumoto, C., 493, 498
Matsuura, F., 211, 213, 222, 231, 241, 242, 243, 248, 252
Maxwell, D. S., 8, 9, 85
Mayer, N., 286, 290
Mazzarella, L., 218, 230, 250
Mead, G. W., 450, 453, 462
Mead, J. F., 457, 463
Mechant, B., 481, 485, 499
Méhés, J., 119, 131
Meier, A. H., 28, 86
Meiselman, H. J., 227, 248
Menzies, R. J., 454, 462
Mersch, F. D., 201, 207
Meyer, H., 343, 355
Meyers, C. D., 492, 497
Meyers, G. S., 450, 452, 461
Miescher, P. A., 466, 474, 499
Miles, H. M., 272, 290
Milic-Emili, G., 304, 355
Millen, J. E., 159, 171, 205, 207, 333, 356
Milokhin, A. A., 110, 131
Mommaerts, W. F. H. M., 155, 170
Mondovi, B., 226, 251
Mookherjii, P. S., 328, 353
Moreau, A., 424, 442
Morgan, R. I. G., 287

Mori, Y., 59, 60, 61, 85, 482, 497
Morita, H., 72, 73, 86
Morita, R. Y., 446, 448, 449, 454, 462
Morita, Y., 26, 86, 95, 96, 98, 106
Morgan, E., 446, 448, 450, 454, 458, 462
Motais, R., 205, 207
Mott, J. C., 124, 131, 133, 135, 141, 149, 150, 158, 159, 160, 167, 171
Moussa, T. A., 367, 370, 371, 410
Müller-Eberhard, H. J., 484, 497
Mugnaini, E., 7, 8, 9, 86
Muir, B. S., 255, 257, 258, 260, 261, 264, 268, 290, 299, 300, 301, 353, 356
Muirhead, H., 211, 219, 220, 250
Muirhead, M., 218, 230, 250
Munk, O., 414, 442
Munro, D. C., 448, 449, 461
Munshi, J. S. D., 365, 370, 371, 410
Murdaugh, H. V., 159, 171, 204, 205, 207, 255, 269, 291, 310, 333, 356
Murray, J., 446, 462
Muschel, L. H., 485, 486, 496
Musnick, R. A., 28, 88
Myhrberg, H., 111, 123, 129, 150, 155, 166, 169

N

Nakajima, Y., 7, 69, 79, 86
Nakano, T., 155, 160, 171, 267, 290
Napora, T. A., 450, 462
Nardi, F., 482, 497
Nashat, F. S., 201, 206
Neil, E., 201, 206
Nelson, D. R., 63, 90
Netter, H., 201, 203, 207
Nevenzel, J. C., 457, 463
Newstead, J. D., 255, 259, 266, 290
Newton, M. S., 448, 463
Ngai, S. H., 317, 318, 319, 325, 358
Niazi, I. A., 74, 86
Nichols, R., 482, 496
Nicholson, C., 50, 51, 87
Nickerson, M., 155, 171
Nicloux, M., 271, 290
Nicol, J. A. C., 109, 110, 117, 126, 128, 131
Nielsen, J. G., 414, 442
Nieuwenhuys, R., 11, 12, 14, 16, 19, 25, 26, 27, 50, 51, 68, 70, 86, 87, 89
Niewisch, H., 223, 250

Nigrelli, R. F., 478, 488, 492, 494, 497
Nishioka, R. S., 71, 75, 76, 82
Nishiyama, T., 459, 463
Noble, E. R., 492, 497
Noble, R., 231, 232, 233, 250
Nolan, C., 210, 250

O

Östlund, E., 110, 111, 112, 116, 128, 129, 130, 131, 132, 150, 155, 166, 168, 169, 171, 267, 290
Oets, J., 141, 171
Ogden, E., 326, 338, 356
Oglialoro, C. M., 363, 410
Oguri, M., 96, 97, 101, 107
Okada, H., 320, 322, 355
Okazaki, T., 222, 243, 244, 252
Oksche, A., 26, 27, 87, 92, 93, 98, 107
Olcott, C., 111, 131
Oldham, J., 222, 245, 246, 250
Olivereau, M., 32, 79, 87, 493, 498
Olthof, H. J., 343, 356
Omura, Y., 96, 97, 101, 107
Osborne, M. P., 62, 63, 85
Otis, A. B., 304, 356
Otorii, T., 111, 131
Overmier, J. B., 20, 21, 82
Owen, R. D., 482, 496
Ozaki, S., 141, 170
Oztan, N., 7, 87

P

Padlan, E., 211, 250
Paganelli, C. V., 254, 264, 265, 289, 344, 355
Painter, B., 485, 491, 496
Palay, S. L., 4, 7, 8, 28, 71, 85, 87, 392, 411
Palayer, P., 96, 100, 107
Palmer, J. L., 475, 494
Palmgren, A., 112, 130
Pang, P. K. T., 102, 107
Papermaster, B. W., 466, 474, 475, 476, 477, 478, 494, 496, 499
Papez, J. W., 41, 54, 55, 87
Pappas, G., 7, 69, 79, 86, 87
Pappenheimer, J. R., 188, 189, 193, 194, 198, 206
Parker, G. H., 126, 131

AUTHOR INDEX

Parkhurst, L., 231, 232, 233, *250*
Parry, G., 254, *290*
Parsons, J. R., 256, *292*
Pasztor, V. M., 264, *290*, 301, *356*
Patterson, T. L., 111, *131*
Pauling, L., 217, *252*
Pavlopulu, C., 222, *248*
Pearcy, W. G., 450, *463*
Pease, D. C., 8, 9, *89*
Pegel', V. A., 456, *463*
Pereira, R. S., 184, *207*
Persson, H., 111, 123, *129*, 150, 155, 166, *169*
Perutz, M. F., 211, 218, 219, 220, 230, *250*
Peter, R. E., 97, 102, *107*
Peters, A., 110, 112, *131*
Peters, H. M., 424, *442*
Petersen, J. A., 363, 364, 365, 366, 368, 369, 376, 377, 379, 380, 383, 386, 387, 389, 391, 399, 400, 405, 406, 407, *409*, *410*
Petit, J. M., 304, *355*
Peyraud, C., 283, *289*, *290*, 350, *356*
Pflugfelder, O., 27, 87, 101, 102, *107*
Phillips, A. M., Jr., 271, *290*
Phillips, D., 211, *249*
Pickford, G. E., 28, 31, 72, *87*
Piiper, J., 159, *171*, 200, *206*, 255, 260, 261, 266, 267, 268, 269, 271, 274, 275, 278, 280, 281, 282, *286*, *290*, *291*, 304, 310, 332, 334, 335, 336, 339, 342, 344, 348, *352*, *356*, 357, 399, *410*
Pleschka, K., 181, 188, 194, 195, 197, 198, 200, 201, *205*, 239, *246*
Pocklington, P., 276, *290*
Polissar, M. J., 446, 448, 458, *462*
Poll, M., *410*
Pollara, B., 476, 481, 490, *496*, *498*
Ponomarev, V. A., 318, *355*
Poole, D. T., 203, *206*
Porter, R., 260, 270, *291*
Potter, D. D., 29, *87*
Potter, M., 481, 485, *499*
Powers, E. B., 327, *356*, 392, *410*
Pressman, D., 468, *498*
Pritchard, A. W., 331, *354*
Prosser, C. L., 38, *89*, *207*, 210, 236, *250*, 294, *356*

Putnam, F. W., 469, 471, 474, 477, *498*, *499*
Pytkowiez, R. M., 449, *461*

Q

Qasim, S. Z., 367, *411*
Qayyum, A., 367, *411*
Quay, W. B., 96, 101, *107*

R

Race, R. R., 467, 487, *498*
Rahmann, H., 9, 49, *87*, *88*
Rahn, H., 186, 201, 202, *207*, 254, 264, 265, 275, 279, *289*, *290*, *291*, 344, 345, 348, *355*, *356*, 386, *409*
Rall, J. W., 155, *172*
Ramel, A., 119, *130*, 414, 431, *442*
Ramon y Cajál, S., 33, 34, *88*
Ramos, J., 216, *249*
Randall, D. J., 116, 123, 124, 125, *130*, *131*, *132*, 136, 137, 138, 141, 142, 143, 144, 145, 146, 149, 150, 151, 155, 156, 157, 159, 160, 166, *168*, *169*, *171*, *172*, 255, 259, 260, 261, 262, 266, 267, 268, 273, 278, 280, 281, 282, 283, 284, 285, 286, *286*, 288, *291*, *292*, 297, 299, 304, 328, 332, 333, 334, 335, 336, 337, 338, 340, 343, 344, 345, 346, 347, 348, 350, 351, 352, *354*, 356, 357, *358*
Rapoport, S., 216, *250*
Rasquin, P., 27, *88*, 93, 94, 99, 100, 102, 104, 105, *107*
Rass, T. S., 447, 453, *463*
Rathschlag-Schaefer, A. M., 240, *247*
Rauther, M., 363, 364, 365, 367, 368, *411*, 417, *442*
Ravitz, M. J., 31, *79*
Raynaud, P., 155, *170*, 349, 350, *357*
Read, J. B., 124, 125, *132*
Rechnitzer, A. B., 447, 460, *461*
Rees, T. A., 475, *495*
Regnard, P., 458, 459, *463*
Reid, P., 483, *498*
Reite, O. B., 372, *409*
Reiter, R. J., 97, *106*
Relkin, R., 97, *107*
Remorov, V. A., 456, *463*
Restieaux, N. J., 56, 57, 58, *88*
Reynafarje, C., 216, *249*

Rhodin, J. A. G., 255, *291*
Richard, J., 427, *442*
Richard, J. D., 63, *90*
Richter, G., 364, *411*
Richter, W., 48, 49, *88*
Ridgway, G. J., 471, 475, 477, 486, 489, 490, 492, 493, *497, 498, 499*
Riggs, A., 210, 213, 220, 221, 222, 223, 226, 227, 228, 236, 237, 238, 245, 246, *247, 249, 250, 251*
Ripplinger, J., 117, 118, *130*
Ritzén, M., 112, *128*, 155, 166, 167, *169*
Rivas, L. R., 94, *107*
Robb, J. S., 134, *171*
Robert, N. L., 491, *498*
Roberts, E., 213, 243, *251*
Roberts, T. D. M., 294, 314, 315, *356*
Robertson Connor, A., 199, *206*
Robertson, E. E., 31, 72, *87*
Robertson, J. D., 6, 7, 8, 9, *88*, 184, *207*
Robertson, K. G., 256, *288*
Robertson, O. H., 136, *171*
Robin, E. D., 159, *171*, 204, 205, *207*, 255, 269, *291*, 310, 333, *356*
Robinson, E., 212, *249*
Robinson, J. S., 457, *463*
Rodegker, W., 457, *463*
Roels, B., 103, *108*
Roggenkamp, P. A., 62, *84*
Rona, M., 213, *251*
Ronald, K., 160, *171*
Root, R. W., 189, 196, 197, 198, *208*, 231, *248, 251*, 273, *291*, 377, *411*, 432, *442*
Rosen, D. E., 450, 452, *461*
Rosengren, E., 121, 124, 126, *129*
Ross, A. J., 492, 493, *498*
Ross, G. D., 485, *497*
Rossi, L., 234, *251*
Rossi-Bernardi, L., 218, 230, *249*
Rossi-Fanelli, A., 210, 226, 230, 232, 233, 234, *251*, 273, *291*
Roth, W. D., 104, *107*
Roughton, F., 193, *208*, 234, *251*
Ruch, T. C., 20, *88*
Rudloff, V., 210, *247*
Rüdeberg, C., 26, 27, *88*, 92, 95, 96, 98, 101, *107*
Rüffel, K., 431, *442*

Rumen, N., 221, 226, 236, 237, 238, *246, 250, 251*
Rushmer, R. F., 148, *171*
Ruska, H., 110, 112, *130*
Rusy, B. F., 304, *354*
Rund, J. T., 209, *251, 291*
Ryback, B., 110, 112, *130*

S

Sadlack, F. J., 25, *80*
Safford, V., 273, *289*
Saito, N., 69, *82*
Salmoiraghi, G. C., 317, 318, 319, 320, 321, 323, 325, *353, 356, 357, 358*
Sanders, B. G., 489, 490, *498*
Sanger, R., 467, 487, *498*
Sano, Y., 71, *88*
Santa, V., 158, *169*
Satchell, G. H., 56, 57, 58, *88*, 116, *132*, 136, 137, 145, 147, 148, 151, 158, 159, 163, *171*, 282, 284, *291*, 322, 323, 326, 327, 338, 340, 348, 349, 350, *357*
Sathyanesan, A. G., 32, *88*
Saunders, R. L., 260, 268, *291*, 294, 297, 298, 300, 301, 304, 329, 333, 334, 337, 338, 343, 344, 347, *357*, 427, *442*
Sauter, V., 482, *498*
Sawaya, P., 184, *207*, 368, 372, 382, *411*
Sawyer, S., 491, *495*
Sawyer, W. H., 28, 31, 72, *87, 88*
Saxena, D. B., 259, *291*, 362, 364, 368, *411*
Schadé, J. P., 18, 34, 55, 56, *89*, 91, *104*, 322, *357*
Scharrer, E., 98, *108*, 392, *411*
Schiffman, R. H., 160, *172*
Schilling, J., 184, *206*
Schlaifer, A., 393, *411*
Schlicher, J., 271, *291*
Schloesing, T., 427, *442*
Schmidt-Nielsen, K., 191, *207*, 227, 244, *249, 251*, 363, 364, 365, 366, 368, 369, 376, 377, 379, 380, 383, 386, 387, 389, 391, 399, 400, 405, 406, 407, *409, 410*
Schnitzlein, H. N., 11, 12, 13, 26, *89*
Schöttle, E., 363, 364, 366, *411*

AUTHOR INDEX

Scholander, P. F., 231, 235, *251*, 278, 289, 421, 422, 426, 427, 428, 430, 431, 432, 433, *441*, *442*, *443*, 448, *463*
Schonherr, J., 99, 102, *108*
Schroeder, W., 213, *251*
Schroeder, W. C., 452, *460*
Schulte, A., 42, *89*
Schultz, R., 8, 9, *89*
Schulz, H., 371, 372, *411*
Schumann, D., 159, *171*, 255, 260, 269, 282, *291*, 304, 310, 332, *356*, *357*, 399, *410*
Schwassmann, H. O., 35, 37, 38, *89*
Schwend, M. J., 431, *443*
Seaman, G. R., 491, *498*
Sears, T. A., 320, 321, *353*, *357*
Sebruyns, M., 372, 373, *409*
Segaar, J., 12, 13, 14, 19, 22, 23, 25, *89*
Seraydarian, K., 155, *170*
Serbenyuk, Ts. V., 323, 327, *357*
Serfaty, A., 155, *170*, 283, 289, 290, 349, 350, *356*, *357*
Severinghaus, J. W., 180, *208*
Seyama, I., 141, *172*
Shapiro, S. M., 42, *89*
Sharma, S. C., 47, *82*
Shealy, C. N., 320, 321, *353*
Shefner, A. M., 477, *495*
Shelden, F. F., 184, *207*
Shelford, V. E., 392, *411*
Shelton, G., 136, 144, 145, 156, 157, *172*, 254, 255, 260, 268, 281, 282, 288, 294, 295, 296, 297, 301, 303, 304, 306, 309, 317, 319, 320, 321, 322, 324, 325, 327, 331, 332, 334, 335, 337, 341, 343, 345, 346, 347, 348, 349, 350, *354*, *356*, *357*
Shepherd, M. D., 94, *105*
Shine, L., 24, 53, 54, *81*
Shino, S. M., 493, *498*
Shishov, B. A., 327, *357*
Shore, V., 211, *249*
Shub, C., 376, 379, 385, 386, 392, 398, *409*, *410*
Shukuya, R., 222, 243, 244, *252*
Sigel, M. M., 466, 476, 478, 485, *494*, *497*, *498*
Siggaard-Andersen, O., 174, 179, 192, *208*
Siniscalco, M., 213, 221, 226, 237, 238, *246*
Sindermann, C. J., 475, 489, 490, 491, 492, 493, *498*, *499*
Singh, B. N., 266, *287*
Sirotinin, N. M., 477, *499*
Skvortsova, N. V., 458, *462*
Slicher, A. M., 32, *79*
Small, L. F., 450, *463*
Small, P. A., 476, *495*
Smith, A. M., 481, 485, *499*
Smith, H. W., 181, 184, *208*, 382, *411*
Smith, J. C., 283, *291*, 340, 350, *356*
Smith, L. S., 136, 160, *171*, *172*, 272, *290*, 297, *357*
Smith, R. T., 466, 474, *499*
Smith, S. W., 392, *411*
Snowdon, C. T., 20, 21, *82*
Solandt, D. T., 150, *170*
Solandt, O. M., 150, *170*
Solomon, S., 184, *206*
Soprunov, F. F., 457, *463*
Spaich, P., 194, *205*
Sperry, R. W., 41, 44, 45, 46, 47, 75, 76, 77, 78, 79, *89*
Sprague, L. M., 487, *499*
Springer, M. G., 321, *358*
Spurway, H., 403, *411*
Stage, D. E., 6, 7, 8, 9, *88*
Standberg, B., 211, *249*
Steen, J. B., 165, 166, *172*, 243, *251*, 257, 259, 266, 267, 268, 274, 277, 286, 287, *292*, 364, 366, 383, 384, 404, 406, *408*, 414, 416, 420, 421, 429, 432, 434, 435, 436, 437, 439, *441*, *442*
Stevens, E. Don, 124, 125, *132*, 136, 137, 144, 145, 150, 155, 156, 157, 158, 159, 160, 161, 162, 166, *171*, *172*, 255, 259, 260, 261, 262, 266, 267, 271, 278, 280, 281, 282, 284, 285, *291*, *292*, 299, 328, 332, 333, 334, 335, 337, 344, 346, 347, 348, 350, 352, *356*, *358*
Stevens, J. M., 491, *499*
Stewart, J. E., 160, *171*
Steyn, W., 93, *108*
Strahan, R., 255, 289, 315, 316, *355*, *358*
Stray-Pedersen, S., 440, *442*
Strickland, J. D. H., 256, *292*

Strittmatter, F., 432, 439, *441*
Studnička, F. K., 91, 93, 95, 96, 98, *108*
Stupfel, M., 180, *208*
Sudak, F. N., 125, *132*, 136, 145, 146, 147, 148, *172*
Sumi, T., 320, *358*
Sundnes, G., 427, 438, *443*
Suran, A. A., 476, 478, *498, 499*
Sussdorf, D. H., 467, *494*
Sutherland, E. W., 155, *172*
Sutherland, N. S., 37, *85*
Sutterlin, A. M., 38, *89*
Svedberg, T., 213, *251*
Sverdrup, H. U., 447, *463*
Swan, H., 273, *292*, 376, 377, *411*
Symmons, S., 158, *169*
Székely, G., 47, *82*

T

Tait, J. S., 427, *443*
Takasugi, N., 70, 71, *79*
Talmage, D. W., *500*
Tarail, M. H., 476, 478, *499*
Tateda, H., 59, 60, *89*
Tavolga, W. N., 63, *89, 90*
Taylor, C., 227, *251*
Taylor, F. H. C., 452, *463*
Taylor, M., 364, 366, *411*
Taylor, W., 341, 347, *358*
Teal, J. M., 163, 164, 165, *169*, 233, *247*, 450, *463, 464*
Tebēcis, A. K., 143, 148, *172*
Teichmann, H., 59, *90*, 312, *358*
Terry, R. J., 92, *108*
Tester, A. L., 59, *90*
Thines, G., 98, *108*
Thoenes, G. H., 491, *496*
Thomas, S. F., 136, *171*
Thompson, A. M., 203, *207*, 278, *288*
Thompson, K., *251*
Thompson, N., 136, *171*
Thornhill, R. A., 62, 63, *85*
Thorson, T. B., 159, 160, *172*
Tilney, F., 91, 92, 93, *108*
Todd, E. S., 363, 366, 382, 383, 407, *411*
Tomataka, S., 474, *499*
Tomlinson, N., 155, 160, *171*, 267, *290*
Torrance, J., 216, *249*
Tretjakoff, D., 100, *108*
Triplett, E. L., 483, *498, 499*

Truesdale, G. A., 256, *292*
Trump, G. N., 477, 487, 489, 495, *499*
Tsuyuki, H., 213, 243, *251*
Tucker, H. H., 189, *206*, 255, 273, *286*
Turitzin, S., 243, *251*
Tyuma, I., 226, *248*

U

Uchida, K., 155, *170*
Uchida, M., 59, 60, 61, *85*
Ueda, K., 15, 17, *83, 90*
Uhr, J. W., 477, *499*
Umezawa, S. I., 303, 329, 333, 339, 340, 350, *354, 355*
Utter, F. M., 490, 493, 497, *499*

V

van Bergeijk, W. A., 67, *90*
Van Brunt, E. E., 94, *105*
Van Citters, R. L., 148, 150, 159, 166, *170*
van Dam, L., 231, 235, *251*, 255, 260, 268, *292*, 295, 304, 331, 332, 336, 338, 343, *358*, 426, 427, 428, 432, 433, *442*, 448, *463*
Van de Kamer, J. C., 92, 93, 95, 96, 97, 100, *108*
van Dyke, H. B., 28, *88*
Vann, D. C., 489, *499*
Vanstone, W., 213, 243, *251*
Vaupel von Harnak, M., 98, *107*
Vecchini, P., 213, *248*
Verriest, G., 205, *206*, 255, 275, 279, 287, 344, *353*
Vesselkin, N. P., 34, *84*
Vivien, J. H., 103, *108*
Voigt, R., 116, *132*
von Baumgarten, R., 25, *90*, 319, 320, 321, 323, 325, *357, 358*
von Euler, U. S., 111, 112, 121, *128, 132*, 155, 166, 167, *169, 172*
von Frisch, K., 98, 99, *108*
von Ledebur, J. F., 439, *443*
von Mecklenburg, C., 111, 123, *129*, 150, 155, 166, *169*

W

Waddell, W. J., 203, 204, *206, 208*
Wagner, H. G., 35, 38, *85, 90*
Wagner, H. H., 160, *169*

AUTHOR INDEX

Walberg, F., 8, 9, *86*
Wald, G., 236, 238, *251*
Waldeck, F., 223, *249*
Wall, J. R., 94, *105*
Walter, F. K., 97, *108*
Wang, S. C., 317, 318, 319, 323, 325, *353, 355, 358*
Warren, C. E., 338, *358*
Warren, J. M., 22, *90*
Warren, J. W., 490, *499*
Warren, W. F., 91, 92, 93, *108*
Washburn, E. W., 256, *292*
Watts, D., 210, *251*
Way, H. K., 326, *357*
Weale, K. E., 448, *464*
Weast, R. C., *292*
Webb, M., 93, *108*
Weiler, I. J., 18, 34, 47, 55, 56, *89, 90*, 322, *357*
Weinheimer, P. F., 491, *495, 500*
Weir, D. M., *499*
Weisbart, M., 102, *108*
Weiser, R. S., 481, 486, *494*
Weiss, E., 204, *207*
Weissback, H. W., 101, *104*
Weitzman, S. H., 450, 452, *461*
West, J. B., 256, 279, *292*
Westerman, R. A., 10, 25, 45, *90*
Whalley, E., 448, 449, *464*
Whitmore, C. M., 338, *358*
Wiener, A. S., 488, *499*
Wikler, M., 474, *499*
Wilber, C. G., 125, *132*
Wild, G., 114, *132*
Wildenthal, K., 110, *129*, 137, 138, 140, 141, 151, 155, 158, 167, *169*, 255, 287
Wilkins, N., 212, 213, 214, 243, *251, 252*
Willem, V., 300, *358*
Williams, C. A., 467, *500*
Willmer, E. N., 273, *292*, 376, 377, 378, 395, 402, *411*
Wilson, J. A. F., 10, *90*
Winn, H. E., 67, *81*
Winterstein, H., 186, *208*
Wisby, W. J., 63, *90*
Wittenberg, B. A., 253, 279, 292, 431, *443*

Wittenberg, J. B., 253, 279, *292*, 428, 431, 438, *443*
Wodinsky, J., 63, *89, 90*
Wolbarsht, M. L., 35, 38, *85, 90*
Woldring, S., 319, 320, *358*
Wolff, T., 447, 451, *464*
Wolsky, A., 119, *131*
Wood, L., 63, *84*
Woodland, W. N. F., 417, *443*
Woskoboinikoff, M. M., 295, *358*
Wrbitzky, R., *206*
Wright, C. I., 256, *292*
Wright, J. E., 489, 490, *498*
Wu, H. W., 365, *411*
Wurtman, R. J., 103, *108*
Wykes, U., 98, *108*
Wyman, J., 213, 221, 226, 230, 237, 238, 240, *246, 248, 252*
Wyman, J., Jr., 210, 228, 229, 230, 231, 233, 234, 237, *252*
Wyman, L. C., 99, *108*, 124, *131*
Wyss, O. A. M., 325, *358*

Y

Yagi, K., 63, 71, 72, 73, 74, *80, 90*
Yamaguchi, K., 211, 243, *252*
Yamaguchi, Y., 231, 241, 242, *248*
Yamanaka, H., 211, *252*
Yamashita, S., 72, 73, *86*
Yamauchi, A., 123, *132*, 141, 150, *172*, 352, *359*
Yazdani, G. M., 300, *359*
Yokota, S., 59, *83*
Yoshioka, M., 222, 243, 244, *252*
Young, J. Z., 63, *90*, 98, 99, 103, *108*, 114, 115, 116, 118, 122, 123, 126, *132*
Young, N., 184, *206*
Young, S., 321, 322, *357, 359*
Yuen, K., 457, *461*

Z

Zagorulko, T. K., 34, *84*
Zotterman, Y., 59, 61, *84, 85*
Zuckerkandl, E., 217, *252*
Zwaardemaker, H., 110, 111, *132*

SYSTEMATIC INDEX

Note: Names listed are those used by the authors of the various chapters. No attempt has been made to provide the current nomenclature where taxonomic changes have occurred.

A

Acipenser, 455
Agnatha, 134, 159
Alosa alosa, 212
Ameiurus
    A. melas, 60, 216
    A. nebulosus, 190
Amia, 264, 372, 374, 375, 382, 383, 385, 393, 402, 403
    A. calva, 93, 362, 368, 376, 392, 396, 400
Amphibia, 47
Amphipnous, 374
    A. cuchia, 365, 367
Anabantidae, 63, 370
Anabas, 259, 370, 374
    A. scandens, 367
Ancistrus, 374
    A. anisitsi, 367
Anglefish, *see Pterophyllum scalare*
Anguilla, 336, 383, 420, 421, 431
    A. anguilla, 364, 366
    A. japonica, 72, 222, 224, 225, 243
    A. vulgaris, 243, 415, 417, 429
Anoplopoma fimbria, 456
Anoptichthys, 99
Aquidens latifrons, 482
Arapaima gigas, 368, 372
Argentina, 121
Astanax, 99
Astronotus, 77
    A. ocellatus, 41, 42, 44, 45, 47, 76
Atlantic Herring, *see Clupea harengus*

B

Barndoor skate, *see Raja laevis*
Barracuda, *see Sphyraena*
Bassogigas, 447
Barbus filamentosus, 482

Bdellostoma, 50
    B. stouti, 111
Betta splendens, 18
Black-spotted barb, *see Barbus filamentosus*
Blue acara, *see Aquidens latifrons*
Bluefin tuna, *see Thunnus thynnus*
Blue-striped grunt, *see Haemulon sciurus*
Bowfin, *see Amia*
Branchydanio rerio, 49
Bramidae, 454
Bream, *see* Bramidae
Brotulidae, 447
Brown recluse spider, *see Lexoceles reclusa*
Bufo marinus, 118
Burbot, *see* Lota

C

California bonito, *see Sarda chiliensis*
Callichthys, 371
Callionymus, 299, 300, 303, 306, 329, 333
    C. lyra, 299
Caranx trachurus, 418
Carassius
    C. auratus, 10, 12, 15, 18, 52, 71, 489
    C. carassius, 43, 48, 52
Cardiodectes, 493
Careproctus, 447
Carp, *see Cyprinus*
    crucian, *see Carassius carassius*
    Japanese, *see Cyprinus carpio*
    Swedish, *see Cyprinus carpio*
Catfish, *see Ameiurus, Parasilurus, Ictalurus*
    channel, *see Ictalurus p. punctatus*
    brown bullhead catfish, *see Ictalurus nebulosus*

Catostomus, 297, 337
Centrolabrus exoletos, 418
Centrolophus sp., 22
Cephaloscyllium uter, 217
Ceratoidea, 493
Chaenobryttus gulosus, 244
Choanichthyes, 134
Chondrichythes, 50, 69, 134, 136, 138, 151, 157, 159
Chondrostei, 14, 159, 372
Clarias, 367, 371, 374, 375
   C. batrachus, 31
   C. lazera, 370, 371
Climbing perch, *see Anabas*
Clupea, 297
   C. harengus, 68, 212, 214, 455, 481
   C. pilchardus, *see Sardina pilchardus*
Clupeadae, 63
Cobitis fossilis, 382, 383
Cobra, *see Naja naja*
Cod, *see Gadus*
Codfish, *see Gadus*
Comephoridae, 451
Conger conger, 299
Coregonus acronius, 427
Coryphoenoides, 422
Cottus scorpius, 64, 67
Crenilabrus, 297
   C. melops, 299, 418
Crossopterygi, 50, 70
Ctenolabrus, 122
   C. rupestris, 418
Cymatogaster aggregata, 483
Cyprinidae, 456
Cyprinodontidae, 458
Cyprinus, 297, 337
   C. carpio, 25, 43, 48, 59, 73, 136, 185, 189, 197, 198, 331, 482

D

Dace, *see* Cyprinidae
Dipnoi, 70, 134, 138, 245, 255
Dogfish, *see Mustelus, Squalus, Scyliorhinus*
   smooth, *see Mustelus canis*
Dragonet, *see Callionymus*

E

Eel, *see Anguilla, Synaphobranchus*
   common, *see Anguilla anguilla*

electric, *see Electrophorus electricus*
nosed, *see Synaphobranchus pinnatus*
Japanese, *see Anguilla japonica*
Electrophorous, 365, 374, 376–380, 383, 385, 402, 405
   E. electricus, 244, 363, 364, 366, 386, 389, 398–400
English sole, *see Parophys vetulus*
Epinephelus morio, 459
Eptatretus stouti, 137, 238, 491
Erythrinus, 395, 402
   E. unitaeniatus, 372
Esox lucius, 97
Euthynnus pelamis, 163, 165

F

Fluke, *see Paralichthys* sp.
Fundulus, 99, 102, 459, 482
   F. heteroclitus, 31, 72, 458

G

Gadus, 297
   G. callarias, 18, 34, 55, 418
   G. morhua, 67, 136, 155
Galeocerdo cuvieri, 7
Gar, *see Lepisosteus*
Gasterosteus aculeatus, 18, 418
Genyonemus lineatus, 488
Gillichtys, 382, 383
   G. mirabilis, 366, 383
Ginglymostoma circiratum, 7, 485
Gobio gobio, 456
Gibius niger, 418
Goldfish, *see Carassius auratus*
Goosefish, *see Lophius*
Gray snapper, *see Lutjanus griseus*
Grimaldichthys, 447
Gudgeon, *see Gobio gobio*
Guppy, *see Lebistes reticularis*
Gymnarchus niloticus, 375

H

Haemulon sciurus, 63
Hagfish, *see Polistotrema, Bdellostoma*,
   Pacific, *see Eptatretus stouti*
Hake, *see Urophysis chuss*
Halibut, *see Hippoglossus stenolepis*
Halmulon album, 478
Hammerhead, *see Sphyrna zygaena*
Hemichromis bimaculatus, 18

SYSTEMATIC INDEX 519

Herring, see *Clupea harengus*
Heterodontus
  H. *francisci*, 159
  H. *portus jacksoni*, 145, 158, 163
Hippocampus
  H. *brevirostris*, 424
  H. *spinosa*, 93
Hippoglossoides platessoides, 276, 489
Hippoglossus stenolepis, 493
Holocentrus ascensionis, 63
Holostei, 14, 159, 372
Hoplias malabaricus, 376, 377
Hoplosternum, 367, 371, 374, 376, 377
Horse mackerel, see *Trachurus*
Hydrolagus colliei, 136, 145, 159
Hypomesus pretiosus, 490
Hypopomus, 364, 374
  H. *brevirostris*, 365, 366

I

Ictalurus, 297, 337
  I. *melas*, 216
  I. *natalis*, 60
  I. *nebulosus*, 60, 190, 478, 489
  I. *punctatus*, 203
  I.p. *punctatus*, 488
Ichthyomyzon, 111
Ide, see *Idus idus*
Idus idus, 456

K

Katsuwonis pelamis, 487, 488
Killifish, see Cyprinodontidae

L

Labrus
  L. *berggylta*, 418
  L. *ossigragus*, 418
Lamnid sharks, see Lamnidae
Lamnidae, 163, 165
Lampetra, 99, 110, 113, 314, 315
  L. *fluviatilis*, 8, 34, 185, 215, 217, 221
  L. *planeri*, 8
  L. *zanandreai*, 8
Lamprey, see *Petromyzon*, *Lampetra*, *Ichthyomyzon*
Latimeria, 245, 246, 288
Lebistes, 101, 102
  L. *reticularis*, 18, 25, 49, 71, 482
Leopard shark, see *Triakis semifasciata*

Lepidosiren, 140, 245, 246, 372, 375, 376, 379–383, 385, 386, 402
  L. *paradoxa*, 135, 217, 368, 387
Lepisosteus, 372, 374, 393
  L. *osseus*, 11
  L. *platyrhincus*, 478
Lepomis, 52
  L. *cyanellus*, 244
  L. *gibbosus*, 486
  L. *macrochirus*, 8, 486
Lernaeoceriformes, 493
Leucaspius delineatus, 49
Leuciscus rutilus L., 141
Lexoceles reclusa, 484
Lingcod, see *Ophiodon elongatus*
Liparidae, 447
Lophius, 29, 31, 126
  L. *piscatorius*, 190
Lota, 16
  L. *vulgaris*, 418
Lungfish, see *Lepidosiren*, *Neoceratodus*, *Protopterus*
  African, see *Protopterus*
  Australian, see *Neoceratodus*
  South American, see *Lepidosiren*
Lutjanus griseus, 478

M

Macropodus, 370
  M. *opercularis*, 22
Margate, see *Halmulon album*
Marine white perch, see *Morene americana*
Megalops, 393
Menidia, 55
Microgadus tomcod, 60
Micropetrus salmoides, 483
Microstomus kitt, 299
Midshipman, see *Porichthys notatus*
Misgurnus fossilis, 367
Molliensia formosa, 483
Monopterus sp., 365, 374
Morene americana, 486
Mormyridae, 63
Mugil, 26, 168
Mustelus, 115, 125, 126, 193, 312, 338
  M. *canis*, 147, 189, 198
  M *mustelus*, 239
Myelus, 376, 377, 379
  M. *setiger*, 376, 377, 379

SYSTEMATIC INDEX

Myxine, 110–114, 185
  M. glutinosa, 7, 8, 93, 110, 138, 238, 315

N

Naja naja, 484
Neoceratodus, 139, 140, 245, 246, 372, 376, 380, 382, 383, 385–392, 394, 400–402, 405
  N. forsteri, 135, 368

O

Oncorhynchus
  O. keta, 241
  O. kisutch, 15, 17
  O. nerka, 26, 99, 101
  O. tshawytscha, 15, 17, 97, 212
Onos
  O. cimbrius, 418
  O. mustela, 418
Ophiocephalus, 374
  O. punctatus, 367
Ophiodon elongatus, 144–146, 149, 156, 157
Opsanus, 329
  O. tau, 189, 197, 198, 433
Oscar, *see Astronotus ocellatus*
Ostariophysi, 55, 62
Osteichthyes, 69, 70, 134

P

Paddlefish, *see Polyodon spathula*
Paradise fish, *see Macropodus opercularis*
Paralabrax nebulifer, 8, 9
Paralichthys sp., 73
Parasilurus asotus, 60
Parophys vetulus, 493
Perca
  P. flavescens, 432
  P. fluviatilis, 190, 417, 418, 494
Periopthalmus, 363, 364, 374
Periopthalmus sp., 366
Petromyzon, 138
  P. marinus, 8, 114, 221, 236–238, 490
  P. planeri, 213
Phoxinus, 98
Pike perch, *see Stizostedion vitreum vitreum*

Pilchard (*Clupea pilchardus*), *see Sardina pilchardus*
Pimelometopon pulcher, 217
Plaice, *see Hippoglossoides*, *Pleuronectes*
Platyfish, *see Xiphophorus maculatus*
Plecostomus, 371, 374
  P. plectomus, 371
Pleuronectes platessa, 299, 455
Plotosus anguillaris, 60
Poecilia formosa, 32
Polistotrema stouti, 110, 217, 238
Polyodon spathula, 476, 485
Polypteriformes, 12, 50
Polypterus, 372, 374, 375
Polypterus sp., 367
  P. senegalus, 392
Pomatomus, 431
Porichthys, 126
  P. notatus, 32, 67
Prionotus, 198
  P. carolinus, 59, 60, 197
Protopterus, 245, 246, 368, 372, 375, 376, 379–383, 385, 386, 402
  P. aethiopicus, 135, 137, 139, 245, 246, 373, 387
  P. annectans, 93, 245, 246
  P. dolloi, 373
Pseudapocryptes lanceolatus, 365
Pterophyllum scalare, 77
Puffers, *see Tetraodontidae*

R

Raja, 22, 35, 70, 72, 119, 311
  R. binoculata, 136, 159, 240
  R. oscillata, 189, 190
  R. laevis, 228
Raniceps raninus, 418
Ratfish, *see Hydrolagus colliei*
Red groupers, *see Epinephelus morio*
Remora, 299
Roach, *see Rutilus*, *Leuciscus*
Roccus americanus, 492
Ruff, *see Centrolophus* sp.
Rutilus, 295

S

Sablefish, *see Anoplopoma fimbria*
Saccobranchus, 365, 374
Salmon, *see Oncorhynchus*, *Salmo*
  Atlantic, *see Salmo salar*

SYSTEMATIC INDEX 521

chinook, see *Oncorhynchus tschawytscha*
chum, see *Oncorhynchus keta*
coho, see *Oncorhynchus kisutch*
landlocked, see *Salmo salar sebago*
sockeye, see *Oncorhynchus nerka*
Salmo, 124, 295, 336, 337
   S. *clarkii henshawi*, 490
   S. *gairdneri*, 116, 136, 141, 142, 151, 152, 154, 161, 194, 198, 212, 333
   S. *gairdneri irideus*, 96, 116
   S. *irideus*, 26, 28, 97, 98
   S. *salar*, 189, 214, 455
   S. *salar sebago*, 189
   S. *trutta*, 455
Salvelinus fontinalis, 61, 331
   S. *namaycush*, 191
Sand bass, see *Paralabrax nebulifer*
Sarda chiliensis, 489
Sardina pilchardus, 212
Sardina pilchardus sardina, 95
Sardine, see *Sardina pilchardus sardina*
Scomber, 197, 198
Scomberomorus, 268
Scorpaenichythes, 213
Sculpin, see *Cottus scorpius*
Scyliorhinus, 11, 26, 198, 311, 312, 349
   S. *canicula*, 194, 339, 458
   S. *stellaris*, 159, 195–197, 200, 239, 336, 339
Scyllium, 115, 116
   S. *canicula*, see *Scyliorhinus canicula*
Sea bass, see *Serranus gabrilla*
Sea horse, see *Hippocampus brevirostris*
Sea Robin, see *Prionotus carolinus*
Sebastes viviparus, 418
Serranus gabrilla, 222, 225
Shad, see *Alosa alosa*
Shark, see *Ginglymostoma, Heterodontus, Scyliorhinus, Scyllium, Squalus*
   horn, see *Heterodontus francisci*
   Port Jackson, see *Heterodontus portus jacksoni*
   nurse, see *Ginglymostoma ciriratum*
   spiny dogfish, see *Squalus acanthias*
Sheephead, see *Pimelometopon pulcher*
Shiner sea perch, see *Cymatogaster aggregata*
Skate, see *Raja*

Skipjack Tuna, see *Euthynnus pelamis, Katsuwonis pelamis*
Sparidae, 63
Sphyraena, 428
Sphyrna zygaena, 7
Spinachia vulgaris, 418
Spratt, see *Sprattus sprattus*
Sprattus sprattus, 212
Squalus, 125
   S. *acanthias*, 116, 136, 147, 159, 190, 191, 198, 204, 226, 338, 490
   S. *lebruni*, 56
   S. *suckleyi*, 136, 159, 239, 336
Squirrelfish, see *Holocentrus ascensionis*
Stickleback, see *Gasterosteus aculeatus*
Stizostedion vitreum vitreum, 454
Sturgeon, see *Acipenser*
Sucker, see *Catostomus*
Sunfish, see *Lepomis*
   bluegill, see *Lepomis macrochirus*
   "green," see *Lepomis cyanellus*
   pumpkinseed, see *Lepomis gibbosus*
Surf smelt, see *Hypomesus pretiosus*
Symbranchus, 364, 374–376, 379, 380, 383, 385, 386, 400, 401, 404, 405
   S. *marmoratus*, 245, 366
Symphysodon discus, 492
Synaphobranchus, 422
   S. *pinnatus*, 421
Syngnathus acus, 93, 418

T

Taragon, see *Megalops*
Tautog, see *Tautogaonitis*
Tautogaonitis, 231
Teleosti, 14
Tench, see *Tinca*
Tetraodontidae, 7
Thunnus albacares, 163, 165
Thunnus thynnus, 93, 163
Tiger, see *Galeocerdo cuvieri*
Tilapia, 21, 22, 73–75
   T. *macrocephala*, 18, 20
Tinca, 118, 121, 295, 337
   T. *tinca*, 212
Toadfish, see *Opsanus*
Tomcod, see *Microgadus tomcod*
Torpedo, 312
   T. *marmorata*, 93
   T. *ocellata*, 93, 239

Trachurus, 297
Triakis semifasciata, 476, 478
Trichogaster, 370, 371
Trigla gurnardus, 418
Trout, *see Salmo, Salvelinus*
  brook, *see Salvelinus fontinalis*
  brown, *see Salmo trutta*
  lahontan, *see Salmo clarkii henshawi*
  rainbow, *see Salmo gairdneri*
Trygon, 35, 115
Tuna
  bluefin, *see Thunnus thynnus*
  skipjack, *see Katsuwonis pelamis, Euthynnus pelamis*
  yellowfin, *see Thunnus albacares*

U

Uranoscopus, 111
  U. scaber, 115, 123, 126
Urophysis chuss, 59

W

Warmouth, *see Chaenobryttus gulosus*
White croachers, *see Genyonemus lineatus*
Whitefish, *see Coregonus acronius*
White perch, *see Roccus americanus*
Wrasse, *see Crenilabrus*

X

Xiphophorus maculatus, 18, 483

Y

Yarrow, *see Erythrinus*
Yellowfin tuna, *see Thunnus albacares*
Yellow perch, *see Perca flavescens*

Z

Zebra fish, *see Branchydanio rerio*
Zoarces viviparus, 7

SUBJECT INDEX

A

Acetozolamide, 191, 437
Acetylcholine, 158, 268, 440
Acid-base balance, 173–205
Acidosis, 199
Acoustic center, 68
Acoustic lobe, 10, 55
Acoustic stimulation, 43, 52, 54, 56
Acoustic threshold, 62, 63
Acoustico-lateralis system, 27
Action potential, 2, 3, 16, 29, 31, 38, 56, 64, 73, 74, 141, 142, 322
  after potential, 29
  all-or-none impulses, 2
Actomyosin, 155, 458
Adenohypophysis, 7, 31, 101
Adenosinetriphosphatase, 458
Adrenaline, 158, 160, 166, 187, 267, 268, 352, 440, 458
Adrenergic nerves, *see* Autonomic nervous system, sympathetic
Adrenocorticotrophic hormone, 32
Agglutinins, 475, 488, 489
Airbladder, *see* Swim bladder
Alanine, 224, 225
Alkalosis, 199
Alpha cells, 32
Alpha rhythm, 18, 34
AMP, 153, 155
Anaerobiosis, 331, 446, 453
Angiotensin II, 168
Anoxia, 50, 116, 339
Anterior commissure, 11, 16
Antibody, 468–479, 482, 483, 485, 486, 489, 491–494
  biosynthesis of, 468, 475
  specificity, 477–479
Antigen, 469, 472, 473, 475, 477, 480, 487–489
  antigen-antibody reactions, 468
  antigenic specificity, 469, 470, 472, 476, 493

Aorta, *see* Circulatory system
Aortic glomi, 340
Apnoea, 318, 350, *see also* Respiratory system
Arginine, 222, 224, 225
Artery, *see* Circulatory system
Aspartic acid, 224, 225, 476
Asphyxiation, 439
ATPase, *see* Adenosinetriphosphatase
Atropine, 111, 116–119, 121–123, 125, 126, 149, 150, 166, 167, 352
Audition, 62–68
Auditory nerve, *see* Cranial nerves
Auriculae cerebelli, 50
Autonomic nervous systems, 19, 23, 109–128
  adrenergic nerves, 111, 112, 114, 118, 121, 123–126, 150, 151, 155, 157, 166–168, 268
  cholinergic nerves, 111, 116, 117, 121–126, 150, 156, 166, 352, 440
  cranial autonomic nerves, 110–112, 114, 122
  in cyclostomes, 109–114
  in gnathostomatous fish, 114–128
  parasympathetic, 27, 115, 117, 127, 128
  spinal autonomic nerves, 112–114, 122–128
  sympathetic, 27, 112, 118, 121–127
Axon, 30–33, 37, 43, 44, 49, 50, 56, 57, 59, 69, 71, 73–77, 95, 320

B

Bactericidins, 491
Baroreceptor, 92, 150, 446
Basihyal, 30
Basophylic cells, 32
Behavior
  abnormal patterns in, 459
  adaptive, 2
  aggressive, 18, 19

avoidance, 21, 23
emotional, 22
erratic, 404
feeding, 23, 24
innate, 19
reproductive
  arousal, 18, 23, 34, 56
  copulation, 18
  courtship, 18, 25, 51
  mating, 18
  nestbuilding, 18
  parental, 19, 51
  sexual, 18, 51
  spawning, 17, 51
respiratory, 292, 392, 393, 400, 403
rheotactic, 37
Beta cells, 32
Blood
  distribution, 160–162
  groups, 466, 467, 478, 486–490
    ABO system, 487, 488, 490, 491
  pressure in arteries and veins, 135–138
  respiratory properties of, 375–380
  transport of CO_2 in, 188–203
  volume, 159, 160
Bohr effect, 218, 223, 228, 230, 231, 234, 235, 239, 240, 242, 243, 245, 249, 274, 280, 376, 378, 379, 431, 432, 434
Boyle's law, 424, 425
Branchial system, see also Gills
  arches, 301, 308, 310, 313–315, 320, 365, 371, 408
  breathing, 393, 394, 396, 399–403
  cavities, 364
  circulation, see Circulatory system
  heart, see Heart
  nerves, 327
  pump, 333, 338, 392, 395
  respiratory rate, 394, 401
  vascular resistance, 405
  vessels
    afferent, 139, 144, 266, 371, 374, 394
    efferent, 266, 374
Breathing, see also Respiratory system
  regulation of, 293–352
  responses to air exposure, 403, 404
    to changes in external gas composition, 393–402
  to mechanical stimuli, 402, 403
Bretylium, 123
Buffer systems, 173, 181–183, 187, 188, 193, 195, 197–199, 201, 202, 204, 235, 344, 381
  action, 181–183, 197, 344
  capacity, 182, 193, 196–200, 203, 275, 279, 376, 380
Bulbar tract, 33
Buoyancy, 72, 453, 457
  equilibrium, 415, 423
  neutral, 414, 425–427
Bursa of Fabricus, 480

C

Carbon dioxide in blood, 342–346
  carbonate-bicarbonate system, 274, 275, 344
  combining curve, 105, 145, 188–196, 199–201
  combining power, 190, 195, 380
  dissociation curve, 188, 199, 271–273, 275, 282, 379, 380
  transport of, 188–203, 215, 229, 380
  Y-bound, 193
Carbonic anhydrase, 190–192, 202, 203, 275, 276, 279, 437
Cardiac arrhythmia, 151, 459, see also Heart
Cardiac function, 119, 458, see also Heart
Cardiac muscle, 111, 138, 141, see also Heart
Cardiac output, 144, 145, 151, 153, 155–160, 283, 284, 340–342, 350, 351, 389, 405, see also Heart
Cardiovascular system, see Circulatory system, Heart
Carotid glomis, 340
Catecholamines, 110–112, 116, 121, 123–126, 136, 150, 151, 155, 156, 160, 166–168, 267, 268, 286, 352, 440, see also Adrenaline, Noradrenaline
Caudal neurosecretory system, see Spinal cord
Central nervous system
  anatomy and physiology of the, 1–90
  integration, 2, 3, 16, 34, 41, 52, 54, 58, 62, 317, 321
  regeneration, 3, 24–26

SUBJECT INDEX

ventricles, 9, 13, 32, 33, 48, 93, 94, 97
ventricular sulcus, 11
Cerebellum, 33, 49–54, 68, 70
  anatomy, 50, 51
  corpus cerebelli, 50, 53
  Purkinje cell layer, 50, 51
  stratum albium, 50
  stratum granulosum, 50
  vestibulo-lateralis lobe, 50
  deficit function of, 51, 52
  stimulation of, 52–54
Cerebrospinal fluid, 43, 49, 69, 70, 92, 97, 100, 191
Chemoreceptors, 31, 58–62, 92, 150, 393, 398, 399
Cholinergic nerves, see Autonomic nervous system, parasympathetic
Chromaffin cells, 112, 125
Chromatophores, 114, 126, 127, 167, 221, 482
Ciliary ganglion, 115
Circulatory system, 133–168, 333, 349, 369, 385, 419, 439, 440, 480
  anterior caudinal vein, 137
  anterior hypobranchial system, 140
  cardinal vein, 419
    posterior, 391
  caudal vein, 163, 168
  coeliac artery, 406, 419, 420
  coronary circulation, 140
  Cuvier, duct of, 138
  dorsal aorta, 135, 136, 138, 150, 158, 163, 166–158, 266–268, 374, 375, 387, 408, 416, 419
  dorsal vein, 416
  gill circulation, see Gill
  hepatic vessels, 419, 420
  intercostal arteries, 420
  jugular vein, 389, 391
  lingual artery, 110
  peripheral circulation, 156
  pneumogastric artery, 419
  pulmonary circulation, 134–137, 140, 385–389, 394, 401, 404–406, 408
  segmental vessels, 158, 163
  systemic circulation, 134–136, 140, 166, 167, 266, 278, 374, 375, 377, 386, 389, 390, 398, 404, 405, 407, 408
  vena cava, 137, 394, 405

ventral aorta, 135–139, 144–149, 159, 166–168, 340, 399, 406
Clasper gland, 128
Cleithrum, 297, 307
Coeliac sympathetic ganglion, 126, 439
Commissural organ, 93
Cortex, 13, 33
Corticotropin, 32
Cranial nerves, 59, 76
  auditory, 55, 64, 65, 123
  facial, 54, 60, 110, 114, 123, 319–321
  glossopharyngeal, 54, 59, 114, 116, 123, 320, 326, 327, 341
  oculomotor, 77, 110, 114–116, 123, 127
  olfactory, 10, 30, 31
  optic, 31, 35, 38, 40, 41, 44–48, 75
  trigeminal, 31, 50, 54, 55, 76, 77, 123, 320
  vagus, 54, 110–114, 116–119, 121–125, 127, 149, 150, 166, 319–321, 326, 349, 350, 403, 439
Creatinine, 193
Crista cerebellis, 320
Curare, 150, 323, 327
Cyanopsin, 40
Cysteine, 61
Cytidine, 49

D

Dahlgren cells, 71, 72, 76
Delta cells, 32
Dendrites, 2, 5, 7, 10, 11, 34, 35, 51, 58, 95
  dendritic potentials, 35
  dendritic summation, 2
  lateral, 6
  primary, 4
Diamox, see Acetozolamide
Dibenamine, see Phenoxybenzamine
Dibenzylene, see Phenoxybenzamine
Diencephalon, 16, 26–28, 93, 96, 98, 99
  diencephalic areas, 14, 17
  diencephalic connection, 14
  epithalamus, 26, 27
  habenular nuclei, 26, 27, 96
  pineal complex, see Pineal complex
Digestive tract, 415, 416

SUBJECT INDEX

E

Ectothermy, 485
EDTA, see Ethylenediaminetetraacetate
EEG, see Electroencephalogram
Electroencephalograms, 18, 34, 56
Eminentiae granulares, 50
Endothermy, 485
Entorhinalis, 13
Epiphysan, 101
Epiphysis, 103
Epsilon cells, 32
Eptatretin, 167
Erythrocytes, 159, 485, 487, 489, 491
Erythrophores, 126
Erythrosinophils, 32
Ethylenediaminetetraacetate, 485
Euphysoclist, 416, 417, 419, 420
Eye
  lateral, 34, 110
  parietal, 34, 93
  vitreous chamber, 49

F

Facial lobe, 10, 55, 319, 320
Facial nerve, see Cranial nerves

G

Gallamine, 323
Gallocyanin, 10
Ganglion cells, 26, 35, 44, 49, 95, 110–114
Gas exchange, 253–285
  between blood and tissues, 278–279
  across gills, 254–276
  across skin, 276–278
  methods of analysis of transfer of gases, 279–283
    capacity rate ratio, 282, 348
    transfer factor, 280, 281, 285, 335, 346
    ventilation-perfusion ratio, 282, 283, 285, 348, 349, 405
Gastrointestinal tract, 123, 371, 374
Geniculate commissure, 42
Genital nerve, 12
GH, see Gonadotropic hormone
Gill, 210, 362, 363, 365, 366, 374, 375, 381, 385, 390, 393, 395, 397, 398, 401, 402, 405, 407, 429, 455, 481, see also Branchial system, Gas exchange
circulation, 134, 135, 156, 163, 165, 166, 340, 374, 375, 401
differential pressure across, 296, 298–301, 303, 305, 311, 312
resistance, 301, 303, 311
slits, 312
ventilation, 261, 294, 295, 298, 303–306, 310, 311, 313, 315, 323, 329, 331, 333, 342–352, 393, 401, 404, 405
  anatomical dead space, 263, 264, 265, 389
  diffusion dead space, 263–265, 344
  distribution dead space, 263–265, 286
buccal cavity, 295–300, 305, 307–309, 315, 340, 341, 364, 365, 398, 406, 407
buccal pump, 299, 305, 309
buccopharyngeal cavity, 383
opercular cavity, 295–300, 305, 307–309, 364–366, 374, 375
opercular pump, 299–301, 305, 309
operculum, 295–297, 300, 303, 307, 308, 333, 366, 374, 402
Glia, 5, 8, 9, 12, 18, 48, 50, 71, 74–76, 95
Globulins, 472
Glossopharyngeal nerve, see Cranial nerves
Glutamic acid, 224
Glutamine pyrollidone carboxylic acid, 476
Glyceryl esters, 457
Glycine, 219, 225
Glycogen, 8, 9, 162
Glycogen phosphorylase, 155
Glycolysis, 162
Gonadotropic hormone, 32
Gonadotropins, 32
Gonosomatic indices, 102
Growth hormone, 32
GSI, see Gonosomatic indices
Guanethidine, 123
Gustatory lobe, 54, 56
Gustatory receptors, 401

SUBJECT INDEX

H

Haldane effect, 199, 200, 203, 239, 376
Heart, 116–118, 123, 124, 134–158, 161, 163, 166, 167, 210, 283, 284, 319, 349–351, 369, 374, 377, 387, 390, 405, 430, 439, 468, 493, *see also* Cardiac functions and properties
  accessory
    cardinal, 158
    caudal, 137, 158, 168
    portal, 137, 138, 141, 143, 158
  atrioventricular plug, 138
  atrioventricular valves, 138, 143, 144
  atrium, 138, 140, 141, 143, 146, 168, 375
  bulbus arteriosus, 118, 136, 138, 143–149
  bulbus cordis, 139, 140, *see also* Heart, conus arteriosus
  conus arteriosus, 136–140, 142, 143, 146–148
  diastole, 136, 144, 146, 147, 149, 151, 153, 157
  electrical properties, 141–143
  ionotropic effect, 117, 150, 151, 153, 155, 156, 166, 167
  mechanical properties, 143–149
  myocardial depression, 155
  pacemaker system, 141, 156, 321, 325
  pericardium, 137, 143, 145, 146, 148, 157, 204
  rate, 149–159, 405, 458, 459
  regulation of cardiac activity, 149–158
    aneural, 151–156
    neural, 149–151
  Starling's law of, 151, 157
  stroke volume
  semilunar valves, 147
  sinoatrial node, 141
  sinoatrial valves, 147
  sinus venosus, 138, 139, 141, 143, 419
  sinus arrhythmia, 151
  systole, 135, 136, 137, 140, 143, 144, 146, 147, 148, 153, 166
  vagal tone, 150, 151, 155, 156, 352
  ventricle, 135, 138, 140, 141, 143–149, 168
Henderson–Hesselbalch equation, 176 179, 190, 204

Henry's law, 177
Hepatic, *see* Liver
Hematocrit, 159, 201, 271, 376, 434
Hemocyanin, 478
Hemoglobin, 135, 137, 144, 160, 182, 190, 192, 197, 200, 202, 205, 272–274, 350, 376, 377, 432, 433, 436, 438, 456, 469, 473
  carbamino, 192, 193, 199, 203, 230
  methemoglobin, 226, 228
  oxygen
    capacity, 376, 377
    combining power, 376
    dissociation curve, 348, 377, 378
    transport, 210, 213, 215, 217, 222–246
  oxygen-hemoglobin equilibrium, 234
  properties of, 209–246
  oxyhemoglobin, 199, 230, 244, 400, 433, 456
  reduced, 199
Herring bodies, 30
Hexamethonium, 111
HIOMT, *see* Hydroxyindole-O-methyl transferase
Histamine, 167
Histidine, 49, 218, 219, 222, 224
Hydroxyindole-O-methyl transferase, 97, 101
5-Hydroxytryptamine, 101, 133, 167
Hyperoxia, 399, 406
Hyperthyroidism, 101
Hypophysis, 31, 32, 71
  hypophysectomy, 31, 32, 73
  pars intermedia, 32
  regeneration of, 31, 32
Hypothalamus, 10, 26, 27, 30, 31, 71
Hypoxia, 136, 137, 151, 271, 278, 282, 283–286, 333–335, 340, 432, 344, 346, 347, 350, 351, 391–395, 398, 399, 405, 406

I

Immunology, 465–494
  adaptive immunity, 490
  experimental, 467
  methods in, 467

immunochemistry, 468
immunodiffusion, 468, 478
immunoelectrophoretic systems, 468, 478
immunogenetics, 468
immunoglobulins, 469–477, 479, 480, 491
  allotypes of 470, 471
  isotypes of, 469–474, 484
  structure of, 470, 473, 477–479
immunological response, 479–483
rejection action, 482
Inderal, 123, 155
Infundibulum, see Neurohypophyseal stalk
Interrenal, 101, 102, 321
Isoprenaline, 166

K

Kidney, 168, 205, 486

L

Labyrinth, 63, 68
Lactic acid, 162, 331
Lagena, 62
Lateral geniculate nucleus, 27
Lateral line system, 50, 54, 55, 67, 451
Lateral ventricle, 11
Lectins, 489
Leucine, 224, 225
Leucocytes, 159, 479
Leptomenix, 71
Limbic system, 13, 22, 97
Liver, 137, 138, 158, 161
  hepatic circulation, 419
  hepatic plexus, 111
  hepatic portal vein, 419, 420
Lymphatic vessels, 168
Lymphocytes, 480, 485
Lymphoid system, 481, 482, 491
Lysine, 224, 225

M

Mandibular, nerve, 76
Mandibularis trigemini, 76
Mastication, 76
Mechanoreceptors, 326, 327, 402
Melanophores, 122, 126, 458, 459
Melatonin, 96, 97

Membrane potential, 30
Memory, dual trace system of, 41, 44
Mercaptoethanol, 475
Mesencephalon, 32–50, 318
  optic tectum, 33
    learning and, 43, 44
    medial and lateral optic tract fibres, 33
    periventricular stratum, 33
    plexiform layer, 33
    pluripotential cells, 9, 48, 49
    regeneration of, 48, 49
    spontaneous activity of the, 33
    stratum opticum, 33
    superior colliculus, 33
  tectal
    commissural cells, 42, 43
    commissure, 33, 42, 43
    event, 40
    evoked response, 38, 39, 40
    flap, 47
    hemisphere, 41, 43, 44
    input, 44
    layers, 37
    neurons, 35, 37
    signals, 38
    transfer, 41
    units, 40, 43
  tegmentum, 32, 34
    tegmental motor nucleus, 35
Metacholine, 166
Metatonin, 97, 101, 103
Methionine, 225
Mouth, 398, 405, 407
Muscarine, 111
Mullers cells, 35, 55, 69
Myocardial contractility, see Heart, ionotropic effect
Myocardial depression, see Heart
Myoglobin, 211
Myotomal muscles, 158, 161, 162, 163, 279

N

Nasal cavity, 10, 15
Neostigmine, 166
Nerve fibers, 16, 17, 25, 27, 33, 35, 37, 44–46, 48, 51, 101, 110–112, 114, 117, 118, 121–124

Neural lobe, 28
Neurocranium, 308
Neuroendocrine system, 28, 29
Neuroglial filaments, 8
Neurohemal organ, 70
Neurohypophysis, 28, 72, 74
Neurohypophyseal hormones, 168
Neurohypophyseal stalk, 32, 70, 71
Neurohypophyseal tract, 29, 30
Neurons, 2, 7–9, 11, 12, 14, 16, 17, 25, 28–31, 33, 35, 38, 45, 46, 48, 49, 56–58, 66, 68–73, 75, 110–112, 117, 121, 126, 317–323, 326
  afferent
  commissural, 43, 69, 70
  efferent
  enteric, 110
  internuncial, 57
  Mauthner, 6, 7, 50, 55, 69
  Muller's, 35
  neurohypophyseal neurosecretory, 7
  neurosecretory, 30, 31, 71, 73–75
  pacemaker, 325
  second-order, 17
  ultrastructure, 4–9, 27–32, 35, 49, 50, 71, 75, 95, 101, 162, 419, 421
Neurosecretion, neurosecretory cells, 30, 32, 33, 70–74, 76, *see also* Spinal cord
  electrical activity, 28–31
  granules, 30, 71
  neurons, *see* Neurons
  secretory activity, 31
Nicotine, 111, 112
Noradrenaline, 121, 124, 127
Norepinephrine, *see* Catecholamines, Noradrenaline
Notochord, 113
Novocaine, 327
Nucleus amygdala, 13
Nucleus lateralis tuberis, 31

O

Oculomotor nerve, *see* Cranial nerves
Olfaction, 14, 17, 92
Olfactory activity, 17
Olfactory bulb, 9, 10, 11, 14–17, 21, 22, 25, 30
Olfactory crura, 23, 25

Olfactory epithelium, 17
Olfactory information, 11, 14, 16
Olfactory integrating centers, 10
Olfactory mucosa, 10
Olfactory nerve, *see* Cranial nerves
Olfactory tracts, 10, 11, 15, 16, 24, 25, 29–31, 33, 61
Olfactory receptors, 401
Opthalmic nerves, 57, 58
Optic, *see* Eye, Vision
Optic chiasma, 37, 38
Optic cysts, 99
Optic hemispheres, 13, 44
Optic nerve, *see* Cranial nerves
Optic tracts, 44, 45
Optic tectum, *see* Mesencephalon, Optic tectum
Osmoreceptor, 31
Osmoregulation, 74
Ovaries, 32
Oxygen, *see also* Hemoglobin
  consumption, 293, 304, 306, 310, 321, 328, 329, 332, 333–336, 338, 342, 343, 346, 350, 381–384, 388, 394, 398, 402, 456
  receptor, 347
Oxytocin, 168

P

Palatal organ, 60–62
Palatine nerve, 59, 60
Paraphysis, 93
Paraphysoclists, 416
Parapineal body, 93
Paraventricular nuclei, 28
Paravertebral ganglia, 122
Pars inferior, 62
Phagocytes, 479, 480
Phagocytosis, 484
Pharyngo-cutaneous duct, 315
Pharynx, 315, 316, 326, 327, 339, 348, 364–371
Phenoxybenzamine, 125, 166
Phenylalanine, 224, 225
Photophores, 452
Phrenic nerve, 318
Physoclists, 416–418, 420, 428, 438
Physostomes, 416, 417, 419, 427–429, 438
Pilocarpine, 166

Pineal body, 91–104
Pineal complex, 26, 27, 28
Pineal gland, 26, 27, 28, 91, 96, 98, 101, 102, 103
  innervation of, 95, 96
  ontogeny of, 92, 93
  organ, 91–104
  photosensitivity of, 28
  physiology, 96–104
  pinealectomy, 97, 99, 100–103
  structure of, 93, 94
  tractus epiphyseos, 96
Pituitary, 28, 30, 32, 97, 99, 104, 483
Plasma proteins, 160, 182, 191, 195, 198, 202, 204, 469, 475, 493
Pneumatic duct, 416, 419, 420
Porphyropsin, 40, 98
Preoptic endocrine cells, 30
Preparavertebral ganglia, 122
Pretectal nuclei, 27
Prolactin, 227
Proline, 224
Proprioceptors, 348
Pseudobranch, 341, 346

Q

Quadrate, 308

R

Rami communicantes, 122, 123
Renin, 168
Reserpine, 155, 167
Respiration, 395, 396, 403
  aerial, 403
  anaerobic, 331
  aquatic, 334, 395, 402
Respiratory system, 294, 317, 318, 321, 325, 380, 385, 391, 392, 395, 403, 418
  activity, 308, 313, 318, 319–326, 329, 347, 371, 372
  cycle, 314, 349
  center, 317–326, 327, 339, 403
    apneustic, 313
    pneumotaxic, 318
    pontine region, 318, 319
    rhythmic activity, 313, 316, 318, 319, 322–326, 327, 381, 402, 408
  coordination, 318
  dependence, 328–331
  gases, see Gas exchange
  homeostasis, 317, 346–347
  muscles
    adductor arcus palatini, 307, 310
    adductor branchialis, 313
    adductor mandibulae, 307, 310, 313, 314
    adductor operculi, 307, 310
    coracobranchialis, 312
    coracohyoideus, 312
    dilator, palato-quadrate, 314
    fasciculus longitudinalis posterior, 316
    hyohyoideus, 307, 308, 310
    hyoid muscle, 76
    hypobranchials, 312
    interarculis dorsalis, 314
    intermandibular muscle, 76
    levator arcus palatini, 307, 310
    levator hyomandibulae, 314
    levator operculi, 308
    levator palato-quadrate, 314
  organ, 365, 369, 371–374, 406, 407, 429
  pump, 294–331
  tract, 315
Rete mirabile, 119, 163–165, 253, 278, 279, 416, 418–423, 429–436, 439, 440
  bipolar, 420
  unipolar, 420
Reticular cells, 321
Reticulomotor system, 56–58
Reticulospinal fibres, 70
Retina, 38, 44–49, 94, 98, 163
  fast potential, 38–40
  ganglion cells, 35
  input, 38
  project, 44
Rhinencephalon, 13, 17
Rohon–Beard cells, 68, 69
Root effect, 231, 245, 376, 379, 431, 432, 434
  root-off-shift, 274, 432, 436, 437
  root-on-shift, 274, 432, 436, 437

S

Sacculus, 62, 67
Schmidt–Lanterman clefts, 7
Semicircular canals, 62
Serine, 224
Serotonin, see 5-Hydroxytryptamine
Somatotrophic hormone, 32
Spinal canal, 9
Spinal cord, 7, 8, 10, 34, 35, 50, 54, 57, 68–76, 112, 114, 123, 319, 320
  anatomy, 67–70
  caudal neurosecretory system and urophysis, 70–72
  regeneration of, 74–76
Spinal nerve, 31, 60, 62, 68, 112–114, 122, 123, 127
Spinotectal tract, 33
Spiracle, 294, 295, 310, 312, 326
Splanchnic nerve, 115, 122, 123, 125, 126
Spleen, 161, 278, 479, 485
Stellate cells, 33
STH, see Somatotropic hormone
Stomach, 116, 117, 488
Stratum opticum, 33
Striatum, 13
Subcommissural organ, 96, 97
Succinylcholine, 111, 166, 304, 323
Sulcus medians, 27
Sulfanilamide, 191
Superior colliculus, 33
Superior commissure, 96
Supraoptic nuclei, 28
Swim bladder, 62, 63, 68, 117–119, 121–123, 125, 167, 191, 231, 235, 278, 367, 368, 371, 372, 374, 375, 413–460
  gas deposition, 424, 425, 427, 433, 438–440
    salting out effect, 431, 433–435, 437
  gas gland, 122, 126, 253, 278, 420, 432
    vessels, 419, 420
  hairpin countercurrent system, 421, 431
Sympathetic system, see Autonomic nervous system
Synapse, 45, 46, 69, 73, 75
  axodendritic potential, 5, 7
  axosomatic synapse, 7, 8
  Bouton terminaux, 5–7, 49
  excitatory postsynaptic potential, 31, 57
  inhibitory postsynaptic potential, 29, 30
  pre- and postsynaptic elements, 2, 3
  postsynaptic potential, 2, 72, 73
  synaptic bar, 5, 7
  synaptic bed, 6
  synaptic cleft, 7
  synaptic potential, 57
  synaptic vesicles, 5–7

T

Tectal commissures, see Mesencephalon
Tectum, optic, see Mesencephalon
Telencephalon, 10–26
  anatomy of, 11–14, 22, 24–26
  electroencephalography of, 17, 18
  regenerative capacity of, 24–26
  specific sensory deficits following ablation of, 19, 20
  stimulation of, 23, 24
Tetraethylammonium, 125
Thalamus, 26, 27
Threonine, 224, 225
Thymus gland, 479, 480, 481, 491
Thyrotropin, 32
Thyroxin, 101
Torus longitudinalis, 48
Trigiminal nerve, see Cranial nerves
Tryptophan, 224
  metabolism of, 96, 97, 100, 104, 205
Tubocurarine, 111
Tyramine, 167
Tyrosine, 222, 224, 225

U

Urinogenital system, 126
Urophysis, 70–73, 76, 168
  urophysial extract, 168
Utriculus, 62

V

Vagal lobe, 10, 65, 320
Vagus nerve, see Cranial nerves
Valine, 218, 222, 224, 225
Vasopressin, 168
Vein, see Circulatory system
Velum, 295
  velar cycle, 316

Ventilation, see Gill ventilation, Gas exchange
Vestibulospinal fibers, 70
Vision, 19, 20, 28, 34–44, 48, 54, 98, 100, 114, 115
  electroretinograms, 38–40
  interocular exchange, 43
  interocular transfer, 20, 41, 42
  monocular vision, 42
  spectral sensitivity, 28, 40, 41
  visual activity, 47
  visual angle, 37, 47
  visual evoked response, 34, 35, 46
  visual field, 35, 36, 42, 46, 92
  visual information, interhemispheric transfer of, 41–43
  visual input, 34–40
  visual pathway, 38, 39
  visual pattern, 43, 47, 48
  visual pigment, 28, 40, 41, 98, 99, 103, 104, 114
  visual projection, 47

W

Weberian ossicles, 62, 68

X

Xanthophores, 126

Z

Zona occludens, 9